STRUCTURE AND EVOLUTION OF THE MAGELLANIC CLOUDS

INTERNATIONAL ASTRONOMICAL UNION
UNION ASTRONOMIQUE INTERNATIONALE

SYMPOSIUM No. 108

HELD IN TÜBINGEN, WEST GERMANY, 5–8 SEPTEMBER, 1983

STRUCTURE AND EVOLUTION OF THE MAGELLANIC CLOUDS

EDITED BY

SIDNEY VAN DEN BERGH

Dominion Astrophysical Observatory,
Herzberg Institute of Astrophysics,
Victoria, B.C., Canada

and

KLAAS S. DE BOER

Astronomical Institute of the University of Tübingen,
West Germany

D. REIDEL PUBLISHING COMPANY

A MEMBER OF THE KLUWER ACADEMIC PUBLISHERS GROUP

DORDRECHT / BOSTON / LANCASTER

Library of Congress Cataloging in Publication Data

Main entry under title:

Structure and evolution of the Magellanic Clouds.

    At head of title: International Astronomical Union, Union astronomique internationale.
    "Symposium no. 108 held in Tübingen, West Germany, 5–8 September, 1983."
    Includes bibliographical references and index.
    1. Magellanic Clouds–Congresses. 2. Bok, Bart Jan, 1906- .
I. Van den Bergh, Sidney, 1929- . II. Boer, Klaas S. de (Klaas Sjoerds), 1941- . III. International Astronomical Union.
QB858.5.M33S77   1984   523.1'12   83-27061
ISBN 90-277-1722-2
ISBN 90-277-1723-0 (pbk.)

Published on behalf of
the International Astronomical Union
by
D. Reidel Publishing Company, P.O. Box 17, 3300 AA Dordrecht, Holland

All Rights Reserved
© 1984 by the International Astronomical Union

Sold and distributed in the U.S.A. and Canada
by Kluwer Academic Publishers,
190 Old Derby Street, Hingham, MA 02043, U.S.A.

In all other countries, sold and distributed
by Kluwer Academic Publishers Group,
P.O. Box 322, 3300 AH Dordrecht, Holland

No part of the material protected by this copyright notice may be reproduced or utilized in any form or by any means, electronic or mechanical, including photocopying, recording or by any information storage and retrieval system, without written permission from the publisher

Printed in The Netherlands

TABLE OF CONTENTS

Bart J. Bok	x
Foreword	xi
Acknowledgements	xii
List of Participants	xiii

MAGELLANIC CLOUD CLUSTERS

S. van den Bergh
Star Clusters in the Magellanic Clouds ... 1

P.W. Hodge
Blue Clusters, Populous Clusters, and Globulars ... 7

L. Searle
The Integrated Spectra of Star Clusters and the History of the Magellanic Clouds ... 13

M. Kontizas and E. Kontizas
Masses and Relaxation Times of Star Clusters in the SMC ... 25

S.M. Fall and C.S. Frenk
An Ellipticity-Age Relation for Globular Clusters in the LMC ... 27

E.H. Geyer and A. Hänel
SIT-VIDICON Surface Photometry of Globular Clusters in the Magellanic Clouds ... 29

P.J. Flower
Intermediate-Age Magellanic Cloud Star Clusters ... 31

B. Nelles and T. Richtler
Metallicities of Young Populous Clusters in the Magellanic Clouds ... 33

E. Schulz-Lüpertz and M. Grewing
The H-R Diagram of NGC 1962-65-66-70 ... 35

J.A. Graham and J.M. Nemec
A Survey of Magellanic Cloud Clusters for RR Lyrae Stars ... 37

J.M. Nemec, M.H. Liller and J.E. Hesser
Period Changes of RR Lyrae Stars in the LMC Globular Cluster NGC 2257 ... 39

J. Andersen, A. Blecha and M.F. Walker
The Electronographic Color-Magnitude Diagram of Hodge 11 ... 41

L.L. Stryker, J.M. Nemec, J.E. Hesser and R.D. McClure
The LMC Globular Cluster Hodge 11 (=SL 868) ... 43

R.M. Rich, J.R. Mould and G.S. Da Costa
Main Sequence Photometry of Kron 3 ... 45

J.E. Hesser, R.D. McClure and W.E. Harris
Color-Magnitude Diagram Morphology of the Oldest Star Clusters in the Large Magellanic Cloud ... 47

G. Alcaino and W. Liller
A BVRI Photometric Study of Star Clusters in the Bok Region of the LMC ... 49

G. Alcaino and W. Liller
A BV Photometric Study of Star Clusters in Two Selected Regions of the SMC ... 51

K.S. de Boer
Far-UV Photometry of Magellanic Cloud Globular Clusters 53
E.H. Geyer and A. Cassatella
UV-observations of Young Populous Clusters in the LMC 55
C. Cacciari
Disentangling Individual Star and Cluster Contributions in IUE
Spectra of Magellanic Cloud Globular Clusters 57
E. Böhm-Vitense and P.W. Hodge
Ultraviolet Studies of O and B Stars in the LMC Cluster NGC 2100,
the SMC Cluster NGC 330 and the Galactic Cluster NGC 6530 59

Panel Discussion on **Star Clusters**
(J.E. Hesser, P.W. Hodge, A. Renzini, L. Searle, S. van den Bergh) 63

EVOLUTION

J. Lequeux
Global Evolutionary Effects in the Magellanic Clouds 67
L.L. Stryker
Field Studies, Luminosity Functions, and Star Formation History in
the Magellanic Clouds 79

J.V. Feitzinger
Stochastic Star Formation and Bubbles in the Large Magellanic Cloud 89
B. Rocca-Volmerange
Past Star Formation History of the Magellanic Clouds 91
J.V. Feitzinger and E. Braunsfurth
The Spatial Distribution of Young Objects in the Large
Magellanic Cloud - a Problem of Pattern Recognition 93
F. Matteucci
Chemical Evolution of the Magellanic Clouds 95
M.T. Brück
Star Counts in Fields Surrounding the LMC 97
P. Linde, A. Ardeberg, H. Lindgren and G. Lyngå
Crowded Field Electronography in the LMC 99
M.R.S. Hawkins and M.T. Brück
The Colour Magnitude Diagram of a Second Field in the SMC Halo 101
M.T. Brück and M.R.S. Hawkins
A Search for Stars in the Magellanic Stream 103
J. Koornneef
Ultraviolet Surface Photometry of Stellar Associations in the LMC 105

DYNAMICS

K.C. Freeman
Kinematics and Dynamics of the Magellanic Clouds 107
M. Fujimoto and T. Murai
The Magellanic Stream and its Related Problems 115

D.S. Mathewson and V.L. Ford
H I Surveys of the Magellanic System   125

N. Martin, E. Maurice, L. Prévot, E. Rebeirot and J. Rousseau
Distribution and Radial Velocities of Late Supergiant Stars
in the LMC and the SMC   137

N.V. Bystrova
A Search for Low-Velocity Neutral Hydrogen in the Magellanic Stream   139

J. Jaaniste
On the Distance of the Magellanic Stream   141

STELLAR POPULATIONS

R.M. Humphreys
The Brightest Stars in the Magellanic Clouds and Other
Late-Type Galaxies   145

M.W. Feast
Magellanic Cloud Cepheids: Abundances, Reddenings, P-L and
P-L-C Relations. A Review.   157

M.S. Bessell
Spectroscopy of Red Variables and Other Luminous Red Stars   171

M. Aaronson
Detection and Photometry of Red Giants in the Magellanic Clouds   183

J. Mould
Red Giant Stars: Comparison of Observations and Theory   195

J.A. Graham
RR Lyrae Stars and Novae in the Magellanic Clouds   207

T. Lloyd Evans
A Survey for Red Variables in the Magellanic Clouds   217

G. Paltoglou, P.R. Wood, M.S. Bessell and K. Ratnatunga
A Search for Red Variable Stars in the LMC   219

D.L. Welch and B.F. Madore
JHK Observations of Magellanic Cloud Cepheids   221

H.A. Smith and L. Connolly
Spectroscopy of Overluminous Cepheids in the Magellanic Clouds   223

C.D. Laney and D.H. McNamara
Spectral Types, Color Indices, and Abundances for Cepheids in the
Magellanic Clouds and the Galaxy   225

H. Deasy and P.A. Wayman
Period Changes in Magellanic Cloud Cepheids   227

M.J. Stift
Magellanic Cloud Cepheids: Intrinsic and Extrinsic Properties as
Inferred from Numerical Simulations of Two-Colour Photometry   229

N. Sanduleak
A More Complete Listing of LMC Planetary Nebulae   231

K. Nandy, G.I. Thompson, D.H. Morgan and L. Houziaux
Spectrophotometry of Early-type Supergiants in the LMC   233

M. Friedjung and C. Muratorio
From What Radius do the Winds of Magellanic Cloud Supergiants
Produce Fe II Emission lines? 235
O. Stahl and B. Wolf
R127: an S Dor Type Variable Intermediate Between Of and WN 237
N.R. Walborn
An Unstable Ofpe Star in the LMC 239
A.P. Cowley, D. Crampton, J.B. Hutchings and R. Remillard
Evidence for a Black Hole in LMC X-3 241

30 DORADUS AND R136

N.R. Walborn
The Stellar Content of 30 Doradus 243

B.D. Savage and E.L. Fitzpatrick
UV Extinction in the 30 Doradus Nebula and the UV Energy
Distribution of R136a 255
M. Rosa, J. Melnick and P. Grosbol
On the Nature of R136, the Central Object of 30 Dor;
a Comparison with the Galactic Cluster NGC 3603 257
Y.-H. Chu
New Optical Observations of R136 259
W. Seggewiss and A.F.J. Moffat
The Object R136 in the Core of the 30 Doradus Nebula 261

Panel Discussion on R136
(Y.-H. Chu, J.V. Feitzinger, B.D. Savage, W. Seggewiss,
Th. Schmidt-Kaler, N.R. Walborn) 263

SUPERNOVA REMNANTS, RADIO CONTINUUM, X-RAY SOURCES

M. A. Dopita
The Magellanic Cloud Supernova Remnants 271
B.Y. Mills and A.J. Turtle
A Radio Continuum Survey of the Magellanic Clouds 283
D.J. Helfand
X-ray Surveys of the Magellanic Clouds 293
J.B. Hutchings
Spectroscopy of Stellar X-ray Sources in the Magellanic Clouds 305

U. Klein, R. Gräve and R. Beck
Radio Continuum Emission from Magellanic-Type and Dwarf
Irregular Galaxies 313
R. Fusco-Femiano and A. Preite-Martinez
Are the Supernova Remnants of the LMC in the Adiabatic Phase? 315
M. W. Pakull
Compact X-ray Sources in the LMC: Optical identifications 317

INTERSTELLAR MATTER

F.P. Israel
Molecules and Dust in the Magellanic Clouds — 319

J. Koornneef
Gas-to-Dust Ratios in the Magellanic Clouds — 333

K. Nandy
Interstellar Extinction in the Magellanic Clouds — 341

R.J. Dufour
The Composition of H II Regions in the Magellanic Clouds — 353

M. Peimbert
The Magellanic Clouds and Planetary Nebulae — 363

K.S. de Boer
Coronae of the Magellanic Clouds — 375

M.V. Copetti, H.A. Dottori, E.L. Bica and M.G. Pastoriza
Age Determination of H II Regions of the LMC and SMC — 383

I.R.G. Wilson and M.A. Dopita
The Effect of Mass-loss on the Evolution of H II Regions in the LMC — 385

J. Caplan and L. Deharveng
Absolute Hα and Hβ Photometry of LMC H II Regions — 389

A. Hänel
Kinematics of the H II Region N11 in the LMC — 391

P. Shull Jr.
The Multiple-Phase Structure of the Interstellar Medium in the LMC — 393

K. Rohlfs, J. Kreitschmann and J.V. Feitzinger
A New 21-cm Line Survey of the LMC — 395

F.F. Gardner
Molecular Line Observations in the Large Magellanic Cloud — 397

M. Rubio, R. Cohen and J. Montani
CO Observations in the Small Magellanic Cloud — 399

R. Cohen, J. Montani and M. Rubio
CO in the Large Magellanic Cloud — 401

G.C. Clayton and P.G. Martin
Interstellar Dust in the LMC — 403

J. Lequeux, E. Maurice, L. Prévot, M.-L. Prévot-Burnichon and B. Rocca-Volmerange
SMC: UV Extinction Curves, Gas to Color-Excess Ratios — 405

E.L. Fitzpatrick
Ultraviolet Interstellar Absorption Towards SK 159 in the Small Magellanic Cloud — 407

FUTURE DIRECTIONS OF RESEARCH IN THE MAGELLANIC CLOUDS

Panel Discussion on Future Research
(I.J. Danziger, K.C. Freeman, J. Frogel, J. Graham, M. Grewing, D.J. Helfand, B. Westerlund) — 409

Index — 419

BART JAN BOK

born in Hoorn, the Netherlands, April 28, 1906
died in Tucson, AZ, U.S.A., August 5, 1983

This Volume is dedicated to the Memory of

BART JAN BOK

who's enthusiasm for the Southern skies, and for the Magellanic Clouds in particular, contributed mightily to the dramatic increase in our knowledge about the Galaxy's nearest neighbours, which is recorded in this symposium.

While he was Director of the Mt. Stromlo Observatory, Bart Bok organized two symposia on the Clouds of Magellan. The first of these, (Kerr and Rodgers, 1964) dealt with both the Galaxy and the Magellanic Clouds, whereas the second (Hindman and Westerlund, 1966 was entirely devoted to the Magellanic Clouds.The only other compendium devoted to the Clouds is the symposium organised in Chile in 1969 by ESO (Muller 1971) to celibrate the dedication of the European Southern Observatory. A fairly complete bibliography of LMC literature prior to 1964 can be found in the introduction to the LMC atlas by Hodge and Wright (1967). Similar data on the SMC up to about 1974 are provided in the introduction to the atlas for the SMC by Hodge and Wright (1977).

During the decade and a half since these symposia, our knowledge of the Clouds of Magellan has been revolutionized by the new generation of 4-meter class telescopes in Chile and in Australia, by the introduction of highly efficient panoramic detectors in optical studies, as well as by infrared measurements, new radio techniques and far-ultraviolet and X-ray studies of the Clouds from spacecraft. The present symposium is intended to review the progress made in the study of the Magellanic Clouds

Sidney van den Bergh

Klaas S. de Boer

Hindman, J.V., and Westerlund, B.E. 1966, Symposium on the Magellanic Clouds, Mt. Stromlo Observatory, Canberra
Hodge, P.W., and Wright, F.W. 1967, The Large Magellanic Cloud, Smithsonian Publication no 4699, Smithsonian Press, Washington D.C.
Hodge, P.W., and Wright, F.W. 1977, The Small Magellanic Cloud, University of Washington Press, Seattle
Kerr, F.J., and Rodgers, A.W. 1964, The Galaxy and the Magellanic Clouds, IAU Symposium no 20, Australian Academy of Science, Canberra
Muller, A.B. 1971, The Magellanic Clouds, Reidel Publ. Co., Dordrecht

ACKNOWLEDGEMENTS

Symposium No. 108 was sponsored by six commissions of the IAU: 28 (Galaxies), 29 (Stellar Spectra), 34 (Interstellar Matter), 37 (Star Clusters and Associations), 44 (Astronomy from Space), and 45 (Stellar Classification). The Scientific Organising Committee consisted of S. van den Bergh (Chairman), I.J. Danziger, J. Einasto, M.W. Feast, J. Graham, P.W. Hodge, K. Kodaira, J. Lequeux, D.S. Mathewson, M. Peimbert, T. Schmidt-Kaler, and B.E. Westerlund.

The meeting was held in the physics complex "Auf der Morgenstelle" of the Universität Tübingen. The Local Organising Committee consisted of K.S. de Boer (Chairman), M. Grewing and G. Krämer. We thank all those who helped in the running of this conference, in particular R.H. Vos for the organisation of the conference secretariat and her help in preparing the index for the proceedings. The Deutsche Forschungsgemeinschaft and the IAU provided substantial financial support for this meeting, which was vital since it enabled us to invite speakers from both Southern and Northern hemispheres to present reviews.

PARTICIPANTS

M. Aaronson	Steward Observatory, Tucson AZ 85721, USA
G. Alcaino	Instituto Isaac Newton, Casilla 8-9, Santiago de Chile, Chile
I. Appenzeller	Landessternwarte Königstuhl, D-6900 Heidelberg, F.R. Germany
M. Azzopardi	E.S.O., Karl Schwarzschildstr 2, D-8046 Garching, F.R. Germany
M. Bessell	Mount Stromlo and Siding Spring Observatories, Private Bag, Woden P.O., ACT 2606, Australia
L. Bianchi	Osservatorio Astronomico, 10025 Pino Torinese, Italy
V.M. Blanco	Cerro Tololo Interamerican Observatory, Casilla 603, La Serena, Chile
E. Böhm-Vitense	Astronomy Deptm., University of Washington, FM 20, Seattle WA 98195, USA
M.T. Brück	Royal Observatory Edinburgh, Edinburgh EH9 3HJ, Scotland
N.V. Bystrova	Special Astrophysical Observatory, Leningrad Branch, 196149 Leningrad, Pulkovo, USSR
C. Cacciari	E.S.A.-VILSPA, Apartado 54065, Madrid, Spain
V. Caloi	Instituto Astrofisica Spaziale, C.P. 67, 00044 Frascati, Italy
A. Cassatella	E.S.A.-VILSPA, Apartado 54065, Madrid, Spain
V. Castellani	Instituto Astrofisica Spaziale, C.P. 67, 00044 Frascati, Italy
Y.-H. Chu	Washburn Observatory, University of Wisconsin, Madison WI 53706, USA
G. Clayton	Deptm. of Astronomy, University of Toronto, Toronto M5S 1A7, Canada
R. Cohen	NASA Institute for Space Studies, 2880 Broadway, New York NY 10025, USA
A.P. Cowley	Deptm. Physics, Arizona State University, Tempe AZ 85281, USA
I.J. Danziger	E.S.O., Karl Schwarzschildstr 2, D-8046 Garching, F.R. Germany
K.S. de Boer	Astronomical Institute Tübingen, Waldhäuserstr 64 D-7400 Tübingen, F.R. Germany
M. de Groot	Armagh Observatory, College Hill, Armagh BT61 9DG, Northern Ireland
L. Deharveng	Observatoire de Marseille, 2 Place le Verrier, F13248, Marseille, France
P. Demarque	Yale University Observatory, P.O.Box 6666, New Haven CT 06511, USA
M. Dennefeld	Inst. d' Astrophysique, 98 bis Blvd Arago, F75014 Paris, France

LIST OF PARTICIPANTS

M.A. Dopita	Mount Stromlo and Siding Spring Observatories, Private Bag, Woden P.O., ACT 2606, Australia
H. Dottori	Inst. de Fisica UFGRS, R. Luiz Englert, 90000 Porto Alegre RS, Brasil
P. Dubois	Observatoire de Strasbourg, 11 Rue de l'Universite, F67000, France
R.J. Dufour	Deptm. Space Physics and Astronomy, Rice Unversity, Houston TX 77251, USA
R.A.W. Elson	Institute of Astronomy, Madingley Road, Cambridge CB3 0HA, England
S.M. Fall	Institute of Astronomy, Madingley Road, Cambridge CB3 0MA, England
M.W. Feast	South African Astronomical Observatory, P.O.Box 9, Observatory 7935, South Africa
J.V. Feitzinger	Astronomisches Institut der Ruhr Universität, Postfach 102148, D-4630 Bochum. F.R. Germany
E.L. Fitzpatrick	Washburn Observatory, University of Wisconsin, Madison WI 53706, USA
P.J. Flower	Deptm. of Physics and Astronomy, Clemson University, Clemson SC 29631, USA
K.C. Freeman	Mount Stromlo and Siding Spring Observatories, Private Bag, Woden P.O., ACT 2606, Australia
M. Friedjung	Institut d' Astrophysique, 98 bis Blvd Arago, F75014 Paris, France
J.A. Frogel	Cerro Tololo Interamerican Observatory, Casilla 603, La Serena, Chile
M. Fujimoto	Deptm. of Physics, Nagoya University, Nagoya 464, Japan
R. Fusco-Femiano	Instituto Astrofisica Spaziale, C.P. 67, 00044 Frascati, Italy
F.F. Gardner	C.S.I.R.O., Division Radiophysics, P.O.Box 76, Epping N.S.W. 2121, Australia
E.H. Geyer	Observatorium Hoher List, D-5568 Daun, F.R. Germany
P.M. Gondhalekar	Rutherford Appleton Lab., Chilton, Didcot OX11 0QX, United Kingdom
J.A. Graham	Cerro Tololo Interamerican Observatory, Casilla 603, La Serena, Chile
M. Grewing	Astronomisches Institut Tübingen, Waldhäuserstr 64, D-7400 Tübingen, F.R. Germany
A. Hänel	Observatorium Hoher List, D-5568 Daun, F.R. Germany
D.J. Helfand	Columbia Astrophysics Lab., 538 W 120th str., New York NY 10027, USA
J.E. Hesser	Dominion Astrophysical Observatory, 5071 W Saanich rd, Victoria BC, V8X 4M6, Canada
P.W. Hodge	Astronomy Deptm., University of Washington, FM 20, Seattle WA 98195, USA
L. Houziaux	Institut d' Astrophysique, 15 Avenue des Tilleuls, B 4000 Liege, Belgium
R.M. Humphreys	Deptm. of Astronomy, 116 Church str SE, Minneapolis MN 55455, USA

LIST OF PARTICIPANTS

J.B. Hutchings	Dominion Astrophysical Observatory, 5071 W Saanich rd, Victoria BC, V8X 4M6, Canada
F.P. Israel	E.S.A., S and D / SP, ESTEC, Postbus 299, 2200AG Noordwijk, the Netherlands
J. Isserstedt	Institut für Astronomie und Astrophysik, Am Hubland, D-8700 Würzburg, F.R. Germany
J. Jaaniste	Tartu Astrophysical Observatory, Toravere 202444, Estonian SSR
T.D. Kinman	Kitt Peak National Observatory, P.O.Box 26732, Tucson AZ 85726, USA
U. Klein	M.P.I.f. Radioastronomie, Auf dem Hügel 69, D-5300 Bonn, F.R. Germany
J. Koornneef	Space Telescope Science Institute, Homewood Campus, Baltimore MD 21218, USA
G. Krämer	Astronomisches Institut Tübingen, Waldhäuserstr 64, Tübingen, F.R. Germany
R.P. Kraft	Lick Observatory, University of California, Santa Cruz CA 95064, USA
B.M. Lasker	Space Telescope Science Institute, Homewood Campus, Baltimore MD 21218, USA
J. Lequeux	Observatoire de Marseille, 2 Place le Verrier, F13248 Marseille, France
P. Linde	Lund Observatory, Box 1107, S22104 Lund, Sweden
T. Lloyd Evans	South African Astronomical Observatory, P.O.Box 9, Observatory 7935, South Africa
M.-C. Lortet	D.A.F., Observatoire de Meudon, F92195 Meudon, France
K. Lundgren	Astronomiska Observatoriet, Box 515, S75120 Uppsala, Sweden
N. Martin	Observatoire de Marseille, 2 Place le Verrier, F13248 Marseille, France
D.S. Mathewson	Mount Stromlo and Siding Spring Observatories, Private Bag, Woden P.O., Act 2606, Australia
F. Matteucci	Instituto Astrofisica Spaziale, C.P. 67, 00044 Frascati, Italy
E. Maurice	E.S.O., Karl Schwarzschildstr 2, D-8046 Graching, F.R. Germany
M.F. McCarthy	Vatican Observatory, I-00120 Vatican City, Vatican City State
D.H. McNamara	Deptm. Physics Astronomy, Brigham Young University, Provo UT 84602, USA
B.Y. Mills	School of Physics, University of Sydney, Sydney N.S.W. 2006, Australia
J. Mould	Deptm. of Astronomy, CALTECH, 1201 E California Blvd, Pasadena CA 91025, USA
K. Nandy	Royal Observatory Edinburgh, Blackford Hill, Edinburgh EH9 3HJ, Scotland
B. Nelles	Observatorium Hoher List, D-5568 Daun, F.R. Germany
M.W. Pakull	Inst. Astronomie Astrophysik, Technische Unversität, Ernst Reuter Platz 7, D-1000 Berlin, F.R. Germany
M. Peimbert	Instituto de Astronomia, A.P. 70264, Mexico 20 DF, Mexico

J.W. Pel	Kapteyn Sterrewacht, Mensingheweg 20, 9301KA Roden, the Netherlands
A. Preite-Martinez	Instituto Astrofisica Spaziale, C.P. 67, 00044 Frascati, Italy
L. Prévot	Observatoire de Marseille, 2 Place le Verrier, F13248 Marseille, France
E. Rebeirot	Observatoire de Marseille, 2 Place le Verrier, F13248 Marseille, France
A. Renzini	Osservatorio Astronomico, C.P. 596, 40100 Bologna, Italy
R.M. Rich	CALTECH 105-24, Pasadena CA 91106, USA
T. Richtler	Observatorium Hoher List, D-5568 Daun, F.R. Germany
B. Rocca-Volmerange	Institut d' Astrophysique, 98 bis Blvd Arago, F75014 Paris, France
K. Rohlfs	Astronomisches Institut der Ruhr Universität, Postfach 102148, D-4630 Bochum, F.R. Germany
M. Rosa	E.S.O., Karl Schwarzschildstr 2, D-8046 Garching F.R. Germany
J. Rousseau	Observatoire de Lyon, F69230 Saint Genis Laval, France
M. Rubio	Deptm. Astronomia, Universidad de Chile, Casilla 36-D, Santiago de Chile
N. Sanduleak	Warner and Swansey Observatory, Case Western Reserve University, Cleveland OH 44106, USA
B.D. Savage	Washburn Observatory, University of Wisconsin, Madison WI 53706, USA
L. Searle	Mount Wilson Observatory, 813 Santa Barbara St, Pasadena CA 91107, USA
W. Seggewiss	Observatorium Hoher List, D-5568 Daun, F.R. Germany
Th. Schmidt-Kaler	Astronomisches Institut der Ruhr Universität, Postfach 102148, D-4630 Bochum, F.R. Germany
P. Shull Jr.	M.P.I.f. Astronomie, D-6900 Heidelberg, F.R. Germany
H.A. Smith	Deptm. Physics and Astronomy, Michigan State Univ., East Lansing MI 48824, USA
Y. Sofue	Nobeyama Radio Observatory, 181 Mitaka, Tokyo, Japan
O. Stahl	Landessternwarte Königstuhl, D-6900 Heidelberg, F.R. Germany
M.J. Stift	Institut für Astronomie, Türkenschanzstr 17, A-1180 Wien, Austria
L.L. Stryker	Deptm. Terrestrial Magnetism, 5241 Broad Branch rd, Washington DC 20015, USA
D.A. VandenBerg	Physics Deptm., University of Victoria, P.O.Box 1700, Victoria BC, V8W 2Y2, Canada
S. van den Bergh	Dominion Astrophysical Observatory, 5071 W Saanich rd, Victoria BC, V8X 4M6, Canada
J. Vigneau	Observatoire de Toulouse, 14 Avenue E. Belin, F31400 Toulouse, France
N.R. Walborn	Code 683, NASA-GSFC, Greenbelt MD 20771, USA

LIST OF PARTICIPANTS

M.F. Walker	Lick Observatory, University of California, Santa Cruz CA 95064, USA
P.A. Wayman	Dunsink Observatory, Castleknock, Co Dublin, Ireland
D.L. Welch	Deptm. of Astronomy, University of Toronto, Toronto M5S 1A7, Canada
B. Westerlund	Astronomical Observatory, Box 515, S75120 Uppsala, Sweden
R.F. Wing	Astronomy Deptm., Ohio State University, 174 W 18th Av., Columbus OH 43210, USA
B. Wolf	Landessternwarte Königstuhl, D-6900 Heidelberg, F.R. Germany
P.R. Wood	Mount Stromlo and Siding Spring Observatories, Private Bag, Woden P.O., ACT 2606, Australia

STAR CLUSTERS IN THE MAGELLANIC CLOUDS

S. van den Bergh
Herzberg Institute of Astrophysics
Dominion Astrophysical Observatory
Victoria, B. C., Canada

I   INTRODUCTION

The evolutionary history of the Magellanic Clouds has been very different from that of the Galaxy. As a result we presently observe major differences between both the stellar content and the cluster populations in the Galaxy and the Clouds.

Perhaps the most striking characteristic of Galactic clusters is the clear-cut dichotomy between massive old globulars and less massive younger open clusters in the galactic disc. No such dichotomy exists in the Clouds where "populous intermediate-age clusters" span the gap between old globulars such as NGC 121 in the SMC and NGC 2257 in the LMC and younger objects similar to Galactic open clusters. A second major difference is that some Cloud clusters, such as NGC 121 and NGC 1978 are quite flattened, whereas Galactic globulars are much more nearly spherical. Finally young clusters in the Clouds (especially those in the SMC) are metal-poor compared to their Galactic counterparts. This difference simply reflects the fact that the Galaxy is a highly evolved system in which the interstellar gas has been more strongly contaminated by stellar ejecta than is the case in the Magellanic Clouds.

II   THE FLATTENING OF CLUSTERS

The flattenings of Large Cloud globulars have recently been measured by Geisler and Hodge (1980) and by Frenk and Fall (1982). These measurements show (van den Bergh 1982) that old LMC clusters are significantly more flattened than are galactic globulars. The strong flattening of NGC 121 suggests that the SMC clusters are similar to those in the LMC rather than to those in the Galaxy. The reason for this difference between the flattenings of Galactic and Cloud clusters remains a mystery. Norris (1983) has made the interesting suggestion that cluster flattening is related to the "second parameter phenomenon". He notes that all Galactic globulars with large ($\varepsilon \sim 0.15$)

flattenings have blue horizontal branches, whereas Galactic globular clusters with red horizontal branches are almost spherical ($\epsilon \sim 0.0$). These systematics, if real, do not seem to apply to Cloud clusters. In the SMC NGC 121 is quite highly flattened but has a red horizontal branch (Tifft 1963, Gascoigne 1966). Casual inspection of the globulars in M31 indicates that they, like their Galactic counterparts, are nearly spherical. Of the Galactic globulars ω Cen ($\epsilon = 0.19$) has the largest intrinsic flattening. It is probably not coincidental that it is also the most luminous Galactic globular. It is of interest to note that four of the six most luminous globulars known in NGC 5128 (Hesser et al. 1984) are noticably flattened. Even in the Clouds cluster flattening seems to be a function of luminosity. Table 1 and Figure 1 show that the most luminous LMC clusters of all ages (van den Bergh 1981) are more flattened (Frenk and Fall 1982) than are less luminous LMC clusters. A Kolmogorov-Smirnov test shows that this difference is significant at the 99% confidence level.

The fact that some relatively young (Searle, Wilkinson and Bagnuolo 1980) luminous Cloud clusters are highly flattened, whereas young Galactic clusters are almost invariably spherical, suggests that differences between cluster formation in the Galaxy and the Clouds persist to the present day. This view is supported by the observation that a number of "populous star clusters" (which do not appear to have a Galactic counterpart) have been formed during the last 1×10^9 yr.

Finally it is noted that the absence of a population of old flattened clusters in the Galaxy implies that our Milky Way System was not formed by the coalescence of numerous Magellanic-type galaxies. Even the merger with a single LMC-like galaxy probably can be excluded by the absence of flattened old clusters.

Table 1
FLATTENING OF LMC CLUSTERS

V	$\langle\epsilon\rangle$	n_{cl}
9.00 - 9.99	0.19	4
10.00 - 10.99	0.14	16
11.00 - 11.99	0.11	15
12.00 - 12.99	0.08	15

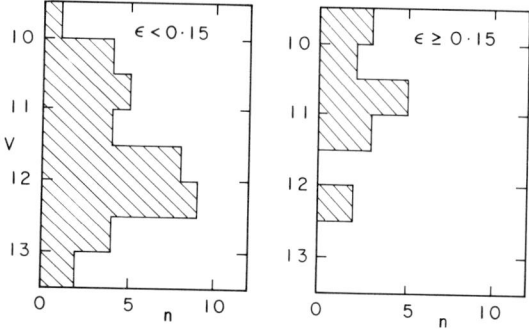

Fig. 1. The magnitude distribution of little-flattened and highly-flattened LMC globulars.

STAR CLUSTERS IN THE MAGELLANIC CLOUDS

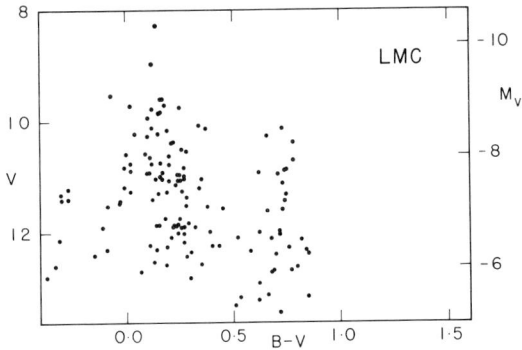

Fig. 2. Integrated C-M diagram of the LMC clusters

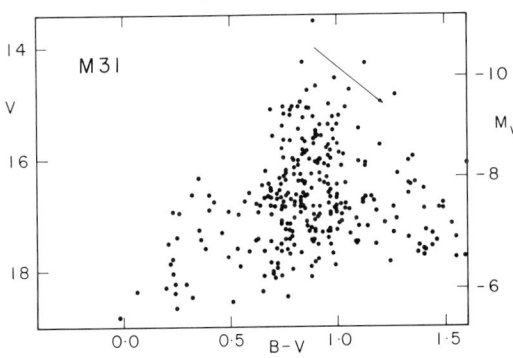

Fig. 3. Integrated CM diagram of M31 clusters

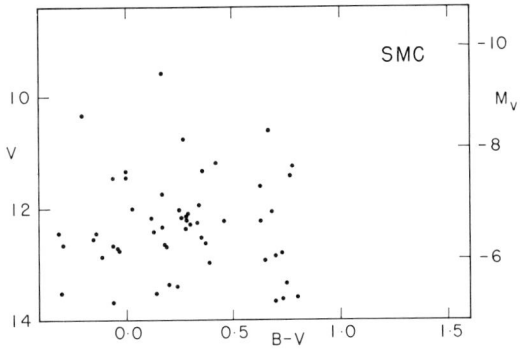

Fig. 4. Integrated Colour-Magnitude diagram for SMC clusters

III THE COLOURS OF CLOUD CLUSTERS

Fig. 2 shows an integrated colour-magnitude diagram (van den Bergh 1981) for all LMC clusters for which photoelectric UBV observations are presently available. The figure shows that (1) blue clusters outnumber red ones by a 5 to 1 margin and (2) the most luminous blue clusters are ~ 2 mag brighter than the most luminous red ones. In this respect the LMC differs drastically from the Andromeda Nebula (see Fig. 3) in which red clusters vastly outnumber blue clusters (Sharov and Lyuty 1982) and in which the most luminous red clusters are much brighter than the most luminous blue clusters. The reason for this striking difference is, no doubt, that the rate of cluster formation has declined drastically over the lifetime of M31, whereas it has remained much more nearly constant with time in the Large Cloud.

The colour-magnitude diagram for Small Cloud clusters (see Fig. 4) is basically similar to that of the Large Cloud except that the SMC contains fewer very luminous blue clusters than does the LMC. For $M_V < -8.0$, for which the photoelectric data should be reasonably complete, the LMC contains 25 blue clusters versus 3 such objects in the SMC. (If the LMC and SMC produced the same number of bright blue clusters per unit luminosity one would have expected to observe 5.7 such objects in the Small Cloud i.e. most of the observed difference is due to the fact that the SMC is intrinsically less luminous than the LMC.)

I hope that the results presented by other speakers at this symposium will throw more light on the origin of the puzzling differences between Cloud clusters and Galactic clusters that have been discussed above.

REFERENCES

Frenk, C.S., and Fall, S.M., 1982. M.N.R.A.S. **199**, 565
Gascoigne, S.C.B., 1966. M.N.R.A.S. **134**, 59
Geisler, D., and Hodge, P.W., 1980. Ap.J., **242**, 66
Hesser, J.E., Harris, H.C., van den Bergh, S., and Harris, G.L.H., 1984. Ap.J., in press
Norris, J., 1983. Preprint
Searle, L., Wilkinson, A., and Bagnuolo, W.G., 1980.
    1980. Ap.J., **239**, 803 1980
Sharov, A.S., and Lyuty, V.M., 1982. Ap.Space Sci., **90**, 371
Tifft, W.G., 1963. M.N.R.A.S., **125**, 199
van den Bergh, S., 1981. Astr.Ap.Suppl., **46**, 79
van den Bergh, S., 1982. Observatory, **102**, 228

DISCUSSION

Fall: A natural explanation for the correlation between ellipticity and luminosity is in terms of the correlation between ellipticity and age. The LMC clusters appear to get rounder with age and they also get fainter with age; hence they appear flattest when brightest and rounder when faintest. Indeed, the correlation between ellipticity and luminosity might be interpreted as support for the correlation between ellipticity and age because the clusters certainly have faded with time.

van den Bergh: First it will have to be established that there is still an age-ellipticity relation _after_ the luminosity-ellipticity relation is removed.

Fall: From measurements by Frenk and myself we find that the significance of the correlation between the ellipticity and the relative ages of populous clusters in the LMC is 97%. This is increased a bit when our large sample is supplemented by the small sample of Geyer and Richtler, which also shows a trend between ellipticity and age.

van den Bergh: I believe that your age versus ellipticity relation is a pseudo correlation that results from the fact that blue LMC clusters are, in the mean, brighter than the old red LMC clusters. A two component analysis will have to be made to see if any statistically significant age-ellipticity relationship remains _after_ the ellipticity-luminosity relation is removed.

Mathewson: The observed flattening of LMC clusters may be due to their forming in the disk rather than in the halo as in the case of our galaxy (see Freeman, these proceedings).

van den Bergh: That does not explain why _young_ clusters in the galaxy are round whereas young clusters in the Magellanic Clouds are flattened. Also bright clusters in the halo of NGC5128 appear to be flattened.

Alcaino: Considering that NGC5128 at a distance whose lower limit is estimated at, I believe, 3 Mpc, is it at all possible to appreciate within some degree of accuracy the flattening of a globular cluster candidate?

van den Bergh: Yes! On CTIO plates taken in good seeing it is possible to see that some of the brightest globulars associated with that galaxy are distinctly non-stellar.

Frogel: Since visual extinction of globulars is a strong function of galactic coordinates, how do you know that your correlation of ellipticity with extinction isn't really due to correlation with location in the Galaxy?

van den Bergh: Tests show that ellipticity of galactic globulars is more strongly correlated with extinction than it is with galacto-centric distance.

Demarque: The galactic globular cluster Omega Cen has the rare property of having a measurable ellipticity. It is also of course very old. Are you suggesting that Omega Cen might be an example of a cluster cannibalized by our Galaxy?

van den Bergh: I think that it is probably significant that Omega Cen is both the most luminous and the most flattened galactic globular cluster.

BLUE CLUSTERS, POPULOUS CLUSTERS, AND GLOBULARS

Paul W. Hodge
Astronomy Department, University of Washington, Seattle, WA

ABSTRACT

The introduction of new photometric techniques for faint objects has tremendously increased our understanding of the MC clusters, especially at the old end of the age spectrum. Many of the well-studied clusters of the past have gone through several transformations, starting as true globulars in 1960, becoming intermediate-age objects in the 1970's, but now returning to the true globular classification in the 1980's. LW 868 is an example of a cluster with such a history. Better photometric accuracy is also greatly improving our understanding of younger clusters, allowing detailed comparisons with theoretical evolutionary models, and a more reliable determination of the age-abundance relationship. The dynamical history of MC clusters can also be examined from the new data. The cluster formation rate, as well as the destruction rate, can be reconstructed from modern, deep cluster surveys.

I. OLD GLOBULAR CLUSTERS

Developing tradition, expanded somewhat by recent experience beyond the local Galactic environment, has held that for a cluster to be a globular cluster it must be luminous, large, massive, very old, and either heavy-element poor or not. Numbers have not yet been set by the IAU for these parameters, but generally all of them are fairly loose, except for age. Most astronomers seem to agree that any cluster younger than about 1.0×10^{10} years is too young to be a genuine globular, regardless of its other properties. Unfortunately, age is the most difficult of these parameters to measure for MC clusters. The only two really reliable methods, among many that have been tried, are searches for RR Lyrae variables and CM diagrams carried out down to the main sequence (MS) turn-off, which unfortunately for the MC's is at $V \sim 22$ to 23, a level unattainable until the last year or two.

For the LMC we now have thorough searches for RR Lyrae variables in most, if not all, of the likely true globulars (Graham and Nemec 1984).

In addition to the well-known outlying cases of NGC 1466, NGC 1841 and NGC 2257, there are well-established RR Lyrae populations in the more central clusters NGC 1786, NGC 1835, and NGC 2210. Nemec (1983) has produced a careful study of those in NGC 2210 and NGC 2257, even finding period changes and several double-mode pulsators among them.

A cluster that apparently does not have RR Lyrae variables but which may belong in this class is SL 868 (sometimes called H11). Originally picked out as a prime candidate for a genuine globular cluster, it was subsequently found to have an anomalous CM diagram (Gascoigne 1966). Later data taken to fainter limits (Freeman and Gascoigne 1977) led to an ambivalent conclusion about its age, while a somewhat later, independent study (Walker 1979) derived an age of only 6×10^8 yr. The latest word, based on much deeper surveys (Hesser et al. 1983 and Andersen et al. 1984), however, indicates a return to the conclusion that it is a genuine globular cluster.

For the SMC, the only certain case of a globular cluster that contains RR Lyraes is still NGC 121, for which Hesser et al. (1983) have a deep CM diagram confirming its old age. K3, long suspected to be old, but with a checkered history not unlike that of SL 868, is not quite old enough to be a true globular, according to the most recent results (Rich et al. 1984). Table 1 is a tentative list of the latest results of which I am aware; it may well need revision very soon.

TABLE 1. Genuine Globular Clusters

Cluster	Criterion		Recent References
	RR Lyraes	CM Diagram	
LMC			
1466*	√	√	Penny 1975, Hesser et al. 1983
1786	√	no	Nemec 1983, Graham et al. 1983
1835	√	no	Graham et al. 1983
1841	√	√	Kinman et al. 1983
2210	√	√	Graham et al. 1983, Nemec 1983, Harris et al. 1983
2257	√	√	Graham et al. 1983, Nemec 1983, Hesser et al. 1983
SL868=H11	no	√	Graham et al. 1983, Hesser et al. 1983
SMC			
121	√	√	Hesser et al. 1983

*Membership questioned (Cowley and Hartwick 1981).

A subject of some considerable disagreement for these clusters (as well as for the blue globulars discussed below) has to do with their projected shape. Geisler and Hodge (1980) found the older clusters of the LMC to be significantly more elliptical than Galactic globulars and Geyer et al. (1983) at least partially confirm this conclusion. Frenk and Fall (1982), however, obtain rather different ellipticities from their measurements. Kontizas et al. (1983) find SMC clusters of both old and young age to be more elliptical than Galactic, or even than LMC, clusters. Because there are interesting possible correlations of shape with age and/or composition (see also Cowley and Hartwick 1982), this is a question that should be cleared up by additional study.

II. BLUE GLOBULAR CLUSTERS

As photometrists have pushed the limits of CM diagrams to fainter and fainter limits, MC globular-like clusters have been found in all age ranges. Among the very young clusters in each Cloud, the most luminous and best studied are NGC 330 in the SMC and NGC 2100 in the LMC. The former was one of the targets of Arp's (1969) pioneer study, and it was investigated again photometrically by Robertson (1974) and spectroscopically by Feast (1972). New, comprehensive, faint-limit photometry is now available from Janes and Carney (1983), who pushed the main sequence photometry down to V = 20. A spectroscopic study in the UV has just been completed for the brighter stars in NGC 330 by Böhm-Vitense et al. (1983a), who find some unexpected properties of the UV extinction and who measure the age to be 2×10^7 years.

NGC 2100, near the 30 Doradus region and enveloped in dust (Cassatella and Geyer 1983), was found by Westerlund (1961) to be a very young cluster, though it is as large (50 pc) and as luminous ($M_V = -9$) as a globular. Robertson (1974) measured its CM diagram and more recently Böhm-Vitense et al. (1983b) obtained high and low dispersion IUE ultraviolet spectra of its brightest members. Using these to measure individual stellar extinctions and comparing UV fluxes with models, they determine an age for the cluster of 1×10^7 yrs.

Integrated UV photometry has also proven helpful in interpreting these clusters (de Boer 1984; Geyer et al. 1983).

The age range between these young values and the ages of genuine globular clusters is becoming fairly well filled out. Table 2 lists a few of the most recent age determinations, finished subsequent to an earlier review (Hodge 1983a). There are still too many data unreduced and too many rich clusters unstudied to reach any firm conclusions about many of the general properties of the system of rich clusters, but three key questions are nearing the point of being definitively answered. First, the age distribution of these clusters is not consistent with a historically uniform rate of formation. There is evidence both for periods of enhanced formation and possibly for relatively rapid destruction of these objects. Second, there is at least a rough inverse correlation of age with heavy element abundance (Cohen 1982, and previous references). Third, most of these objects belong to a system that has the kinematics of a disk population (Freeman et al. 1983, Cowley and Hartwick 1982).

Table 2. Some Very Recent Studies of Blue Globulars

Cluster	Age (10^6 yr)	[Fe/H]	Reference
LMC			
NGC 1831	190	–	Hodge & Morris 1983
NGC 1846	intermediate	–	Aaronson et al. 1983
NGC 1856	80	mildly depleted	Hodge & Lee 1983
NGC 1866	86	–0.4	Becker & Mathews 1983, Flower 1982a
NGC 1978	700	–0.5	Olszewski 1982
NGC 2058	120	–	Flower 1982b
NGC 2065	120	–	Flower 1982b
NGC 2100	10	–	Böhm-Vitense et al. 1983b
NGC 2121	400	–1.3	Flower et al. 1983
NGC 2133	130	–1	Hodge & Schommer 1983
NGC 2134	110	–1	Hodge & Schommer 1983
NGC 2190	intermediate	–	Aaronson et al. 1983
NGC 2203	intermediate	–	Hesser et al. 1983
E2	intermediate	–	Aaronson et al. 1983
23 clusters	various	–	Flower 1983
SMC			
NGC 330	20	–	Böhm-Vitense et al. 1983a, Janes & Carney 1983
NGC 339	intermediate	–	Aaronson et al. 1983
NGC 419	intermediate	–	Aaronson et al. 1983
NGC 643	intermediate	–	Hesser et al. 1983
K3	5000	–	Rich and Mould 1983
L 113	4000	–1.4	Mould & DaCosta 1983

III. OTHER MC CLUSTERS

Less attention has been paid to the thousands of clusters in each Cloud that are the obvious counterparts of our Galaxy's open clusters. The Edinburgh group (e.g., Brück 1975, Kontizas 1983, and many other references) has studied the cluster system of the SMC, providing catalogues, CM diagrams, and structural parameters for many open clusters. Less exhaustive open cluster surveys have been carried out for a few SMC fields at the University of Washington (Hodge 1983b, van Duine 1983).

For a series of deep 4-m plates of the LMC, Hodge (1983b) and Olszewski (1982) have made exhaustive searches that should include all possible star clusters of all possible ages, up to the Hubble time.

As expected from dynamical considerations, the number of open clusters decreases dramatically with age. Somewhat surprisingly, considering the very different dynamical environments, the LMC cluster destruction rate is almost identical to that of the Galaxy (Wielen 1971).

I want to thank the many astronomers who sent me information and preprints in advance of publication. There are many important papers relevant to my topic that I could not include because of space limitations and I apologize to their authors. I thank NASA for supporting my IUE spectroscopy of MC cluster stars.

REFERENCES

Aaronson, M., Olszewski, E. and Schommer, R.: 1983, in preparation.
Andersen, J., Blecha, A., and Walker, M.F.: 1984, this volume, p. 41.
Becker, S. A. and Mathews, G. J.: 1983, Astrophys. J. 270, 155.
Böhm-Vitense, E., Hodge, P. W., and Proffitt, C.: 1983a, preprint.
Böhm-Vitense, E., Hodge, P. W., Nemec, J. and Proffitt, C.: 1983b, preprint.
Brück, M.: 1975, Mon. Not. Royal Astron. Soc. 173, 327.
Cassatella, A. and Geyer, E. H.: 1983, Proc. Third Europ. IUE Conf., 523.
Cohen, J. G.: 1982, Ap. J. 258, 143.
Cowley, A. P. and Hartwick, F. D. A.: 1981, Astron. J. 86, 667.
Cowley, A. P. and Hartwick, F. D. A.: 1982, Astrophys. J. 259, 89.
de Boer, K. S.: 1984, this volume, p. 53.
Feast, M. W.: 1972, Mon. Not. Royal Astron. Soc. 159, 113.
Flower, P. J.: 1982a, Publ. Astron. Soc. Pacific 94, 122.
Flower, P. J.: 1982b, Publ. Astron. Soc. Pacific 94, 894.
Flower, P.: 1983, preprint.
Flower, P., Geisler, D., Hodge, P., Olszewski, E. and Schommer, R.: 1983, Astrophys. J., in press.
Freeman, K. and Gascoigne, S. C. B.: 1977, Proc. Astron. Soc. Australia 3, 136.
Freeman, K., Illingworth, G. and Oemler, G.: 1983, preprint.
Frenk, C. S. and Fall, S. M.: 1982, Mon. Not. Royal Astron. Soc. 199, 565.
Gascoigne, S. C. B.: 1966, Mon Not. Royal Astron. Soc. 134, 59.
Geisler, D. and Hodge, P. W.: 1980, Astrophys. J. 242, 66.
Geyer, E. H., Hopp. U., and Nelles, B.: 1983, preprint.
Graham, J. and Nemec, J.: 1984, this volume, p. 37.
Hesser, J., Nemec, J., Stryker, L. and McClure, R.: 1983, preprint.
Hodge, P. W.: 1983a, Astrophys. J. 264, 470.
Hodge, P. W.: 1983b, paper in preparation.
Hodge, P. W. and Lee, S.-O.: 1983, Astrophys. J., in press.
Hodge, P. W. and Morris, V.: 1983, preprint.
Hodge, P. W. and Schommer, R.: 1983, preprint.
Janes, K. and Carney, B.: 1983, preprint.
Kinman, T. D., Stryker, L. L., and Hesser, J. E.: 1983, preprint.
Kontizas, M.: 1980, Astr. Ap. Suppl. 40, 151.
Kontizas, M.: 1983, preprint.

Kontizas, E., Dialetis, D., Prokakis, Th., and Kontizas, M.:
   1983, preprint
Mould, J. and Da Costa, G.: 1983, preprint.
Nemec, J.: 1983, Ph.D. dissertation, University of Washington.
Olszewski, E.: 1982, Ph.D. dissertation, University of Washington.
Penny, A. J.: 1975, Mon. Not. Royal Astron. Soc. 172, p. 65
Rich, R. M., Mould, J. and Da Costa, G.: 1984, this volume, p. 45.
Robertson, J. W.: 1974, Astron. Astrophys. Suppl. 15, 261.
Stryker, L., Butcher, H. and Jewell, J.: 1981, in IAU Colloq. No. 68
   (Schenectady, L. Davis Press), p. 267.
van Duine, D.: 1983, preprint.
Walker, M. F.: 1979, Mon. Not. Royal Astron. Soc. 188, 735.
Westerlund, B.: 1961, Ann. Uppsala Astron. Obs. 5, No. 1.
Wielen, R.: 1971, Astron. Astrophys. 13, 309.

DISCUSSION

Searle: Do your age distributions for clusters refer to a magnitude-limited sample?
Hodge: Yes. The open clusters in the sample are those with members brighter than about $M \simeq 4.0$. This should be, therefore, a complete sample of clusters that are younger than the universe. There is also a strong correlation between age and physical size for LMC clusters. Young clusters show a wide range of sizes, while the old clusters are almost all very small.
Mould: It is necessary to distinguish between a limiting magnitude for your Cloud cluster sample and a limiting mass, because the clusters fade with time. Was this taken into account in calculating the formation history for the clusters?
Hodge: Yes. The sample is magnitude limited, which translates into a mass limit in the sense that a cluster so small in population as to have no stars brighter than $M \simeq 4.0$ will not be detected. By analogy with the Galaxy, I do not believe such clusters exist, as they are disrupted, and thus our sample is probably complete.
van den Bergh: Do you feel that there might be a conflict between your observed distribution of cluster ages in a complete sample and the notion that there was a great burst of star formation in the LMC a few Gyrs ago!
Hodge: No. I believe the age data need to be made more complete and reliable before the clusters can be used to test the burst hypothesis.

THE INTEGRATED SPECTRA OF STAR CLUSTERS AND THE HISTORY OF THE
MAGELLANIC CLOUDS

Leonard Searle
Mount Wilson and Las Campanas Observatories of the
Carnegie Institution of Washington

ABSTRACT

This paper reviews the attempts that have been made to derive the ages and compositions of star clusters from studies of their integrated light. It discusses what can be learned by such methods regarding the history of the Magellanic Clouds. Finally, it reviews what is known about the age-spread and abundance-range in the cluster-systems of other galaxies and considers the history of the Magellanic Clouds in this context.

1. THE INTERPRETATION OF INTEGRATED LIGHT

The simplest problem in the study of the integrated light of stellar systems is that of inferring the age and chemical composition of a star cluster from the character of its integrated spectrum. If we can't do that, how can we hope to understand the spectra of galaxies, or interpret their spectral evolution?

The Magellanic Clouds are the ideal laboratory for investigating such problems. They contain many populous clusters of different ages and compositions, and they are sufficiently close that inferences from integrated light can be checked by spectroscopy and photometry of individual stars. What is learned in this way about the interpretation of integrated spectra can be applied to more remote galaxies, making possible the intercomparison of the systematic properties of cluster systems.

It is reasonable to begin with the assumption that the main factors determining the integrated spectrum of a star cluster are its age and its chemical composition. It is not difficult to imagine other factors that may play a role but, with an important exception that I shall discuss in a moment, there is no direct evidence that any other factors actually do modify integrated spectra. If we are mistaken in this assumption we shall find out soon enough! The important exception that

I referred to is the undoubted importance of stochastic fluctuations in the the populations of rare but luminous stars. Persson and his colleagues (1983) have shown that in a typical intermediate-age Cloud cluster about half the bolometric luminosity is radiated by two or three carbon stars. The infrared spectra of such clusters are inevitably dominated by stochastic effects. Similar effects are likely to be important in the far UV. Stochastic effects will be minimal in those spectral regions dominated by the light of numerous faint stars. For clusters with turnoff effective-temperatures between 5000 and 10000K, the spectral region around 4000Å appears to be optimum. At this wavelength light-profiles of populous clusters are quite smooth, showing that this light comes from large numbers of faint stars.

If two parameters, age and abundance, determine the blue spectra of star clusters, their spectral classification will evidently require a two-dimensional scheme. Older classifications of cluster spectra were one-dimensional, and are therefore quite incapable of describing the relations among clusters that formed in galaxies with essentially different chemical histories. I shall not review in any detail these one-dimensional classification schemes here. For globular clusters the best schemes have beem photometric. Particularly important are those of Zinn (1980) and of Aaronson and his colleagues (1978) for the clusters of the Galaxy, and that of Frogel, Persson, and Cohen (1980) for the clusters of M31; these schemes provide useful abundance rankings. No convincing two-dimensional classification of the spectra of Galactic globular clusters has yet been achieved, although it is clear from the existence of the "second parameter" phenomenon that such a classification is needed. The age classification of the spectra of open clusters is less well studied. Again, the best schemes are photometric and can provide useful age estimates for sufficiently populous clusters [see, for example, Sandage (1963), Searle, Sargent, and Bagnuolo (1973), and Larson and Tinsley 1978)]. Stochastic effects are important for most open clusters and have been explored by Barbaro and Bertelli (1977). No two-dimensional work has been attempted, so far as I know.

The realization that there is a need for a two-dimensional classification scheme for the integrated spectra of star clusters arose from intercomparison of the spectra of old clusters in the Clouds with those of Galactic globulars. Danziger (1973) obtained the first significant quantitative data on the spectra of old Cloud clusters and attempted a first classification. His ordering was essentially one by metal-line strength and it plainly failed to order all the line-indices that he had measured. While puzzling over this data, I luckily happened upon a reference to Gelfand's (1969) thesis: "Seriation of Multivariate Observations through Similarities." Application of Gelfand's algorithms quickly brought out the real order to be found in Danziger's data. This order is not one by metal-line strengths but rather by hydrogen-line strength. Along the sequence of clusters ordered in this way the metal-line strengths smoothly vary, but their behavior is not monotonic. This work was published, along with a supporting photometric classification, in a paper by Althea Wilkinson, Bill Bagnuolo, and myself (SWB 1980).

A brief word about this SWB classification. It is not wedded to
the photometric system that I happened to use; it can be reproduced in
your favorite system, so long as that generates two independent
reddening-free blanketing measures. In the case of the Clouds, where
reddening seems to be negligible, even the two-color plane of UBV photometry suffices, as was first realized by Frenk and Fall (1982). As
I shall show, the SWB classification is also a spectral classification.

The natural classification plane for the integrated spectra of old
star clusters has the strength of the Balmer lines on one axis and the
strength of the metal-lines on the other. The globular clusters of the
Galaxy and the old clusters of the Clouds lie on sequences in this plane,
but the sequences are different. Rabin (1982) termed this classification
diagram the Hydrogen-Metals diagnostic diagram, or HMD. In a careful
study of the integrated spectra of 16 old star clusters in the Magellanic
Clouds, he both confirmed its importance for the classification of integrated spectra, and made clear its physical interpretation. From models
based on stellar evolutionary tracks, and on model-atmospheres of a
cluster's component stars, Rabin was able to estimate quantitatively the
behavior of hydrogen and metal line strengths in the integrated light of
star clusters. He convincingly demonstrates that the hydrogen-line
strengths in the integrated spectra of a star cluster are, and are
expected to be, strongly dependent upon the cluster's age.

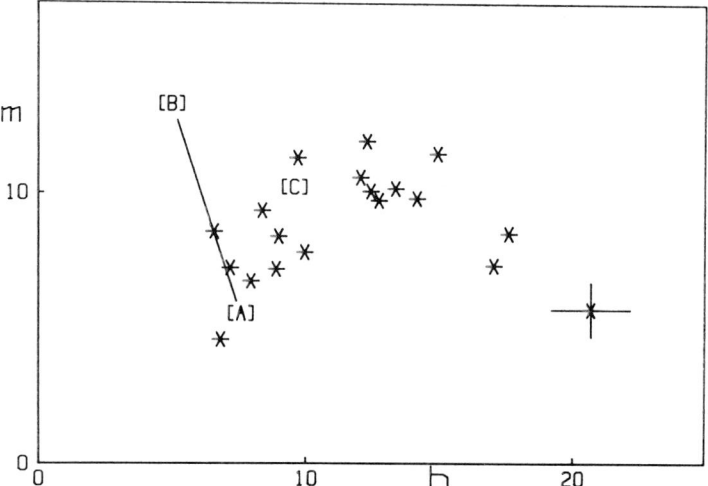

Figure 1. The HMD for integrated spectra of old clusters
in the Large Cloud. The straight line is the locus of
Galactic globular clusters. The points A, B, and C are
for the clusters M15, 47 Tuc, and NGC 2158 respectively.

In Figure 1, I have plotted the HMD for some star clusters in the Large Cloud, together with the locus of the Galaxy's globulars. The data are taken from observations of integrated spectra obtained with Shectman's (1981) photon-counting spectrograph on the du Pont telescope at Las Campanas. The data are from a study in progress by Horace Smith, Armando Manduca, and myself. The quantity h is proportional to the equivalent widths of the Balmer lines, while m is proportional to the equivalent width of strong metallic features in the spectra. The errors in the determination of these quantities arise not from photon statistics but from stochastic effects in the subtraction of the spectrum of the star field in which the cluster is embedded. It will be difficult to beat these errors down. These new results confirm Rabin's conclusion that, in the HMD, almost all the old cloud clusters lie apart from the sequence of Galactic globular clusters.

In Figure 1 the LMC clusters define a sequence; the scatter about it is no greater than that expected from the precision achieved. In Figure 2, I have drawn zones in the HMD that contain clusters of different SWB type. The point here is that the spectra insist on the same classification as that derived from the SWB photometry. The advantage of a classification in terms of the HMD, rather than in terms of photometric indices, is that the physical interpretation of the classification becomes clear.

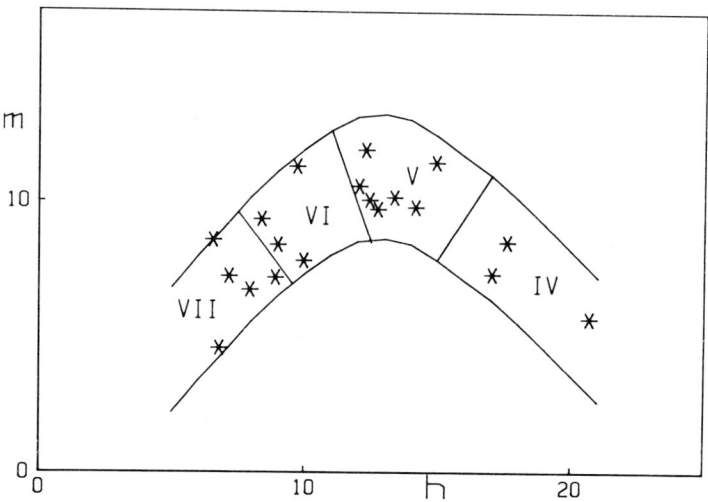

Figure 2. The SWB classification in the HMD.

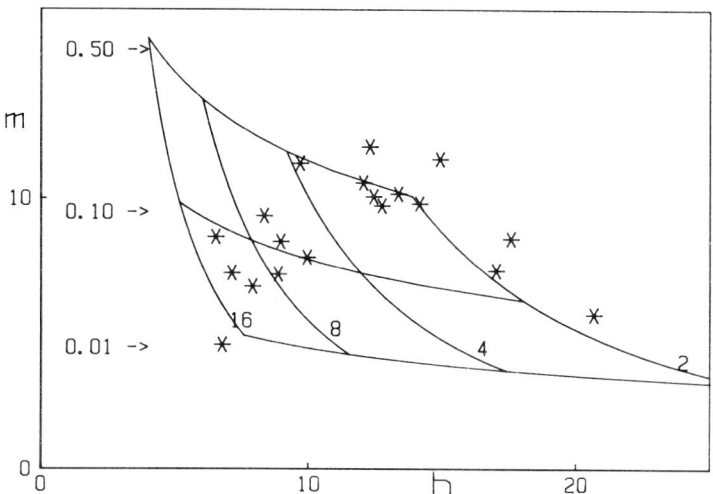

Figure 3. Preliminary classification of the HMD after Manduca. Vertical grid lines are labelled by age in gyr. Horizontal grid lines are labelled by metal-abundance in solar units.

Manduca has recently extended and improved on Rabin's model-making using the synthetic spectra of Bell and Gustafsson. In Figure 3, I show a preliminary version of Manduca's calibration of the HMD. The problem of calibration is difficult and fundamental, and I do not believe that it can be solved by model-making alone. Clearly, a few reliable ages and abundances for calibrating clusters are necessary before one can have confidence. But, Manduca's work is a big step forward, and I think it is now possible to draw reliable inferences from integrated spectra concerning the relative ages and compositions of clusters.

How do the results compare with those of others? Forming the average of the values of h and m for the clusters of SWB types V, VI, and VII respectively, I obtain ages of 2, 5, and 10 gyr respectively for those types. The corresponding values of $\log z$ (in solar units) are -0.3, -0.7, and -1.3. These ages are somewhat lower than those suggested by Rabin for types V and VI and in fair agreement with the median age at a given type in the compilation of ages inferred from color-magnitude diagrams by Hodge (1983). The compositions for a given type are also in fair agreement with the results of Cohen (1982) who derived abundances from spectroscopy of individual cluster giants.

It is comforting to see convergence, but possibly we shall learn more by focussing on disagreements. Let us consider some of these. Some clusters were undoubtedly misclassified in SWB. A good example is NGC 1831. Frenk and Fall (1982) pointed out, and spectroscopy confirms, that this is a type IV, and not a type V as SWB photometry suggested.

Another type of disagreement arises, I suspect, from misinterpretation of color-magnitude arrays; an example may be NGC 416. In Hodge's (1983) compilation NGC 416 and NGC 419 are assigned the same age of 0.6 gyr. Danziger's work shows that these two clusters have similar metal-line strengths but that their hydrogen-line strengths are very different. Rabin's spectra confirm this; the equivalent widths of the Balmer lines are 2 or 3 times greater in NGC 419 than in NGC 416. Neither cluster has a blue horizontal-branch. How can this difference in hydrogen line strengths arise if not from a big difference in age and in turnoff temperature? Rabin's Figure 5 shows that an age difference of something like 4 gyr is required. The ages assigned to these two clusters by Smith, Manduca, and myself from our new spectroscopy confirm Rabin's conclusions; we find ages of 2 and 6 gyr for NGC 419 and NGC 416 respectively. So here is a nice test case.

2. THE HISTORY OF THE CLOUDS.

When the calibration of the HMD is put on a sound empirical basis it will directly yield the age-abundance relations for both Clouds. Until then, no firm conclusions can be drawn. In the present circumstances, a consideration of what the HMD implies about the history of the Clouds is still useful, however, since it helps to define problems for future work.

Notice, in Figures 2 and 3, the small abundance range that is indicated for the type V clusters. Chemical homogeneity at the present epoch is a feature of Magellanic irregulars, in contrast to spirals (Pagel and Edmunds 1981, Webster and Smith 1983). In this respect, the Large Cloud, 2 gyr ago, appears to have been typical. It would be valuable to set good limits on the abundance range among type V clusters in the Large Cloud by differential study of the spectra of their red giants.

In Figure 4, I show two sequences in the HMD based on the predictions of the simple model of galactic evolution (i.e., homogeneous evolution with constant yield, see e.g., Searle and Sargent 1972). These represent hypothetical histories of the Large Cloud and are based on present-day abundance that is 0.75 solar, a present gas fraction of 12 percent, and an age of 16 gyr for the oldest stars. The upper sequence supposes a uniform rate of star formation. The lower one superposes a Gaussian burst of star formation on a uniform background rate. This burst is supposed to have peaked 3 gyr ago, and to have a dispersion of 2 gyr. In the model illustrated the number of stars formed in the burst is just twice the number formed by the uniform background rate of star formation. Comparing such tracks with the observed location of clusters in the HMD, I tentatively conclude, on the basis of the present calibration, that the rate of star formation in the Large Cloud in the recent past was greater than the past average rate but that the hypothetical recent burst formed fewer than 50 percent of its stars. More important than this shaky inference, Figure 4 emphasizes

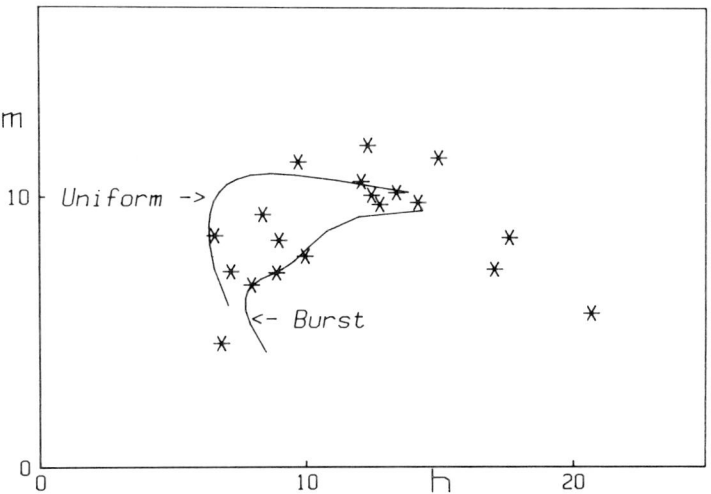

Figure 4. Hypothetical histories in the HMD. See text.

that the only clusters that contain significant information about the chemical history of the Clouds are those of SWB type VI. It would be really important to determine accurately the age and composition of some of these. NGC 1978 is probably the best candiate.

In the HMD the sequence of the Galaxy's globular clusters lies along a line of constant age. The sequence of old Cloud cluster appears to cut right across the age lines, even for cluster of SWB type VII; this is a point made by Rabin. The search for the "true globular clusters" in the Clouds is an old one, and the list of candidates shrinks as knowledge grows. Perhaps the search is futile. The clusters of the Clouds seem to be telling us that the age-abundance relation is different in different galaxies. Whether the age-abundance relations of different galaxies have points of intersection in the remote past seems to me to be a crucial question. We ought not to assume the answer.

Finally, in the HMD the clusters of the Small Cloud show a small displacement from those of the Large. On the basis of the calibration in Figure 3, Small Cloud clusters with ages between 5 and 10 gyr appear to be about twice as metal-rich as Large Cloud clusters of the same age. Since the interstellar medium of the Small Cloud today is metal-poor compared to that of the Large Cloud (Pagel and Edmunds 1981), this apparent reversal is surprising and interesting. It would be important to try to confirm or refute this inference; a differential spectroscopic comparison of red-giant spectra could provide a direct test.

3. CLOUD CLUSTERS IN CONTEXT

Many of the clusters in the Clouds have no obvious counterparts in the Galaxy. It may be that, located in the murk of the Galactic plane, we are poorly placed to make the comparison. The question I want to consider is: Do the types of Cloud clusters that have been recognized, have counterparts in other nearby galaxies?

Certainly populous blue clusters do. These are the Cloud clusters of SWB type II and III. The brightest of these have absolute visual magnitudes between -8 and -9. Christian and Schommer (1982) have discussed their occurrence in M33. Recently, Judy Cohen, Eric Persson, and I have found that in all spectroscopic and photometric characteristics Hiltner's (1960) clusters a and c in M33 closely resemble the type II and III clusters in the Clouds. The absolute magnitudes are also similar using the new distance to M33 derived by Sandage and Carlson (1983). Similar clusters also exist in M31. In his classic study, van den Bergh (1969) recognized a number of Hubble's "globular clusters" as being blue and young. Hubble 5 is one of the brightest; it lies near the boundary of the type II and type III clusters in the Q-Q plane according to my spectrophotometry, just like the clusters a and c of M33. It has an absolute visual magnitude of -8, but is probably somewhat reddened.

Nor are intermediate-age clusters unique to the Clouds. Christian and Schommer (1982) have an interesting discussion of this point. NGC 1783 is one of the more remarkable intermediate-age clusters in the Large Cloud. It is a type V, with an age near 2 gyr according to the HMD, and has an absolute visual magnitude of -7.8. This is less than 2 magnitudes fainter than the brightest younger clusters. If a cluster has a Salpeter mass function it fades about 5 magnitudes as it ages from .01 gyr to 1 gyr. The existence of such a cluster as NGC 1783 might, therefore, be taken to support the notion of a special epoch of star formation in the Large Cloud a few billion years ago. Do clusters like NGC 1783 exist in other nearby galaxies?

I think that they do. In absolute magnitude and location in the Q-Q plane (i.e., in SWB type), Hiltner's cluster f of M33 is a close match to NGC 1783. The same problem exists for it too, it cannot have evolved from clusters like the younger clusters now seen, if the mass-functions are of Salpeter type. Similar clusters, with similar problems appear among the "globular clusters" of M31; numbers 68 and 137 from Vetesnik's list (1960) are the brightest found in my unpublished survey. Whether type VI clusters, like NGC 1978, exist in M31 or M33 is an important unanswered question. Such clusters exist in the peripheral regions of the Galaxy's disk. NGC 2158, whose location in the HMD is illustrated in Figure 1, is a well known example (Arp and Cuffey 1962, Hardy 1981). In any case, the existence of blue and intermediate-age globular clusters in not a privilege unique to the Clouds.

A real and important difference between the Clouds and the Galaxy is that the latter contains old metal-rich clusters. All systems contain old metal-poor clusters; they differ in the abundance range of old clusters, and in the fraction of old clusters that are metal-rich. As a readily visible signature of such differences among galaxies, we might take the presence or absence of clusters at least as old as 47 Tucanae and at least as metal-rich. Such clusters can easily be recognized photometrically and spectroscopically. The existence of such old, metal-rich clusters is, of course, evidence for rapid enrichment. No such clusters are known in the Clouds. None are known in M33 either, although the study of the clusters in this galaxy is very incomplete. In the Galaxy some 10 percent of the globular clusters are of this type, while in M31 about 30 percent of its clusters meet the requirements. There appears to be a significant trend here, perhaps related to the prominence of the spheroidal population.

The facts that I have reviewed suggest some caution in accepting catastrophist accounts of the evolution of the Clouds. The cluster contents of the Clouds are not so unusual as once was thought, and the undoubted differences between the chemical history of the Large Cloud and that of the Galaxy may find their explanation within the regular systematics of galactic chemical evolution.

REFERENCES

Aaronson, M., Cohen, J.G., Mould, J., and Malkan, M.: 1978, Astrophys. J. 223, 824.
Arp, H.C., and Cuffey, J.: 1962, Astrophys. J. 138, 356.
Barbaro, G., and Bertelli, G.: 1977, Astron. Astrophys 54, 243.
Christian, C.A., and Schommer, R.A.: 1982, Astrophys. J. Suppl. 49, 353.
Cohen, J.G.: 1982, Astrophys. J. 258, 143.
Danziger, I.J.: 1973, Astrophys. J. 181, 641.
Frenk, C.S., and Fall, S.M.: 1982, M.N.R.A.S. 199, 565.
Frogel, J.A., Persson, S.E., and Cohen, J.G.: 1980, Astrophys. J. 240, 785.
Gelfand, A.E.: 1969, Thesis, Stanford University, University Microfilms, Ann Arbor, Michigan.
Hardy, E.: 1981, Astron. J. 86, 217.
Hiltner, W.: 1960, Astrophys. J. 131, 163.
Hodge, P.W.: 1983, Astrophys. J. 264, 470.
Larson, R.B., and Tinsley, B.M.: 1978, Astrophys. J. 219, 46.
Pagel, B.E.J., and Edmunds, M.G.: 1981, Ann. Rev. Astron. Astrophys. 19, 77.
Persson, S.E., Aaronson, M., Cohen. J.G., Frogel, J.A., and Matthews, K.: 1983, Astrophys. J. 266, 105.
Rabin, D.: 1982, Astrophys. J. 261, 85.
Sandage, A.R.: 1963, Astrophys. J. 138, 863.
Sandage, A.R., and Carlson, G.: 1983, Astrophys. J. (Letters) 267, L25.
Searle, L., and Sargent, W.L.W.: 1972, Astrophys. J. 173, 25.
Searle, L., Sargent, W.L.W., and Bagnuolo, W.G.: 1973, Astrophys. J. 179, 427.

Searle, L., Wilkinson, A., and Bagnuolo, W.G.: 1980, Astrophys J. 239, 803.
Shectman, S.A.: 1981, Ann. Report of the Director, Mount Wilson and Las Campanas Observatories, 1980-81, p. 586.
van den Bergh, S.: 1969, Astrophys. J. Suppl. 19, 145.
Vetešnik, M.: 1962, Bull Ast. Inst. Czechoslovakia 13, 180.
Webster, B.L., and Smith, M.G.: 1983, M.N.R.A.S. 204, 743.
Zinn, R.J.: 1980, Astrophys. J. Suppl. 42, 19.

DISCUSSION

McCarthy: Could you illustrate the hydrogen line fluctuations in the ranking of metal-line strengths? What is the magnitude of such fluctuations?
Searle: The answer to your question is contained in Figure 1. Very roughly you may take the quantity to be the equivalent width of Hγ. It ranges from 5 to 20Å with a measurement uncertainty of about 1Å. In the ordering by hydrogen-line strength the metal-line strength behaves regularly but not monotonically.
Graham: How closely do the integrated spectra of the oldest, metal-poor clusters in the Galaxy and the Magellanic Clouds compare?
Searle: I think you can see the answer from Figure 1. I have the impression that most type VII clusters have significantly stronger hydrogen lines than the galactic globulars. Rabin already made this point. It seems likely that the age-metallicity relations of the Clouds and the Galaxy began to disagree very early.
Frogel: Would you comment on D. Burstein's claim that M31 globulars have systematically stronger H-line strengths than Galactic globulars? Do the M31 clusters differ from Milky Way clusters in the same way as LMC clusters do?
Searle: I believe that Rabin (in his thesis) concluded that the phenomenon you refer to existed for Hβ but not for Hγ and Hδ. He attributed it to blending, I believe. In any case, I doubt that it is in any way very connected with the hydrogen line strength-age relation found in the Clouds.
Mould: Your (m,h) diagram suggested a lower dispersion about their mean line for the galactic globular clusters than those of the Clouds. Is that real? Is there a reason for it?
Searle: I think it results from observational techniques. The background subtraction problem is more difficult for the clusters in the Clouds.
Peimbert: Do you think that the positions of the two youngest clusters of the LMC plotted in the (m,h) diagram are significant?
Searle: Menduca's grid (from the work we do as mentioned in the text) refers to clusters older than 2 Gyr - it would be unsafe to extrapolate it to young clusters. In particular, I do not think it would be safe to infer from Figure 3 that the clusters you refer to are metal deficient compared to the type V's. I doubt that very much. More work is needed on the integrated spectra of young clusters.

Flower: Let me point out a potentially serious problem with Rabin's (1982 Ap.J. 261, p 85) cluster age estimate based on Balmer-line strengths from integrated spectra of Magellanic Cloud star clusters. He used integrated light models for star clusters which were constructed from a grid of evolutionary tracks that did not include core-helium-burning giants (no published grid includes these stars). However, color-magnitude diagrams of populous (red or blue) Magellanic Cloud star clusters are dominated by red giant (core-helium-burning) stars. Color-magnitude diagrams of NGC2121 and NGC1978 (both c-m diagrams suggest ages of less than 10 Gyr) show red clump giants (B-V $\approx$ 1.0) that are a magnitude or more brighter than the brightest main-sequence stars. These red giants are as bright in the blue as are the brightest main-sequence stars. Thus, the Balmer line strengths from integrated cluster spectra are clearly contaminated by these bright red giants. The brightness of core-helium-burning red giants relative to the brightest main-sequence stars depends on both the chemical composition and the number of giants. Thus chemical composition variations and stochastic effects in the population of red giants from cluster to cluster will tend to invalidate current integrated light ages of Magellanic Cloud star clusters.

Searle: The answer is, of course, that red giants are relatively rare. The smoothness of scans over these clusters shows that their blue light is dominated by contributions of intrinsically faint stars. Stochastic effects are not important here. Danziger's, Rabin's, and our new spectra confirm that the integrated light of NGC1978, for example, closely resembles that from Kron 3, which has an age of 7 Gyr. If NGC1978 has an age an order of magnitude smaller, as you suggest, the value of intergrated spectra as an age indicator would, indeed, be brought into question. But it would take better color-magnitude diagrams than those currently available to cause me to have serious doubts. I suggest the situation may be similar to Hodge 11, for which c-m diagrams gave erroneous conclusions. The great age of Hodge 11, insisted upon from its integrated light, has been confirmed by the better color-magnitude diagrams now available.

MASSES AND RELAXATION TIMES OF STAR CLUSTERS IN THE SMC

M. Kontizas and E. Kontizas

University of Athens, Observatory of Athens

ABSTRACT

Plates taken with the 1.2m U.K. Schmidt Telescope and the 3.8 AAT Telescope have been used in order to derive the dynamical parameteres of 43 various clusters of the SMC by means of star counts. The clusters are divided into two main categories : (i) the disk, "blue" and "intermediate" in colour, young, mainly globulars and (ii) the halo, "red", old globular clusters. The disk clusters have been found to be more massive and older than the galactic open clusters, whereas the halo clusters are at least 10 times less massive than the galactic globulars. The relaxation times of the disk clusters are larger than their evolutionary age while the observed density profiles always show evidence of well relaxed systems.

OBSERVATIONS

The observational material was taken by the 1.2m U.K. Schmidt Telescope in Australia and by the 3.8m AAT Telescope on IIaD and IIaO emulsion plates. The star count reductions and the derived density profiles are discussed and given elsewhere (Kontizas, Danezis, Kontizas, 1981 ; Kontizas and Kontizas, 1982).

DYNAMICAL PARAMETERS

By means of star counts it is possible to derive the stellar density distribution and therefore the tidal radii, r_t, of the star clusters (King, 1962). Assuming that all clusters are approximately spherical a fit with theoretical models gives us the core radii, r_c, and the concentration parameters, $\log \frac{r_t}{r_c}$.

These values and the location of each cluster in the SMC permit us to derive their masses and relaxation times. The integrated colours,

existing c-m diagrams and previous classifications of the SMC clusters were the criteria of dividing the clusters into two basic categories :
   (i) the disk clusters which are the young, "blue" and intermediate in colour clusters and
   (ii) the halo, "red" old clusters

The disk clusters were found to have large tidal radii compared to the galactic open clusters with values accumulated at ~30pc and concentration parameters higher than their galactic counterparts. Their masses are also found at least 10 times higher than those of our galaxy. Their relaxation times, much higher than their evolutionary ages mean that the clusters are not relaxed, although their density profiles favour well relaxed systems. The LMC, "blue", clusters were found to behave in the same way (Freeman, 1974 ; Geyer and Hopp, 1982).

The "halo" clusters are found to have tidal radii and concentration parameters similar to those of the galactic globulars. The masses are at least 10 times less massive than their galactic counterparts and the M/L ratios rather small for old clusters. Their relaxation times are smaller than their evolutionary ages (where exist) showing that the old SMC clusters are well relaxed systems and being in agreement with their density profiles.

CONCLUSION

The disk clusters of the SMC are more massive and older than the galactic open clusters whereas the halo SMC clusters are less massive and younger than the galactic globulars.

ACKNOWLEDGEMENTS

The authors would like to express their sincere thanks to the 1.2m, U.K. Schmidt Telescope Unit.

REFERENCES

Freeman, K.C. (1974). ESO/SRC/CERN Conference on Research Programs for the New Large Telescopes, Geneva, May, 1974.
Geyer, E.H. Hopp. U. (1982). IAU Colloquium No. 68, 235.
King, I.R. (1962), Astron. J., 67, 471.
Kontizas, M., Danezis, E., Kontizas, E. (1982). Astron. & Astroph. Suppl. Ser. 49, 1.
Kontizas, E. and Kontizas, M. (1983). Astron. & Astroph. Suppl. Ser. 53, 143.

AN ELLIPTICITY-AGE RELATION FOR GLOBULAR CLUSTERS IN THE LARGE
MAGELLANIC CLOUD

S. Michael Fall
Institute of Astronomy, Cambridge, England

Carlos S. Frenk
Astronomy Department, Berkeley, California

We have measured the ellipticities of 52 globular clusters in the
LMC and 93 in the Galaxy by eye from polaroid enlargements of the sky
surveys (Frenk & Fall 1982). In most cases, the measurements pertain to
regions between $(1-2)r_h$ where r_h is the median radius of a cluster; i.e.
the radius containing half of the light in three dimensions. These were
compared with determinations based on star counts for 12 members of the
LMC sample and 19 members of the Galactic sample. We found no systematic difference between the ellipticities from the two methods and
concluded that the eye-measurements are free of any major bias. They
are also in reasonable agreement with the measurements by Geyer &
Richtler (1981) and Geyer, Hopp & Nelles (1983), who used the **Agfa**
contourfilm technique. The ellipticities measured by Geisler & Hodge
(1980) from microdensitometer scans are systematically large in comparison
with our results and those of Geyer and his associates. Since the scans
cover only a small part of each cluster, a few bright stars can cause
spurious elongations in the fitted contours.

Our measurements show that the globular clusters in the LMC are
stochastically flatter than the globular clusters in the Galaxy. We
also found a correlation between the shapes of the LMC clusters and
their classification in the scheme devised by Searle, Wilkinson &
Bagnuolo (1980). An approximate dating of this sequence, based partly
on Hodge's (1983) compilation of turnoff photometry, gives $\log(\text{age/yr}) \approx$
$0.5(\text{SWB type}) + 6.6$. Young clusters are flatter on average than old
clusters with most of the change between SWB types III and IV and therefore at ages of a few times 10^8 yr. This relation has much scatter but
the general trend is significant at the 97 per cent level. Another
indication that the effect is real comes from the correlation between
the ellipticities and luminosities of clusters in the LMC (van den Bergh
1983). Since the clusters fade with time, the dependence of their
shapes on age probably explains why they appear flattest when brightest
and roundest when faintest.

One process that can change the shape of a rotating cluster is
evaporation by two-body diffusion. As Agekian (1958) pointed out, stars
that escape in the direction of rotation carry away more angular momentum
per unit mass than stars that escape in the opposite direction. He

computed this effect for a Maclaurin spheroid with an isotropic velocity distribution and found that its ellipticity decreased on a timescale of order $10^2 \tau_r$ where τ_r is the local relaxation time. Shapiro & Marchant (1976) applied this formalism to Galactic globular clusters, using central values of τ_r, and suggested that evaporation alone could explain why they are so round in comparison with elliptical galaxies. In the regions where ellipticities are measured, however, two-body diffusion is characterized better by the reference relaxation time τ_{rh} which is defined as the value of τ_r at the mean density within r_h (Spitzer 1975). Almost all of the globular clusters in the Galaxy have 10^8 yr $\lesssim \tau_{rh} \lesssim 10^{10}$ yr and those in the LMC have slightly smaller values of τ_{rh}. Since the time-scale $10^2 \tau_{rh}$ is greater than the ages of these clusters, evaporation probably plays only a minor role in the evolution of their shapes.

A more promising explanation for the relation between ellipticity and age also involves two-body diffusion but operates on faster timescales. If the clusters form by aspherical collapse and violent relaxation, they will generally be distorted by anisotropic velocity distributions when they first reach dynamical equilibrium. Two-body diffusion will then isotropize the stresses within r_h on small multiples of the reference relaxation time τ_{rh} before evaporation is important. The clusters will then become rounder until they reach a state that is consistent with their total angular momenta and isotropic velocity distributions. We have studied this process in some detail by means of N-body simulations and find that the changes in shape can be quite dramatic (Fall & Frenk, in preparation). For example, the true ellipticities in one series of models evolved from 0.4 to 0.1 on a timescale of a few τ_{rh} with no evaporation. A detailed comparison with the empirical relation between ellipticity and age is difficult because the data needed to estimate τ_{rh} are available for only a few globular clusters in the LMC. Nevertheless, the agreement is as good as could be expected and strongly suggests that changes in the velocity distributions of the clusters are responsible for much of the observed changes in their shapes.

REFERENCES

Agekian, T.A. (1958) Sov. Astron - AJ, 2, 22.
Frenk, C.S. & Fall, S.M. (1982) Mon.Not.Roy.Astr.Soc., 199, 565.
Geisler, D. & Hodge, P.W. (1980) Astrophys. J., 242, 66.
Geyer, E.H., Hopp, U. & Nelles, B. (1983) preprint.
Geyer, E.H. & Richtler, T. (1981) in "Astrophysical Parameters for Globular Clusters", eds. Davis Philip, A.G. & Hayes, D.S. (L. Davis Press, Schenectady) p.239.
Hodge, P.W. (1983) Astrophys. J., 264, 470.
Searle, L., Wilkinson, A. & Bagnuolo, W.G. (1980) Astrophys.J., 239, 803.
Shapiro, S.L. & Marchant, A.B. (1976) Astrophys. J., 210, 757.
Spitzer, L. (1975) in "Dynamics of Stellar Systems", ed. Hayli, A. (Reidel, Dordrecht) p.3.
van den Bergh, S. (1983) preprint.

SIT-VIDICON SURFACE PHOTOMETRY OF GLOBULAR CLUSTERS IN THE MAGELLANIC CLOUDS

Edward H. Geyer, Andreas Hänel
Observatorium Hoher List der Universitätssternwarte Bonn, D-5568 Daun, FRG

INTRODUCTION

The globular clusters in the Large Magellanic Cloud offer an unique possibility for the study of their spatial structures. Their ages show a very wide range from $6 \cdot 10^7$ to $>5 \cdot 10^9$ years which is unknown for their galactic counterparts. The latter ones are all very old and therefore dynamically relaxed systems, whereas the young globular clusters of the Magellanic Clouds (age $<10^8$ years) have still preserved the dynamical state with which they were formed (Geyer et al. 1979). Furthermore, these young globulars show in their Hertzsprung-Russell diagram morphology the massive and luminous stars still on the main sequence with only a few evolved red supergiants. As in general the luminosity of the individual stars depends on their evolutionary state, the total mass to total luminosity ratio of a stellar cluster represents also its evolutionary state.

Furthermore, if during the forming process of the cluster the spatial mass distribution is anisotropic or in the case of old systems mass segregation has occured by the dynamical evolution, this should be seen in the radial variation of the mass/luminosity ratio within the clusters.

We have started an observing program for star counts and surface photometry of a sample of young and old LMC globular clusters to study the spatial star and luminosity distribution to search for these effects. We present here first preliminary results about a red (NGC 1806) and a blue (NGC 1818) cluster. The star counts of these objects have been reported earlier by Geyer et al. (1979 and 1982).

The spatial density distribution is described with a polytrope and the constants are derived from the surface density distribution (counts and photometry) as described by Geyer et al. (l.c.).

OBSERVATIONS

The surface photometry in the UBV and RGU colour systems was carried out for 22 LMC globulars with a panoramic detector system ("OMA 2" - Optical Multichannel Analyzer, EG&G Corp.) based on a SIT vidicon at the 1m telescope of the European Southern Observatory. The whole equipment was described by us (1982). The field of the detector-telescope system is 2.9' x 2.9' with a pixel size of 3.45" x 3.45" (50x50 pixels for the whole frame). The SIT vidicon detector was cooled with dry ice and we used integration times of 0.9 and 5.8 minutes. The data were digitzed and stored on 8" flexible disks using a LSI 11 based microcomputer.

REDUCTIONS

All data have been corrected for dark current and flat field variations. As the microcomputer of the detector system has only limited programming facilities, we transfered the data to a HP 9835A desktop microcomputer for further reductions:
1) The center of the cluster is determined in fitting a Gaussian to the row and column sums (marginal distributions) of the array describing the image frame. This is used to center a diaphragm simulating one as used for photoelectric photometry (e.g. from literature). the intensity of the pixels is summed up in this diaphragm and with the photometric magnitude value converted to astronomical units (mag/arcsec^2/pixel). These have been used for calibrated contour plots.
2) To remove the influence of single bright stars which would considerably contaminate the strip functions, a simple filtering algorithm was applied.
3) These corrected row and column sums are fitted with the strip functions to determine the cluster center (the least square fits use the Marquardt algorithm).
4) Finally all values are made symmetric in respect to the center yielding the radial surface density distribution. These are fitted to the strip function $L(r)$ with the results for the constants describing the spatial distribution.

PRELIMINARY RESULTS FOR NGC 1806 AND 1818

From the spatial star and luminosity distribution it can be seen that the average luminosity per cluster member has its maximum in the center and drops radially outward except for NGC 1818 in B. If this finding holds true for the other clusters it would indicate that the massive stars are already more concentrated during the clusters' formation.

We gratefully acknowledge the support by the Deutsche Forschungsgemeinschaft, Bonn, and the generously granted observing time by ESO.

REFERENCES

Geyer, E.H.; Hänel, A.: 1982, ESO Messenger 29, 5
Geyer, E.H.; Hopp, U.: 1982, Astrophys. Space Sci. 84, 133
Geyer, E.H.; Hopp, U.; Kiehl, M.; Witzigmann, S.: 1979,
    Astron. Astrophys. 77, 61

INTERMEDIATE-AGE MAGELLANIC CLOUD STAR CLUSTERS

Phillip J. Flower
Clemson University

Ages of six intermediate-age Large Magellanic Cloud star clusters have been estimated using the time dependent behavior of the luminosity of stellar interior models of red giants. All clusters studied, NGC 1783, NGC 1868, NGC 1978, NGC 2121, NGC 2209, and NGC 2231, were found to have ages $< 10^9$ yr. It is concluded that there is currently no substantial evidence for a major cluster population of large, populous clusters $> 10^9$ yr in the Large Magellanic Cloud.

The distributions of red giants on the six cluster color-magnitude diagrams were compared to a grid of 33 stellar evolutionary tracks, evolved from the main sequence through core-helium-exhaustion and up the asymptotic giant branch, spanning the expected mass range (2-3 solar masses) and metallicity range ($-0.2 \leq$ [Fe/H] ≤ -1.2) for intermediate-age Large Magellanic Cloud clusters. The faintest model core-helium-burners clearly decreased in luminosity with decreasing mass; thus, model red giant luminosities decreased with age.

Although Cannon (1970) indicates that the mean M_V of the red "clump" giants reaches a limiting magnitude of +1, many galactic intermediate-age clusters exhibit red clump giants much fainter; e.g., the faintest red giants in NGC 559 reach $M_V = +2.5$, in NGC 752, NGC 1245, NGC 2477, and NGC 3496 they reach $M_V \simeq +2$. These clusters clearly substantiate the age-luminosity trends of the model core-helium-burning red giants.

Since the current main sequence photometry is generally still too inaccurate to obtain reliable Magellanic Cloud cluster ages with main sequence turnoffs, the red giant models have been used to estimate cluster ages (the red giant photometry is more accurate than the necessarily fainter main sequence photometry). These red giant ages are compared in Table 1 to the main sequence termination ages of Hodge (1982), Olszewski (1983), and Flower et al. (1983), the AGB ages of Mould and Aaronson (1982), and the integrated spectra ages of Rabin (1982). The red giant ages agreed with the main sequence termination ages; both techniques are based on cluster color-magnitude diagrams (CMD).

TABLE 1   LMC CLUSTER AGE ESTIMATES (10^9 yr)

Cluster	Red Giants	MS Termination	Mould and Aaronson	Rabin	[Fe/H] Estimate	SWB Type
NGC 2209	0.8	0.7	3	1.5-2.5	-1.1	III-IV
NGC 1868	0.75	0.3	-	-	-1.2	-
NGC 2231	0.55	1.2	-	-	-1.3	V
NGC 1783	0.46	>0.2	3	4	-0.5	V
NGC 1978	0.45	0.7	2	>6	-0.5	VI
NGC 2121	0.39	0.4	4	5	-1.0	VI

In every instance of an age estimate for a cluster by both CMD and AGB dating techniques, the CMD ages are always significantly (factor of 3 or more) lower. The AGB ages are, however, very sensitive to the choice of the mass loss parameter used in the AGB evolutionary models. Increasing it from 0.45 (Mould and Aaronson 1982) to 1.0 will reduce all AGB age estimates to less than 10^9 yr. The AGB ages are also susceptible to statistical fluctuations in the luminosity function (in many clusters only one star defines the age).

Because the integrated spectra ages (Rabin 1982) are based on integrated light models that ignore the contribution of red clump giants, the extremely large ages (based on the strengths of the Balmer lines) are probably entirely unreliable. The large number of red giants in the clusters may contribute enough flux on the blue to fill in the H-line contribution by the fainter main sequence stars, thus mimicking the effects of a weakening main sequence contribution in the blue due to cluster aging.

REFERENCES

Cannon, R.: 1970, Monthly Notices Roy. Astron. Soc. 150, pp. 111.
Flower, P.J., Geisler, D., Hodge, P., Olszewski, E., and Schommer, R.: 1983, Ap. J. December 15.
Hodge, P.: 1982, in *Astrophysical Parameters for Globular Clusters*, IAU Colloq. No. 68, A.G. Davis and D.S. Hayes, eds. (L. Davis Press: Schenectady) pp. 207.
Mould, J. and Aaronson, M.: 1982, Ap. J. 263, pp. 629.
Olszewski, E.: 1983, Ph.D. Thesis, University of Washington.
Rabin, D.: 1982, Ap. J. 261, pp. 85.

METALLICITIES OF YOUNG POPULOUS CLUSTERS IN THE MAGELLANIC CLOUDS

B. Nelles, T. Richtler
Observatorium Hoher List der Universitätssternwarte Bonn
D-5568 Daun, Fed. Rep. Germany

INTRODUCTION

We present metallicities for the young LMC clusters NGC 1818, 1866 2157, 2136 and 2214. This sample is increased by taking over additional values from literature, so for NGC 1850, 1994, 2004, 2100 from the list of Hodge (1981). The intention is to compare these values with metallicities found in a field population of comparable age. Such a population may be represented by bright Cepheids, for which metallicities based on Washington photometry (Harris 1983) are available.
The only young SMC cluster, for which we could estimate a metallicity, is NGC 330. Therefore, a comparison with SMC Cepheids, whose metallicities are also quoted by Harris (1981), can lead to uncertain statements only, until further observations of young SMC clusters will have been made.

OBSERVATIONS

Strömgren photometry has been obtained for stars in the Magellanic Cloud clusters mentioned above. Furthermore, stars in the galactic clusters M 67, 47 Tuc, M 79 and NGC 288 were observed to build up a reference frame of metallicities, in which the Magellanic Cloud stars can be inserted. The observations were performed with the ESO 1m telescope, single channel photometer and ESO standard filters.
In addition, IDS-spectra have been obtained from stars in NGC 1866 and NGC 330 as well as for 6 late type SMC supergiants and from stars in the above mentioned galactic clusters. Instruments were the ESO 3.6m and 1.5m telescopes. From these spectra, abundance indicators were synthesized following Canterna et al. (1982).
As a second photometric approach, stars in NGC 1818, 2004, 2100, 2157, 2214 and in NGC 330 have been observed in the Walraven system. The measurements were made with the Walraven photometer at the Dutch 91cm telescope on La Silla.

RESULTS

The metallicities we derived in the Strömgren system for the LMC clusters NGC 1818, 1866, 2157, 2136 are -1.6, -1.2, -0.6, -1.2, -0.4 dex respectively. The internal mean error is about 0.2 dex. The spectroscopic value for NGC 1866 is in agreement with the photometric metallicity, although the calibration of the metallicity indicators must be considered as preliminary. For the clusters NGC 1818, 2004, 2100, 2157, 2214 the Walraven data yield -0.5, -1.3, -1.4, -0.9, -1.0 dex, respectively. But the calibration is somewhat unsatisfying (perhaps due to a strong luminosity effect) so we rely on the Strömgren data alone.

A comparison with the metallicity distribution of the LMC Cepheids leads to the result, that only one cluster from the whole sample (NGC 1994 with -0.2 dex) falls in the metallicity range covered by the Cepheids. All other clusters are less abundant (also in the Walraven system). Beside all the uncertainties due to questionable cluster membership, errors in the calibration, inhomogeneous observational material etc. it seems surprising, that there is so little overlapping.
One may conclude, that young populous clusters in the LMC are preferably formed from low enriched material. In this sense a possible condition for the existence of these clusters lies in the low metal content of regions of active star formation.

The situation in the SMC remains unclear. For NGC 330 the Strömgren system yields a metallicity of -1.8 dex. This is supported by the spectroscopic method (although the Walraven data give -0.9 dex). The 6 supergiants seem to have metallicities between -1.2 and -1.5 dex. But they still lie in the range of the SMC Cepheids. Since there are much more low abundant Cepheids than in the LMC (Harris (1981) has found his most metal poor objects at about -1.5 dex) one cannot decide wether there are systematic metallicity differences between the cluster and field population until further data about young SMC clusters will become available. However, NGC 330 may be taken as a hint, that such an effect is present in the SMC too.

REFERENCES

Canterna, R. et al. : 1982, Astrophys. J. 258, 612
Harris, H. : 1981, Astron. J. 86, 1192
Harris, H. : 1983, Astron. J. 88, 507
Hodge, P. : 1981, in "Astrophysical Parameters for Globular Clusters"
            IAU Coll.68 , eds. A.G. Davis Philip and D.S. Hayes,
            L. Davis Press Inc.,Schenectady, p.205

THE H-R DIAGRAM OF NGC 1962-65-66-70

E. Schulz-Lüpertz (1,2) and M. Grewing (1)
(1) Astronomisches Institut der Universität Tübingen
(2) Space Sciences Laboratory, University of California, Berkeley

The complex NGC 1962-65-66-70 is one of the many giant filamentary shell nebulae in the LMC. The nebulosity, also known as N144 was studied originally in H-alpha by Henize (1956). It has an angular size of 698" x 620" and surrounds an association of very young blue stars. The cluster has been studied first photographically in V and B by Westerlund (1961).

We have observed the brightest members of the association with the IUE satellite and obtained both low and high resolution spectra in the ultraviolet region. Some results on these UV observations, especially concerning the stellar energy distribution of HD 269546, as well as our studies of the hot galactic coronal gas have been published earlier (Grewing and Schulz-Lüpertz, 1980, Schulz-Lüpertz and Grewing, 1982). Here we present results of direct photography, photometry, and spectroscopy. The observations have been carried out at the European Southern Observatory, La Silla, Chile. The UBV magnitudes of 68 stars in the field of NGC 1962-65-66-70 have been measured. The association contains a few very luminous evolved supergiants, but most of the stars belong to the upper main sequence of the cluster. Some galactic foreground stars can clearly be seperated by making use of the two-colour diagram. From this diagram we also derive a total interstellar reddening of $E(B-V) = 0.10 \pm 0.01$ for the LMC members in our sample.

All observed supergiants in the cluster have composite spectra, and both our optical and IUE spectrograms show strong emission features for most of these stars indicating mass loss by stellar wind. A more detailed investigation of these evolved stars on the basis of high resolution spectra from different wavelength bands is currently under way. Also the stars at the upper end of the main sequence show clearly Of emission characteristics. From our optical low resolution spectra we could derive the spectral types of some of these stars. We find a mean spectral type of O8(f) for the main sequence turnup region.

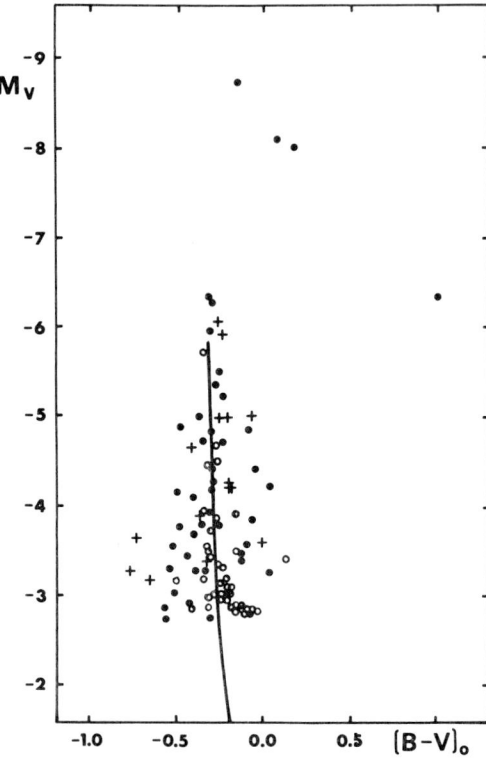

Fig. 1: Observed colour-magnitude diagram of the bright stars in NGC 1962-65-66-70.
● : cluster members observed in the present investigation;
o : additional cluster members measured by Westerlund (1961);
+ : possible LMC field stars.
The curved line represents the mean colour-magnitude relation for luminosity class V (Schmidt-Kaler, 1982).

By combining our spectroscopic and photometric data and converting them into physical parameters we are able to construct the observed H-R diagram of the cluster. This is reproduced in Fig. 1. If we compare these data with theoretical model calculations for stellar evolution which consider the mass loss effects in the most massive stars (e.g. Maeder 1981) we find an age of 3.5 - 4 million years for NGC 1962-65-66-70.

A more detailed paper on these investigations will be published elsewhere (Schulz-Lüpertz and Grewing, 1983).

REFERENCES

Grewing, M., Schulz-Lüpertz, E. 1980, Proc. 2nd Europ. IUE Conf., ESA SP-157, 357
Henize, K.G. 1956, Astrophys. J. Suppl. $\underline{2}$, 315
Maeder, A. 1981, Astron. Astrophys. $\underline{99}$, 97, and Astron. Astrophys. $\underline{102}$, 401
Schmidt-Kaler, Th. 1982, Landolt-Börnstein Vol.2b, eds. K.Schaifers, H.H.Voigt
Schulz-Lüpertz, E., Grewing, M. 1982, Mitt. Astron. Ges. $\underline{55}$, 123
Schulz-Lüpertz, E., Grewing, M. 1983, Astron. Astrophys. (submitted)
Westerlund, B. 1961, Uppsala Astronom. Obs. Ann. $\underline{5}$, no. 1

A SURVEY OF MAGELLANIC CLOUD CLUSTERS FOR RR LYRAE STARS

J.A. Graham
Cerro Tololo Inter-American Observatory
La Serena, Chile

J.M. Nemec
Dominion Astrophysical Observatory
Victoria, B.C. Canada

ABSTRACT

Preliminary results are presented of a search for RR Lyrae variable stars in old clusters in the Magellanic Clouds. RR Lyraes have been found in NGC 1786 and NGC 2210. With the exceptions of NGC 339 (SMC) and NGC 2019 and Hodge 11 (LMC), all clusters classified as type VII by Searle, Wilkinson and Bagnuolo (1980) contain RR Lyrae variables. On the other hand, none have yet been found in clusters of type VI.

At CTIO, a survey is underway to look for variable stars in the globular clusters of the Magellanic Clouds. Initially, we have used as an observing list those clusters which were classified as types VI and VII by Searle, Wilkinson and Bagnuolo (1980). Searle et al. showed that it is possible to order the integrated spectra of the Magellanic Cloud star clusters in a one-dimensional sequence. The character of the spectral variations along the sequence suggested that both cluster age and chemical composition vary as the sequence is traversed so that clusters of type VII are the oldest and most metal-poor while those of type VI are younger and have higher metal abundances on the average.

For our search, we used the Yale 1m telescope with an RCA 33063 image-tube camera which has a field of 13'. With baked IIIa-J plates, a limiting magnitude of $20^m.5$ is obtained with exposures of only five minutes so that many clusters can be observed a number of times during each night of observation. We have found RR Lyrae stars in the type VII clusters NGC 1786 and NGC 2210. We searched but we did not find RR Lyrae stars in NGC 339, NGC 2019 and Hodge 11 (type VII) or in NGC 416, NGC 2121 and NGC 2155 (type VI).

TABLE I

Magellanic Cloud Clusters with RR Lyrae Stars

Cluster	No.	$<B>$	$\overline{P_{ab}}$	Ref.	Comment
NGC 121	4	$19^m.7$	$0^d.55$	(1)(2)	SMC
NGC 1466	43	19.1	0.53	(3)(4)(5)	foreground Galactic?
NGC 1786	11	-	0.69		LMC, crowded.
NGC 1835	23	19.4	0.57	(6)	LMC, crowded.
NGC 1841	12	19.9	0.66	(7)	background?
NGC 2210	25	-	0.64		LMC
NGC 2257	42	19.5	0.58	(8)(9)	LMC

References

(1) Tifft, W.G.: 1963, MNRAS 125, p. 199.
(2) Graham, J.A.: 1975, PASP 87, p. 641.
(3) Wesselink, A.J.: 1971, MNRAS 152, p. 159.
(4) Norris, M.V.: 1973, in IAU Colloquium No. 21 ed. J.D. Fernie (D. Reidel, Dordrecht) p. 113.
(5) Cowley, A.P. and Hartwick, F.D.A.: 1981, A. J. 86, p. 667.
(6) Graham, J.A. and Ruiz, M.T.: 1974, A. J. 79, p. 363.
(7) Kinman, T.D., Stryker, L.L. and Hesser, J.E.: 1976, PASP 88, p. 393.
(8) Alexander, J.B.: 1960, MNRAS 121, p. 97.
(9) Nemec, J.M. et al.: 1984, this volume, p. 39.

Table I summarizes those clusters which are now known to contain RR Lyraes. All are of type VII and appear to be as old as the oldest clusters known in our Galaxy. No RR Lyrae variables have yet been found in clusters of type VI. Those assigned to NGC 1978 by Thackeray and Wesselink (1953) are now thought to be LMC field stars (Olszewski 1982).

REFERENCES

Olszewski, E.W.: 1982, Dissertation, Univ. of Washington.
Searle, L., Wilkinson, A. and Bagnuolo, W.G.: 1980, Ap. J. 239, p. 803.
Thackeray, A.D. and Wesselink, A.J.: 1953, Nature 171, p. 693.

PERIOD CHANGES OF RR LYRAE STARS IN THE LMC GLOBULAR CLUSTER NGC 2257

James M. Nemec
Dominion Astrophysical Observatory, and
Department of Astronomy, University of Washington

Martha H. Liller and James E. Hesser
Dominion Astrophysical Observatory

The period changes of RR Lyrae stars can be compared with models of horizontal branch stars as a means of investigating the physical properties of the stars themselves, and of the stellar systems in which they are found (Smith and Sandage 1981). The present study is the first in which period change rates of extragalactic RR Lyraes have been estimated.

The RR Lyraes in NGC2257 were first investigated by Alexander (1960). Since 1971, over 100 plates have been taken of NGC2257 at CTIO to study the cluster and field RR Lyraes (see Hesser, Nemec and Ugarte 1980). Their properties include (Nemec, Hesser and Ugarte 1983, in preparation):

	$\langle P_{ab} \rangle$	s.e.	N	$\langle P_c \rangle$	s.e.	N	P_{trans}	$P_{ab,max}$	$\langle B \rangle$	
Cluster	0.589	0.08	14	0.351	0.03	21	0.493	0.686	19.40	0.10
Field	0.563	0.08	12	0.362	0.04	9	0.423	0.728	19.60	0.15

Note that NGC2257 now appears to lie ~4-5 kpc in front of the LMC, rather than ~9 kpc as we reported earlier. The revision is due to an improved radial background light correction.

By combining new PDS photometry of 45 of the Radcliffe plates, taken from 1953-1961, with the photometry of the CTIO plates, period change rates have been estimated for 19 of the 41 RR Lyraes in NGC2257. Eight plate groups were formed, light curves were plotted with JD2440000.0 as zero phase, and phases at maximum light were measured from the light curves. Linear and parabolic regressions were fit to the data, plotted as a function of time (Fig.1). Upward curves correspond to increasing periods (2 stars), downward curves to decreasing periods (10 stars), and straight lines to constant periods (7 stars), with period change rates ranging from -1.85 to +0.98 d/Myr. Thus for these extragalactic RR Lyraes decreasing periods appear to be more frequent than increasing periods, and the rates of change are similar to those of most Galactic RR Lyrae stars. Only with continued observations will it be possible to distinguish between abrupt vs. continuous period changes.

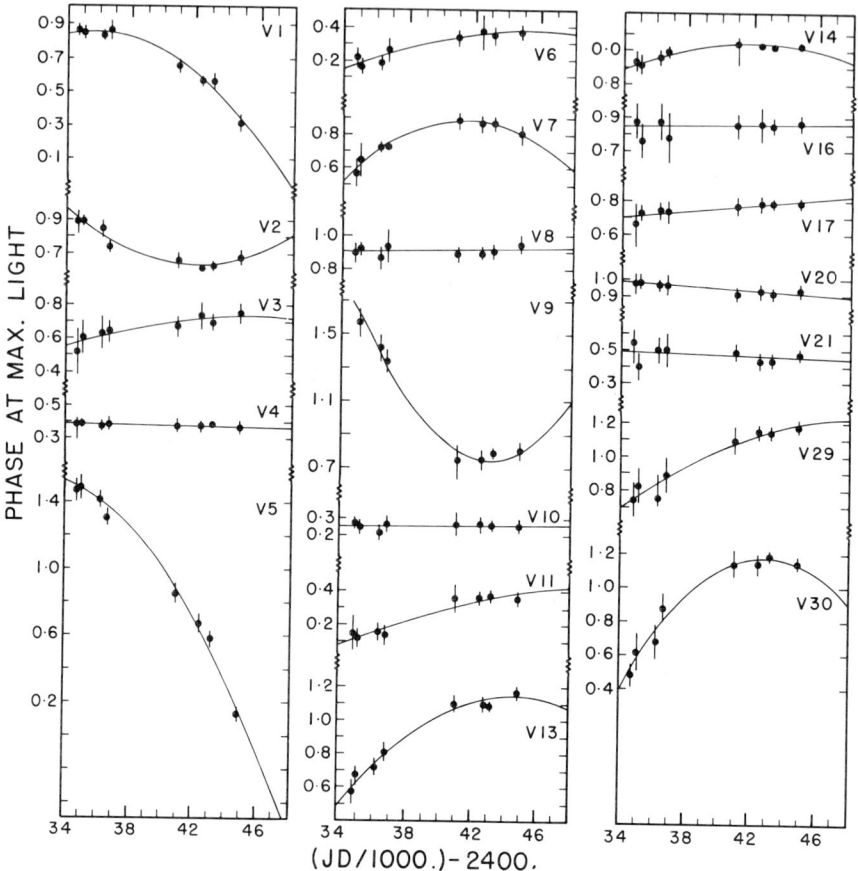

Figure 1: The phase-shift diagrams for 19 RR Lyrae stars in NGC 2257

A few Blazhko effect stars, and several possible double-mode RR Lyrae stars also were found. One such star, with $P_1 = 0.4065$ and $P_1/P_0 = 0.7463$ implying $M/M_0 = 0.65$, is similar to the double-mode RR Lyrae stars in M15 (Cox, Hodson and Clancy 1983) and the Draco dwarf galaxy (Nemec 1983, 1984). Metal abundances and radial velocities for the NGC2257 cluster and field RR Lyrae stars will be useful for evaluating Freeman, Illingworth and Oemler's (1983) claim that the LMC does not possess a halo population.

REFERENCES:
Alexander, J.B.: 1960, M.N.R.A.S. 121, p.97.
Cox, A.N., Hodson, S.W. and Clancy, S.P.: 1983, Ap.J. 266, p.94.
Freeman, K.C., Illingworth, G. and Oemler, A. 1983, preprint.
Hesser, J.E., Nemec, J.M. and Ugarte P., P.: 1980, in Star Clusters, I.A.U. Symp. 85, ed. J.E. Hesser (Dordrecht:Reidel), p.347
Nemec, J.M.: 1984, in I.A.U. Symp. 105, in press.
-----. 1983, Ph.D. Dissertation, University of Washington.
Smith, H.A. and Sandage, A.: 1981, A.J. 86, p.1870.

THE ELECTRONOGRAPHIC COLOR-MAGNITUDE DIAGRAM OF HODGE 11

J. Andersen
Copenhagen University Observatory
A. Blecha
Geneva Observatory
M. F. Walker
Lick Observatory

A program of electronographic photometry of star clusters in the Magellanic Clouds is being carried out using the McMullan camera on the Danish 1.5-m telescope at La Silla. The observations are being reduced using the Geneva microdensitometer and data reduction programs. Results are shown for the LMC cluster Hodge 11 (=SL 868), based on measurements of two exposures in both yellow and blue-light, each three hours in length, recorded on Ilford L4 emulsion.

In the reductions, areas of 250x250μ, 340x340μ or 380x380μ centered on each star were raster-scanned with a scanning aperture of 14 or 20μ. X-Y raster steps of 12, 16 or 18μ were used, producing a 21x21-point matrix. For each area, the local sky background level was determined by fitting a third-degree polynomial. Standard star profiles were determined for each film using images of bright stars, fitted to modified Gaussian profiles having three shape parameters allowing for seeing and image elongation and orientation. A multiple-peak profile, capable of resolving up to nine stars per scan-area, was then fitted to each sub-field scan and the volume of the density solid of each star image found by numerical integration.

Measures were made of 413 sub-fields in and near the cluster. Volumes of the density solids of the measured stars were converted to the BV systems using a least-squares fit to the previous electronographic photometry by Walker (1979) for those stars of V < 19.50 in yellow and B < 20.00 in blue light. Analysis of other data indicates no significant divergence of the McMullan camera color system from B-V colors; color errors are $< \pm 0.02$ B-V/mag. Eliminating double stars and stars in badly crowded regions and considering only the 108 uncrowded stars measured on all four exposures and having radial distances of $38" \leq R \leq 80"$ from the cluster center, the color-magnitude diagram shown in Figure 1 was obtained. This diagram, extending to V $\simeq$ 22, indicates that Hodge 11 is an old, metal-poor globular cluster, similar to M 92 and thus having [Fe/H] $\simeq$ -2.1, in agreement with the integral photometry by Danziger (1973) and by Searle, et al. (1980). Fitting the horizontal branch of Hodge 11 to that of M 92 and assuming E_{B-V} = 0.06, the distance modulus of the cluster is $(m-M)_o = 18.1$ if M_v (HB) = +0.9 in M92 as found by Sandage (1970) from main-sequence fitting, or 18.4 if M_v (HB) = +0.6 (Harris 1976).

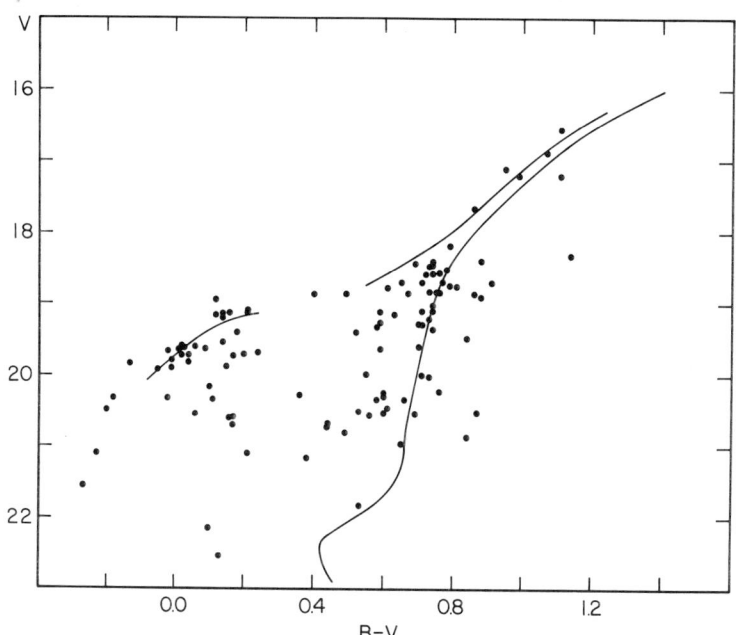

Figure 1. Color-Magnitude Diagram of Hodge 11. Line represents the diagram for M 92, superimposed as indicated in the text.

The observations of Hodge 11 have also been reduced at the Lick Observatory using: (1) scans of the star profiles with a Joyce-Loebl microdensitometer as employed by Walker in earlier studies (Walker 1979 and references therein) and (2) iris-diaphragm measures with a Sartorius astrophotometer using electronographically observed stars as standards and a calibration curve that varies with radial distance from the cluster center. These tests indicate that useful results can be obtained by these methods, though with larger random error than obtained with the Geneva system; for $21.00 \leq B \leq 21.99$, the standard deviation of a single measure is: Geneva, $\pm$ 0.26 mag; Joyce-Loebl, $\pm$ 0.33 mag; Sartorius, $\pm$ 0.36 mag.

References

Danziger, I.J.: 1973, Astrophys. J., 181, p. 641.
Harris, W.E.: 1976, Astron. J., 81, p. 1095.
Sandage, A.R.: 1970, Astrophys. J., 162, p. 841.
Searle, L., Wilkinson, A., Bagnolo, W.G.: 1980, Astrophys. J., 239, p. 803.
Walker, M.F.: 1979, Monthly Notices Roy. Astron. Soc., 186, p. 767.

THE LMC GLOBULAR CLUSTER HODGE 11 (=SL 868)

L.L. Stryker, J.M. Nemec, J.E. Hesser and R.D. McClure
Dominion Astrophysical Observatory
Herzberg Institute of Astrophysics

The age of the star cluster H11 has been controversial for a number of years. The color-magnitude diagram (CMD) of Walker (1979) to V=21.5 was interpreted as an "...evolved main-sequence, whose termination point corresponds to an age of about 0.6 Gyr, but with a giant branch which is displaced blueward by about $\Delta(B-V)_o=0.4$ from the positions of the giant branches of open clusters of similar age in our Galaxy." On the other hand, the integrated colors are similar to those of metal-poor globular clusters in the Galaxy (Freeman and Gascoigne 1977, and references therein), and "...incompatible with an age of say 0.3 Gyr." Searle, Wilkinson and Bagnuolo (1980) classify it as Group VII, the oldest group. The system has no RR Lyrae stars (Graham and Nemec 1984).

To settle the age question, deep CCD exposures in B and V to V=23.5 were taken of the cluster and of a neighboring field with the CTIO 4-m telescope in Dec. 1982 and Feb. 1983. CMDs have been measured in the 3x5' fields with a point-spread-function fitting technique (Mould and Shortridge 1983, private communication) at the DAO for a pair of frames containing the cluster at one end ($\gtrsim$1500 stars) and for a pair covering the field 30' N of the cluster ($\gtrsim$1000 stars). The frames were calibrated with observations of standard fields. The upper row of Fig. 1 contains the CMDs for: a) the portion of the frame containing the cluster; b) the field portion of that frame; and c) the difference between a) and b). Fig. 1c is morphologically like metal-poor Galactic globulars with pronounced blue horizontal branches. ΔV(horizontal branch to turnoff) is >3.0 mag, which is also indicative of great age. Walker's low inferred age arose naturally from: 1) his shallower CMD, and 2) the lack of data for a comparison field in this crowded region. The latter data particularly reveal that the field star component was easily mistaken for the upper main sequence of a much younger cluster.

The lower row of Fig. 1 contains CMDs from the field 30' N for: d) the entire 3x5' area; e) the upper half of the frame; and f) the lower half of the frame. There is general similarity between Figs. 1b and 1d, with both indicating the presence of a wide range in ages, presumably comprised of stars younger than NGC7789 to stars older than

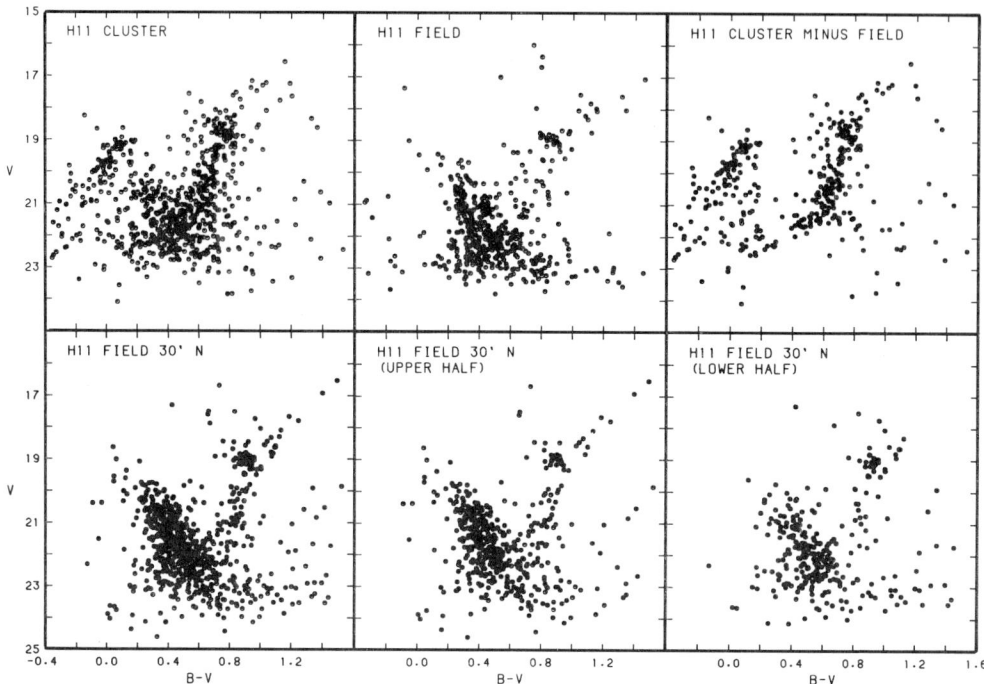

Fig. 1. Preliminary CMDs for the cluster (upper) and field (lower) frames.

those in M67/NGC188 in the Galaxy. The spread in color of the giants seems to indicate a range in [Fe/H], as well. The sharply defined giant branch in the field region of Fig. 1e is quite remarkable. At the same time the differences between Figs. 1e and 1f, which occur over ~2' (or ~30 pcs), are reminders of the care necessary when generalizing to global properties of the Magellanic Cloud field star populations from studies of small fields.

We are very grateful to J.R. Mould, E.W. Olszewski, H. Yee, D. Tsai and K. Haukaas for invaluable assistance with the image processing.

REFERENCES

Freeman, K.C. & Gascoigne, S.C.B.:1977, Proc.Astron.Soc.Aust. 3, p.136.
Graham, J.A. & Nemec, J.M.: 1984, this volume, p. 37.
Searle, L., Wilkinson, A. & Bagnuolo, W.: 1980, Ap.J. 239, p.803.
Tody, D.: 1980, in SPIE Symp. "Applications of Digital Image Processing to Astronomy", ed. D.A. Elliott, p.121.
Walker, M.F.: 1979, Mon.Not.Roy.Ast.Soc. 186, p.767.

MAIN SEQUENCE PHOTOMETRY OF KRON 3

R.M. Rich and J.R. Mould
Palomar Observatory, California Institute of Technology
G.S. Da Costa
Yale University Observatory

ABSTRACT

The globular cluster Kron 3 in the Small Magellanic Cloud presents an interesting opportunity to study stellar evolution in a population of intermediate age populous enough to contain many stars in short-lived phases of their evolution. The cluster lies approximately two degrees west of the body of the SMC and just outside the tidal radius of the Galactic globular cluster 47 Tuc.

THE COLOR-MAGNITUDE DIAGRAM

We note the following significant features:

1. Kron 3 has a distinct turnoff population starting at R = 21.5.

2. The giant and subgiant branches are well delineated.

3. The horizontal branch (HB) has a "C" shape which opens towards the giant branch. The HB qualitatively resembles a Sweigart and Gross track for a HB star with $0.20 < Y < 0.30$, $Z = 0.001$, and $M = 0.6 - 0.8\ M_\odot$. The width of the blue turning point constrains the dispersion in HB masses (and mass loss on the first giant branch) to $\sigma < 0.01 M_\odot$, and the narrow luminosity range constrains the cosmic dispersion in Y to be $\sigma(Y) < 0.01$.

4. There may be a gap below the main sequence turnoff. It is clearly visible in the high quality sample of 169 stars, persists in the large sample of 459 stars, and also appears in the observed luminosity functions of all samples. If it is real, and analogous to the hydrogen core exhaustion gaps in old open clusters, it is very interesting because theory does not predict such a gap in this range of age and composition.

5. The SMC background has well delineated subgiant and horizontal

branches, both of which coincide with Kron 3 within 0^{m}01 in R and B-R. We conclude that the Kron 3 and local SMC background Fe-peak element abundances differ by no more than 0.1 dex.

6. The background has an old turnoff population fainter than that of Kron 3; at R = 22, there is an increase in the luminosity function and a pronounced Hertzprung gap.

7. There is a sequence of blue stars in the field sample which are more luminous than the turnoff population. The widespread occurence of young populations in the SMC confirms that this is a young main sequence.

THE AGE

Isochrone fits were made using the Yale tracks for [Fe/H] = -1.3 and Y = 0.20 transformed to the observational plane using Kurucz model atmospheres. The tracks were reddened by A_R = 0.09 and E(B-R) = 0.064, and the fit placed the turnoff at R = 22.2. Values for the SMC distance modulus reported in the literature cluster about two means of 18.8 and 19.3, and fits yielded ages of 8 and 6 Gyr, respectively. These ages are corrected to Y = 0.25, in agreement with recent determinations for the SMC, as well as the primordial He abundance. The turnoff of the SMC field is fainter than that of Kron 3. Considering that this region of the SMC contains many other older objects, including NGC 121, the background field may be the oldest population in the SMC.

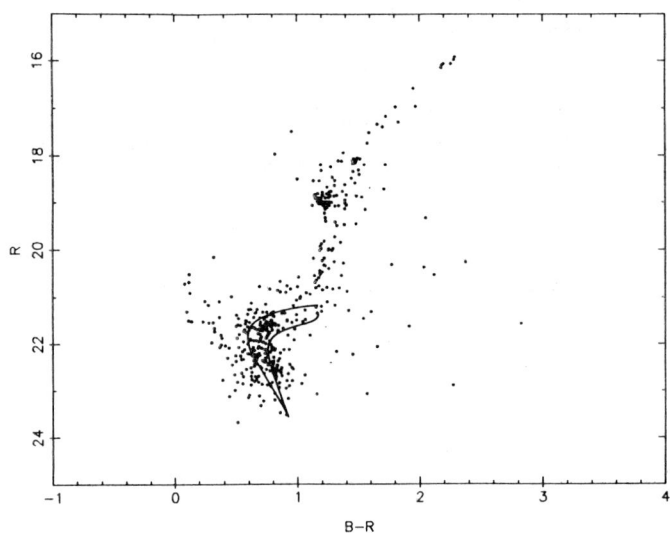

Figure 1
The Color Magnitude Diagram of Kron 3 with
6 and 8 Gyr isochrones superposed.

COLOR-MAGNITUDE DIAGRAM MORPHOLOGY OF THE OLDEST STAR CLUSTERS IN THE
LARGE MAGELLANIC CLOUD

James E. Hesser and Robert D. McClure
Dominion Astrophysical Observatory
Herzberg Institute of Astrophysics

William E. Harris
McMaster University

We summarize B,V imagery results from the CTIO 4-m telescope and two electronic detectors: (1) The SIT vidicon at the R-C focus yielded 0".19 pixels in the 0.82 or 1.1 arcmin^2 frames (Harris, Hesser and Atwood 1983). (2) KPNO's P-F CCD (used by McClure and Hesser) gave 0.6 pixels in 3x5' fields. (The latter data are being analyzed in cooperation with Stryker and Nemec, who observed some of the same objects.) Zero points for the CCD data are set by observations of E region standards and of giants in Galactic star clusters. Our results from point-spread-function profile fitting are:

NGC2210 (~4° E):. SIT data to $\underline{V}$ ~ 21.5 support previously noted similarities to Galactic globular cluster color-magnitude-diagram (CMD) morphology (Hesser, Hartwick and Ugarte P. 1976). CCD data are being reduced.

H11 (~4° E): CCD data to $\underline{V}$ ~ 22.5 presented elsewhere in this volume (Stryker, et al. 1984) resolve the controversy surrounding the interpretation of the CMD in favor of its similarity to Galactic globulars with pronounced blue horizontal branches.

LW4 (~5° SW): Hodge (1983, private communication) suggested that this cluster might be an overlooked member of the sparse "red globular" group; but CCD photometry (Fig. 1a) reveals a CMD strikingly similar, except for the lack of blue stragglers, to that of NGC7789, a Galactic cluster a few billion years old.

NGC2203 (~6° SSE): Gascoigne (1978, private communication) suggested this as another candidate for the "red globular" group. Alas, SIT data reaching $\underline{V}$ ~ 22.5 also resemble those of NGC7789. Deeper CCD data are being reduced.

NGC1466 (~8° SW): SIT observations to $\underline{V}$ ~ 21.5 support Gascoigne's (1965) and Penny's (1975) conclusions that the CMD morphology is similar to Galactic globulars. NGC1466's membership in the Clouds remains problematical, however.

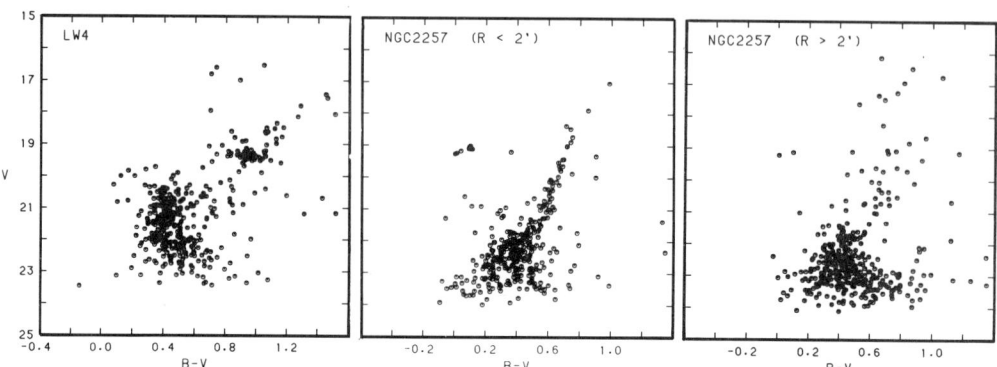

Fig. 1: Preliminary CMD's for LW4(left); NGC2257, r $\lesssim$ 2' (center); and NGC2257, r $\gtrsim$ 2' (right).

NGC2257 (~8° NE): Extensive SIT and CCD data define a CMD like that of the metal-poor Galactic globular, M92. The CCD data reveal (Fig. 1b) the upper 1.5 mags of the main sequence for the first time in an extragalactic globular. The turnoff appears at V ~ 22.4, and $\Delta \underline{V}$(horizontal branch to turnoff) = 3.3 ± 0.1 mag is similar to Galactic globulars, thereby confirming and extending Stryker's (1983a) inferences from a photographic plate pair. Fig. 1c shows the CMD of the region beyond ~2', i.e., outside the cluster region of Fig. 1b. There is again a cluster-like turnoff at $\underline{V}$ ~ 22.4. Although these are undoubtedly LMC stars, a large and puzzling spread in color exists among the giants (see also Stryker 1983b). It seems unlikely that the spread can be accounted for entirely by differences in age, since the turnoff stars indicate a very old population with few, if any, stars younger than 10^{10} years. Perhaps in the NGC2257 direction the LMC halo has a spread in metal abundance comparable to that seen in the Galactic globular cluster system.

REFERENCES

Gascoigne, S.C.B.: 1966, Mon.Not.Roy.Ast.Soc. 134, p.59.
Harris,W.E.,Hesser,J.E. & Atwood,B.A.: 1983, Pub.Ast.Soc.Pac.,submitted.
Hesser,J.E.,Hartwick,F.D.A. & Ugarte,P.,P.: 1976, Ap.J.Suppl.32, p.283.
Penny,A.J.: 1975, Mon.Not.Roy.Ast.Soc. 172, p.65P.
Stryker,L.L.: 1983a, Ap.J. 266, p.82.
_____ :1983b, ibid., submitted.
Stryker,L.L.,Nemec,J.M.,Hesser,J.E. & McClure,R.D.:1984,this vol. p. 43.

A BVRI PHOTOMETRIC STUDY OF STAR CLUSTERS IN THE BOK REGION OF THE LMC.

Gonzalo Alcaino and William Liller
Instituto Isaac Newton, Ministerio de Educación de Chile,
Santiago, Chile.

The large number of star clusters in the Large Magellanic Cloud provides us with an excellent opportunity to understand well this neighboring galaxy. The large variety of integrated cluster colors present suggests a wide range in evolutionary stages, and several studies have established vague relationships between their location in the LMC and properties including color, total luminosity and metallicity.

We are at present deriving color-magnitude diagrams for a selection of clusters located in the so-called Bok region situated in the northwestern section of the Bar of the LMC. This region relative to the LMC is identified in Figure 1, and a magnification of the region itself is shown in Figure 2. As can be seen in Figure 2, this zone is densely populated with clusters, among the most conspicuous being NGC 1847, NGC 1850, NGC 1854 and NGC 1856.

The photographic material consists of 4 B plates, 4 V plates and 2 R plates each covering 1° x 1° obtained with a Pickering-Racine wedge at the 2.5 m du Pont telescope of Las Campanas, ($\Delta m \sim 5.1$ mag, scale 10.8 arc sec mm^{-1}), and three I plates obtained using the triplet configuration at the 3.6 m telescope at ESO/La Silla (scale 18 arc sec mm^{-1}). The plates are being calibrated with a BVRI photoelectric sequence to $V \sim 15.3$, obtained by us using the 1 m telescope at ESO/La Silla (Alcaino and Liller 1982).

Reference

Alcaino, G., and Liller, W.: 1982, Astron. Astrophys. 114, p. 213.

Figure 1. The Bok region in the NW section of the LMC Bar   (next page)
Figure 2. Enlargement of a section of Figure 1.   (next page)

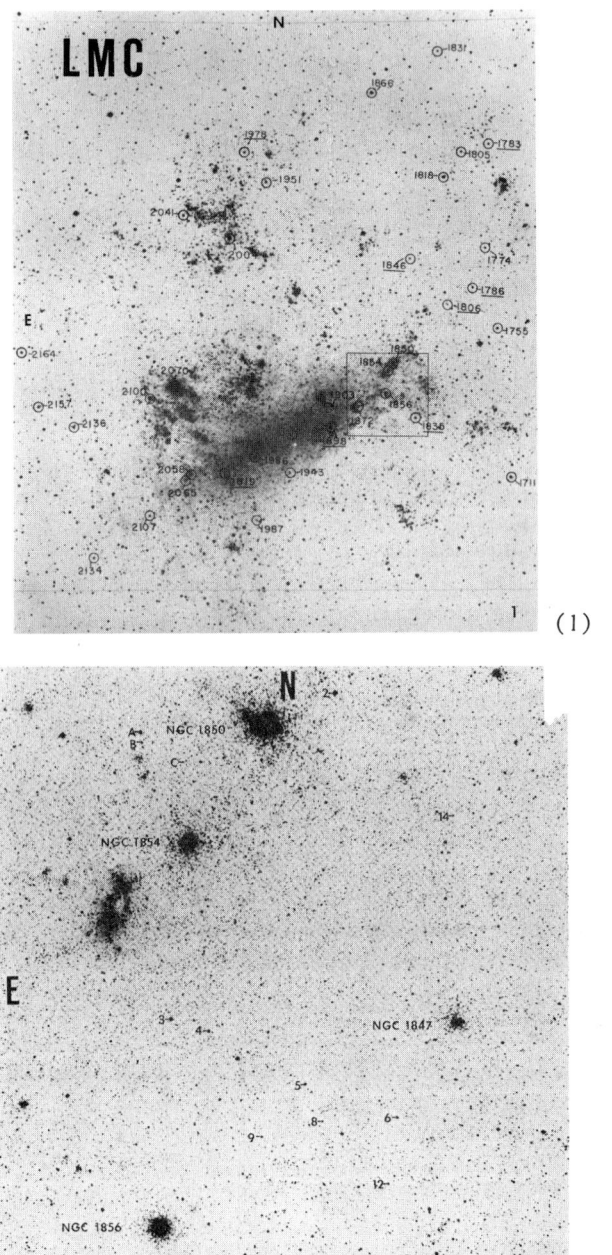

(1)

(2)

A BV PHOTOMETRIC STUDY OF STAR CLUSTERS IN TWO SELECTED REGIONS OF THE SMC.

Gonzalo Alcaino and William Liller
Instituto Isaac Newton, Ministerio de Educación de Chile,
Santiago, Chile.

We are deriving BV color-magnitude diagrams of star clusters in two selected regions in the SMC. These zones, characterized by the presence of a high density of star clusters, are centered at the 1981 coordinates for region 1: (RA: $1^h 10^m.33$, Dec $-73°08'$), and for region 2: (RA: $1^h 0^m.33$, Dec $-73°00'$). See Figure 1 for their identification relative to the SMC.

For region 1, some of the most conspicuous clusters are: NGC 376, NGC 416, NGC 419, NGC 456, NGC 460 and NGC 465. For region 2: NGC 290, NGC 292, NGC 294, NGC 299, NGC 306, NGC 330, NGC 346, NGC 376, NGC 416 and NGC 419.

Large size photographic plates (20 x 20 inches) have been obtained with the 2.5 m du Pont telescope at Las Campanas. They cover an area of $1°.5 \times 1°.5$, have a plate scale of 10.8 arc sec mm^{-1}, and have been taken with a Pickering-Racine wedge ($\Delta m \sim 5.1$ mag). The plates are now being calibrated with electronographic sequences (Walker 1972), as well as with other existing photoelectric sequences.

Reference

Walker, M. F.: 1972, Mon. Not. Roy. Astron. Soc. 159, p. 379.

Figure 1. Identification chart of the SMC clusters studied (next page)

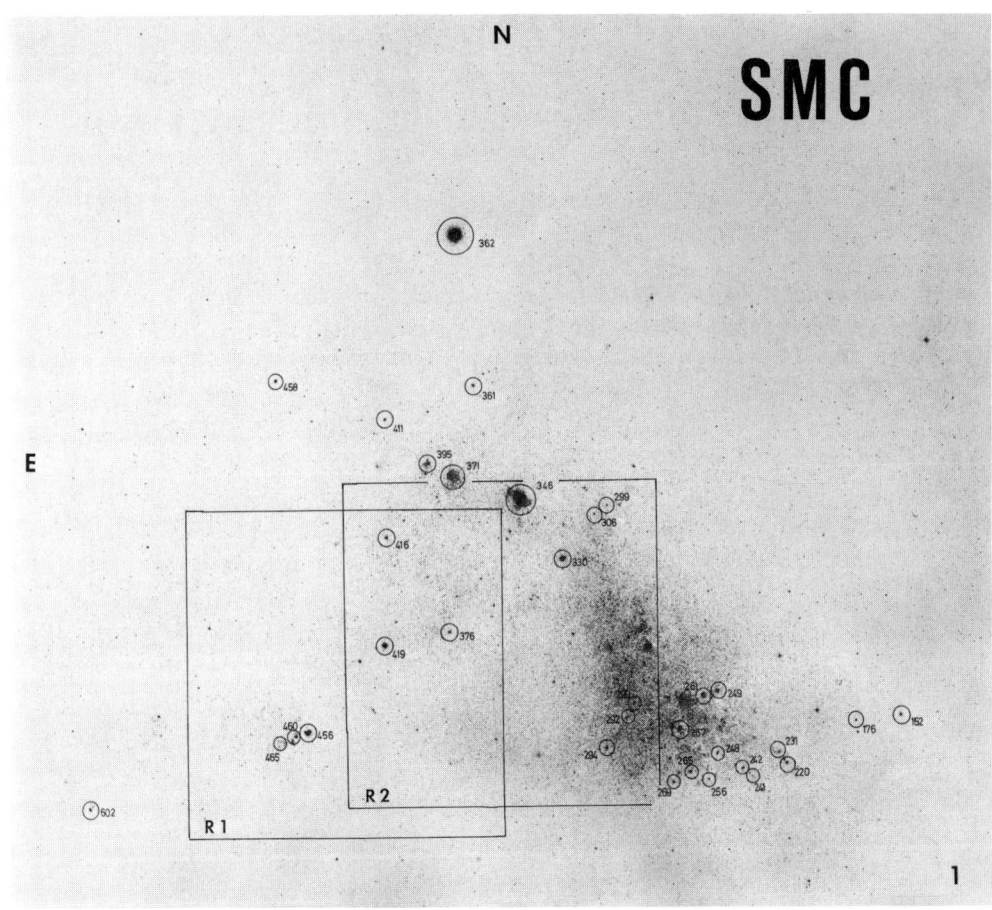

Figure 1.

FAR-UV PHOTOMETRY OF MAGELLANIC CLOUD GLOBULAR CLUSTERS

Klaas S. de Boer
Astronomisches Institut Tübingen, D-7400 Tübingen, F.R.Germany

The International Ultraviolet Explorer (IUE) has been used by me since January 1979 to observe globular clusters in the LMC, under various IUE programs in collaboration with Art Code. A first report was presented at the NASA 2nd IUE conference in 1980 (de Boer 1981a). A comparison of the UV colors for 11 MC clusters with the colors of galactic globular clusters was presented during IAU Colloquium 68 (de Boer 1981b), in a review which accumulates also all references to literature on far-UV photometry of galactic globular clusters. Using the VILSPA data base Cacciari et al (1982) repeated the comparison. The spectra of three clusters with good signal were presented by Code (1982). Data for a few other clusters were reported by Cacciari and Fusi-Pecci (1981) and by Cassatella and Geyer (1982; and this symposium).

Spectra of the redder clusters are extremely weak, in part due to the limited extent of the IUE aperture (see de Boer 1981b), and in spite of hours long integration times. Special extraction methods, working on the original image, are needed to address the noise problem properly (de Boer et al 1981), to get the best spectral resolution from these wide spectra (de Boer et al 1982), and to solve for spatial structure (de Boer et al 1981; Cacciari in this symposium).

Various spectral features can be seen in the spectra of the MC clusters. The resonance lines in part are of interstellar origin (see for average Magellanic Cloud interstellar line strengths de Boer and Savage 1980); the MgII blend at 2800Å surely is interstellar in the spectra of the bluest (hottest-youngest) clusters, but becomes gradually dominated by stellar absorption when going to spectra of the reddest clusters. The bluest clusters show easily recognizable stellar lines at the shortest wavelength range (SiIII, CIV, etc) while towards the redder clusters lower ionization stages become noticable, but for those the spectra unfortunately but not unexpectedly have much smaller S/N.

Full details of the spectral analysis will be reported eslewhere.

References

Cacciari, C., Fusi-Pecci, F. 1981, in "Astrophysical Parameters for Globular Clusters", IAU Coll.68, p 217, Ed A.G.D. Philip and D.S. Hayes, Davis Press, Schenectady NY
Cacciari, C. et 9 altera. 1982, in "Third European IUE Conference", ESA SP-176, p 519
Cassatella, A., Geyer, G.H. 1982, in "Third European IUE Conference", ESA SP-176, p 523
Code, A.D. 1982, COSPAR XVIII, in press
de Boer, K.S. 1981a, in "The Universe at Ultraviolet Wavelengths", NASA-CP 2171, p 527, Ed R.D. Chapman
de Boer, K.S. 1981b, in "Astrophysical Parameters for Gobular Clusters", IAU Coll. 68, p 3, Ed A.G.D. Philip and D.S. Hayes, Davis Press, Schenectady NY
de Boer, K.S., Koornneef, J., Meade, M.R. 1981, in "The Universe at Ultraviolet Wavelengths", NASA-CP 2171, p 771, Ed R.D. Chapman
de Boer, K.S., Preussner, P.-R., Grewing, M. 1982, A. Ap. 115, p 128
de Boer, K.S., Savage, B.D. 1980, Ap. J. 238, p 86

UV-OBSERVATIONS OF YOUNG POPULOUS CLUSTERS IN THE LARGE MAGELLANIC CLOUD

Edward H. Geyer[1], Angelo Cassatella[2]

[1] Sternwarte der Universität Bonn -
Observatorium Hoher List, D-5568 Daun, FRG

[2] European Space Agency Astronomy Division, Villafranca
Tracking Station, Madrid, Spain

INTRODUCTION

The young populous star clusters give evidence for the 'explosive' star formation in the Magellanic Clouds which took place in the time interval $5 \cdot 10^6$ yrs to $< 10^8$ yrs agoe. They are also key objects for the understanding of the formation of massive stellar clusters, because they are still situated close to their 'birthplace' in the parent galaxy and are dynamically not relaxed (Geyer et al. 1979). Their HRD-morphology shows most of the member stars in the upper Main Sequence range with only a few massive yellow and red supergiants. The lower massive stars are still in the pre-main-sequence evolution phase ('T-Tauri state'), which cannot be observed at the MC's distances. Thus in the uv-spectral range the blue stars with (B-V) < 0.1 on the upper MS contribute to the uv-fluxes. In the optical spectral regions the bright 'blue' globular clusters seem not be embedded in remanent interstellar matter, though neighbouring loose stellar aggregates of similar age are in many cases surrounded by dense HII-regions. This rises the questions wether the starformation process in such massive clusters was so efficient that no remanent matter was left over, or was this material blown away by the uv-radiation of the numerous OB-member stars?

OBSERVATIONS AND RESULTS

In the last two years we have obtained low resolution uv-spectra of the central parts of the four young LMC-globular clusters NGC 1818; 1866; 2004 and 2100 with the International Ultraviolet Explorer. From star counts on ESO-Schmidt plates we estimate that on the average about 150 stars per cluster contribute to the integrated spectra through the large entrance aperture (10 x 20 arc sec^2) used with the IUE-spectrometer.

The clusters NGC 1818 and NGC 1866 are situated about $3.5°$ northwest of the LMC Bar center where no dust lanes can be traced on direct plates. NGC 2004 is situated in the very young Shapley III-constellation with faint dust patches visible, and the position of NGC 2100 is on the east outskirts of the 30 Dor complex with dust patches in its vincinity.

The observed integrated uv-spectra reveal the young character of the stellar content of the clusters according the absorption line features and the energy distribution:

To the absorption line features of these low resolution spectra contribute the cluster stars and the interstellar medium (i.m.) of the galactic foreground and the LMC. These are the lines with the low excitation potential about zero (e.g. OI, MgI, CII, SII, AlII, MgII and FeII). Strong resonance line features of SiIV, CIV and Al III as seen in NGC 2004 and 2100 get gradually weaker in NGC 1818 and 1866. As this finding is correlated with spectral gradients, these lines are more cluster intrinsic than to originate from the ISM. Actually, high resolution IUE spectra of NGC 2004, taken recently by one of us (Cassatella et al. 1983), show that the SiIV and CIV contribution from the i.m. is marginal.

The interstellar $\lambda\lambda 220$nm dust extinction is strong in the spectra of NGC 2100, weak in those of NGC 2004 getting marginal in NGC 1818 and 1866. Thus individual cluster reddening in the LMC must be taken into account.

The observed uv-fluxes were de-reddened by a trial and error method for the adoption of a constant galactic foreground-absorption and the individual cluster reddening in the LMC making use of the different extinction laws for the Galaxy and the LMC.

The relevant extinction laws for the Galaxy and the LMC where taken from Seaton (1979) and Nandy et al. (1981), respectively. The accuracy of the corrected fluxes depends mainly on the relevant extinction law curves beeing of the order of 15 to 20%.

For the determination of the unknown E_{B-V}^{G} and E_{B-V}^{LMC} we first adopted reasonable values and changed them until the $\lambda\lambda 220$nm depression was "ironed out". As this cannot be done without ambiguity, we made use of a 'uv-two spectral gradient' $\Delta\Phi_{\lambda 1}$, $\Delta\Phi_{\lambda 2}$-diagram. This t.s.g.-diagram was calibrated according to the uv-spectral gradients of unreddened B2 to A0 MS stars derived from published IUE-spectra (Silko et al. 1981). With the two methods we finally arrived at consistent results for the E_{B-V}. The de-reddened $\Delta\Phi^{o}$ for the four clusters show a perfect linear correlation with the reddening free colour index differences Q (Searle et al., 1980).

Furthermore, the $\Delta\Phi^{o}$-spectraltype correlation for MC-stars allows to determine 'effective' integrated spectral types for the clusters. As these i.s.t. are a measure for the luminosity functions of the upper MS of the individual clusters, and therefore also for their evolutionary state, a relative age-sequence for the clusters can be established:

$7 \cdot 10^6$ yrs $\sim t_{2004} < t_{2100} < t_{1818} < t_{1866} \sim 8 \cdot 10^7$ yrs.

ACKNOWLEDGEMENT:
E.H. Geyer acknowledges the support of this investigation by the Deutsche Forschungsgemeinschaft.

REFERENCES
Cassatella, A., Geyer, E.H., Pettini, M.: 1983, paper in preparation
Geyer, E.H., Hopp, U., Kiehl, M., Witzigmann, S.: 1979, Astron. Astrophys. 77, 61
Nandy, K., Morgan, D.H., Willis, A.J., Wilson, R., Gondhalekar, P.M.: 1981, M.N.R.A.S. 196, 955
Searle, L., Wilkinson, A., Bagnuolo, W.G.: 1980, Astrophys. J. 239, 803
Sitko, M.L., Savage, B.D., Meade, M.R.: 1981, Astrophys. J. 246, 161
Seaton, M.J.: 1979, M.N.R.A.S. 187, 73

DISENTANGLING INDIVIDUAL STAR AND CLUSTER CONTRIBUTIONS IN IUE SPECTRA
OF MAGELLANIC CLOUD GLOBULAR CLUSTERS

Carla Cacciari
ESTEC Astronomy Division, Villafranca, Spain

The globular clusters NGC 1786 in LMC and NGC 121 and 330 in SMC have
been observed in UV with IUE, both in the long and short wavelength
range, in the low dispersion mode. The spectra show stellar component(s)
superimposed on the cluster contributions.

New test extraction software, recently developed at the ESA-IUE
Tracking Station in VILSPA, with a spatial resolution two times better
than the resolution provided by the standard extraction software, has
been applied to the spectra in order to disentangle the stellar and
cluster contributions. A multiple-gaussian fitting technique then was
applied to the line-by-line spectra averaged over bands 100Å wide. The
FWHM of the gaussian profiles representing the stellar contributions
were fixed on the basis of previous studies of the point spread
function (Barbero and Cassatella 1983). An exemple of the fitting for
each cluster is shown in Figure 1.

The energy distributions for the stars and clusters have been derived
at 100Å intervals. A short summary of the estimated stellar characteristics is given in Table 1. These estimates have been obtained by
comparing the observed flux distributions with Kurucz' (1979) model
atmospheres taking into account the results by Zinn (1974) on
Population II UV-bright stars (for NGC 121) and by Robertson (1974)
(for NGC 330). No reddening correction has been applied to the derived
energy distributions, since they are consistent (in first approximation)
with a reddening value smaller than 0.05 mag.
For more details reference should be made to the main paper, in preparation.

REFERENCES

Barbero J. and Cassatella A., 1983, Report to the IUE 3-Agency Meeting,
    London.
Kurucz R.L., 1979, Astrophys. J. Suppl. 40, pp. 1.
Robertson J.W., 1974, Astron. Astrophys. Suppl. 15, pp. 261.
Zinn R., 1974, Astrophys. J. 193, pp. 593.

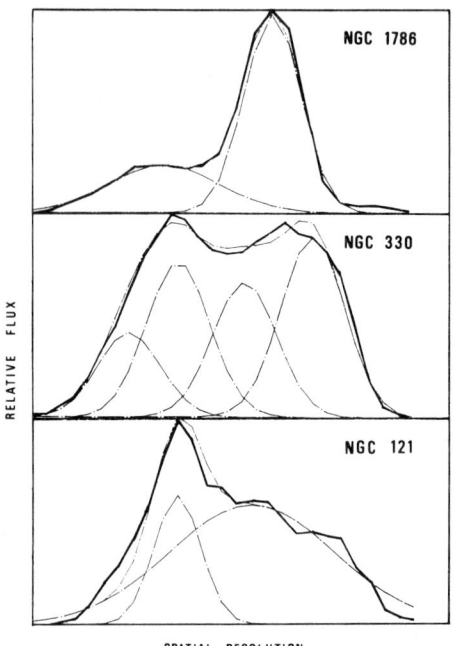

Figure 1. Multiple-gaussian fitting applied to the line-by-line spectra. The thick lines represent the observed cross-sections of the spectra, the thin lines represent the total reconstructed profiles and the profiles of the single components.

Table 1

Star ID	Sp.Type	Teff	log g	V	Comments
1786	F0-F5 IV-V	7000	4.5	12.0	Field star
121	UV-bright	30000	4.0	19.0	Member?
330 # 1	A1 I	12000	2.5	13.0	Member
330 # 2	B9 I	13000	3.0	12.7	Member
330 # 3	B1-B2 I	15000	2.5	13.4	Member
330 # 4	B0 I	20000	3.0	14.2	Member

ULTRAVIOLET STUDIES OF O AND B STARS IN THE LMC CLUSTER NGC 2100, THE SMC CLUSTER NGC 330 AND THE GALACTIC CLUSTER NGC 6530

Erika Böhm-Vitense and Paul Hodge
Astronomy Department, University of Washington, Seattle, WA

ABSTRACT

We have studied high a low resolution IUE spectra of O and B stars in the LMC cluster NGC 2100, the SMC cluster NGC 330, and the young Galactic cluster NGC 6530. Temperatures and luminosities were determined. In the LMC and SMC clusters the most luminous stars are evolved stars on the "horizontal" supergiant branch, while in NGC 6530 the stars are all still on the main sequence.

Extinction laws were determined. They confirm the known differences between LMC and galactic extinctions.

No mass loss was detected for the evolved B stars in the LMC and SMC clusters, while the high luminosity stars in NGC 6530 show P Cygni profiles.

INTRODUCTION

We try to determine the mass loss for early-type cluster stars for which distances are well known and for which the evolutionary stage can be well established. In the populous LMC and SMC clusters we have the opportunity to study stars along the evolutionary track for essentially one mass. We therefore may hope to see the mass-loss change during stellar evolution. A comparison of different clusters with different metal abundances may teach us something about the driving mechanism for the mass loss.

Figure 1 shows the measured color magnitude diagrams for the clusters studied here. The vertical scales were shifted so as to take into account the different distances and average extinctions. These diagrams are based on measurements by Robertson (1974), Walker (1957), and Chini and Neckel (1981). Colors for the LMC and SMC clusters were also measured by Westerlund (1961), Lucke (1974), Nemec (1982), and Olszewski (1983), Janes and Carney (1983). The color-magnitude diagrams of the three clusters look very similar.

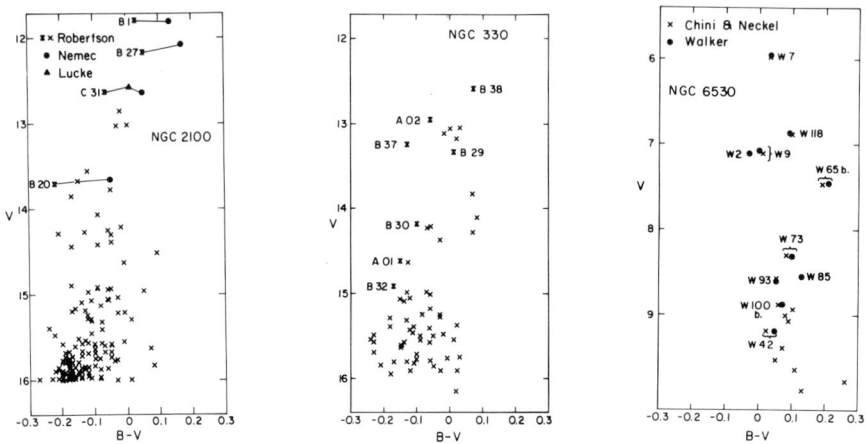

Figure 1: The color magnitude diagrams for the LMC, SMC, and Galactic clusters are shown.

THE T_{EFF}, LUMINOSITY DIAGRAMS

The conversison from color magnitude to T_{eff}, luminosity diagrams is seen in Figure 2. We used distance moduli $m_V - M_V$ = 11.5, 18.6, and 19.0 for NGC 6530, NGC 2100, and NGC 330 respectively. In Figure 2 we also show the evolutionary tracks for stars with 10 to 40 solar masses according to the calculations of Brunish and Truran (1982) and Flower (1976).

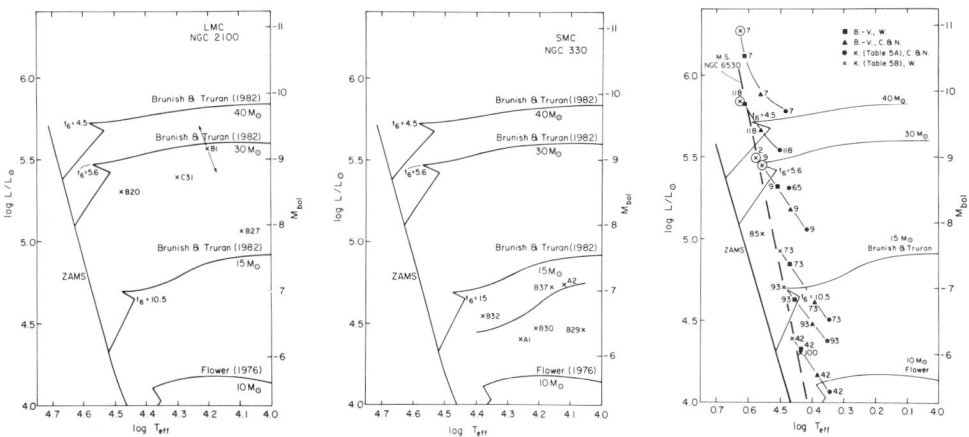

Figure 2 shows the T_{eff}, luminosity diagrams for the three clusters which look quite different.

The T_{eff}, luminosity diagrams for the three clusters look very different, mainly because of the very different reddening corrections, which are E(B-V) = 0.35 for NGC 6530, 0.18 for NGC 2100, and 0.02 for NGC 330. The ages t of the clusters come out to be $t=5\cdot10^6$ years for NGC 6530, $t=10^7$ years for NGC 2100, and $t=2\cdot10^7$ years for NGC 330.

EXTINCTION DETERMINATIONS

Extinction laws were derived for the LMC stars and NGC 6530. For NGC 330 the E(B-V) is too small to be measured reliably, we can only give $A_\lambda - A_V$. For the LMC stars we see steep increase in the extinction for $\lambda < 1800$ Å. An increase in the UV extinction was also found for other LMC stars, for instance by Nandy et al. (1981), Hutchings (1982), and Koornneef (1982).

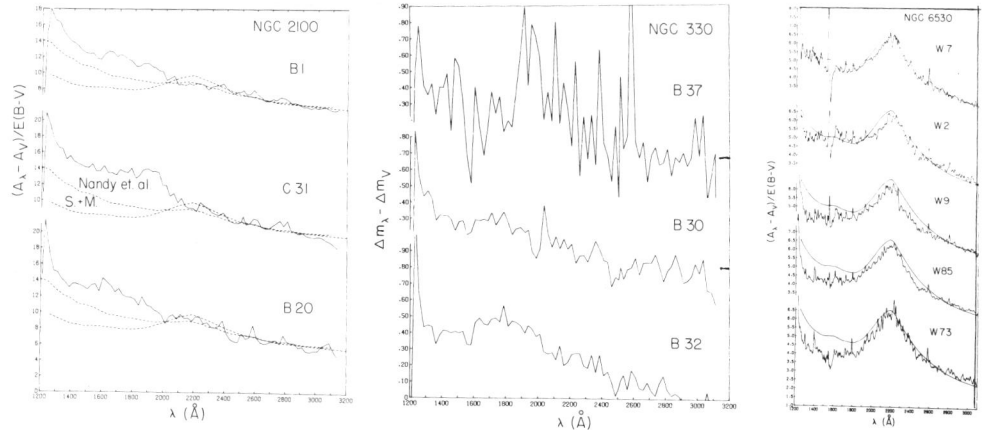

Figure 3 compares the extinction curves for the stars in the LMC and SMC clusters and in the Galactic cluster. The well known differences are obvious, but our UV extinction values are generally higher than those by Nandy et al. for instance. The average galactic extinction curve from Savage and Mathis 1979 is shown for reference.

MASS LOSS IN THE CLUSTERS

We did not detect any P Cygni profiles or profiles with extended short wavelength wings in the LMC and SMC stars, while the stars with $\log L/L_\odot \geq 5$ in NGC 6530 did show mass loss. As can be seen in Figure 4 the LMC stars are in a T_{eff}, L domain where Galactic stars show strong mass loss. If the LMC stars lose mass like the galactic stars the mass loss should be visible. The SMC stars are in a T_{eff}, L domain where the galactic stars show only weak mass loss, which could have escaped detection on our low resolution IUE spectra.

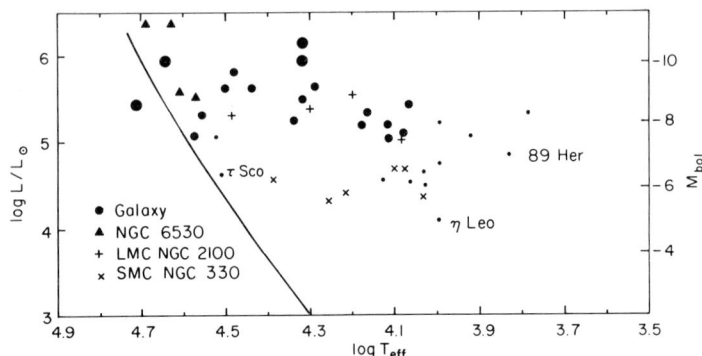

Figure 4. The positions of galactic stars with mass loss in the T_{eff}, luminosity diagram according to Cassinelli (1979). The size of the circles is a measure for the mass loss. ▲ NGC 6530 stars with strong mass loss. + NGC 2100 stars with no visible mass loss. x NGC 330 stars with no visible mass loss.

We are grateful to the staff of the IUE observatory for their continued help with observations and data reduction and to the Space Science Data Center for supplying the previously available data. This research was supported by a NASA grant NAG5-207 which is gratefully acknowledged.

REFERENCES

Brunish, W. M., Truran, J. W.: 1982, Astrophys. J. Suppl. 49, 447.
Cassinelli, J. P.: 1979, Ann. Rev. Astron. Astrophys. 17, 275.
Chini K. and Neckel, Th.: 1981, Astron. Astrophys. 102, 171.
Flower, P.: 1975, Thesis University of Washington
Hutchings, J. B.: 1982, Astrophys. J. 255, 70.
Koornneef, J.: 1982, Astron. Astrophys. 107, 247.
Lucke, P.: 1974, Astrophys. J. Suppl. 28, 73.
Janes, K. and Carney, B.: 1983, private communication.
Nandy, K., Morgan, D. H., Willis, A. J., Wilson, R., Gondhalekar, P. M.: 1981, Monthly Not. Roy. Astr. Soc. 196, 955.
Nemec, J.: 1982, private communication.
Olszewski, E.: 1983, private communication.
Robertson, J. W.: 1974, Astronomy and Astrophys. Suppl. 15, 261.
Savage, B. D. and Mathis, J. S.: 1979, Ann. Rev. Astr. Astrophys. 17, 73.
Walker, M.: 1957, Astrophys. J. 125, 637.
Westerlund, B. 1961, Upp. Ast. Obs. Ann. 5, 1.

PANEL DISCUSSION ON STAR CLUSTERS

Given below are the texts of the introductory statements by panelists J.E. Hesser, P.W. Hodge, A. Renzini, L. Searle and S. van den Bergh. The transcription of their remarks was prepared from a very low-quality tape recording. The Editors wish to apologize for any misinterpretations that may have been made of the speaker's actual remarks.

J.E. Hesser
Dominion Astrophysical Observatory

After listening to the talks this morning, I should like to reiterate Searle's point concerning the deliterious effects of stochastical background fluctuations on both the corrected integrated colors and spectra of Magellanic Cloud clusters and the dominance that a few very blue and/or red stars may have on the observed integrated cluster properties. At the same time I feel that we should not lose sight of the fact - as I think we tended to do today - that the Cloud clusters are incredibly valuable probes of stellar evolution: Rich cloud clusters contain stars in unusual regions of the HR diagram which are very difficult to sample by studying less populous Galactic star clusters. I feel that this whole field is going to change very much in the near future with the advent of CCD detectors and the application of the profile fitting techniques which have been developed for stellar magnitude determination. These new techniques are going to constrain theories of stellar evolution very powerfully through much higher quality data on the Magellanic Cloud star clusters. Many more color magnitude diagrams will be required to take full advantage of these improved techniques.

Leonard Searle made another good point about the importance of the Magellanic Cloud star clusters for understanding light from composite systems. The Clouds are the only other galaxies in which one can observe the individual stars and synthesize system properties (in integrated light) for comparison with observed integrated cluster properties. I really think that this is terribly important: We will not be able to understand the integrated light of galaxies if we can't understand the composite light of Cloud star clusters. I am again very optimistic. The CCD technology in particular will make it possible to obtain extremely good color-magnitude diagrams of Cloud clusters (even with ground-based telescopes and certainly with Space Telescope) to extremely faint absolute magnitudes. But I think we should not lose sight of the fact that the CCDs may reveal that we are not in very good control of the Galactic globulars as comparison objects. I would not be surprised if we find that an awful lot of color-magnitude diagrams that have been constructed in the past 10 years or so turn out to be rather "ropey" at the faint end. We're going to find a lot of new, exciting results after we tighten up their color-magnitude diagrams with data of 1 or 2% precision, and much will be learned about stellar evolution. At the same time we will be in a position to predict the luminosity functions much better, and hence to model synthetic spectra and colors for integrated systems.

From an observational viewpoint, I am, concerned about the application of CCDs to C - M diagram determinations. When I think back to the photoelectric UBV photometry that I used to do, I wouldn't have considered a night to have been well calibrated if I had only 8, 9 or 10 standard star observations without any idea of extinction. I am very concerned that we observers may not be thinking enough about how we are going to set the zero points in all this magnificent new CCD photometry. We can only make correct deductions when we compare objects in the Galaxy with those in the Magellanic Clouds, or compare our observations with theory, if we get the zero-points right. To take advantage of the magnificent evolutionary tracks that Don VandenBerg, among others, has been calculating will require great observational care in this matter.

Finally I would like to emphasize the point that John Graham just made on Ken Freeman's talk. I think the RR Lyrae stars offer an enormous potential to understand more about the halo of the Magellanic Clouds. I am perhaps not quite as confident as he is that we know the age-range of the RR Lyrae's that well, but there is certainly no doubt that they are a probe of the oldest component. Those people who are attempting to study RR Lyrae stars in clusters ought not to concentrate their efforts entirely on the clusters, but should also look at the fields around them - they might turn out to be as interesting as the clusters themselves.

P.W. Hodge
University of Washington

Firstly I think the importance of the discrepant ages arrived at by different means shouldn't be ignored and I hope very much that a solution to this discrepancy will be found in the next year or so. Additionally, I am concerned about 2 clusters; NGC 1978, which is very old by spectroscopic techniques, perhaps not quite as old by some people's preliminary CCD techniques, and rather young by other people's photometric results. Similarly, NGC 2021 is a key cluster to reexamine to see why it gives such different results. A second area where there is a lot of disagreement in detail, and some disagreement in the general picture, is that of stellar metallicities. The first thing there is not to argue about the question of whether metallicities correlate with age but to argue about how you define metallicity indices since different definitions yield different results. Thirdly the question of age distributions is a very important one. The problem of the age distribution for open clusters and of very massive clusters is a question that hasn't been resolved. In that general connection I feel very puzzled by the lack of large numbers of old NGC 1866's and other such intermediate-age clusters. And then finally I mention the age-chemical composition relation because we have heard a lot about this today. It is an extremely difficult question; people get different results using different calibrations. It is perhaps something for 20 to 50 years from now.

A. Renzini
University of Bologna

The Magellanic Cloud globulars are, in a broad sense, of interest to the theory of stellar evolution for several reasons, and I will just single out a couple of them. In the first place, to study advanced stages of stellar evolution, one needs a large assemblage of stars in order to find one star in an advanced stage of stellar evolution. This is particularly true for the so called asymptotic-branch stars, which are in the double shell burning stage. One needs a total of something like 10^5 stars to have one star in such a state. (These figures will change a little bit with the age of the system). The problem with even the richest Galactic clusters is that they contain only something like 1,000 stars. I would also like to draw the attention of observers to the very bright stars that exist in the Magellanic Cloud clusters. It would be important to get reliable bolometric magnitudes and high dispersion spectra of these objects that could become K-type giants or supergiants. Not only do these stars contribute substantially to the models of stars, but they also provide a variety of interesting elements - not only carbon, nitrogen and oxygen. Finally, my impression is that all the evolutionary population synthesis models attempted so far are wrong in the sense that they have forgotten important evolutionary phases.

L. Searle
Mount Wilson and Las Campanas Observatory

I have no ability to prognosticate the future and I have no idea what the direction of research will be a year or a few years from now. I can't even predict my own research on a six-month time scale. So I will confine my remarks to things that struck me during today's presentations. And I should like to emphasize the point that Paul Hodge made. I think that we saw a very clear confrontation between different techniques for estimating the ages of clusters. And some striking disagreements. I think that this represents a remarkable opportunity for future research because it is very clear that these discrepancies must be resolved. And in the direction of resolving them I think that it is not just a matter of more work of the same kind but it is of upgrading the entire level of instrumental and intellectual effort that is put into these problems. In the poster session in the work of Rich,

Mould and Da Costa one sees a whole new level of the work on color-magnitude arrays of globular clusters which gives us some idea of what will be coming along in the next few years.

One other point that I would like to make is on the absence of something that was not talked about today. It seems to me that while it is important to age-date clusters accurately, it is also important to understand the chemical evolution of galaxies and also to determine abundances with accuracy. It was a strange thing that there was no review at this symposium concerning spectroscopic determinations of abundances from studies of individual stars in globular clusters. It is clear that Judy Cohen made a remarkable and important start on that problem. It just scratches the surface and there is a tremendous opportunity for individual spectroscopic studies of stars belonging to these clusters. A comparative study, for example, of the clusters in the Large and Small Magellanic Clouds. And if I were to select a field that is opportune for advancing our knowledge of the evolution of galaxies I think that following Judy Cohen's lead would be such a field.

      S. van den Bergh
      Dominion Astrophysical Observatory

I think there are perhaps still a number of things which could be mentioned that have not been discussed by previous panelists. One of the important clues, I think, to cluster formation is the fact that the Magellanic Clouds contain significant numbers of populous clusters - whereas such objects appear to be rare or absent in giant spirals such as the Galaxy and M31. It would be interesting to hear from theoreticians what suggestions they might have to account for this observation. Many of the rich clusters which we see now must, in the past, have been even more spectacular objects than they are at present; objects comparable to (or more luminous than) the cluster presently located in the center of the Tarantula Nebula. And it seems quite likely that populous intermediate-age clusters also contained objects that were once similar to R136 near their centers. One can ask one's self to which extent the evolution of such clusters might have been affected by such massive objects in their centers. Finally, I think that after hearing the papers this morning, one perhaps worries about the fact that there appears to be considerable evidence from observations of stars for a burst of star formation that took place two or three billion years ago. It is puzzling (but possibly not significant) that the evidence from clusters for such a burst of star formation is, at best, much weaker. I think that this is a situation which we really should try to straighten out with observational techniques that are already available. One would certainly like to think that bursts of star formation go hand in hand with a high rate of cluster formation. Although this need not, of course, be the case as is shown by the example of the dwarf galaxy NGC 1613 which is presently forming stars quite vigorously but (according to Baade) does not contain a single star cluster.

GLOBAL EVOLUTIONARY EFFECTS IN THE MAGELLANIC CLOUDS

James LEQUEUX
Observatoire de Marseille, 13248 Marseille CEDEX 4, France

ABSTRACT: Following Larson and Tinsley the integrated colours of the MCs from the UV to the red can be used to obtain the ratio of the present rate of star formation to the total amount of stars formed. Other tracers of recent star formation (number of bright stars, of supernova remnants, Hα emission, etc...) will be used to determine the present rate of star formation and to obtain some information on the initial mass function. The recent (t < a few 10^7 years) history of star formation in the MCs will then be discussed. Finally, it is found that the ratio present rate/total amount of stars ever formed is of the order of 0.1 Gyr^{-1} for both Clouds, implying a rather uniform average rate of star formation if the Cloud ages are the order of 10^{10} years. The results will be confronted with the metallicity-age relation and the age distribution of stars and clusters.

This review considers mainly each Magellanic Cloud as a whole and uses global properties to try to shed light on its evolution. Such an approach is relatively new and no trace of it can be found in the earlier symposia on the Magellanic Clouds. Not only is this method interesting *per se*, but its application to the Magellanic Clouds for which many data are available allows to assess its use for more distant galaxies in which one can only observe global parameters.

I will first discuss the present initial mass function (IMF) of stars and the present star formation rate (SFR) in the Clouds. Then I will go back in time and address the problem of the short-term history of star formation in the Clouds. Finally I will use their global properties to try to derive the past history of star formation in the Clouds, by comparison with the predictions of models of photometric and chemical evolution ; some comparisons will also be made with the results of detailed population studies.

I. THE PRESENT INITIAL MASS FUNCTION AND STAR FORMATION RATE IN THE MAGELLANIC CLOUDS

The direct determination of the IMF requires the consideration of a complete sample of stars in the Hersprung-Russell (H-R) diagram. Assuming a constant star formation rate (SFR) and using theoretical stellar evolutionary tracks, one adjusts the IMF until a match is achieved between theoretical and observed relative numbers of stars in the various parts of the HR diagram. Unfortunately those star catalogues covering all, or almost all, the face of the MCs are reasonably complete only down to $B \simeq 13$ and 14 for the LMC and the SMC respectively (see e.g. Vangioni-Flam et al., 1980). They contain only the very brightest zero-age main-sequence (ZAMS) stars ; the least massive stars which appear in complete samples have masses of the order of 15 $M_\odot$ (however there are deeper star surveys in some small areas of the Clouds, as we will see later). For the determination of the mass function, the stars have to be binned into mass intervals, thus requiring evolutionary tracks which are still uncertain : in addition, evolution times along the tracks are needed to derive the birthrate. Large uncertainties also arise from the necessary transformation between the M_{bol}, $\log T_{eff}$ diagram and the V, B-V diagram.

Dennefeld and Tammann (1980) have first attempted to derive the IMF of the Clouds using this method, for masses $m > 9\ M_\odot$. They found the IMF of the LMC to be $dn(m)/dm \propto m^{-2.0}$, and that of the SMC to be flatter : $dn(m)/dm \propto m^{-1.39}$. The errors on the exponents are unspecified, but they are obviously very large. They also derive the present rate of star formation for those stars and find it to be about the same per unit *total* mass in the LMC and in the solar neighbourhood (SN), and about twice larger in the SMC. Unfortunately these results depend very critically on the evolutionary tracks *after* the main sequence, which themselves depend considerably on the assumed stellar mass loss rates, convection overshooting and internal mixing. Recently, Meylan and Maeder (1982) have made a detailed comparison of the upper HR diagrams of very young *clusters* in the Galaxy, the LMC and the SMC with the predictions of models. They find that the relative frequency of yellow and red supergiants increase with decreasing metallicity, a result which can be understood as the result of a decreasing mass-loss rate and is also visible for field stars (Maeder et al., 1980 ; Maeder, 1981b ; Meylan and Maeder, 1983). While this effect might in principle be taken into account in the derivation of the IMF and SFR if the mass-loss rates were known, another problem turns out to be worse : too many stars are observed outside the theoretical MS band, a discrepancy noted by several authors after Stothers and Chin (1977). Everything occurs as if the MS was extending to effective temperatures as low as $\log T_{eff} = 4.0 - 3.9$, thus including A stars. Whatever the cause of this extension (internal mixing, or opacity of stellar winds as in Wolf-Rayet stars) it is clear that it is still premature to try to derive the IMF from the present star samples. The SFR is perhaps less sensitive to these problems : the similarity between the luminosity functions and also between the complete parts of the HR diagrams of the SN, the LMC and the SMC

(Vangioni-Flam et al., 1980) might be taken as an indication that they can be used to infer relative SFRs, as done in Table 1. However these similarities might be due to a compensation of factors, given the probable differences in the stellar mass loss rates (Prévot et al., 1980 ; Hutchings, 1982).

Some indirect tracers which deal with main-sequence stars cause less problems, since the M-S evolution is less sensitive to abundance than the post M-S one. These tracers are :

a) The flux of Lyman continuum photons

These photons are essentially produced by O stars, which are all on the main-sequence, and induce ionization of the interstellar gas ; recombination produces photons in the visible, in particular $H\alpha$ photons, whose flux is proportional to the flux of ionizing photons. Integral $H\alpha$ photometry exists only for the SMC (Schmidt, 1972). From an integration of the $H\alpha$ map, a colour excess $E(B-V) = 0.08$ and the usual relations, I find assuming a distance of 70 Mpc a flux of Lyman continuum photons absorbed by the gas in the SMC $N'_c = 4.6 \, 10^{51}$ ph s^{-1}. In the SN Güsten (1981) and Abbott (1982) estimate independently a flux of ionizing photons emitted by the stars $N_c = 3 \, 10^{50}$ ph s^{-1} kpc^{-2}, of which according to Güsten (1981) half are actually used to ionize the gas : hence $N'_c = 1.5 \, 10^{50}$ ph s^{-1} kpc^{-2}. This number is combined with that for the SMC to derive the relative SFR in Table 1.

Table 1. Basic data (from Vangioni-Flam et al., 1980) and relative rates of massive star formation from 4 indicators

	Distance	$M_{gas}(M_\odot)$ (1)	$M_{tot}(M_\odot)$	L_B/M_{tot} (2)	N_*/M_{gas} (3)	N'_c/M_{gas}	L_{1690}/M_{gas}	N_{SNR}/M_{gas}
				(Solar units)				
Solar neighbourhood (SN)								
1 kpc^2	-	6.0 10^6	9.0 10^7	-	1	1	1	1
LMC	52 kpc	7.0 10^8	6.1 10^9	1.5	1.2 - 1.6	-	1.3	1.8
SMC	70 kpc	6.5 10^8	1.8 10^9	2.3	0.15 - 0.27	0.28	0.30	0.4 - 0.6

(1) $M_{HI} \times 1.3$ to take helium into account.

(2) These low values may raise problems. They might mean that the total mass is somewhat underestimated.

(3) From Vangioni-Flam et al. (1980).

b) The far-UV fluxes

They are dominated by *main sequence* B stars, thus are not very sensitive to the mass-loss rate. I have re-evaluated the fluxes at 1690 A from Vangioni-Flam et al. (1980) using better extinction corrections, and find L_{1690} = 3.5 10^{39} and 7.3 10^{38} erg s^{-1}Å^{-1} for the LMC and the SMC respectively. These figures are used in Table 1 together with the calculated flux of 2.26 10^{37} erg s^{-1}Å^{-1} kpc^{-2} in the SN, in order to estimate relative SFRs.

Other tracers of SFR have been discussed by Lequeux (1979). The most reliable one appears to be the number of supernova remnants (SNR), since at least the supernovae of type II result from the evolution of the core of massive stars ; this evolution is almost unaffected by mass loss, according to Maeder (1981a). The catalogue of Mathewson et al. (1983) contains 17 SNRs with diameter less than 32 pc in the LMC, and 3 to 5 in the SMC while there are 0.08 such SNRs per kpc^{-2} close to the Sun according to Clark and Caswell (1976). Ignoring the difficulty in distinguishing between remnants of SN of type I and type II, and assuming the evolution of SNRs to be similar in the Galaxy and in the Clouds, this yields the SFRs given in Table 1.

The determination of the relative SFRs given in the 4 last columns of Table 1 are in rather good agreement with each other considering the large uncertainties in the parameters. The good agreement between the ratios N'_c/M_{gas} and L_{1690}/M_{gas} for the SMC is interesting since it gives an indication that the IMF for massive stars should not differ much between the SMC and the SN. We see that the SFR *per unit mass of gas* is about 1.5 times larger in the LMC than in the solar neighbourhood, and 0.3 times in the SMC. Per *unit total mass*, these relative SFRs become 2.7 and 1.6 respectively for the LMC and the SMC. It is to be remembered that the gas masses in the Clouds are probably *underestimates* because of the unknown saturation of the 21-cm line which is used to derived HI masses, and because of the presence of molecular hydrogen in unknown quantities. As to the total masses, they are probably uncertain by a factor 2 at least, especially for the SMC.

Finally, one should remember that these SFRs refer to massive stars, more massive than about 4 $M_\odot$, and that we know nothing of the present global SFR for less massive stars. However Butcher (1977) and Stryker and Butcher (1982) have built the MS luminosity function in two small fields of the LMC down to $M_V \simeq 5$ and find that it is close to that of the solar neighbourhood between $0 \lesssim M_V \lesssim 3$ or 4 suggesting a similar IMF down to masses slightly larger than 1 $M_\odot$. They interpret the break at $M_V \simeq 3-4$ as an age effect ; I will come back to this point later.

II. THE RECENT STAR FORMATION IN THE MAGELLANIC CLOUDS
(< a few 10^7 years)

The recent history of star formation can be considered from two complementary points of views : a) global variations of the SFR ; b) spatial variations and propagation of star formation.

The first type of study involves unbiased statistics of stars or star clusters of different ages. An early attempt has been made by Hodge (1973) who considered the 509 clusters of the LMC in which the brightest star is brighter than V = 15.5. He considered these stars as giving the approximate turning point of the main sequence turn-off for each cluster, hence obtaining a rough estimate of the age. As acknowledged by the author himself, this method is very crude. Statistics on the ages yield a roughly constant rate of production of clusters from 14 to $4\ 10^6$ years ago of about $4\ 10^{-5}$ per year. There is a lack of clusters younger than $4\ 10^6$ years, but it is probably only apparent given the large difficulty in dating very young clusters. No such study exists for the SMC. A similar attempt has been made by Ardeberg (1976) using supergiant stars, that he places between theoretical isochrones on the HR diagram. However the stars are observed only on limited portions of the evolutionary tracks which correspond in practice to post main-sequence evolution, and one must take into account the lifetime of the stars over each of these portions. These lifetimes are extremely model-dependent (we encounter here the same difficulty as when we wanted to know the IMF and SFR for massive stars). For this reason, I do not think that his result - that the bulk of recent massive star formation took place about $7.5\ 10^6$ years ago - can be taken too seriously. The question must still be considered as unsettled.

However the relatively good agreement between the SFRs of Table 1, which refer to stars of different mass ranges hence of different mean ages, suggests a relatively constant SFR in both Clouds, together with an upper IMF not too different from the galactic one - unless there is by chance some compensation between these two factors - There has been also an argument, based on the well-known apparent absence of carbon stars with $M_{bol} \lesssim -6.5$ in the Clouds, that little star formation has occurred during the last $2.5\ 10^8$ years : this time is the lifetime of a star of $3\ M_\odot$, a mass which is supposed to be the lower limit of the mass of the progenitors of the "missing" bright carbon stars ; however younger stars certainly exist in vast quantities, e.g. young Cepheids noticed by Becker (1982), in the fields where carbon stars have been investigated.

The study of spatial variations in star formation has given more convincing results. A pioneer in the field was C. Payne-Gaposhkin (1972) who published maps of the distribution in the LMC of Cepheids of different periods, hence of different mean ages (see also Schmidt-Kaler, 1977). She showed that the sites of star formation have moved considerably within the last 10^8 years. In particular, the Bar has been the site of an active formation of massive stars about $5 \ 10^7$ years ago, while it has not been particularly active for the last $3 \ 10^7$ years ; this is confirmed by the study by van den Bergh (1981) of the space distribution of young clusters (for which one should rather use the age calibration of Hodge, 1983). Similar studies spanning a shorter past have been made on LMC clusters by Hodge (1973) and on LMC supergiants by Ardeberg (1976) and more recently by Isserstedt (1983) and Prévot and Vigroux (1983). Although details differ somewhat, probably because it is rather difficult to assign absolute ages to clusters and individual stars due to uncertainties in their evolution, all these studies agree in that massive star formation occurs in important local bursts of size $\simeq$ 1 to 2 kpc. This agrees with the idea of stochastic star formation (Feitzinger et al., 1981). However it is not obvious that star formation is generally contagious unless it propagates at extremely large velocities, of the order of 100 km/s ; these velocities seem too large to be physically meaningful.

The SMC has been much less studied in this respect. Brück (1975) has noticed that young clusters of various (very uncertain) ages are not distributed in the same way ; the study by van den Bergh (1981) yields a more detailed picture. While old clusters are scattered everywhere, young clusters (22 - 200 10^6 years according to the calibration by Hodge, 1983) are found preferentially in the Bar, while the present (< $30 \ 10^6$ years) cluster formation takes place mainly in the NE tip of the Bar and in the wing. This is also apparent in the Hα surface photometry of Schmidt (1972) which yields indirectly the distribution of O stars ; the distribution of Hα also reveals some very recent star formation in the SW part of the Bar. Brück (1981) considers that the SMC central bar is the seat of a rather continuous star formation while the surrounding "arm" regions rather proceed by bursts ; this difference does not seem very obvious to me in view of the studies mentioned above.

Another way of looking at the recent star formation is to compare the distribution of young stars with that of the gas from which they form. In the LMC massive star formation seems to take place mainly on the edge of HI complexes or between them, yielding an overall poor correlation between the distributions of gas and young stars (Martin et al., 1976); in some cases like the Shapley III superassociation it might be that the gas has been pushed out by newly-born stars. In the SMC, there is apparently a better correlation between gas and young stars (Sanduleak, 1969 ; Schmidt, 1972 ; Azzopardi and Vigneau, 1977). However this correlation is only apparent. When stars and gas in the same range of radial velocities are compared, it becomes as poor as for the LMC (Martin et al., this Symposium), although the SMC stars have a lesser efficiency to disturb the gas (Table 1 and Tarrab, 1983).

III. THE PAST HISTORY OF STAR FORMATION IN THE CLOUDS

The past history of star formation in a galaxy is reflected in its colours and in some parameters like the chemical composition of gas and stars and the mass fraction in interstellar matter. Many papers have been written about this problem and I will review here only those dealing specifically with the Magellanic Clouds. For the moment, I will ignore the structure of the Clouds and their obvious complexity and inhomogeneity and consider them as single entities.

Even a casual inspection of the global properties of the Clouds reveals differences in their past evolution. The SMC contains a higher mass fraction of gas than the LMC, presently forms 5 times less stars per unit mass of gas (Table 1) and contains 4 times less heavy elements : the SMC is clearly a less evolved galaxy than the LMC. As to the LMC, it is more similar to the SN although there remain important differences as we will see later. The first attempt to study quantitatively the evolution of the MCs is due to Olson and Peña (1976) who treated simultaneously the evolution of the SN and of the MCs. They built a computer code yielding as a function of time for a given IMF the gas fraction, the gas chemical composition and the UBV colours of the system. A law of star formation as a function of the mass (or density) of the gas had to be assumed as well as a rate of infall of external gas, which was supposed to be primordial, without heavy elements. Olson and Peña note correctly that the blue B-V colours of the Clouds imply a higher proportion of massive stars than in the SN, hence a SFR decreasing less rapidly with time and perhaps even nearly constant. This in turn implies that if the SFR is assumed to vary as a power K of the mass of gas, K must be smaller than 1. This conclusion might at first sight seem at variance with the finding by various authors (e.g. Sanduleak, 1969 ; Hamajima and Tosa, 1975 ; Martin et al., 1976 ; Azzopardi and Vigneau, 1977 ; Brück, 1981) of a positive correlation of the surface density of young, massive stars or of HII regions with that of neutral hydrogen (Schmidt's law) : however Olson and Peña remark that this concerns the *spatial* behaviour of star formation at the present epoch and has nothing to do with the *global* SFR as a function of time. Moreover we have already seen that there are local *anticorrelations* between young stars and gas in the LMC, e.g. in the region of Sh III : Martin et al., 1976 ; there are similar phenomena in the SMC : Martin et al., this Symposium. Once the slope of the IMF and a rate of infall are chosen, the chemical composition and gas mass/total mass ratio can be used to set the lower mass limit of the IMF. One problem that Olson and Peña encountered is that their models are still not blue enough, especially for the SMC ; they had to invoke a recent enhancement in star formation in order to produce bluer colours.

After this paper, Larson and Tinsley (1978) made a general study of the colour evolution in galaxies. They showed that for a given IMF, (assumed to be constant in time) the position of a galaxy on the common line in the U-B, B-V diagram is "almost uniquely determined by the SFR per unit (total) mass averaged over the past 10^8 years". To be more precise, this position depends only on the ratio $R(t_0)$ of the "present" SFR (averaged over a few times 10^7 years) over the integrated past SFR

$$R(t_0) = SFR(t_0) / \int_0^{t_0} SFR(t) \, dt.$$

Rocca-Volmerange et al. (1981) have extended this property to *any* combination of two colours, using their models of photometric evolution. They also showed that the colours, and in particular B-V, are sensitive to metallicity. This effect was not taken into account by Olson and Peña, and is apparently sufficient to explain why they found a too red B-V for the SMC. Rocca-Volmerange et al. used their model to try to reproduce the full spectrum of the Clouds from 1690 Å to the red. They succeeded by chosing $R(t_0)$ = 0.10 to 0.14 Gyr^{-1} for the LMC and 0.045 to 0.11 Gyr^{-1} for the SMC. These values are dependent on the choice of the IMF (which is also assumed to stay constant in time). The heavy element abundances can give constraints to the IMF. It excludes an IMF as flat as Salpeter's IMF ($dn(M)/d \ln M \propto M^{-x}$ with x = 1.35) which produces too much heavy elements ; a slope $x \simeq 2$ appears more appropriate. However this conclusion will not hold if there is a very large rate of infall, since a large heavy-element production could then be partly or almost entirely balanced by dilution with the accreted gas. Ignoring this problem for the moment, the quoted range of values of $R(t_0)$ corresponds to extreme possible IMFs in the case of no infall. The Salpeter's IMF (with large infall) would yield a smaller $R(t_0)$. It is possible to try to go farther and to reproduce simultaneously the colours, the integrated luminosity, the mass of gas and the total mass. For the LMC, Rocca-Volmerange et al. (1981) find that a good fit is obtained with a uniform SFR, hence an age of $R(t_0)^{-1} \simeq 7$ to 10 Gyr. For the LMC, a slightly decreasing SFR $\propto \exp - 0.15(t/10^9 \text{ yr})$ seems more appropriate, the age being of the same order. However this assumes no infall, and on the other hand the total masses of the Clouds are so uncertain that these conclusions cannot be taken without care. The previous study is presently extended to the infrared (Rocca-Volmerange, **in preparation**), but the results will be *a priori* less secure due to uncertainties in the evolution of the red giants which dominate the IR emission, due to our poor knowledge of their mass losses.

One interest of the previous studies is to show that there is no need of invoking a recent burst of star formation to account for the global properties of the Magellanic Clouds. They give no evidence that the Clouds may have a small age, a complex past history, etc. This evidence has to be searched for by other means, and I will come back to this point. However the above studies already raise a big problem which is perhaps clearer in the case of the LMC : how could the SFR have been roughly uniform in the past while the amount of gas available to form stars has decreased by more than a factor 10 ? As discussed by Rocca-Volmerange et al. (1981) there are at least two ways out of this difficulty. One is to suppose that a large rate of accretion compensates for the SFR, keeping the mass of gas roughly constant. Another one is that the present SFR (integrated over the IMF) has been overestimated :

if we impose a low-mass cut-off of say 1-2 $M_\odot$ to the IMF, the *present* lifetime of the gas against star formation is raised from 1 to 4 Gyr. In this case, the low-mass stars which are certainly present and contribute much to the luminosity must have been formed mainly at an early epoch. Although speculative such a bimodal star formation is not excluded by theory (see e.g. Silk, 1980). The presence in the LMC of a halo and of a bar mainly made of low-mass stars might be taken as a possible evidence for this speculation (remember however that there has been relatively recent star formation in the haloes : see later).

It is of interest, before ending this section, to make a short review of the other information on the past SFR in the Clouds.

First of all, it appears that some globular clusters of the Clouds are very old. From the main-sequence turn-off points, the ages of NGC 121 and Linsay 1 in the SMC are estimated as 13 ± 5 and $9 \; 10^9$ yr respectively, and that of NGC 2257 in the LMC as about $14 \pm 2 \; 10^9$ years (see Hodge, 1983). There is however a remote possibility that the latter object is a galactic globular cluster captured by the LMC (Stryker, 1983), in which case star formation could have started later. From the ages and metallicities of globular clusters and individual stars it is possible to infer the chemical evolutionary history for the Clouds. Searle, Wilkinson and Bagnuolo (1980) first noticed that age and chemical composition of LMC clusters are highly correlated. Cohen (1982), Cowley and Hartwick (1982) and Butler et al.(1982) further studied this correlation (the latter also included individual stars). They agree that in both Clouds the heavy element enrichment has been relatively slow, a result already foreseen by van den Bergh (1975). This progressive evolution differs markedly from that of the Galaxy, where a phase of fast enrichment was followed by little or no enrichment at all. The observed correlations for both Clouds are not inconsistent with the predictions of a closed model with a uniform SFR and constant IMF (see Cohen, 1982); however errors in the ages and metallicities of clusters are very critical in this kind of comparison, and are still too large to be able to trace the evolution in more detail.

Some "anomalous" heavy element abundances may indirectly give a hint on the past SFR. According to Dufour et al.(1982) the C/O ratio is noticeably smaller in HII regions of the Clouds than in the Galaxy, by a factor as high as about 5 in the SMC. The weakness or absence of the 2200 A feature in the extinction laws of the Clouds is another evidence for this deficiency in carbon, since this feature is supposed to be due to graphite. Carbon can be produced in rather low-mass stars in the red-giant phase, and a smaller C/O ratio can be interpreted as the consequence of a relatively smaller integrated death -rate of such stars, compared to the Galaxy. This is another possible indication of a relatively smaller SFR in the past. However Foy (1981, 83) finds that iron-peak elements in SMC stars are less deficient than oxygen, when compared to the Galaxy. Since these elements, like carbon, are often considered as being produced by smaller-mass stars than oxygen (see e.g. Tinsley, 1979), this result seems in contradiction with the strong underabundance of carbon; I cannot offer an explanation for this discrepancy.

Direct information on the past SFR could be derived if statistics of stellar ages could be obtained. This is unfortunately a most delicate topic. Butcher (1977) and Stryker and Butcher (1982) find in two fields of the LMC a deficiency in stars less bright than $M_V \simeq 3-4$ with respect to the solar neighbourhood. They suggest that the bulk of star formation started in the LMC only 3-5 10^9 years ago. However the lack of low-luminosity, hence low-mass stars, might also be a property of the IMF : see above the discussion of a possible bimodal star formation. In this model, stars with $M \gtrsim 1 M_\odot$ formed early in the evolution would have evolved from the MS (this would not be inconsistent with the HR diagram presented in Fig. 3 of Butcher, 1977). But independent evidence for a large amount of star formation 3-5 10^9 years ago in the LMC has been presented. Mould and Aaronson (1982) suggest a peak in the age distribution of LMC globular clusters at $\simeq 4\ 10^9$ years, but a decreasing distribution from 12 10^9 years ago to now for the SMC ; however their age scale does not agree with that of Hodge (1983), and moreover Mould and Aaronson consider that their age distribution might be biased by luminosity evolution of the clusters and is not inconsistent after all with a constant SFR. The age distribution of SMC clusters looks different from that of the LMC and may imply a SFR decreasing with time ; these results agree with the suggestions of Rocca-Volmerange et al.(1981). It should be emphasized however that a strong increase of SFR in the LMC 3-5 10^9 years ago would not necessarily be in contradiction with the global properties of this galaxy.

In the previous discussion, the Clouds have been considered as homogeneous objects. It is well known however that both show structure : the Bar, a disk, and a halo with different properties in the LMC, and a main body, a Wing and a halo in the SMC. Surprisingly, star formation seems to have occurred fairly recently in some parts of the halos: around NGC 2257 in the LMC (Stryker, 1982), and in the region of Kron 3 in the SMC (Hawkins and Brück, 1982). In the latter region, star formation has taken place about $3 \pm 1\ 10^9$ years ago : the presence of older stars is also suggested. Clearly the history of the various components of the Clouds has been different ; another intriguing related problem is the apparently different kinematic behaviour of the old ($\gtrsim 1-2\ 10^9$ years) and young ($< 10^9$ years) globular clusters of the LMC : Cowley and Hartwick, 1982 ; Freeman et al., 1983.

IV. CONCLUSIONS

Although an enormous observational and intellectual investment has been put into the problem of the evolution of the Magellanic Clouds, we are still far from having obtained a definite picture. A simple-minded model consistent with most observations is that of a relatively smooth evolution at a nearly constant rate since about 10 billion years. However some observations (luminosity function for relatively faint stars) do not seem to fit, and it is clear that at small scales in time and in space the evolution is not smooth. It may be that the Ockam's razor does not work for the Magellanic Clouds after all, and that their evolution has indeed been complex. This is not unexpected, since the Clouds are likely to form an interacting system with our Galaxy ; each encounter might produce a big effect on their evolution, with mass exchange etc.. The problem is certainly quite interesting but extremely complex. The possible lines of approach are multiple, and several of them are followed at the present time. I hope that their results will allow to draw a more convincing picture of the Magellanic Cloud evolution at the next specialized symposium.

REFERENCES

Abbott, D.C.: 1982, *Astrophys. J.* 263, 723.
Ardeberg, A.: 1976, *Astron. Astrophys.* 46, 87.
Azzopardi, M., Vigneau, J.: 1977, *Astron. Astrophys.* 56, 151.
Becker, S.A.: 1982, *Astrophys. J.* 260, 695.
Brück, M.T.: 1975, *Mon. Not. R. Astron. Soc.* 173, 327.
Brück, M.T.: 1981, *Astron. Astrophys.* 87, 92.
Butcher, H.R.: 1977, *Astrophys. J.* 216, 372.
Butler, D., Demarque, P., Smith, H.A.: 1982, *Astrophys. J.* 257, 592.
Clark, D.H., Caswell, J.L.: 1976, *Mon. Not. R. Astron. Soc.* 174, 267.
Cohen, J.G.: 1982, *Astrophys. J.* 258, 143.
Cowley, A.P., Hartwick, F.D.A.: 1982, *Astrophys. J.* 259, 89.
Dennefeld, M., Tammann, G.A.: 1980, *Astron. Astrophys.* 83, 275.
Dufour, R.J., Shields, G.A., Talbot, R.J,Jr.: 1982, *Astrophys. J.* 252, 461.
Feitzinger, J.V., Glassgold, A.E., Gerola, H., Seiden, D.E.: 1981, *Astron. Astrophys.* 98, 371.
Freeman, K.C., Illingworth, G., Oemler, A.,Jr.: 1983, preprint.
Foy, R.: 1981, *Astron. Astrophys.* 103, 135.
Foy, R.: 1983, *The Messenger* 31, 24.
Güsten, R.: 1981, unpublished Ph. D. thesis, University of Bonn.
Hamajima, K., Tosa, M.: 1975, *Publ. Astr. Soc. Japan* 27, 561.
Hawkins, M.R.S., Brück, M.T.: 1982, *Mon. Not. R. Astron. Soc.* 198, 935.
Hodge, P.W.: 1973, *Astron. J.* 78, 807.
Hodge, P.W.: 1983, *Astrophys. J.* 264, 470.
Hutchings, J.B.: 1982, *Astrophys. J.* 255, 70.
Isserstedt, J.: 1983, *The Messenger,* 31, 14.
Larson, R.B., Tinsley, B.M.: 1978, *Astrophys. J.* 219, 46.
Lequeux, J.: 1979, *Astron. Astrophys.* 71, 1.
Maeder, A.: 1981a, *Astron. Astrophys.* 101, 385.
Maeder, A.: 1981b, *Astron. Astrophys.* 102, 401.
Maeder, A., Lequeux, J., Azzopardi, M.: 1980, *Astron. Astrophys.* 90, L 17.
Martin, N., Prévot, L., Rebeirot, E., Rousseau, J.: 1976, *Astron. Astrophys.* 51, 31.
Mathewson, D.S., Ford, V.L., Dopita, M.A., Tuohy, I.R., Long, K.S., Helfand, D.J.: 1983, *Astrophys. J. Suppl.* 51, 545.
Meylan, G., Maeder, A.: 1982, *Astron. Astrophys.* 108, 148.
Meylan, G., Maeder, A.: 1983, *Astron. Astrophys.*, 124, 84.
Mould, J., Aaronson, J.: 1982, *Astrophys. J.* 263, 629.
Olson, G.L., Pena, J.H.: 1976, *Astrophys. J.* 205, 527.

Payne-Gaposchkin, C.: 1972, *IAU Coll. No 17* "L'âge des étoiles",
    ed. G. Cayrel de Strobel et A.M. Delplace, p. III 1.
Prévot, L. et al.: 1980, *Astron. Astrophys.* **90**, L 13.
Prévot, L., Vigroux, L.: 1983, in preparation.
Rocca-Volmerange, B., Lequeux, J. Maucherat-Joubert, M.: 1981,
    Astron. Astrophys. **104**, 177.
Sanduleak, N.: 1969, *Astron. J.* **74**, 47.
Schmidt, Th.: 1972, *Astron. Astrophys.* **16**, 95.
Schmidt-Kaler, Th.: 1977, *Astron. Astrophys.* **54**, 771.
Searle, L., Wilkinson, A., Bagnuolo, W.G.: 1980, *Astrophys. J.* **239**, 803.
Silk, J.: 1980, in *Star Formation*, 10th Advanced Course, Saas Fee,
    ed. Geneva Observatory.
Stothers, R., Chin, C.W.: 1977, *Astrophys. J.* **211**, 189.
Stryker, L.L.: 1982, *Publ. Astr. Soc. Pac.* **94**, 760.
Stryker, L.L.: 1983, *Astrophys. J.* **266**, 83.
Stryker, L.L., Butcher, H.R.: 1982, *IAU Coll. No 68*, ed. A.G.D. Philip,
    D.S. Hayes, L. Davis Press, Schenectady, NY., p. 255.
Tarrab, I.: 1983, *Astron. Astrophys.* in press.
Tinsley, B.: 1979, *Astrophys. J.* **229**, 1046.
Van den Bergh, S.: 1975, *Ann. Rev. Astron. Astrophys.* **13**, 217.
Van den Bergh, S.: 1981, *Astron. Astrophys. Suppl.* **46**, 79.
Vangioni-Flam, E., Lequeux, J., Maucherat-Joubert, M., Rocca-Volmerange,
    B.: 1980, *Astron. Astrophys.* **90**, 73.

DISCUSSION

Danziger: If it was known that the slope of the IMF was a continuing function of the metallicity would this affect your conclusions concerning the differential evolution of the SMC and the LMC?
Lequeux: Most of the conclusions will be affected by changes in the IMF, as I was careful to state in my review. However the fact that the SMC is less evolved than the LMC is purely observational (from metallicity and M(gas)/M(tot). There are indeed several lines of evidence showing that either the IMF slope flattens or the upper mass cut-off increases when metallicity decreases (Serra, Puget, etc.).
Danziger: Nevertheless Melnick and Terlevich have presented some evidence recently for this dependence of slope of IMF on metallicity!

FIELD STUDIES, LUMINOSITY FUNCTIONS, AND STAR FORMATION HISTORY IN THE MAGELLANIC CLOUDS

L.L. Stryker
Dominion Astrophysical Observatory
Herzberg Institute of Astrophysics
Victoria, B.C. Canada, V8X 4M6

I. INTRODUCTION

One of the most fundamental questions we might ask about galaxies is, Do all galaxies have the same age? A less general question, and one which we can surely succeed in answering is, Are the Magellanic Clouds (MCs) the same age as the Galaxy? We must also make clear what is meant by the same age if, in fact, star forming activities in these systems have proceeded along different timescales. The age of a system can be masked if the strongest star-forming epoch was not coincident with the initial epoch. Deep colour-magnitude diagrams (CMDs) and luminosity functions (LFs) have had to wait until the advent of large southern telescopes, sensitive emulsions and detectors, and accurate methods of measuring crowded images.

Our knowledge of the stellar population structure in the Clouds as of 1970 was mainly based on observations to V<18 by Arp, Bok and collaborators, Hodge, Tifft, Westerlund, and Woolley (Westerlund 1970; and the ESO volume, Muller 1971). Most of these studies were made relatively close to the inner regions of the Clouds, or near associations or clusters. Many showed that clusters and their nearby fields had closely related populations. Even so, CMDs really become interesting at V ~19, where one can begin to learn something about the horizontal branch and giant clump.

Harvard workers, Shapley, Bok, and Hodge, produced star counts and LFs down to V=17.5 (M_V=-1.5) in a number of areas in the Clouds. Their investigations indicated that both Clouds had LFs that were similar to the van Rhijn Solar Neighbourhood function but with a possible excess of bright supergiants in the LMC (Hodge 1961).

Very little was known of the halo-type populations in either Cloud other than the globular cluster CMDs of Gascoigne (1966) and that of Tifft's (1963) field near NGC 121. Equally unstudied was the LMC bar. It had been suggested (Tifft 1971) that "the older stellar components of the MCs are significantly younger, on the average, than the old star component of the Galaxy", and that the oldest population appeared only sparsely in the Clouds.

This paper will attempt a) to outline the advances made since 1970 in faint field CMDs and main-sequence LFs, and b) to derive a general notion from these studies of how star formation (SF) may have proceeded in the Clouds.

II. FIELD STUDIES

II.1. The Large Magellanic Cloud.

The first published field CMDs typically showed strong, blue main sequences and varying proportions of red giants. In the last decade many deeper studies have appeared (c.f. Table I). The discussion here will emphasize studies of the Bar and outlying areas.

Tifft and Snell (1971=TS) studied the Bar West (BW) 1°5 NW of Bar centre, down to V~17.5. Many of their conclusions have been superceded by a study to V~21 by Hardy et al. (1983), who found the main giant clump at V=19.2, B-V=0.8, with an asymptotic giant branch (AGB) tip at V=16, B-V=2.2 -- suggesting a strong intermediate-age population (contrary to TS). No blue horizontal branch was found. From their CMD, LF, and from evolutionary models of clump red giants, they concluded that: 1. SF continues in the bar; 2. old Pop. II is not a significant contributor; 3. most SF started between 1 to 3.10^9 years ago (and has since been continuous); and 4. the Bar population is considerably younger and more metal-poor than that in the Solar Neighbourhood.

Frogel and Blanco (1983=FB) produced an infrared CMD of M giants in the same BW region. They found two well-separated AGBs, indicating two discrete epochs of SF, the main one having taken place "a few 10^9 years ago". This latter result is in agreement with conclusions found in other regions (Butcher 1977; Stryker and Butcher 1981; Hardy et al. 1983), and strengthens the idea that a global increase in the star-formation rate (SFR) occurred in the LMC during that epoch. The secondary episode of SF was ~10x less-efficient than the first and took place ~10^8 years ago. Between episodes, the SFR was much lower, suggesting non-continuous SF, contrary to the suggestion of Hardy et al. (1983). Integrated colours for Area 5 (=BW) in Hardy's (1978) study of colours in and around the Bar, suggest that BW is typical of the entire Bar. However, ~11-12' S of the Bar, colours become significantly redder.

These studies indicate that the Bar has been a place of quite active SF, at least up to ~10^8 years ago. Since the dust lane signature is lacking and the LMC is neither regular nor massive, there is no evidence thus far that the presence of a bar in any way drives SF.

From a study near the old globular cluster, NGC 2257, Stryker et al. (1981), and Stryker (1983), find that even in the outer (~9 kpc) regions SF must have taken place in relatively recent times. The field CMD is that of a more metal-rich population, showing a strong red horizontal branch. Also there is no old, extremely metal poor (in the galactic globular cluster sense) component in the field, other than the RR Lyrae stars (Nemec et al. 1984), which themselves may show a range in metallicity.

These field studies show a mixture of young and old stars in all areas of the LMC. V_B, the apparent magnitude of the brightest blue evolved stars (B-V <0.4) in each field CMD, can be used to get a rough notion of ages (Hodge 1983). Figure 1 shows V_B plotted versus radial distance from the LMC centre. (We ignore the Shapley Constellations to concentrate on the older, background population.) Approximate relative ages, from Hodge (1983, fig. 1), appear on the right ordinate. Fig. 1 shows a radial drop-off in "youngest stars".

TABLE I
Recent Field Studies in the Magellanic Clouds

RA 1975. Dec	Dir	Near cluster	CMD Ref.[a]	V_B	M_V	Age
LMC						
05 09.5 -68 57	1°.4 NW	N1850	(1,2,3)	14.5,16	-4.1,-2.6	$5 \cdot 10^7$
05 19.7 -70 59	1.5 S	---	(4)	18	-0.6	$2-3 \cdot 10^8$
05 48.5 -71 30	2.9 SE	N2121	(5)	18.5	-0.1	
05 28.4 -66 15	3.3 NE	N1978	(6)	17,18,19	-1.6,0.4	$1-4 \cdot 10^8$
05 21.8 -73 31	4.0 S	---	(7)	20	1.4	$< 10^9$
05 13.0 -65 30	4.1 N	N1866	(8)	18.5	-0.1	
04 59.0 -66 00	4.2 NW	N1783	(9,10)	19.5	0.9	$6 \cdot 10^8$
06 14.0 -69 20	4.4 E	H11	(11)	19.5-20	1.2	$< 10^9$
05 14.3 -63 59	5.6 N	N1868	(12)	19	0.4	$4 \cdot 10^8$
06 10.0 -73 50	5.7 SE	N2209	(13)	19.3	0.7	$5 \cdot 10^8$
06 30.0 -64 10	8.5 NE	N2257	(9,14,15)	20.5	1.9	$\sim 10^9$
SMC						
01 07.6 -72 42	1.2 E	N419	(16)	17.5-18	-1.5	$1-2 \cdot 10^8$
00 27.0 -72 52	1.8 W	K3	(17)	20.2	1.0	$8-9 \cdot 10^8$
01 09.0 -71 42	1.8 NE	---	(18)	18.5	-0.7	$2 \cdot 10^8$
01 22.0 -72 02	2.4 NE	---	(18)	20	0.8	$6-7.18^8$
00 25.7 -71 41	2.4 NW	N121	(19)	~ 22	2.8	$\sim 10 \cdot 10^9$
01 06.0 -75 30	2.7 S	N339	(20)	~ 20.8	1.6	$1 \cdot 10^9$
00 09.8 -73 36	3.5 SW	---	(21)	20	0.8	$6-7 \cdot 10^8$
01 49.0 -73 51	4.2 E	L113	(22)	~ 22.2	3	$\sim 10 \cdot 10^9$

[a] Refer to the Reference list, which includes these codes.

Evidently, from Fig. 1 and the studies mentioned above, the Bar has had several episodes of SF right up to several 10^7 years ago. The farthest fields, however, show no star-forming activity after $1-2 \cdot 10^9$ years ago (though tidal forces may have dispersed the stars away from their original connections with gas and molecular clouds). For the fields listed in Table I, the ratio of blue to red stars (for stars with V <20) is roughly constant to $\sim 4.5-5°$ from the Bar, beyond which it falls steadily. In many CMDs, a typical Pop. I giant "clump" (age $<5 \cdot 10^9$ years) is seen, in contrast to, say, a Pop. II red horizontal branch (age $>10 \cdot 10^9$ years). In no field region studied thus far has a convincing blue horizontal branch been seen.

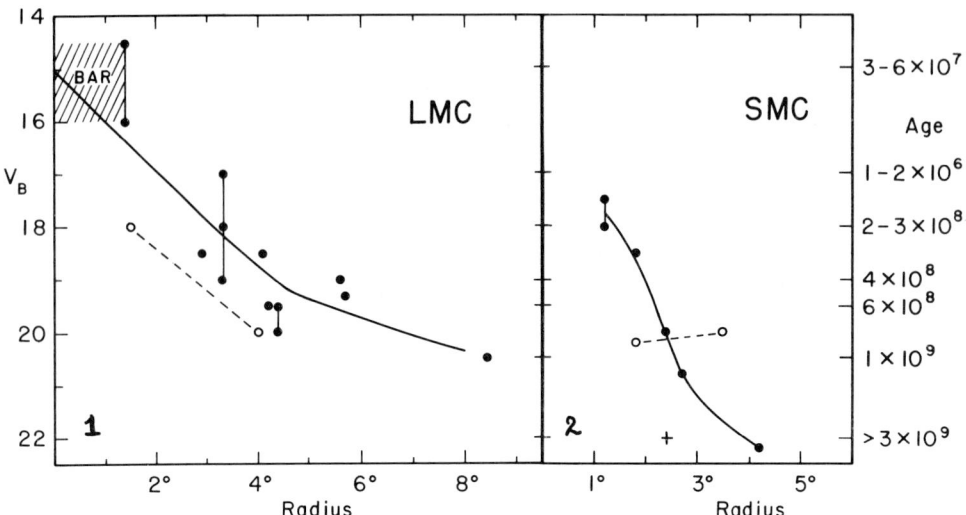

Fig. 1,2. V_B vs. radial distance. LMC: (●) for fields in BW, E, and N; (o) for fields S. SMC: (●) for fields E, N, and S; (o) for fields W. (+) is for the field near NGC 121.

II.2. The Small Magellanic Cloud.

Following the fine review of the population structure of the SMC by Brück (1982), we outline the regions by increasing age (decreasing SF activity).

Bar, Arms, and Wing. current or recent SF 10^6-10^7 yrs ago, no intermediate-age clusters, planetary nebulae in Bar. SF began in Bar ~10^9 yrs ago (Pop. I), Arms and Wing show evidence of recent SF bursts (extreme Pop. I).

Outer Arm, Outer Wing, and Bridge: SF 1-6.10^7 yrs ago, nothing since. No populous clusters in Wing.

Disc: Pop. I objects in dense regions, older objects spread over entire area. SF 5.10^7 to 10^8 yrs ago.

Halo: Objects much younger than galactic halo members, large density of RR Lyrae stars (similar to that in the galactic disc), large intermediate-age component ~3.10^9 yrs old.

The Edinburgh Group have made extensive star counts (Brück 1978, 1980) and field CMDs to V~21.2 (see Table I). Brück and Marsoglu (1978) find, in fields in the outer arm and halo near NGC 458, evidence of layers of SF superposed on a population >10^9 years old. Hawkins and Brück (1982), near K3, confirm a 3.10^9 year old component without ruling out an older constituent. Their latest work (Hawkins and Brück 1984) is much farther out (3°5 SW) and similar results are obtained.

Hardy and Durand (1983) studied fields in the inner wing near NGC 419 and found a probable subgiant branch, indicating a median age of SF older in the SMC than in the LMC. Stryker and Nemec (1983), working down to V ~22.5, find the majority of stars in a field S of NGC 339, to be older than the nearby intermediate-age cluster NGC 643. Stryker et al. (1983), find the field near the old cluster NGC 121, to be $\sim 10.10^9$ years old, and Mould et al. (1983) find the same result for their field 4° E of the Bar, near L113. Thus, from analyses to date, the background sheet of stars in the SMC appears to be of a similar age as the galactic disc. Fig. 2 shows the radial distribution of V_B for the SMC. Note that the SMC displays an essentially older stellar population than appears in the LMC, even after allowance for differences in metallicity.

III. MAIN SEQUENCE LUMINOSITY FUNCTIONS

According to Salpeter (1955), the observed Solar Neighbourhood LF depends on: 1. the relative probability for the creation of stars of mass M at time t (i.e., the initial mass function, IMF(M)); 2. the rate of creation of stars as a function of time since the formation of the Galaxy (i.e., the SF rate, SFR(t)); and 3. the evolution of stars of different masses after they leave the main sequence. Assumptions were that mass-loss is important only in the later stages of evolution; the IMF is time-independent; and the SFR is constant over the last 5.10^9 years (the lifetime of the Pop. I Solar Neighbourhood). Salpeter found a "knee" or change in slope in the local LF at M_V~+4, consistent with an age of the disc, $\sim 10.10^9$ years. Since it is believed that other galaxies condensed and formed at about the same time, their LFs should also exhibit either a discontinuation (elliptical galaxies) or a change in slope (spirals and irregulars) at M_V~+4. What is found from the faint LFs in the Magellanic Clouds?

Many problems are encountered in actually deriving a main sequence LF in the Clouds: getting data to sufficiently faint magnitudes where things begin to get interesting (V >21.5); doing accurate photometry in crowded fields; making corrections for the loss of faint stars, due partly to obscuration by brighter stars, to achieve a complete starcount; eliminating non-main sequence stars and foreground/background contaminators. Added to these problems are the uncertainties in the distance modulus ($\sim 0\overset{m}{.}2$), the chemical composition, the original shape of the IMF, and the SF history or rate.

The first derivation of a main sequence LF was discussed by Butcher (1977) for a field ~4° N of the Bar. Later, Stryker (1981) and Stryker and Butcher (1981) discussed a second region 4°.5 NW of the Bar. These authors deconvolved partially merged images, and studied the error distributions and completeness with the use of artificial images. Their major finding was that in the LMC the change of LF slope occurs at M_V=+3, a full magnitude brighter than in the Solar Neighbourhood. The difference in chemical composition between the Galaxy and LMC contributes only $\sim 0\overset{m}{.}1$, assuming similar helium content. The interpretation of this magnitude difference has been that a major epoch of SF occurred in the LMC at ~ 2-5.10^9 years ago, instead of 10.10^9, as expected. Also, when the younger "age" is assigned the LMC, the correction to the initial LF reproduces the Salpeter initial LF.

Concern remained because these results were obtained from photometry done at the very limits of photographic plates and because of the possibility that the LMC mass function may be deficient at the low mass end (thus causing a brighter knee to be observed in the LF).

In an attempt to understand better the LMC LF and its interpretation, Stryker and VandenBerg (1983=SV) are constructing LF models for chemical compositions [Fe/H]= 0, -0.5, and -1, and IMF slopes of 0 and 1.35. Their preliminary models use the set of isochrones generated by VandenBerg (1983), for masses from $3M_\odot$ to $0.5M_\odot$ (lifetimes from $0.25 - 10.10^9$ years). Evolutionary tracks are found to be sensitive to the metallicity adopted (as demonstrated by the convective "hooks" along the main-sequence). SV are also testing several different SFRs, such as sporadic bursts, constant rate, and continuous SF in equal intervals of log time (power law). The IMF used is the usual simple power law $dn/dm \propto m^{-(1+x)}$. A SF burst occurs with IMF slope x and burst strength specified. Stars are counted until hydrogen-shell burning begins.

Models for [Fe/H] = -1, x = 0 and 1.35 have been compared to observations in three SF cases: 1. sporadic star bursts, 2. constant SF (Δt), and 3. constant SF ($\Delta \log t$). For Case 1, SV find a reasonably good fit with bursts (in 10^9 years, followed by burst strength) at 5(1), 4(1), 2(0.5), 0.25(0.25), 0.05(0.05). Exchanging 10(0.25) for 4(1) fits equally well. Figs. 3-4 show Case 2 models; Figs. 5-6, Case 3 models. Regardless of IMF slope or type of SF chosen, all models require a strong intermediate-age component.

SV conclude that a significant amount of SF 10.10^9 years ago cannot be ruled out, but its strength must have been less than that $\lesssim 5.10^9$ years ago, the age of the majority of stars. Large factors of a $6,8,10.10^9$ year old population move the knee towards M_v=+4. It is realized that field CMDs show local variations, but the data used (Stryker and Butcher 1981) are nearly identical to those of Butcher (1977). Also, IMFs in the Galaxy and LMC are probably similar. Case 3 agrees with Rocca-Volmerange (1983) who, from UV colours of Irregular galaxies, finds an active present SFR, a low mean past SFR, and similar IMFs.

The strength of the 10.10^9 year old population can be best ascertained by Space Telescope observations, since it shows up at $M_v \gtrsim +4$. The slope of the IMF and any low-mass deficiency can be determined as well with observations $+4 < M_v < +6$. Ground-based CCD observations can aid considerably in defining the LF with observations in many fields to V=24.

IV. FUTURE WORK

Is the LMC halo spherical or flat? What are the chemical compositions of halo stars, RR Lyrae stars, and intermediate-age giants? Are the oldest clusters as old as the galactic globulars? More deep CMDs and LFs to V=24 are needed in many more areas of both Clouds. The

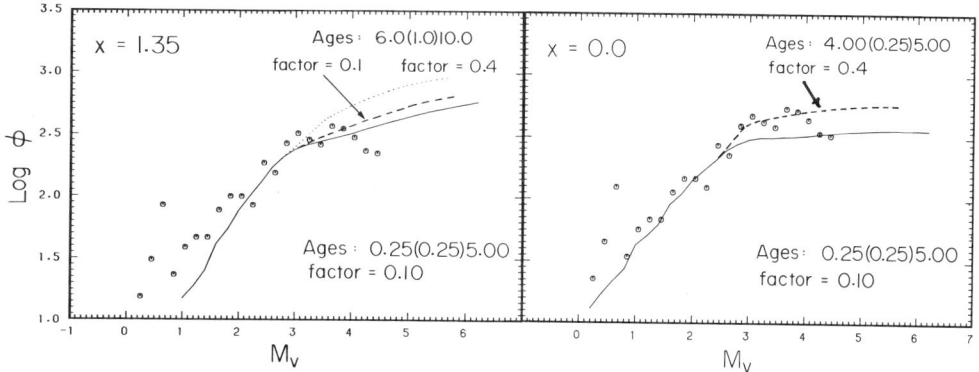

Fig. 3. Model LF with constant SFR, $x = 1.35$, ages $0.25-5.10^9$ in time-steps of $0.25.10^9$ (equal factors). Dashed line shows ages $6-10.10^9$ added, (1/4 weight); dotted line, same (full weight).

Fig. 4. Same as Fig 3. but with $x = 0$. Dashed line is result after addition of strong intermediate-age component.

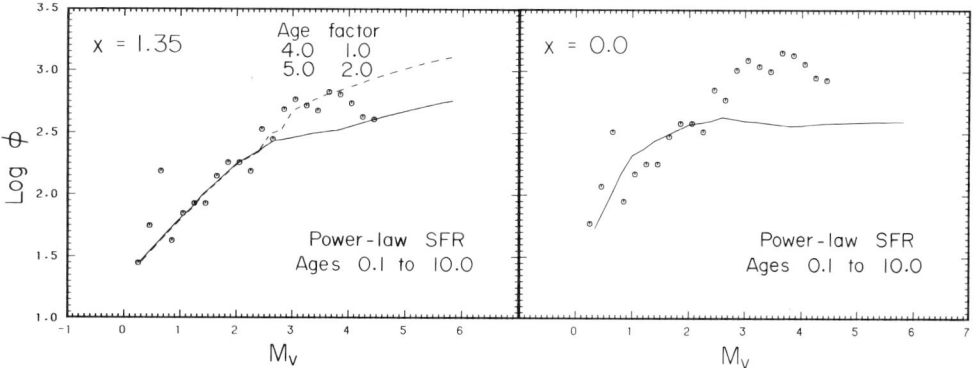

Fig. 5. Model LF, power law SFR ages $0.1-10.10^9$, $x = 1.35$. Dashed line shows addition of very strong intermediate-age component.

Fig. 6. Same, as Fig. 5 but with $x = 0$.

Space Telescope will not only be able to reach $M_V \sim +7$, but will reduce considerably the problem of crowded images. We should then be able to define the LF shape past $M_V \sim +4$, learn more about the oldest population component in the Clouds, and be able to find out whether the Clouds are deficient in low-mass stars compared to the Solar Neighbourhood. In addition, more studies of galactic halo and disc fields would be useful for comparison.

The author wishes to thank M. Brück, J. Frogel, E. Hardy, J. Mould, E. Olszewski, and B. Rocca-Volmerange for sending information pertinent to this paper, and J. Hesser, S. van den Bergh, and H. Harris for comments on early versions. The Deutsche Forschungsgemein-shaft (DFG) and the IAU provided support for travel to this symposium.

REFERENCES

Brück, M.T.: 1978, Astron. Astrophys. 68, 181.
Brück, M.T.: 1980, Astron. Astrophys. 87, 92.
Brück, M.T.: 1982, in Compendium in Astronomy, eds. E. G.
    Mariolopoulos et al. (Dordrecht: Reidel), p. 297.
Brück, M.T., & Marsoglu, A.: 1978, Astron. Astrophys. 68, 193.   (18)
Butcher, H.R.: 1977, Astrophys. J. 216, 372.   (8)
Flower, P., Geisler, D., Hodge, P., & Olszewski, E.: 1980,
    Astrophys. J. 235, 769.   (12)
Flower, P., Geisler, D., Hodge, P., Olszewski, E., & Schommer,
    R.: 1983, Astrophys. J., (in press).   (5)
Frogel, J.A., & Blanco, V.C.: 1983, Astrophys. J. Letters,
    (in press). (FB)   (3)
Gascoigne, S.C.B.: 1966, Monthly Notices Roy. Astron. Soc. 134, 59.
Hardy, E.: 1978, Pub. Astron. Soc. Pacific 90, 132.
Hardy, E., & Durand, D.: 1983, (preprint).   (16)
Hardy, E., Melnick, J., & Rheault, C.: 1980, in IAU Symp. 85,
    ed. J.E. Hesser (Dordrecht: Reidel), p. 343.   (13)
Hardy, E., Buonnano, R., Corsi, C., Janes, K., & Schommer, R.:
    1983, Astrophys. J., (in press).   (2)
Hawkins, M.R.S., & Brück, M.T.: 1982, MNRAS 198, 935.   (17)
Hawkins, M.R.S., & Brück, M.T.: 1984, this volume, p. 101.   (21)
Hodge, P.: 1961, Astrophys. J. Suppl. 6, 235.
Hodge, P.: 1983, Astrophys. J. 264, 470.
Lindgren, H., Ardeberg, A., Linde, P., & Lyngå, G.:1980, in Proc. of
    ESO Workshop on 2-Dim. Photometry, eds P. Crane & K. Kjar, p.155 (4)
Mould, J.R., Da Costa, G.S., & Crawford, M.D.: 1983, (preprint)   (22)
Muller, A.B.: 1971, ed "The Magellanic Clouds", (Dordrecht:Reidel).
Nemec, J.M., Liller, M.H., & Hesser, J.E.: 1984, this volume, p. 39.
Olszewski, E.: 1982, Ph.D. Thesis, University of Washington.   (6)
Rocca-Volmerange, B.: 1983, "First Stellar Generations", Frascati
    Workshop, Mem. della Soc. Ast. Italiana, (in press).
Salpeter, E.: 1955, Astrophys. J. 121, 161.
Stryker, L.L.: 1981, Ph.D. Thesis, Yale University.   (9)
Stryker, L.L.: 1984, Astrophys. J., (in press).   (15)
Stryker, L.L.: 1983, (in preparation).   (7)
Stryker, L.L., & Butcher, H.R.: 1981, in IAU Coll. 68, eds.
    A.G.D. Philip & D.S. Hayes (Schenectady: Davis), p. 255.   (10)
Stryker, L.L., & Nemec, J.M.: 1983, (in preparation).   (20)
Stryker, L.L., & VandenBerg, D.A.: 1983, (in preparation). (SV)
Stryker, L.L., Butcher, H.R., & Jewell, J.L.: 1981, in IAU Coll. 68,
    eds. A.G.D. Philip & D.S. Hayes (Schenectady:Davis), p.267   (14)
Stryker, L.L., Da Costa, G.S., & Mould, J.R.: 1983, (in prep.)   (19)
Stryker, L.L., Nemec, J.M., Hesser, J.E., & McClure, R.D.:
    1984, this volume, p. 43.   (11)
Tifft, W.: 1964, IAU/URSI Symp. 20, eds F.J. Kerr & A.W. Rodgers
    (Canberra: Austral. Acad. Sci.) p. 349.
Tifft, W., & Snell, C.: 1971, MNRAS 151, 365. (TS)   (1)
VandenBerg, D.A.: 1983, Astrophys. J. Suppl. 51, 29.
de Vaucouleurs, G., & Freeman, K.: 1972, in Vistas Astron. 14, 163.
Westerlund, B.E.: 1970, in Vistas Astron. 12, 335.

DISCUSSION

Flower: Your composite color-magnitude diagram of 10 fields in outlying regions appears to be very similar to the color-magnitude diagrams of NGC2121 and NGC1978; i.e. the brightest main sequence stars in your fields are near $V \simeq 20.3$ (vs $V = 20.0$ for NGC2121 and $V = 20.3$ for NGC1978) and the clump of red giants (or red horizontal branch stars) is at $V \simeq 19.2$ (vs $V = 19.0$ for NGC2121 and $V = 19.3$ for NGC1978). The relatively bright main sequence and red giants in these clusters suggest cluster ages less than about 1 Gyr. Thus your fields appear to be rather young.

Stryker: I would be surprised if these 10 outlying fields really did look like NGC1978 or NGC2121. The fields show a well-defined red horizontal branch, not an intermediate-age clump. My point is that there are several ages present in the fields, the youngest at about 1 to 2 Gyrs old; the oldest about 10 Gyrs. The oldest population is stronger. The other point was that although the near cluster (NGC2251) is more than about 14 Gyrs old and very metal-poor, the field seems not to have a significant portion of similar stars.

Lequeux: I am prepared to accept your conclusion from the luminosity function that intense star formation has started only about 4 Gyrs ago in the LMC; by the way, this is not inconsistent with what can be derived from global properties. However this conclusion rests entirely on the assumption that there are and have been stars with $M \lesssim 1\ M_\odot$ formed in the LMC. Unfortunately nothing in the theory or observations can prove this statement, and it may well be that the lack of stars with $M_V \gtrsim +3$ in the luminosity function is not an age effect but the result of a break or cut-off in the IMF at low masses.

STOCHASTIC STARFORMATION AND BUBBLES IN THE LARGE MAGELLANIC CLOUD

J. V. Feitzinger
Astronomisches Institut der Ruhr-Universität, Bochum, BRD

Nearly all places in the LMC where ring nebulae or shell structures in the neutral or ionized interstellar medium are observed, an OB association and/or WR-stars can be located (Braunsfurth, Feitzinger, 1983). Several mechanisms have been propsoed to generate shell or bubble structures: stellar winds, supernovae explosions, evolving HII regions, sequential starformation, collapsing hydrogen clouds interacting with stellar winds and radiation pressure. Ordered motions resulting in a shell or bubble structure are the result of almost any point like energy injection into the interstellar medium. Therefore all the mechanisms result in similar morphological structures, thus similar shapes can have heterogeneous origins.

For the diameter evolution we might expect time to be the dominating evolution parameter. However the large scatter of diameters of emission regions at a given age indicates that this is not so (Fig. 1a, b). It is evident from these figures that the other parameters: e.g. mechanical wind power, ionizing flux, initial energy of the objects and the mean density of the surrounding medium are the main factors determing the shape of the HII regions. Wind power ($\dot{E}_W$) and density (N_0) have opposite effects and are presented as one parameter $\dot{E}_W/N_0$. From the diameter age diagrams we conclude that this parameter covers 4 orders of magnitude and is a primary factor for the diameters of emission regions. Only the initial conditions determine the positions of the emission regions in the diameter age diagram.

Feitzinger et al. (1981) discussed the bubble formation process as a consequence of selfpropagating stochastic starformation. The bubble and shell structures are a natural byproduct of the starforming process. Fig. 2 shows the cold gas disk structured by a multitude of bubbles and shells. The supergiant shells (Meaburn, 1980) are the result of a sequential succession of starforming events. Stochastic selfpropagating starformation is the driving mechanism. Those large gas structures then are not caused by one star generation, but by a quick succession of stellar generations, spreading out from one ignition region to the enormous ionized ring-like gas filaments like Shapley III.

Braunsfurth, E., Feitzinger, J.V., 1983, Astron. Astrophys., in press
Feitzinger, J.V., Glassgold, A.F., Gerola, H., Seiden, P., 1981,
   Astron. Astrophys. 98, 371
Meaburn, J., 1980, Mon. Not. Roy. astr. Soc., 192, 365

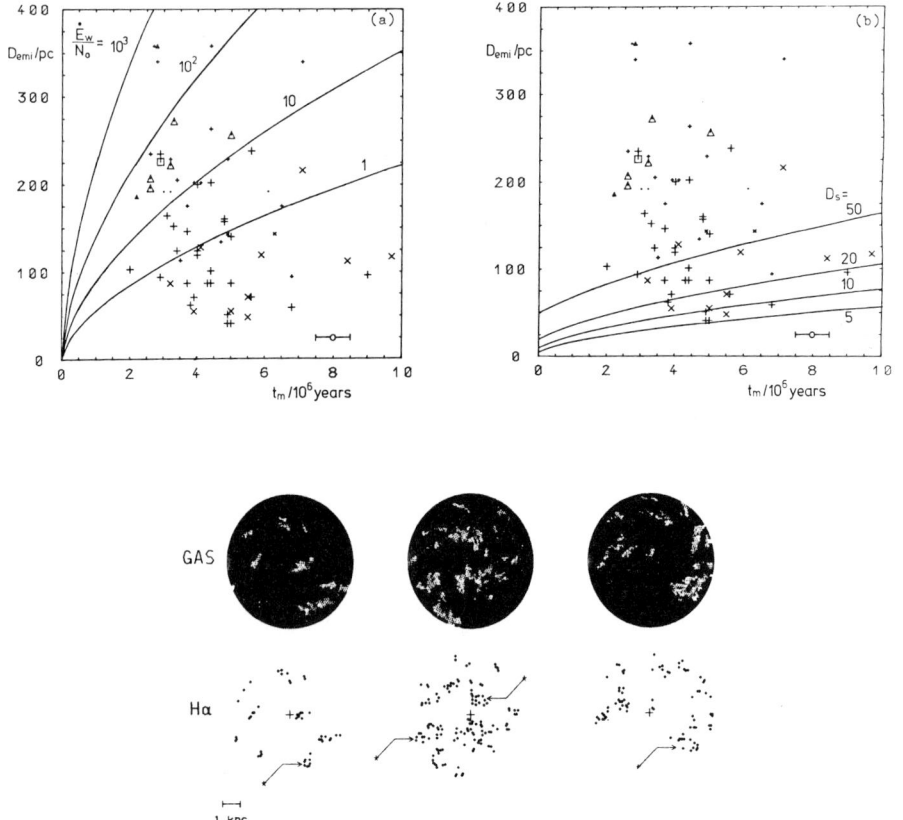

Fig. 1 Diameters of emission regions (D_{emi}) coinciding with associations in the LMC versus age (t_m) of the association. The symbols +, x, Δ, $\Box$ mark the association classes: very certain ... uncertain (star clouds). The symbol size is proportional to the weight given to the coincidences of emission regions and associations.
  a) Evolutionary tracks of wind driven bubble diameters with the mechanical stellar wind power $\dot{E}_w/10^{36}$ erg/s and the density of the ambient medium N_0/cm^3
  b) same as (a) for expanding HII regions; D_s is the initial Strömgren diameter

Fig. 2 One model of the gas distribution of the LMC and the simulation of an Hα picture at three different time steps. A range of 20 in gas density is shown; the white areas mark the regions of lower gas density. In the Hα pictures some shells are marked by arrows (compare Feitzinger et al., 1981)

PAST STAR FORMATION HISTORY OF THE MAGELLANIC CLOUDS

B. Rocca-Volmerange
Institut d'Astrophysique, Paris

Abstract: From far-UV to visible integrated photometry of the Magellanic Clouds we give an estimate of the past star formation parameters. Most of our results can be applied to Irregular galaxies.

1. PAST STAR FORMATION PARAMETERS
1.1 Observational data
Visible data and UV data from the D2B-satellite are available from Maucherat-Joubert et al. (1980), the reddening for gas and stars was derived by Vangioni-Flam et al. (1980), and the gas content and total mass are taken from Lequeux et al. (1979). All data are given in Table 1.

Table 1. Collected Observational Parameters for LMC and SMC

	$m_{1690}-V$	$m_{2200}-V$	$m_{3100}-V$	B-V	V-R	E(B-V)	M_{gas}	M_{tot}
LMC	−0.69	−0.18	0.12	0.39	0.10	0.18	0.9	6.1
	−0.95	−0.41	−0.04	0.30	0.006	0.37	\multicolumn{2}{c}{$(10^9 M_\odot)$}
SMC	−0.75	−0.17	0.23	0.37	0.10	0.08	0.8	1.5

1.2 Model Results
From our synthetic population model evolving with time, any UV or visible color gives a good estimate of the R parameter (defined as the ratio of the present over the time integrated star formation rate) as long as the initial mass function and the metallicity are given. A good fit of the luminosities, colors and gas content (Fig. 1), taking into account the respective metallicities of the clouds $Z(LMC) = 8\ 10^{-3}$ and $Z(SMC) = 3\ 10^{-3}$, leads to the star formation parameters given in Table 2.

From our results: a) The MCs are not presently undergoing a global burst of star formation. b) The global star formation rate stays about constant in the LMC, is lower and lightly decreasing with time in the SMC. c) Their age is about 9-10 Gyr and the IMF is standard (Rocca-Volmerange et al. 1981).

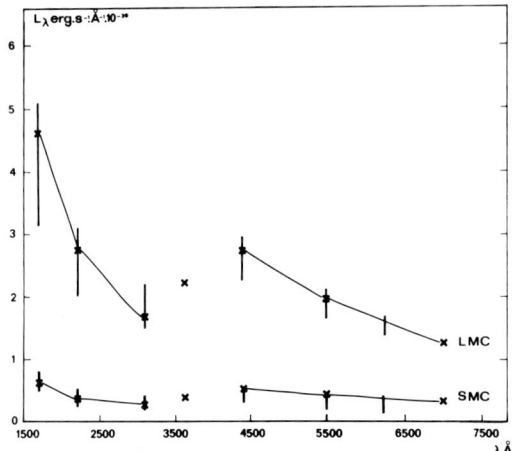

	Mgas/Mtot	
	Observ.	Calculat.
LMC	0.13	0.15
SMC	0.42	0.38

Fig.1. Calculated luminosities for 1690, 2200, and 3100 Å, and for U, B, V, and 6250 Å, compared to observational values with error bars.

2. EXTENSION TO IRREGULAR MAGELLANIC GALAXIES

From integrated photometry of Im and Sd galaxies in far-UV (Code and Welch 1982) and visible (de Vaucouleurs et al. 1976) bands, such irregular galaxies show similar colors (with very low dispersion) and similar metallicities ($Z = 10^{-3}$ to 10^{-2}). Our fit of the Magellanic Cloud photometry can be applied to such irregular galaxies. These colors may be interpreted in terms of a model in which the reddening is not too high and a similar history of star formation (Rocca-Volmerange 1983), see Fig. 2.

Tab. 2		R(T) Gyr^{-1}	Present¹ Past SFR	Present Age T Gyr	Present¹ SFR M$_\odot$ Gyr^{-1}	Average¹ Past SFR/M$_\odot$ Gyr^{-1}
Irregular Galaxies	SMC	0.04 - 0.11	0.4-1	9	0.04	∿ 0.07
	LMC	0.10 - 0.14	0.9-1.2	9	0.12	∿ 0.11
Our Galaxy Disk	Solar Neighbourhood (10 kpc)	0.018	0.18	10	0.05	0.29
	Ring 5 kpc	0.02	0.2	10	0.046	0.26

1 Relative to 1M$_\odot$ of galaxies.
Results from : Rocca et al.,1981 ; Guiderdoni and Rocca, 1982, divided by the ξ parameter

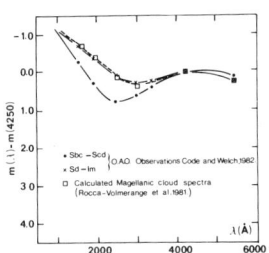

References:
Code, A.D., Welch, G.A.: 1982, Astrophys. J. 255, 1
de Vaucouleurs, G., de Vaucouleurs, A., Corwing, G.: 1976, 2nd Reference Catalogue of Bright Galaxies, University of Texas Press
Guiderdoni, B., Rocca-Volmerange, B.: 1982, A & A, 109, 393
Lequeux, J., Peimbert, M., Rayo, J.F., Serrano, A., Torres-Peimbert, S.: 1979, A & A, 80, 155
Maucherat-Joubert, M., Lequeux, J., Rocca-Volmerange, B.: 1980, A & A, 86, 299
Rocca-Volmerange, B.: 1983, in "The first Stellar Generations", Vulcano Workshop, Mem. Soc. Astr. Italiana, 54, 161
Rocca-Volmerange, B., Lequeux, J., Joubert, M.: 1981, A & A, 104, 177
Vangioni-Flam, E., Lequeux, J., Maucherat-Joubert, M., Rocca-Volmerange, B.: 1981, A & A, 90, 73

THE SPATIAL DISTRIBUTION OF YOUNG OBJECTS IN THE LARGE MAGELLANIC CLOUD - A PROBLEM OF PATTERN RECOGNITION

J.V. Feitzinger, E. Braunsfurth
Astronomisches Institut der Ruhr-Universität, Bochum, BRD

Methods used in pattern recognition and cluster analysis are applied to investigate the spatial distribution of OB associations and emission regions in the LMC. For our analysis we used the catalogue of associations of Lucke and Hodge (1970) and the catalogue of emission regions of Davies et al. (1976).

Several clustering algorithms (Anderberg 1973) were applied to the two-dimensional distribution of the young objects. The linkage in circle clustering method (Fig. 1 and 2) shows the conectedness of the objects at a given scale length. All objects are linked with center to center distances less than a given distance, i.e. the radius of a circle around an object. Fig. 3 and 4 show the dependence of the mean number of cluster members divided by the total number of objects versus the linkage circle. It reminds of a phase transition curve defined by a critical distance for clustering. The slope of the curve is a measure for order in the system. The hierarchical clustering method leads to three orders of clustering for both associations and emission regions (Fig. 5 and 6). A comparison with the distribution of OB and WR stars and the UV emission (see Fig. 4a, Martin et al. 1976) shows the one-to-one correspondence with the hierarchical clusters. In the same way the 2nd and 3rd order clusters coincide with the supergiant shells found by Meaburn (1980, 1981), see his Fig. 2.
The cluster analysis gives an objective procedure for a hierarchical partition of the system. The scales range from 0.06 deg at the 0th order (the median diameter of associations and emission regions) to 8 deg for the whole system. A grand design structure of the system (spiral arms) should be visible in the last hierarchical order. The LMC as a late type galaxy shows no grand design. However, the spiral arm filaments being responsible for the flocculent appearances of spiral galaxies are identical with our intermediate scale structures found in the cluster analytic partition of the system. They are, too, identical with the spiral arm filaments described by Schmidt-Kaler (1977) and Feitzinger (1980).

Anderberg, M.R., 1973, Cluster Analysis for Applicants, Academic Press, New York
Davies, R.D., Elliott, K.H., Meaburn, J., 1976, Mem. Roy. Astr. Soc. 81, 89
Feitzinger, J.V., 1980, Space Science Rev. 27, 35
Lucke, P.B., Hodge, P.W., 1970, Astron. J. 75, 171
Martin, N., Prevot, L., Rebeirot, E., Rousseau, J., 1976, Astron. Astrophys. 51, 31
Meaburn, J., 1980, MN 192, 365
Meaburn, J., 1981, in: Investigating the Universe, ed. by F.D. Kahn, Reidel Pub. Comp., Dordrecht, p. 61
Schmidt-Kaler, Th., 1977, Astron. Astrophys. 54, 771

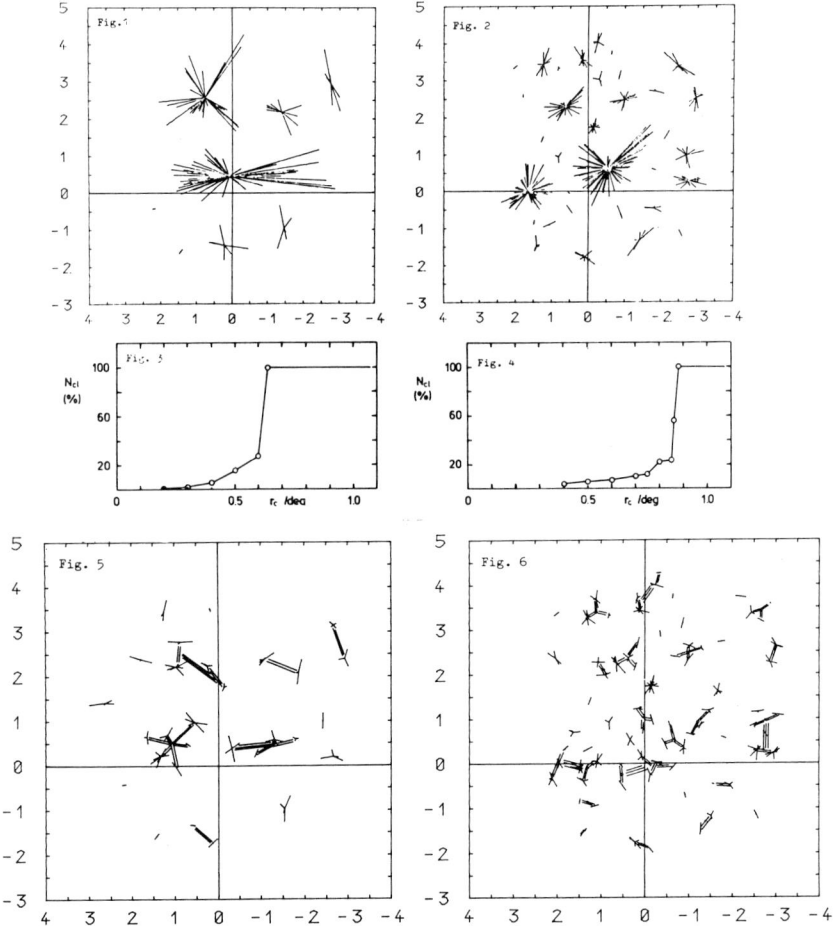

Fig. 1 Linkage in circle clusters of associations, r_c=0.666 deg.
Fig. 2 Linkage in circle clusters of emission regions, r_c=0.272 deg.
Fig. 3 Mean number of cluster members/total number of points versus circle radius/mean next neighbour distance for associations
Fig. 4 As Fig. 3 for emission regions
Fig. 5 Hierarchical clusters of associations; three orders are shown
Fig. 6 As Fig. 5 for emission regions

CHEMICAL EVOLUTION OF THE MAGELLANIC CLOUDS

Francesca Matteucci
Istituto di Astrofisica Spaziale, Frascati, Italy.

1. INTRODUCTION. According to the stochastic self propagating star formation theory(SSPSF,Gerola and Seiden,1978), the star formation process in galaxies changes from a fluctuating but continuous mode to a bursting one, when the size of the system becomes relatively small($R \lesssim 3Kpc$). Then, due to its size, the LMC should be a system undergoing fluctuating but continuous star formation activity, whereas the SMC should be in the region between continuous and bursting modes(Gerola et al.1980). In order to look better inside this problem a model of chemical evolution of the two Clouds, which takes into account the stochastic star formation rate, has been built. For the SMC both the continuous and the bursting modes of star formation have been considered. As a result we find that the different chemical histories of the two Clouds may be related to the fact that SMC has undergone several bursts of star formation(between 50 and 60), while a continuous star formation activity was present in LMC.

2. MODEL RESULTS. The basic assumptions of the model are: single zone description and complete instantaneous mixing of the gas; delay in the metal enrichment of the ISM due to stellar lifetimes(leading to a drop of the instantaneous recycling approximation); detailed description of the variation over the galactic lifetime of a set of elements(H,He,C+O,Si+Fe), due to stellar nucleosynthesis, stellar mass ejection and inflow of primordial gas. The basic equations of the temporal variation of the fractionary mass of each element can be found in Matteucci and Chiosi,1983.
The stellar birthrate function is defined as: $Bm(t)=B(t)\varphi(m)$, where $\varphi(m)$ (the initial mass function) is assumed the same for the two Clouds($x=1.35$ for $m \leq 2m_\odot$ and $x=2.0$ for $m > 2m_\odot$, Lequeux 1979). The rate of star formation is given by: $B(t)=\langle\eta\rangle G(t)$, where: $\langle\eta\rangle$ is the average fractionary number of cells undergoing star formation during the galactic lifetime, as predicted by SSPSF theory; $G(t)$ is the current fractionary mass of gas; ν is an efficiency parameter(i.e. the rate of star formation per unit mass of gas involved in the star formation process). The value of ν has been estimated

by means of observational properties of the two Clouds,(see Matteucci and Chiosi,1983),under the assumption that a fluctuating but continuous mode of star formation had taken place in both the galaxies. We find that a value of $\gamma = 4 \cdot 10^{-9}$ yrs fairly fits the main features of SMC,whereas $\gamma = 10 \cdot 10^{-9}$ yrs reproduces those of LMC. The different value of γ weakens the universality of the SSPSF theory, since other physical mechanisms

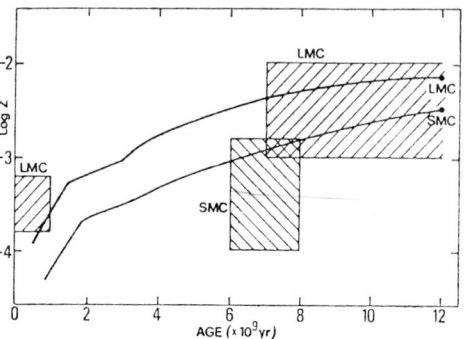

Fig.1—Metallicity vs. age relationships for the MC. The shaded area visualizes the data of Barbaro(1982).

have to be invoked to explain the increase of γ from SMC to LMC. Anyhow,another possible explanation exists. Since SMC lies in the transition region,according to the SSPSF theory,the number of star forming events occurred in it is not well known. If SMC has undergone several bursts during its lifetime,$\langle \eta \rangle$ should be different from the assumed value of 0.03. Then,starting from $\gamma' = \gamma \langle \eta \rangle = 0.12 \cdot 10^{-9}$ yrs,which well fits the features of SMC, and keeping $\gamma = 10 \cdot 10^{-9}$ yrs as in LMC,a value of $\langle \eta \rangle = 0.012$ is derived. At this stage we can predict how many star forming events have taken place in SMC,due to the functional relationship between $\langle \eta \rangle$ and the total number of bursts Nb. Assuming the bursts to have a triangular shape with a typical duration of $50 \cdot 10^6$ yrs and maximum intensity of 0.1,according to the SSPSF theory,we find Nb$\simeq$58. In this way,the rate of star formation per unit mass of gas involved in the star forming process is constant for the two Clouds($\gamma = 10 \cdot 10^{-9}$ yrs),and their main features can be reproduced by varying only the number of star formation events. Figure 1 shows the Z vs. age relationship for clusters in the MC,accordingly to Barbaro(1982).The metal abundances in the present ISM, as given by Peimbert and Torres-Peimbert(1977),are also indicated for comparison(full dots) See that,although many uncertainties exist in both theory and observations,the theoretical relationships from our models seem to fit fairly well the observational trends. As a conclusion, the SSPSF theory is able to reproduce the main observed properties of the Magellanic Clouds,withouth invoking other physical processes as it seems the case for dwarf irregular galaxies(Matteucci and Chiosi,1983).

References.
Barbaro,G.:1982,Astrophys.and Space Sci. 83,143.
Gerola,H.,Seiden,P.,L.:1978,Ap.J.223,129.
Gerola,H.,Seiden,P.L.,Schulman,L.S.:1980,Ap.J.242,517.
Lequeux,J.:1979,Astron.Astrophys.80,35.
Matteucci,F.,Chiosi,C.:1983,Asrton.Astrophys.123,121.
Peimbert,M.,Torres-Peimbert,S.:1977,M.N.R.A.S. 179,217.

STAR COUNTS IN FIELDS SURROUNDING THE LMC

M.T. Brück
Department of Astronomy
University of Edinburgh
Royal Observatory, Edinburgh EH9 3HJ

ABSTRACT

Star counts in a number of fields near the Large Magellanic Cloud have shown a maximum extension of about 10° for the faint stars. The luminosity function in the outer regions of the LMC is consistent with an intermediate age population.

OBSERVATIONS

Observational material consists of five UKST IIIaJ plates of standard survey fields (Fields 30, 31, 32, 56 and 86) measured by the COSMOS automatic measuring machine as described in earlier papers (Brück 1978, 1980). Two plates cover the north-east and south-west quadrants of the main body of the LMC; the other three extend westward from the LMC to the wing of the SMC. The inner fields reach 18^m (B) and contain photometric standards by Butler (1976) and Martin (1977) to the same limit. One of the other fields, that which includes the SMC wing, has photometric calibration due to Kunkel (1980) which reaches 22^m. The measured areas of the inter-Cloud plates overlap slightly so that it has been possible to estimate magnitudes approximately in the uncalibrated fields.

LUMINOSITY FUNCTIONS

Counts in half magnitude bins from 13^m to 18^m in the inner parts of the LMC (Figure 1a) are equivalent to cumulative counts increasing at the rate of 0.40 in $logN$ per magnitude. Among the faint stars in the outer periphery (5° - 6° from the centre in the south-west direction) a characteristic luminosity function emerges (Figure 1b) with an abrupt jump in numbers at around 19^m where horizontal branch stars in the LMC intermediate age population are known to be numerous (Butcher 1977; Stryker and Butcher 1981). A possible zero-point error in magnitudes due to transferred photographic photometry does not alter the unambiguous nature of this LF.

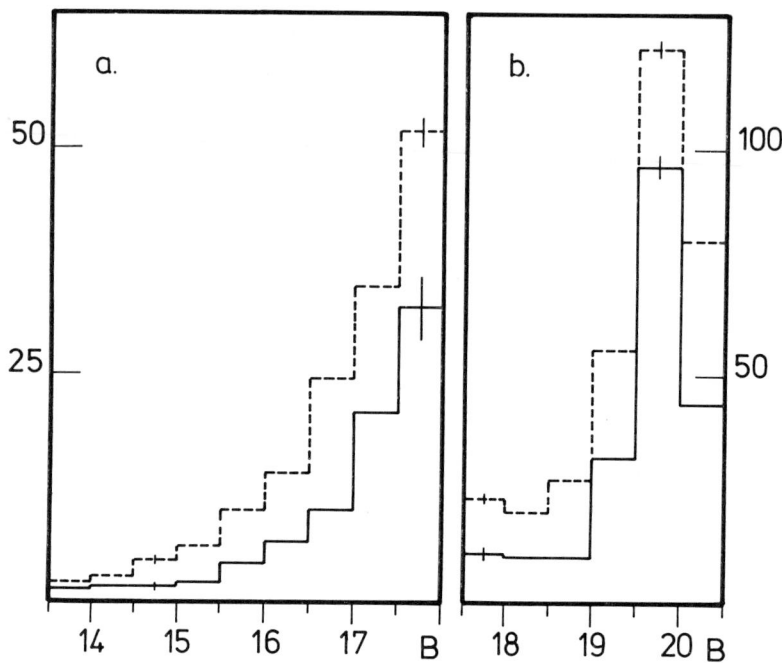

Figure 1. Luminosity functions in the LMC (a) inner disk and (b) halo. Dotted lines are observed total counts; solid lines are counts after subtraction of galactic foreground stars. Numbers are given per unit area 11.'2 x 11.'2. Multiply by 2870 to obtain numbers per square degree.

ACKNOWLEDGEMENT

The author thanks Dr W. Kunkel for generously supplying charts and data of his sequence.

REFERENCES

Brück, M.T.: 1978, Astron. Astrophys. 68, 181
Brück, M.T.: 1980, Astron. Astrophys. 87, 92
Butcher, H.: 1977, Astrophys. J. 216, 372
Butler, C.J.: 1976, Dunsink Obs. Publ. 1, No.6
Kunkel, W.: 1980, IAU Symp. 85 (ed. J.E. Hesser) p.353
Martin, W.L.: 1977, Mem. Roy. ast. Soc. 83, 69
Stryker, L. and Butcher, H.: 1982, IAU Colloq. 68 (ed. A.G. Davis Philip and D.S. Hayes) p.255

CROWDED FIELD ELECTRONOGRAPHY IN THE LMC

Peter Linde, Arne Ardeberg, Harri Lindgren and Gösta Lyngå
Lund Observatory, Box 1107, S-221 04 Lund, Sweden
AA and HL also at ESO, La Silla, Chile

OBSERVATIONS

An area in the LMC situated 1.2 degrees from the centre of the Bar has been studied with the ESO 3.6 m telescope using electronography and photoelectric measurements for calibration. Coordinates for 1950.0 are 5^h20^m, $-71°$. Six exposures have been used, with exposure times ranging from 8 to 90 min. Typical seeing was 1.5 arcseconds FWHM. This investigation is a continuation of our earlier study with the ESO 1.5 m telescope (Lindgren et al., 1980).

REDUCTIONS

To deal with the difficulties of making precision photometry in crowded fields, software has been developed along the following principles:
1) A numerical two-dimensional stellar profile (point spread function, PSF) is formed by merging a number of non-overlapping star images.
2) Stars and conglomerates of stars are pin-pointed interactively on high contrast, multi-coloured displays.
3) A least squares fit is iterated between each conglomerate and a model group using the PSF (step 1) and the initial coordinates (step 2). During this process saturated and disturbed pixels are disregarded. Free parameters are intensity, position and background level.
4) Local background positions are interactively selected for each conglomerate.

RESULTS

A. In figure 1 is shown the resulting colour-magnitude diagram, based on 312 stars. The right and top scales, used by ZAMS, are M_v and $(B-V)_o$. The main features are:

1) A well developed giant branch with a narrow red tip, typical of an old population.
2) A clump on the red horisontal branch, similar to clumps in old galactic clusters.
3) A main sequence which also contains young stars.

4) Red giants typical of a young population.
5) An unusual number of stars at the population I subgiant position.

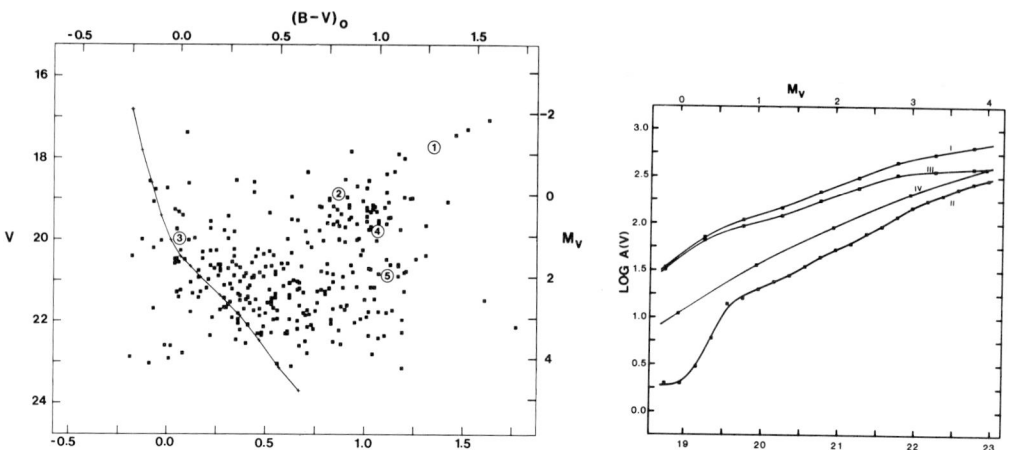

Figure 1. Colour-magnitude diagram.   Figure 2. Accumulated star counts.

B. Figure 2 compares the following four graphs of log A(V) values:

I. The counts in our field multiplied by a completeness factor derived from experiments with insertion of synthetic stars.
II. The star counts obtained by Stryker and Butcher (1981) near NGC 1783 corrected for the difference in field sizes.
III. Our counts minus those of Stryker and Butcher. This should approximately cancel the halo contribution to our counts and leave only the LMC disk stars.
IV. Nearby stellar luminosity function reduced to a volume corresponding to the area studied in the LMC with a depth of 600 parsecs. This value has been chosen so as to give the same value for V=23 as log A(V) of graph III.

The main features are:
1) Our field is more than three times richer in stars brighter than V=22 than the field of Stryker.
2) Particularly stars of about V=19 are much more abundant in our field. Figure 2 shows that some of these are blue main sequence stars with ages less than 100 million years.
3) Graph III flattens out at about V=23, showing again the behaviour of a young population. The comparison with the solar neighbourhood graph (IV) reinforces this point.

REFERENCES

Lindgren, H., Ardeberg, A., Linde, P., Lyngå, G.: 1980, in ESO Workshop on two dimensional photometry, Noordwijkerhout, p. 155
Stryker, L.L., Butcher, H.R.: 1981, in IAU Colloquium No. 68, Schenectady, N.Y., p. 255

THE COLOUR MAGNITUDE DIAGRAM OF A SECOND FIELD IN THE SMC HALO

M.R.S. Hawkins and M.T. Brück
Royal Observatory, Edinburgh EH9 3HJ, U.K.

ABSTRACT

A colour magnitude diagram of a field in the south-west periphery of the SMC reveals a population similar to that obtained earlier in another field, with an age of 3×10^9 years. The present field is $3°.5$ from the SMC centre, more than a degree more distant than the first one. The similarity in stellar content between the two emphasises the widespread nature of this population.

OBSERVATIONS

The field examined is centred on $(00^h\ 09^m.8,\ -73°36')$ (1980). Four electronographs each in B and V were obtained using the 8 cm McMullen camera with the 1.5 Danish telescope at La Silla, Chile. Exposure times were all 60 minutes on L4 emulsion. Six standard stars were observed photoelectrically by R.D. Cannon using the Danish 0.5 m telescope. The electronographs were reduced following the procedures already described (Hawkins and Brück 1982).

RESULTS

The colour magnitude diagram (Figure 1) has the same general morphology as that obtained for the first field (Hawkins and Brück 1982) at $(0^h\ 26,\ -73°.0)$ (1950), which we interpreted as signifying an intermediate age population, with an age of $3 \pm 1 \times 10^9$ years. The present observation refers to a field $3°.5$ from the SMC centre and well away from the Bar. It was intended, in choosing this field, to decide whether there was evidence of a different, older, population in the outer regions of the SMC which would be expected to become relatively more conspicuous the further one moved outwards from the centre. The similarity in the morphologies of the CM diagrams suggest that the intermediate age population is still dominant in the second field and is indeed widespread in both Clouds. Reductions of similar observations of a third field at a still greater distance from the SMC are currently being carried out.

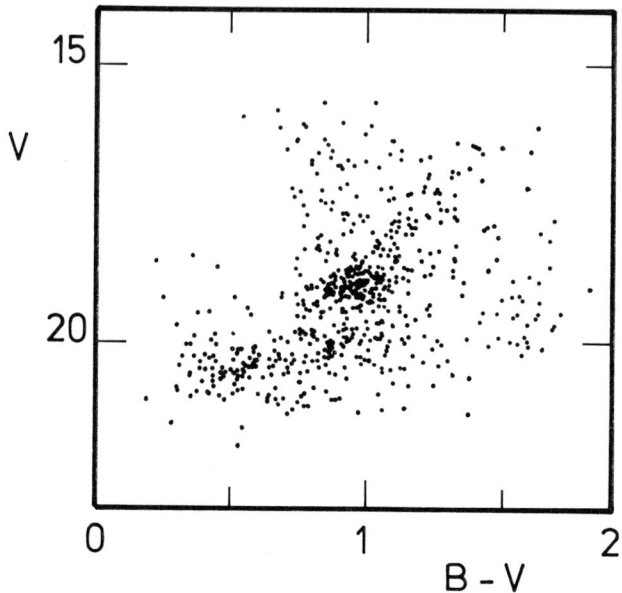

Figure 1. Colour-magnitude diagram of the SMC halo field.

REFERENCE

Hawkins, M.R.S. and Brück, M.T.: 1982, Mon. Not. R. astr. Soc. 198, 935

A SEARCH FOR STARS IN THE MAGELLANIC STREAM

M.T. Brück and M.R.S. Hawkins
Department of Astronomy, University of Edinburgh,
Royal Observatory, Edinburgh EH9 3HJ, U.K.

ABSTRACT

A search for faint stars (to $20^m.5$ in B) in a bright part of the Magellanic Stream designated MSI by Mathewson (1976), carried out from automatic scans of UKST photographs by means of the COSMOS machine gives a negative result as far as the main HI cloud is concerned, but reveals a small surplus of images in a patch 1 degree away from the main HI concentration.

OBSERVATIONS

The observational material consisted initially of a set of UKST plates on IIIaJ (blue) emulsion covering several degrees of sky on either side of the narrow MSI ridge. A photoelectric-electronographic sequence to 22^m in B was set up to calibrate the plates. Various considerations resulted in the decision to terminate counts at $20^m.5$ and also to confine our final analysis to an area of 24 square degrees which was observed on two and in part on three plates. Counts were made in squares of 1 cm^2 on the plates (11'.2 x 11'.2).

RESULTS

Average counts were ~ 60 per square. A few squares had counts in excess of 80, all but one of which could be attributed by visual inspection to clusters of galaxies. Only one extended (~ 0.1 square degree) patch at ($0^h 38^m$, $-44°.1$) 1950 did not coincide with any obvious cluster of galaxies. The patch is $1°$ east of the main HI concentration and the excess equivalent to about 700 objects per square degree. At the actual position of the brightest HI contour there are in fact slightly fewer than average numbers. The result for MSI is therefore negative; the observed patch of surplus images constitutes at most a tentative identification.

DISCUSSION

It is well established that the Large and Small Magellanic Clouds contain a high proportion of intermediate age stars (3×10^9 years) (e.g. Hawkins and Brück 1982; Brück 1984; Stryker and Butcher 1982). It can be argued that if the Magellanic Stream were in fact a tail drawn out from one of the Clouds during a relatively recent encounter with our Galaxy, as many theories suggest, the stars of the intermediate age group might well have accompanied the gas and might therefore be present in the Stream. Our failure to find such stars there is not a decisive result; it is paralleled by the absence of similar stars in the HI bridge joining the Large and Small Clouds (Brück 1982).

CONCLUSION

A small patch with a low level of surplus faint images found 1° from MSI is the only possible location we have been able to identify of a stellar component in this part of the Magellanic Stream. The result is ambiguous; the patch could be a genuine part of the Stream at an undefined distance, in which case the stars and gas have become separated; an extended cluster of faint galaxies; or a fairly large fluctuation in local galactic star numbers. Only a colour magnitude diagram combined with a larger scale photograph will decide the nature of this group of images.

A detailed account of the work is in press (Brück and Hawkins 1983).

REFERENCES

Brück, M.T.: 1982, in "Structure du Petit Nuage de Magellan", Comptes Rendus sur les Journées de Strasbourg, 4, 56
Brück, M.T.: 1984, this volume, p. 97.
Brück, M.T. and Hawkins, M.R.S.: 1983, Astron. Astrophys. (in press)
Hawkins, M.R.S. and Brück, M.T.: 1982, Mon. Not. R. astr. Soc. 198, 935
Hawkins, M.R.S. and Brück, M.T.: 1984, this volume, p. 101.
Stryker, L. and Butcher, H.R.: 1982, IAU Coll. 68, (ed. A.G. Davis Philip and D.S. Hayes) p.255

ULTRAVIOLET SURFACE PHOTOMETRY OF STELLAR ASSOCIATIONS IN THE
LARGE MAGELLANIC CLOUD

Jan Koornneef
Space Telescope Science Institute
Homewood Campus,
Baltimore, MD 21218, USA

Observations of the Large Magellanic Cloud obtained with the 10-channel ultraviolet photometer of the OAO-2 are presented. The aperture was circular and 10 arcmin in diameter whereas the wavelength coverage was from 4250 to 1430Å. A total of 50 fields has been measured. Using photometric criteria, the data fall into three groups which are also spatially separated. The different characteristics are easily understood in terms of population and reddening differences, but only if the previously reported "LMC extinction law" (Koornneef and Code 1981; Nandy et al. 1980) is adopted.

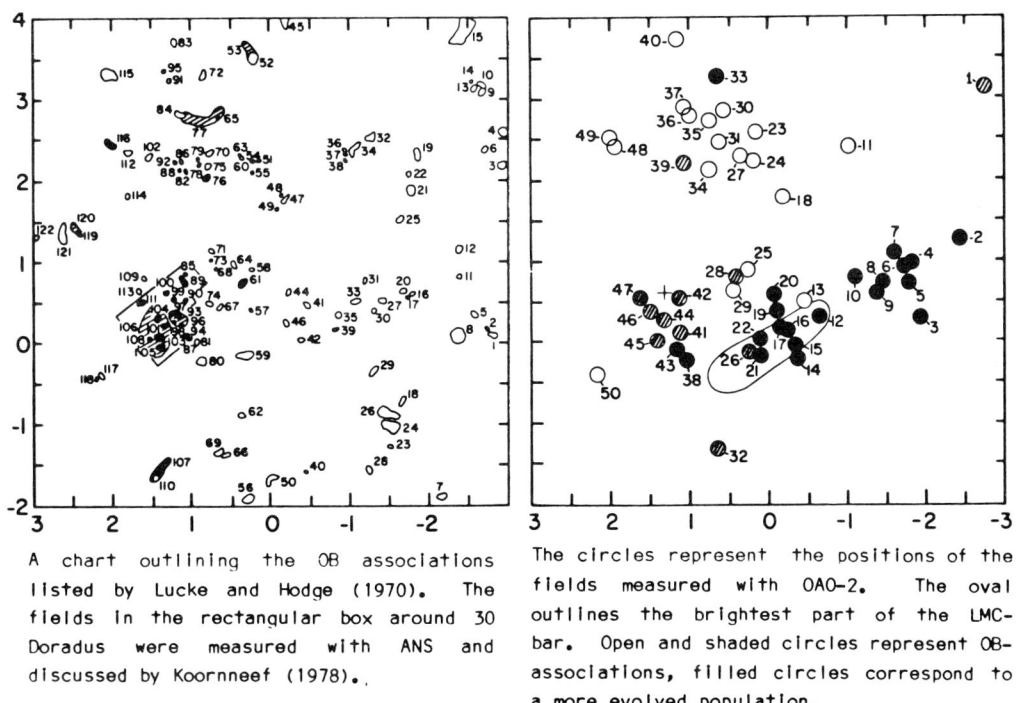

A chart outlining the OB associations listed by Lucke and Hodge (1970). The fields in the rectangular box around 30 Doradus were measured with ANS and discussed by Koornneef (1978).

The circles represent the positions of the fields measured with OAO-2. The oval outlines the brightest part of the LMC-bar. Open and shaded circles represent OB-associations, filled circles correspond to a more evolved population.

The bluest group comprises fields of the Northern Constellation Sh. III. These fields are only moderately reddened ($E_{B-V} \approx 0.15$) and the observed energy distribution is consistent with a Sandage-Salpeter Initial Luminosity Function (ILF).

The same ILF has been adopted for the stellar associations in the 30 Doradus region. But the ultraviolet colours observed here are significantly redder which is explained by a higher amount of dust.

The third group is dominated by the fields observed in the "Bar" of the LMC. The far-ultraviolet slope observed here is similar to the two other categories, but the far to near ultraviolet flux ratios are much redder. Also, this group exhibits a more pronounced Balmer discontinuity. A slightly reddened solar-neighbourhood luminosity function is found to be appropriate.

A full account of this work will be given elsewhere.

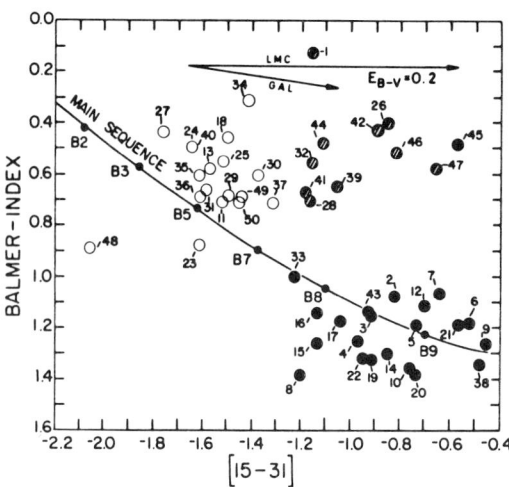
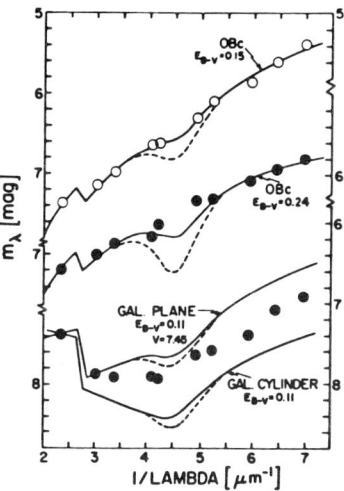

The ordinate is an OAO-2 photometric parameter which mimics the strength of the Balmer - discontinuity. The colour-index on the abcissa is a measure of the slope shortward of the Balmer-jump. Note the gap in this diagram between the OB-association and the evolved population indicating spatial separation. Galactic and LMC reddening vectors are shown.

The final results of the model fluxes as fitted on the overall flux distributions observed for the three categories. The broken line indicates the predicted 2200A feature for a galactic reddening law.

REFERENCES

Koornneef, J. 1978, Astron. Astrophys. **64**, 179
Koornneef, J., and Code, A.D. 1981, Astrophys. J. **247**, 860
Lucke, P.B., and Hodge, P.W. 1970, Astron. J. **75**, 171
Nandy, K., Morgan, D.H., and Carnochan, D.J. 1980,
    Mon. Not. Roy. Astr. Soc. **186**, 421

KINEMATICS AND DYNAMICS OF THE MAGELLANIC CLOUDS

K.C. Freeman
Mount Stromlo and Siding Spring Observatories
Research School of Physical Sciences
The Australian National University

1. INTRODUCTION

Why are the kinematics and dynamics of the Magellanic Clouds worth studying ? Some of the reasons are:

1. The Clouds are the closest examples of Magellanic systems. These asymmetric systems give some interesting dynamical problems. Because the Clouds are so close, a unique amount of information can be obtained on the kinematics of objects of all ages. This should be very helpful for understanding the dynamics.

2. The Clouds and the Galaxy are interacting. This produces complex kinematics of the gas in and between the Clouds, and also the Magellanic Stream. Again, very detailed information can be derived. We would like to know enough about the gas dynamics of interacting galaxies, to be able to explain the kinematics produced by this interaction.

3. The interaction will affect the star formation and chemical evolution in the Clouds. As new results are obtained on the star formation history and the chemical evolution, it is important to follow in parallel the dynamical history of the system, to see if the dynamics, star formation and chemical evolution can be tied together.

New results on the HI kinematics of the LMC and SMC, and on the dynamics of the interaction, have been reviewed by others. I will concentrate mainly on the dynamics of individual Magellanic systems, with particular application to the LMC and SMC. I will also discuss some new results on the kinematics of the globular cluster system of the LMC; this is interesting, because it includes objects of all ages.

2. STRUCTURE AND DYNAMICS OF MAGELLANIC SYSTEMS

For detailed reviews, see de Vaucouleurs and Freeman (1973: dVF) and Feitzinger (1980: JF). These reviews include photographs and diagrams to illustrate many of the points made below.

2.1 Basic Structural properties

Magellanic systems frequently come in pairs. The LMC/SMC is just one example. NGC 4618/25 and NGC 4027/4027A are among others that are strikingly similar to the LMC/SMC in their appearance and gross properties. This supports the view that the LMC/SMC is probably a relatively longlived binary pair.

The basic structure shared by these barred Magellanic systems is the strong asymmetry of the spiral structure about the bar axis. The bar appears displaced from the center of the outer isophotes. In this respect, the SBm systems are analogous to the "lopsided" normal spirals like M101, in which the spiral structure appears very asymmetric about the nucleus. (See Baldwin et al.(1980) for a recent discussion.)

This largescale asymmetry is seen also in the rotation curves. The LMC, NGC 4027 and NGC 55 are just a few examples in which the center of the rotation curve is displaced from the center of the bar by several hundred parsecs. JF has compiled data on the displacement of the bar from the isophotal center and from the rotation center, for a sample of Magellanic systems. The two displacements are indeed very similar.

2.2 Intrinsic Flattening

It seems clear from their appearance and rotation that the Magellanic systems are disklike. Surface photometry shows that their disks have the exponential surface brightness distribution that is so characteristic for disk galaxies of all types. The intrinsic flattening of the Magellanic systems is particularly interesting. Heidmann et al (1972) showed how the intrinsic flattening of disk galaxies increases monotonically from S0 to Sd, where it takes a maximum value (major to minor axis ratio) of about 12. From Sd towards later types, the intrinsic flattening then _decreases_ abruptly, to about 6 at Sm and 5 at Im.

Why should the Magellanic systems be so much less flat than the Sd spirals ? The reason may have to do with the asymmetry described above, which begins to appear at stage Sd. The intrinsic flattening is the ratio of the isophotal major to minor axis for an edge-on system. The major axis is determined mainly by the galaxy's exponential lengthscale, which in turn is closely correlated with its absolute magnitude. The minor axis, for these late-type pure disk (ie almost bulgeless) galaxies, is defined by the thickness of the disk, which in turn is determined mainly by disk heating processes. These disk heating processes are not yet fully understood (see for example the review by Wielen and Fuchs, 1984). In particular, the primary heating mechanism remains uncertain. However, in the asymmetric Magellanic systems, the asymmetry itself provides _another_ source of disk heating, through the resonant excitation of motion perpendicular to the galactic plane (Binney 1981). This mechanism acts in addition to the heating processes that operate in the more symmetric disks, and it could explain why the Magellanic systems are less flat.

It would be interesting to make a study of the vertical (z) structure of Magellanic systems. Do they show the characteristic $sech^2(z/z_o)$ density profile seen in earlier type spirals (van der Kruit and Searle 1982) ? What vertical structure would we expect if the resonant heating by the asymmetry is the dominant mechanism for heating the Magellanic disks ? The large amount of kinematical information that can in principle be obtained for stars of all ages in the LMC would be very useful for understanding the time-dependent vertical structure of Magellanic systems.

2.3 Dynamical Studies of Isolated Magellanic Systems

Not much work has been done so far on the dynamics of these asymmetric galaxies, so the subject is in a fairly primitive state. The early work was reviewed by dVF, and we will just summarise it briefly here. The purpose of these dynamical models was to explain the structural properties of Magellanic systems (the asymmetric spiral structure and rotation curves, and the wave observed in the rotation curve of NGC 4027: see dVF for details). The first step was to adopt a background gravitational potential. This potential should include the basic features of SBm systems which distinguish them from the more symmetric barred galaxies. A typical SBm system has an exponential disk, which provides most of the light, and a small bar. However the centers C_b of the bar and C_d of the disk do not coincide, but are separated by a distance Λ which is typically between 0.5 and 1 kpc. To represent this asymmetrical situation in a simple time-independent way, Freeman and Harrington (1968: see also dVF) introduced a potential based on Figure 1a. The bar rotates around the center of the disk, such that the line $C_d C_b$ is always normal to the major axis of the bar. The potential field for this model is shown in Figure 1b. For parameters appropriate to SBm systems, the total potential (gravitational + centrifugal) has one stable neutral point M_5 and one unstable neutral point M_4. The stable neutral point M_5 acts to trap matter circulating around it. Christiansen and Jefferys (1976) used this potential in a particle orbit study of the SBdm galaxy NGC 4027. They were able to explain the large wave observed in the rotation curve of this galaxy: it results from the trapping and circulation of particles around M_5.

The particle orbit work is only a first step. A proper hydrodynamical treatment of the gas response and motions in this Magellanic potential is needed. Colin and Athanassoula (1983) have presented a preliminary report of such a study, which shows how a stationary gas response develops.

2.4 The LMC-SMC-Galaxy Interaction

The dynamics of isolated SBm systems need to be understood but, on their own, will not help us to understand the complex kinematics in the

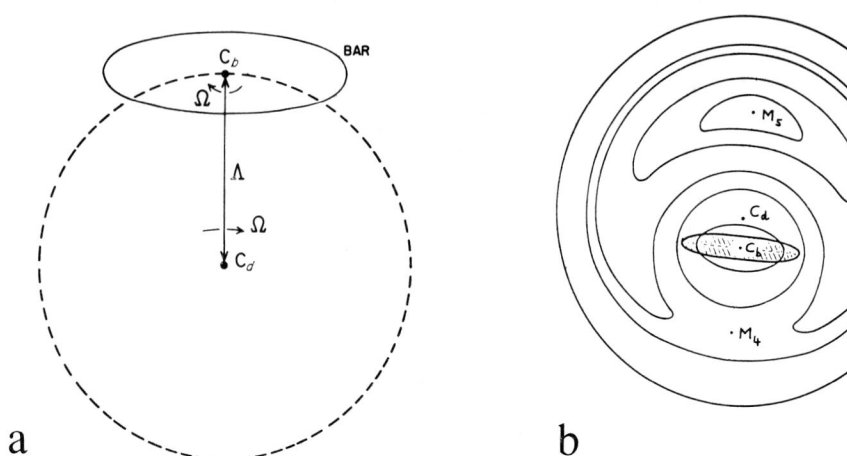

Figure 1a. Motion of the bar relative to the disk center C_d in the SBm model. The broken circle is the orbit of the bar center around C_d.
Figure 1b. Equipotentials of the gravitational + centrifugal potential for the model of Figure 1a, with typical parameters (see dVF). M_4 and M_5 are the neutral points.

LMC and SMC, except perhaps in the innermost parts of the LMC. The problem is, of course, that the LMC/SMC is apparently an interacting binary system which is also interacting with the Galaxy. Observationally there seems no doubt about the important influence of the LMC-SMC-Galaxy interaction on the HI kinematics. Mathewson et al (1979) showed that the velocity field of the HI between the Clouds has a very regular structure. The "isovels" are approximately parallel to the LMC-SMC line, in the region between the Clouds and also to the West of the SMC and the East of the LMC. The disturbing effect of the LMC's rotation on this pan-Magellanic velocity field can be seen in the isovels, but only within about 5 degrees of the LMC's rotation center.

The HI structure and kinematics of the LMC and SMC are discussed in the contributions by Feitzinger, Kreitschmann and Rohlfs, and by Mathewson and Ford, in this volume. However it seems clear that the motions of the HI in the outer parts of the LMC, and over much of the SMC, will be significantly affected by the pan-Magellanic flow field. This makes it very difficult to derive a reliable circular velocity curve and velocity field for the LMC itself, except within a few degrees of the rotation center. In particular, there seems to be no immediate hope of determining from the kinematics whether or not the LMC has a massive dark corona.

For the SMC, the situation is even more difficult. The SMC has an absolute blue magnitude of -16.2, so we expect from the Tully-Fisher relation a maximum rotational velocity of about 35 km s^{-1}. This small rotation would be almost lost in the pan-Magellanic velocity field of Mathewson et al.(1979).

If this direct observational evidence for the effects of the interaction is not enough, recent theoretical work shows how large these kinematical disturbances are expected to be. For example, Murai and Fujimoto (1980) made a study of the dynamics of the LMC and SMC orbiting around a model Galaxy with a massive halo. They showed how the Magellanic stream results, in their model, from SMC material torn off by the tidal force of the LMC. SMC material also forms a bridge between the SMC and the LMC. The interaction produces velocities that are comparable with those observed by Mathewson et al.(1979).

Again, a hydrodynamical treatment of the motion of gas in the interacting system would be valuable. However there is still some disagreement about the orbit of the LMC/SMC around the Galaxy: even the sense of rotation of the orbit is still in dispute (see for example Tanaka 1981).

There seems no way that we can give a good dynamical description of the LMC and SMC until the effects of the LMC-SMC-Galaxy interaction are properly understood. This must be a top priority item for dynamical studies of the LMC and SMC. The LMC/SMC pair is not the only example of interacting Magellanic systems. Weliachew et al.(1978) made an HI study of the interacting pair NGC 4631/NGC 4656, and Combe (1978) was able to make a successful dynamical model for the gas motions and distribution in this system.

2.5 Dark Matter in Magellanic Systems

Although it will not be easy to determine whether the LMC and SMC have massive dark coronas, some progress has been made for the more isolated Magellanic system NGC 3109. This system lies between the LMC and SMC in absolute magnitude: M_B = -16.2, -17.3, -18.2 for the SMC, LMC and NGC 3109 respectively. Carignan (1983) has made surface photometry for this galaxy and, in a combined Fabry-Perot and 21-cm study, has measured its rotation curve to a radius of about 10 kpc. The rotation curve continues to rise, to about 60 km s^{-1}. Carignan calculated the rotation curve to be expected, if the mass distribution follows the light distribution (Kalnajs 1983). This expected rotation curve fits well in the inner 2 kpc, but then drops rapidly. To fit the rotation data, Carignan required, in addition, a massive dark corona (isothermal sphere) with a core radius of 3.3 kpc and a velocity dispersion of 45 km s^{-1}. Within 10 kpc, this dark corona is 7 times more massive than the luminous disk. NGC 3109 is one of the most unambiguous examples of a galactic dark corona: it suggests that other Magellanic systems, like the LMC and SMC, may be similarly endowed.

3. KINEMATICS OF THE GLOBULAR CLUSTERS OF THE LMC

Most of the information about the kinematics of the LMC and SMC comes from the extreme population I (HI, young stars, HII region) which, as suggested above, may be kinematically affected by recent events in the LMC-SMC-Galaxy interaction. A recent study by Freeman et al (1983) compares the kinematics of LMC globular clusters of all ages. (This study owes much to Drs Cowley Hartwick Searle and Smith, who allowed us to use their data before publication.) We made rotation solutions for young clusters and old clusters separately. Because the total number of clusters with good velocities was only 59, we did not attempt to derive the rotation curve: rather, we assumed that it was flat, and solved for the rotation velocity V_m, the systemic velocity V_o (relative to the galactic center), the position angle θ_o of the line of nodes, and the dispersion σ about the rotation solution. The results are summarised in the table below. The age groups are similar to those defined by Searle et al (1980). All velocities are in km s^{-1}.

Group	Age	N	pa(θ_o)	V_m(rot)	V_o(sys)	σ
I-III	< 3.10^8 y	24	1 ± 5	37 ± 5	40 ± 3	15
IV-VII	3.10^8 -> 10^{10}	33	41 ± 5	37 ± 3	27 ± 2	17
VII	> 10^{10}	9	44 ± 6	54 ± 7	38 ± 4	16

The solution for the young clusters is fairly similar in position angle, rotational velocity and systemic velocity to the solutions for the HI and HII components (see JF). It seems clear that the young clusters are moving with the gas from which they recently formed. The older clusters are also apparently in a disklike distribution, with a similar rotation amplitude and an intrinsic velocity dispersion of only 17 km s^{-1}. (This small dispersion was noted by JF, and corresponds to a vertical scale height of only 600 pc). However the position angle for the old cluster line of nodes is very different (41°) from the position angle for the young clusters (1°). We see from the table that even the oldest clusters (group VII) appear as part of the same old disk; their dispersion is only 16 km s^{-1}, so there is no evidence for a kinematic halo population among the globular clusters of the LMC.

Why should the position angles of the old cluster and young cluster disks be so different ? (This difference remains in solutions that include a transverse motion of 300 km s^{-1} for the LMC). We suggest that the old clusters delineate the true old disk of the LMC, while the kinematics of the young clusters and the gas have again been affected by recent events in the interaction of the LMC-SMC-Galaxy system. This view is supported by the recent HI study of the LMC by Feitzinger et al (this conference); they find a position angle of about 28° for the HI in the inner few degrees of the LMC, where the effects of the interaction would be least. Qualitatively, it fits in with the interaction picture of Murai and Fujimoto (1980). They find that the LMC and SMC suffered a close approach about 2.10^8 y ago, with a 3 kpc separation of the two systems; this is by far the closest approach in the entire history of the interaction. If their interaction picture is correct in concept, then this recent severe interaction is presumably the one responsible for the disturbed state of the gas kinematics in the LMC (see JF for details). We might then expect that the young cluster system (ages less than 3.10^8 y) will reflect the disturbed kinematics of the gas from which it formed, while the kinematics of the older cluster system reflect the more sedate dynamical history of the period before this recent close approach.

This recent close encounter fits well also with the evidence by Frogel and Blanco (preprint) for an epoch of enhanced star formation at a time corresponding to the age of the young clusters. We recall the suggestion by Gunn (1980) that the young clusters themselves (which have no counterparts in the Galaxy) formed in a shock-induced star formation episode excited by recent interaction between the two Magellanic Clouds.

It would be worth studying the kinematics of other LMC objects that cover a wide range of ages, to find out if the kinematical effects shown by the old and young clusters are seen again. Planetary nebulae are such objects: however their kinematics do not conform to either of the cluster solutions, young or old. More work would be welcome here. Bessell, Wood and I have begun a kinematical study of the LMC long period variables, which again straddle the relevant age range.

4. SUMMARY

1. For the dynamics, the important property of Magellanic systems is their structural and rotational asymmetry.

2. Magellanic systems are significantly less flat than Sd galaxies. This may be due to disk heating associated with the largescale asymmetry.

3. The effects of the LMC-SMC-Galaxy interaction need to be properly understood before much progress can be made on the dynamics of the individual Magellanic Clouds.

4. At least one Magellanic system, NGC 3109, shows very compelling evidence for a dark massive corona.

5. Both the young and the old globular clusters of the LMC lie in kinematically defined rotating disks. However the two disks have very different lines of nodes. We suggest that the old clusters represent the true old disk of the LMC, and that the kinematics of the young clusters and the gas are affected by the LMC-SMC-Galaxy interaction. There is no evidence for a kinematical halo population in the LMC.

REFERENCES

Baldwin, J.B., Lynden-Bell, D., Sancisi, R. 1980. M.N.R.A.S. 193,313.
Binney, J. 1981. M.N.R.A.S. 196,455.
Carignan, C. 1983. Australian National University PhD thesis.
Christiansen, J., Jefferys, W.H. 1976. Ap.J. 205,52.
Colin, J., Athanassoula, E. 1983. Internal Kinematics and Dynamics of Galaxies (IAU Symposium 100), ed. E. Athanassoula (Reidel, Dordrecht), 239.
Combe, F. 1978. Astron. Astrophys. 65,47.
de Vaucouleurs, G., Freeman, K.C. 1973. Vistas in Astronomy, ed. A. Beer, (Pergammon, Oxford), 14,163. (dVF).
Feitzinger, J.V. 1980. Space Science Reviews 27,35. (JF).
Freeman, K.C., Harrington, R. 1968. Bull. Astronomique, 3e Série 3,269.
Freeman, K.C., Illingworth, G.D., Oemler, A. 1983. **Ap.J.** 272,488
Gunn, J. 1980. Globular Clusters, ed D. Hanes and B. Madore (Cambridge University Press,Cambridge) 301.
Heidmann, J., Heidmann, N., de Vaucouleurs, G. 1972. Mem.R.A.S. 75,85.
Kalnajs, A. 1983. Internal Kinematics and Dynamics of Galaxies (IAU Symposium 100), ed. E. Athanassoula (Reidel, Dordrecht), 87.
Mathewson, D.S., Ford, V.L., Schwarz, M.P., Murray, J.D. 1979. The Largescale Characteristics of the Galaxy (IAU Symposium 84) ed W.B. Burton (Reidel, Dordrecht), 547.
Murai, T., Fujimoto, M. 1980. P.A.S.Japan 32,581.
Searle, L., Wilkinson, A., Bagnuolo, W. 1980. Ap.J. 239,803.
Tanaka, K. 1981. P.A.S.Japan 33,247.
van der Kruit, P., Searle, L. 1982. Astron. Astrophys. 110,61.
Weliachew, L., Sancisi, R., Guélin, M. 1978. Astron. Astrophys. 65,37.
Wielen, R., Fuchs, B. 1984. Presented at IAU Symposium 106, "The Milky Way Galaxy", to be published.

DISCUSSION

Walborn: You showed some interesting cases of pairs of asymmetrical irregulars, but also similar asymmetries in apparently single objects. Are the observed spatial structures of the Magellanic Clouds believed to be due to mutual interaction?
Freeman: No. There are many examples of relatively isolated Magellanic systems. I was just trying to say that pairs of Magellanic galaxies, like the LMC/SMC are not rare.

THE MAGELLANIC STREAM AND ITS RELATED PROBLEMS

M. Fujimoto and T. Murai
Department of Physics
Nagoya University
Nagoya Chikusa 464

ABSTRACT

A brief survey is made of recent 21-cm and optical observations of the Magellanic Stream(MS). The space orientation of the Magellanic Clouds is touched upon in relation to modelling the MS. After summarizing a variety of models for the MS, we show that if our Galaxy is massive with a huge dark halo, a tidal model is most suitable for reproducing its characteristic structure and high-negative radial velocity. Past orbits of the Large and the Small Magellanic Cloud (LMC and SMC) are determined uniquely for the last 2×10^9 yr, if we postulate that the LMC and SMC are bound together for 10^{10} yr: Highly-noncircular motion of the SMC around the LMC could give a clue to understand some peculiar features associated with the Magellanic Clouds.

1. THE MAGELLANIC STREAM

The Magellanic Stream(MS) is a narrow band of neutral hydrogen gas extending along a great circle from the Magellanic Clouds past the South Galactic Pole to $l=90°$, $b=-30°$ (Wannier and Wrixon 1972; Mathewson et al. 1974). The radial velocity of the gas varies systematically with angular distance along the MS, $V_r=-240\sin(\theta+2°)$ km s^{-1}, where θ is measured from the origin at $l \approx 298°$, $b \approx -69°$ toward the northern tip of the Stream.

As shown in figure 1 and references therein, recent high sensitive 21-cm surveys reveal large number of narrower filaments and elongated cloudlets of $0.5° \times 0.3°$ aligned along the MS proper. Two parallel substructures are found along the MS near the Magellanic Clouds. Large amplitude fluctuations of more than 50 km s^{-1} are superimposed locally on the smooth velocity features V_r (Haynes 1979).

Giovanelli(1980, 1981) and Mirabel(1981a and b) have observed and noted a wide-spread population of high-velocity clouds in the first and second quadrant in the southern galactic hemisphere. They consider that some of them are of the same origin as the MS: In fact, the radial velocity of the high-velocity clouds at the extension of or in the neighborhood of the MS connects smoothly with that of the MS.

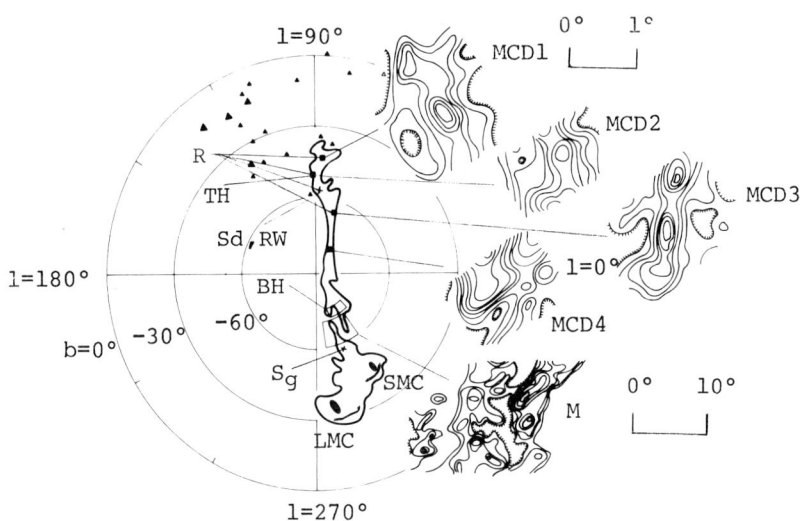

Fig. 1. Outline of the Magellanic Stream and some recent higher-sensitive 21-cm surveys. Six clouds MSI-VI by Mathewson et al.(1977) are not given, but the steep HI gradients are indicated closely along the leading edges of the LMC and SMC opposite to the MS. The selected areas for 21-cm and optical observations are shown: MCD1-4=Mirabel et al.(1979), M=Morras (1983), R=Recillas-Cruz(1982), TH=Tanaka and Hamajima(1982), Sd=Sanduleak(1980), BH=Brück and Hawkins(1983), Sg=Songaila (1981), RW=Richer and Westerlund(1983). The triangles ▲▲▲ are high velocity clouds near the northern tip of the Stream (Giovanelli 1980, 1981; Mirabell 1981a, 1981b). Other 21-cm observations by Haynes(1979) and Cohen(1982), not given here, also reveal that the MS consists of large number of small scale but large-amplitude fluctuations in the velocity and density.

2. OPTICAL OBSERVATIONS OF THE MAGELLANIC STREAM

Songaila (1981) detected visual absorption lines, Ca K and Na D, on the background continuum of the Seyfert galaxy, Fairall-9, at Sg in MS in figure 1. The metalicity of the MS gas is determined as 0.01 to 1.78, depending on the calcium depletion. Although it ranges still widely from metal-poor globular cluster values to those of the LMC and SMC, Songaila's observations may shed some light on the origin of the MS material.

Following negative searches for faint stars in the MS by Philip (1976a and b) and by Mathewson et al.(1979), Sanduleak (1980) found a carbon star at Sd in figure 1, suggesting that it might belong to the MS. Although it seems to await confirmation, the distance to the MS is estimated as 15 kpc or 50 kpc depending on its absolute magnitude M_V=0.4 or -2.5. In their near-infrared survey of carbon stars in galaxies of the Local Group, Richer and Westerlund(1983) identified a carbon star at the same place as Sanduleak. If it belongs to the MS, the distance to the northern tip of the MS is 38 kpc, but its radial velocity differs from that of the MS by 150 km s^{-1}.

Faint A type stars of apparent magnitude around m_V=18.5 are searched in several selected areas in the directions R in figure 1(Recillas-Cruz 1982). The number of stars of this magnitude is a factor 10 lower than what we should expect if the tip of the Stream is located as close as 10 kpc from the sun. A similar search is made by Tanaka and Hamajima (1982) by means of three-colour photometry whose limiting magnitude is 17.5(TH in figure 1). Again no candidates for A type star have been found associated with the MS. These two results suggest strongly that the distance to the MS is far greater than 15 kpc and comparable to that to the Magellanic Clounds.

Brück and Hawkins(1983) performed star counts to 20.5 in B magnitude on an extended area in the MS. The amplitude of excess of stars is not so large as expected and their positional correlation with the maximum HI density is not present or just marginal. Since the excess stars to 20.5^m in B were expected to consist of intermediate age objects, Brück and Hawkins(1983) consider that the age of the MS is older than 2×10^9yr or the distance to the MS is greater than that to the Clouds.

To sum up, the optical search for faint stars in the MS has been negative, suggesting that the distance to the MS is far greater than 15 kpc.

3. SPACE ORIENTATION OF TRANSVERSE MOTION OF THE MAGELLANIC CLOUDS

The angular size of the LMC is finite that a small fraction of its transverse motion would be seen as radial velocity superimposed on the intrinsic rotational motion. Taking into account this and the apparent distribution of HI, supergiants etc., Feitzinger et al.(1977) concluded that the transverse component of the LMC motion is about 275 km s^{-1} relative to the LSR, toward the galactic plane and parallel to the MS. Lin and Lynden-Bell(1982) applied a similar kinematics to the LMC-SMC system whose angular separation is so large as 21° that one quarter of the transverse motion of the Magellanic Clouds would be seen as a part of radial velocity of the SMC. Assuming a circular motion of the SMC around the LMC, Lin and Lynden-Bell(1982) derived the same transverse motion of the Magellcnic Clouds as Feitzinger et al.(1977).

Mathewson et al.(1977) suggested already the above motion in their hydrodynamical model for the MS (section 4). As shown schematically in figure 1, the steep HI gradients along the leading edges of the LMC and SMC are considered as due to the ram pressure of the galactic halo gas through which the Magellanic Clouds pass toward the galactic plane and in pararell with the MS.

4. MODELS FOR THE MAGELLANIC STREAM

Three kinds of models — tidal, primordial and hydrodynamical — have been proposed for the MS. They attempt to explain the narrow and long-extended structure of the MS, together with its smooth sinusoidal and highly-negative radial velocity features.

4.1. Tidal Interaction Models

Figures 2a and 2b summarize the tidal models for the MS, with references and adopted parameters. Orbits of the Magellanic Clouds are approximately on a plane perpendicular to the sun-galactic center line: They are seen from the direction of l=180°, b=0°. The Magellanic Stream is simulated by some hundreds of test particles. [See Toomre(1972) for earlier studies on the tidal disruption of the Magellanic Clouds at their close approach to the Galaxy. The present paper deals with the tidal models published after the discovery of the Magellanic Stream.]

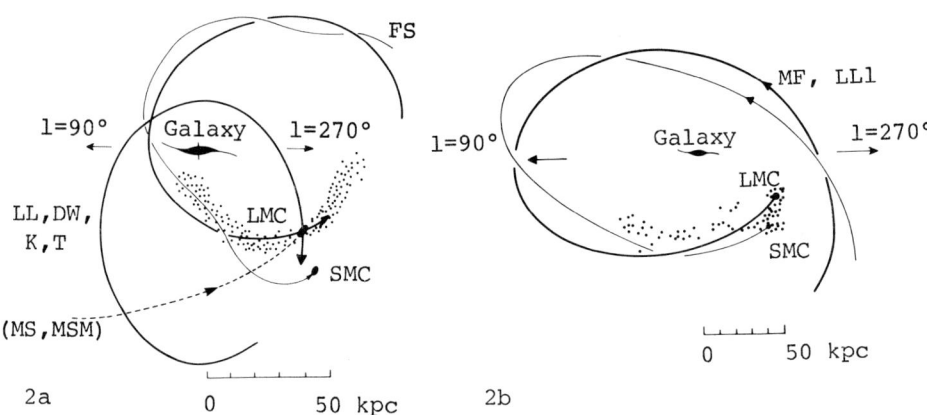

Figs.2a and 2b. Tidal models for the Magellanic Stream. FS=Fujimoto and Sofue(1976, 1977), LL=Lin and Lynden-Bell (1977), DW=Davies and Wright(1977), K=Kunkel(1979), MF=Murai and Fujimoto(1980), T=Tanaka(1981), LL1=Lin and Lynden-Bell (1982). The thin lines of FS and MF indicates the SMC orbits bound to the LMC. In figure 2a, the perigalactic distance of the orbit is 10 to 20 kpc and the total mass of the Galaxy is $1.2-2.8 \times 10^{11} M_\odot$. In figure 2b, the perigalactic distance is

50 kpc, and the Galaxy is assumed to be massive with a huge dark halo. The total mass of the Galaxy within a radius of 50 kpc is $7\times10^{11}M_\odot$. The dashed lines, MS and MSM, are due to Mathewson and Schwarz(1979) and Mathewson et al.(1978) for the LMC orbit of the primordial and the hydrodynamical model.

The orbits, LL, DW, and K, are clockwise as seen from the sun in an opposite direction to that derived in section 3. The MS is a gaseous (test particles) bridge torn from the LMC and leading toward the Galaxy. The apparent features and the high-negative radial velocity of the MS are reproduced qualitatively. However, the northern part of the model Stream covers too large an area of the sky because the test particles approach the sun so close as 10 kpc. Although its amplitude is reasonable, the radial-velocity distribution does not resemble the observed one V_r. It is also to be noted that these models do not take into account the SMC: A tail or a counterpart to the bridge appears unavoidably in the two body system, the Galaxy and the LMC. As seen in figures 1 and 3, the MS misses the tail and is asymmetric with respect to the Magellanic Clouds. The model T considers the LMC and SMC which are bound together for 10^{10}yr. The radial velocity at the tip of the model Stream is, however, -140 km s^{-1}, too large by 80 km s^{-1} compared with the observed value -220 km s^{-1}.

The parameters adopted for the model FS in figure 2a are the same as T except for the direction of the revolution around the Galaxy. It is counterclockwise as seen from the sun. A narrow band of test particles can be obtained on the sky, but the observed radial velocity V_r can not be reproduced. This model FS is open to drastic improvement.

In order to overcome these difficulties, a model Galaxy with a massive halo is introduced, in which the LMC and the SMC are postulated to be bound together for 10^{10}yr. The binary state like MF and LL1 in figure 2b is realized when the perigalactic distance is 40-50 kpc. A good

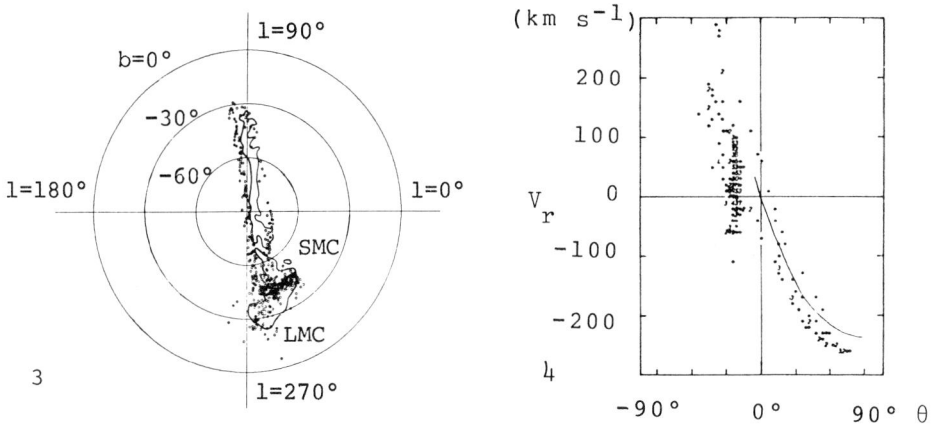

Figs. 3 and 4. The model Stream and its radial velocities are superimposed on the observed ones.

example of our particle simulations is given in figures 3 and 4, where the apparent feature and radial-velocity distribution of the MS is reproduced, much better than the models in figure 2a. The Magellanic Clouds revolve around the Galaxy in the direction consistent with that in section 3. The asymmetric structure of the Stream, a tail with no bridge, is understood by taking into account the SMC: A bridge from the SMC enters into the LMC and captured by it.

Since the massive halo assumption of the Galaxy is very reasonable (Rubin et al. 1978; Blitz 1979), the tidal models, MF and LL1, in figure 2b can be regarded as most realistic at present. It is interesting to note that such a similar orbit and a similar picture for the origin of the Stream have emerged through various investigators who have previously different opinions like LL and FS in figure 2a.

4.2. Other Theoretical Models

The primordial and the hydrodynamical (or wake) model have been introduced to produce the high-negative radial motion of the MS, without disrupting tidally the LMC-SMC system at its close approach to the Galaxy (Mathewson and Schwarz 1976; Mathewson et al. 1977). They explained also the observed inhomogenities in the density and velocity which can not be treated by the particle simulation in the tidal theories. Fujimoto and Sofue(1977) tried to find a mechanism to stretch a primordial gas cloud left from the condensation of the LMC and SMC. A drag force is assumed between the intergalactic gas and this gas cloud. However, it is not yet successful to reproduce a long-extended and gapless structure of the MS.

Bregman(1979) studied vortices in the wake which arise behind the Magellanic Clouds passing through a tenuous and high-temperature halo gas. The viscosity, thermal conduction and radiative cooling are considered in a realistic unstable gas. Bregman concluded, also through his hydrodynamical simulation, that the cooling time is not much less in the wake than the surroundings and consequently too large an amount of cooled (neutral) gas must be found downstream after the Magellanic Clouds. This feature is not observed in the MS.

The present authors consider that these non-tidal models await more extensive theoretical work, and also that the presently available data is not so well understood as to support them more strongly than the tidal models.

5. PAST ORBITS OF THE LMC AND SMC

When we postulate a bound state between the LMC and SMC in the tidal models, their past orbits are rather uniquely determined for the last, say, 2×10^9 yr (Murai and Fujimoto 1980; Fujimoto and Murai 1984).

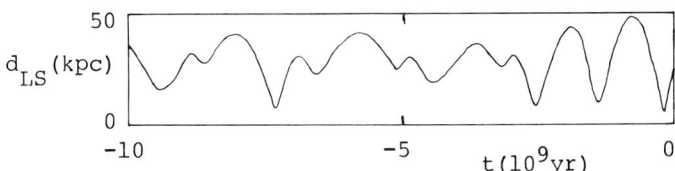

Fig. 5. The distance between the LMC and SMC, $d_{LS}(t)$, for the model MF in figure 2a.

Figure 5 shows a binary orbit of the LMC and SMC in the form of the separation (in kpc) between them, $d_{LS}(t)$. One finds that d_{LS} is not constant but varies with time from 2-7 kpc at $t \approx -2\times10^8$, -1.5×10^9 and -2.6×10^9yr to 50 kpc at $t \approx -8\times10^8$ and -2×10^9yr and so on.

Since this tendency in $d_{LS}(t)$ is found in most cases of our computed binary orbits of the LMC and SMC, we consider that the "collision" such as d_{LS}=2-7 kpc is not an accident but a dynamical event repeated in the history of the Magellanic Clouds. The last collision occurred 200 million years ago. Thereby the LMC and SMC must have been much disturbed and its aftereffect could be seen in some peculiar features in the Magellanic Clouds. In order to see them, we would like to refer to the spatial distributions of HI gas(Hindman et al. 1963), of H_α nebulosities(Johnson et al. 1982) and of planes of optical polarization of stars(Mathewson and Ford 1970; Schmidt 1970 and 1976). As shown in figure 6, they are observed along the relative orbit of the SMC to

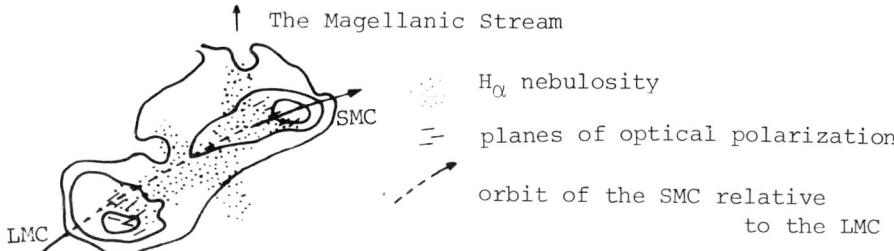

Fig. 6. Spatial distributions of HI gas, H_α nebulosity and planes of optical polarization of stars. The relative orbit of the SMC is given to the LMC.

the LMC, as if the SMC has stretched and stimulated the common magnetized gas as it departed from the LMC after the collision.

REFERENCES

Blitz, L.: 1979, Astrophys. J., 231, L115.

Bregman, J.N.: 1979, Astrophys. J., 229, 514.
Brück, M.T. and Hawkins, M.R.S.: 1983, to be published in Astron. Astrophys..
Cohen, R.J.: 1982, Mon. Not. R. astr. Soc., 199, 281.
Davies, R.D. and Wright, A.E.: 1977, Mon. Not. R. astr. Soc., 180, 71.
Feitzinger, J.V., Isserstedt, J. and Schmidt-Kaler, Th.: 1977, Astron. Astrophys., 57, 265.
Fujimoto, M. and Sofue, Y.: 1976, Astron. Astrophys., 47, 263.
Fujimoto, M. and Sofue, Y.: 1977, Astron. Astrophys., 61, 199.
Fujimoto, M. and Murai, T.: 1984, to be published in IAU Symposium 106 (ed. H. van Woerden).
Giovanelli, R.: 1980, Astron. J., 85, 1155.
Giovanelli, R.: 1981, Astron. J., 86, 1468.
Haynes, M.P.: 1979, Astron. J., 84, 1173.
Hindman, J.V., Kerr, F.J. and McGee, R.X.: 1963, Australian J. Phys., 16, 570.
Johnson, R.G., Meaburn, J. and Osman, A.M.I.: 1982, Mon. Not. R. astr. Soc., 198, 985.
Kunkel, W.E.: 1979, Astrophys. J., 228, 718.
Lin, D.N.C. and Lynden-Bell, D.: 1977, Mon. Not. R. astr. Soc., 181, 59.
Lin, D.N.C. and Lynden-Bell, D.: 1982, Mon. Not. R. astr. Soc., 198, 707.
Mathewson, D.S., Cleary, M.N. and Murray, J.D.: 1974, Astrophys. J., 190, 291.
Mathewson, D.S. and Schwarz, M.P.: 1976, Mon. Not. R. astr. Soc., 176, 47P.
Mathewson, D.S., Schwarz, M.P. and Murray, J.D.: 1977, Astrophys. J., 217, L5.
Mathewson, D.S., Ford, V.L. and Schwarz, M.P.: 1979, IAU Symposium 86 (ed. W.B. Burton) 547.
Mathewson, D.S. and Ford, V.L.: 1970, Astrophys. J., 160, L43.
Mirabel, I.F.: 1981a, Rev. Mexican Astron. Astrof., 6, 245.
Mirabel, I.F.: 1981b, Astron. J., 250, 528.
Mirabel, I.F., Cohen, R.J. and Davies, R.D.: 1979, Mon. Not. R. astr. Soc., 186, 433.
Morras, R.: 1983, to be published in Astron. J..
Murai, T. and Fujimoto, M.: 1980, Publ. Astron. Soc. Japan, 32, 581.
Philip, A.G.D.: 1976a, B. A. A. S., 8, 352.
Philip, A.G.D.: 1976b, B. A. A. S., 8, 532.
Recillas-Cruz, E.: 1982, Mon. Not. R. astr. Soc., 201, 473.
Richer, H.B. and Westerlund, B.E.: 1983, Astrophys. J., 264, 114.
Rubin, V.C., Ford, W.K., Jr. and Thonnard, N.: 1978, Astrophys. J., 225, L107.
Sanduleak, N.: 1980, Publ. Astron. Soc. Pacific, 92, 246.
Schmidt, Th.: 1970, Astron. Astrophys., 6, 294.
Schmidt, Th.: 1976, Astron. Astrophys. Suppl., 34, 357.
Songaila, A.: 1981, Astrophys. J., 243, L19.
Tanaka, K.I.: 1981, Publ. Astron. Soc. Japan, 33, 247.
Tanaka, K.I. and Hamajima, K.: 1982, Publ. Astron. Soc. Japan, 34, 417.
Toomre, A.: 1973, not published. See Mirabel, I.F. and Turner, K.C.: 1973, Astron. Astrophys. 22, 437.
Wannier, P. and Wrixon, G.T.: 1972, Astrophys. J., 173, L119.

DISCUSSION

Fall: How close does the LMC-SMC pair get to the Milky Way in your simulation? When was the last perigalactic passage?
Fujimoto: 200 million years ago the Galaxy and the Magellanic Clouds approached each other to about 50 kpc.

HI SURVEYS OF THE MAGELLANIC SYSTEM

D. S. Mathewson and V. L. Ford
Mount Stromlo and Siding Spring Observatories
The Australian National University

THE MAGELLANIC SYSTEM

The global distribution of HI in the Magellanic System is shown in Figure 1. The gas covers some 1500 square degrees of sky and has a mass of 1.8×10^9 $M_\odot$. There are four main components: the LMC, the SMC, the inter-Cloud region and the Magellanic Stream. The integrated HI of the first three components is mapped in Figure 2 with the Parkes 64-m radio telescope which has a resolution of 15 arc min. The previous surveys of McGee and Milton (1966), Hindman (1967), Mathewson et al. (1979) have been combined with a recent survey by Mathewson et al. (1983) of the outer regions of the System to give this large-scale picture of the gas distribution. The last two surveys were made with a velocity resolution of 4.12 km s^{-1} and a minimum detectable signal of 0.2 K. The long spurs extending from the LMC and SMC and the bridge joining the two galaxies with prominent spurs pointing to the Magellanic Stream are all compelling evidence for tidal interaction between the LMC and SMC (Mathewson 1976a, Murai and Fujimoto 1980). The detailed velocity field of the HI is given in Mathewson et al. (1983). Its large-scale features are shown in Figure 5 of Mathewson et al. (1979) which indicate that the radial velocities of the inter-Cloud region form a velocity continuum with those of the LMC and SMC. This plus the continuity of the general velocity gradient across the entire Magellanic System strongly suggest that the two galaxies are bound.

THE LARGE MAGELLANIC CLOUD

The radial velocity field of the LMC is shown in Figure 3 from observations with the 18-m Parkes radio telescope. The effect of a translational motion of 250 km s^{-1} of the LMC along the great circle of the Magellanic Stream in the direction of the galactic plane has been removed from the observed velocities (Mathewson et al. 1977; Feitzinger et al. 1977; Lin and Lynden-Bell 1982). This has the effect of moderating the anomalously high velocities observed along the eastern edge of the LMC. The velocity field is peculiar and it is clear that

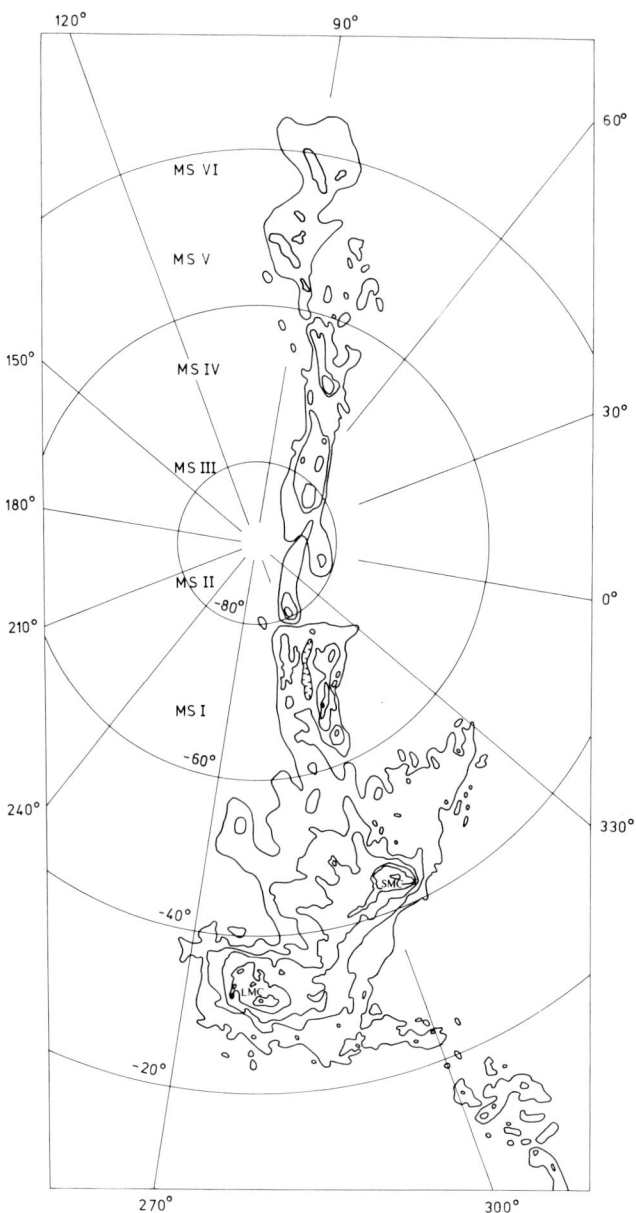

Fig. 1 The global distribution of HI in the Magellanic System in galactic co-ordinates. The outer contour is the 10^{19} atoms cm^{-2} isophote and the inner contours have been selected to show the main features of the components. The Magellanic Stream is divided into six HI concentrations, MSI-VI, (Mathewson et al. 1977). Data have been taken from Mathewson (1976b), Mathewson et al. (1979), Mathewson, Ford and Fisher (1983), McGee and Milton (1966), Hindman (1967), Wannier and Wrixon (1972), Haynes, (1979), Mirabel (1981), Cohen (1982), and Morras (1983).

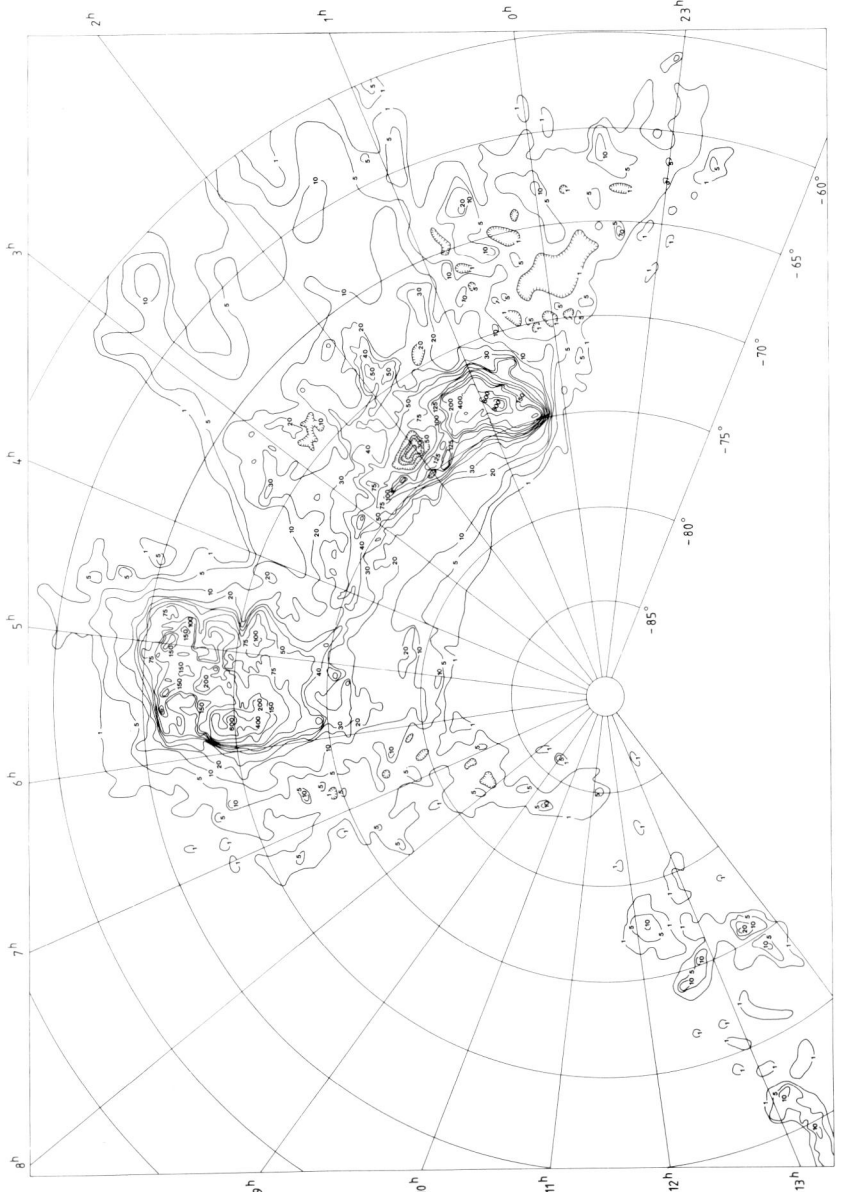

Fig. 2 The integrated HI surface density contours of the region of the LMC and SMC obtained with the 64-m Parkes radiotelescope. The contour unit is 10^{19} atoms cm^{-2}. Data have been taken from Mathewson et al. (1979), Mathewson, Ford and Fisher (1983), McGee and Milton (1966), and Hindman (1967).

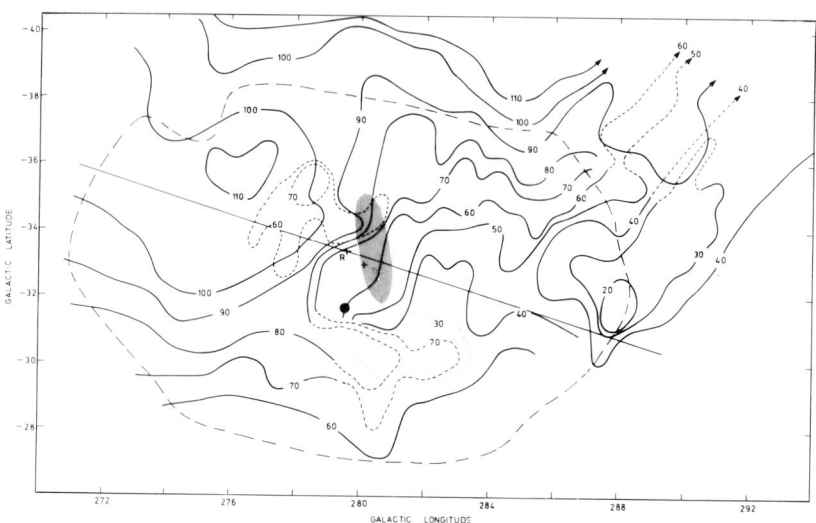

Fig 3 The radial velocity field of the LMC in galactic co-ordinates obtained using the 18-m Parkes radiotelescope. The contour numbers are in km s^{-1} and represent the V_{GSR} of the HI peaks (calculated using $V_\odot$ = 225 km s^{-1}). The dotted contours indicate regions where 2 peaks occur in the HI profile. The effect of a translational motion of 250 km s^{-1} of the Magellanic Clouds has been removed. The dot is 30 Doradus and the shaded region is the Bar. The dashed line is the outer limit of the LMC clusters.

tidal interaction between the LMC and SMC has produced substantial warping of the plane of the LMC, complex z structure and non-circular motions (Feitzinger 1980; Rohlfs et al. 1984). The velocity field of the LMC is similar in many respects to M83 which Rogstad et al. (1974) modelled successfully by a series of rings having a progression in their angle of inclination and position angle of the line of nodes with increasing radii.

THE SMALL MAGELLANIC CLOUD

A well-known characteristic of the SMC HI profiles is their double peaks separated by 30-40 km s^{-1} (Hindman 1967; McGee and Newton 1981; Bajaja and Loiseau 1982). Hindman proposed that this velocity structure originates in three expanding shells of gas. This explanation is untenable as the double peaks occur throughout the SMC and are not confined to the areas covered by Hindman's shells. Figure 4 maps the velocity structure along the major axis of the SMC obtained with the 64-m reflector. This clearly shows the splitting of the gas into two components with similar velocity gradients of about 8 km s^{-1} per degree.

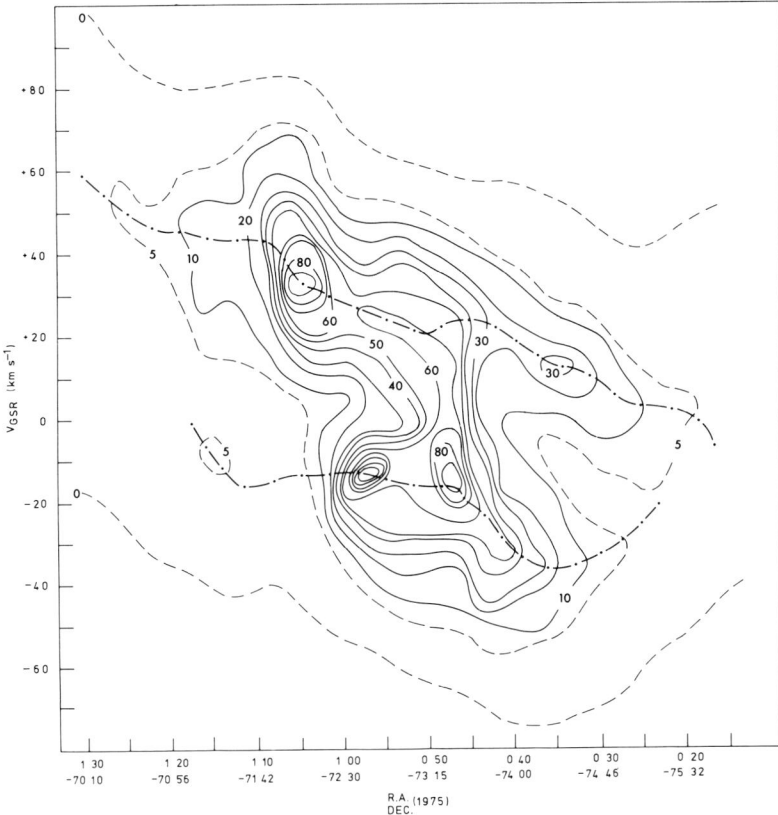

Fig. 4 Radial velocity structure along the major axis of the SMC obtained using the 64-m Parkes radiotelescope. V_{GSR} is calculated using $V_\odot = 250$ km s^{-1}. Contour numbers are the brightness temperatures (°K) of the HI. The dot/dashed lines indicate the ridges of the HI peaks.

Figure 5 shows that stars, HII regions and planetary nebulae also display this dichotomy in their radial velocities. Thus the two HI velocity components in the SMC delineate quite separate entities each with their own nebular and stellar population. Therefore we conclude that the SMC has been badly torn during its near collision with the LMC 2×10^8 years ago (Murai and Fujimoto 1980; Fujimoto 1984) and a large fragment, the Mini-Magellanic Cloud (MMC), is now separating from the remnant of the SMC (SMCR) at about 30 km s^{-1}. They are about 6 kpc apart. The CaII absorption line velocities (Feast et al. 1960) are marked along each velocity base-line in Figure 5. They invariably fall under the lower velocity HI profile even though the stars are about equally divided between the two velocity components. Hence the low velocity component must be in front of the high velocity component (Hindman 1964). It is assumed that the lower velocity component is the SMCR as it is the more intense. The SMCR is more compact than the MMC which extends further to the north-east. The red globular clusters

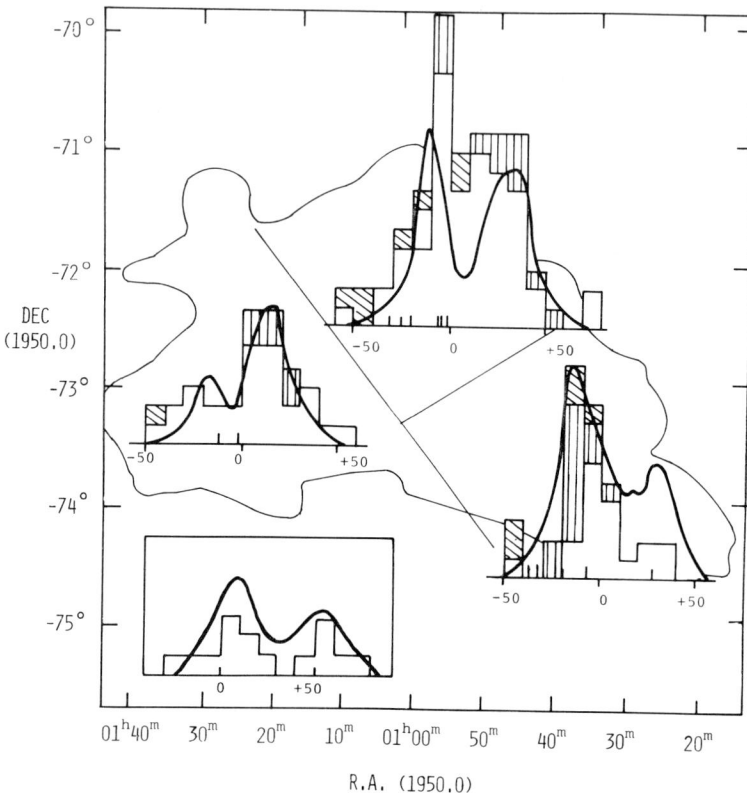

Fig. 5 Synoptic histograms showing the radial velocity distribution (V_{GSR}) of stars (unshaded), HII regions (vertical lines) and planetary nebulae (diagonal lines) drawn on the region of the SMC which they represent indicated by the straight boundary lines. The inset is for the bridge region. (V_{GSR} is calculated using $V_\odot = 250$ km s^{-1}). A HI profile representative of each region is superimposed on each histogram. The CaII absorption line velocities are marked on each velocity baseline. Data taken from Carozzi (1974), Dubois (1975), Feast et al. (1960), Feast (1968), Ardeberg and Maurice (1979), Westerlund and Glaspey (1971), Smith and Weedman (1973), and Hindman (1967).

concentrate around the SMCR at the centre and south end of the Bar whilst the blue clusters extend well to the north-east (Brück 1975).

The splitting of the SMC into two smaller "galaxies" provides a simple explanation for the stellar observations of Ardeberg and Maurice (1979), Florsch et al. (1981) and Welch and Madore (1984) which indicate that the SMC has a large extension along the line of sight. It also explains why the intense regions of HI as shown in the integrated HI map of Hindman (1967) are displaced with respect to the intense HII regions. Positional agreement becomes good when HI maps of the same velocity as the HII regions are used (McGee and Newton 1981; Bajaja and Loiseau 1982).

THE INTER-CLOUD REGION

The dominating feature of the inter-Cloud region is the intense bridge of gas connecting the SMC with the LMC, which is clear evidence for tidal interaction between the two galaxies. Its HI profiles show the characteristic double peaks separated by 30-40 km s^{-1} of the gas in the SMC (Mathewson et al. 1979). This indicates that both the SMCR and the MMC are the sources of the bridge gas. The velocity maps labelled -11 km s^{-1} and +21 km s^{-1} in Figures 1a and 1b by Bajaja and Loiseau (1982) show the points of origin in these two "galaxies".

Stars have been identified in the HI condensations in the bridge as far out from the SMC as $3^h15^m - 74°00'$. The radial velocities of these stars are divided into similar groups to the gas indicating that they originate in both HI streams (Fig. 5). This stellar component suggests that the bridge is not diffuse halo gas but that each component is probably only a few kpc thick.

The long spurs extending northwards from the bridge towards the Magellanic Stream are a noticeable feature of the inter-Cloud region. These spurs must be the result of tidal forces between the LMC and SMC and are the start of the Magellanic Stream. The gas in the inter-Cloud region has a mass of 5×10^8 M$_\odot$ and is loosely bound to the Magellanic Clouds. Some interaction between halo and Magellanic Cloud gas appears to be taking place as deep Hα photography by Johnson et al. (1982) shows weak, diffuse Hα emission (EM~25) between the Magellanic Clouds. In addition, the steep HI gradients along the eastern edge of the LMC and the southern edge of the inter-Cloud region are probably the result of ram pressure of the halo. Mathewson et al. (1977) took this as observational evidence that the direction of the orbit of the Magellanic Clouds is counter-clockwise as viewed from the Sun. It is estimated that a gas density of about 10^{-4} in the halo of our Galaxy would produce the observed compression of the HI. If the orbital velocity of the Magellanic Clouds is about 300 km s^{-1} and if the masses of the LMC and SMC are 10^{10} M$_\odot$ and 2×10^9 M$_\odot$, respectively, a halo gas density of 4×10^{-4} would start stripping HI from the inter-Cloud region.

THE MAGELLANIC STREAM

Figure 6 shows the HI surface density and velocity contours of MSI obtained using the Parkes 64-m reflector. The narrow elongated structures containing condensations about 30-50 arc min in size with FWHM of 18-25 km s^{-1} are characteristic of the Magellanic Stream. Mirabel (1981) has surveyed MSV and MSVI using the Arecibo radio telescope with the very high angular resolution of 3.3 arc min. These observations show much fine structure (~5 arc min) and were interpreted as the ongoing disintegration of the tip of the Magellanic Stream as it approached the Galaxy. The very high negative velocity HI clouds (VHVCs) observed in the first and second quadrants of the southern galactic hemisphere were assumed to be fragments of this disintegration

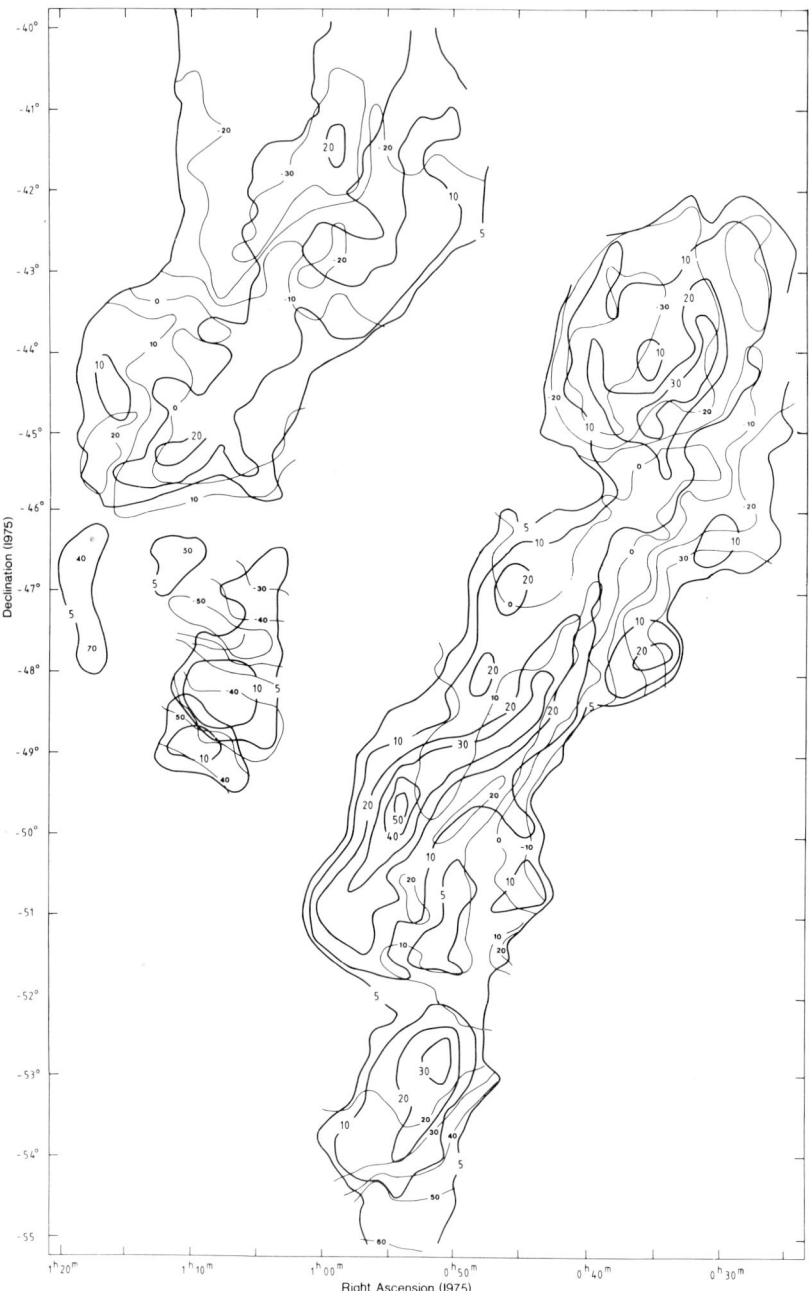

Fig. 6 The thick lines are the HI surface density contours of MSI. The contour unit is 10^{19} atoms cm^{-2}. The thin lines are the radial velocity contours. The contour numbers are V_{GSR} in km s^{-1} calculated using $V_\odot = 225$ km s^{-1}. Data were obtained using the 64-m Parkes radio telescope.

scattered over a wide area. Giovanelli (1981) is attracted to this unified theory of HVCs but there are the following difficulties with this currently popular hypothesis:

1) This hypothesis is based on the models of Davies and Wright (1977) and Lin and Lynden-Bell (1977) which place the tip of the Stream 10-15 kpc from the galactic centre. As Lin and Lynden-Bell (1982) have pointed out, these models are incorrect and the tip of the Stream is about 60 kpc from the centre of our Galaxy (Murai and Fujimoto 1980).
2) There are a number of other well-known, unrelated phenomena besides gravitational forces that accelerate HI to high velocities e.g., supernova remnants (Giovanelli and Haynes 1979; Mathewson et al. 1983); gas stripping of galaxies, active galactic nuclei and stellar winds are other examples. The prominent high-latitude galactic radio spurs which have been identified with SNRs, (Berkhuijsen et al. 1971) occupy the first and second quadrants and have the major HVC complexes A, C, and AC (Hulsbosch 1975) lying along their boundaries thus suggesting an association (see also Weaver 1979).
3) The centre of mass of the Local Group of galaxies lies in this region of sky so that intergalactic gas clouds would be expected to lie in these directions. It is likely that some of the VHVCs will be such clouds particularly as a number of intergalactic clouds have now been identified in other groups (Mathewson et al. 1975; Briggs et al. 1980; Hart et al. 1980; Mirabel and Cohen 1979; Schneider et al. 1983).
4) It is not clear how orbits can be generated by the fragmentation of the tip of the Stream which will produce the observed spatial and velocity distribution of the VHVCs, i.e., a fairly uniform $V_{GSR} \sim -250$ km s^{-1} over $\ell = 0^\circ$-190° at low to medium galactic latitudes.
5) MSI-IV may have similar fine structure to MSV and VI but unfortunately the Arecibo reflector cannot observe them.

An example of a group of VHVCs that appear to have a SNR origin is the HI complex near $\ell = 165^\circ$ b = -45°. Observations by Cohen (1981) show a 25° long HI streamer at $V_{LSR} = -110$ km s^{-1} coincident with a -10 km s^{-1} filament in the local gas at a distance of 100-300 pc. Cohen suggests that the high velocity stream is colliding with the galactic HI disk and that the -10 km s^{-1} filament is shocked galactic gas. In a later paper, Cohen (1982) maps a VHVC at $V_{LSR} = -190$ to -360 km s^{-1} which is the same size as the lower velocity filaments. In fact if Figure 3 of Cohen (1981) is superimposed on Figure 4 of Cohen (1982) the similarity between the shape of the 3 filaments is obvious and the high velocity filament lies along the northern boundary of the lower velocity filaments. The whole complex coincides with the strongest region of radio continuum emission from the Cetus Arc, a nearby SNR (Large et al. 1962). The close spatial relationship between the three HI streams and the SNR almost certainly implies an association and not, as Cohen argues, that the VHVC streamer is debris from the break-up of the tip of the Magellanic Stream. In addition, the intensity of the high velocity component is much greater than MSVI of which it is supposed to be a small fragment. Therefore it is our opinion that the Magellanic Stream is as illustrated in Figure 1.

Murai and Fujimoto (1980) obtained a good fit both to the radial velocities and the spatial structure of the Stream when they introduced their test particles at the beginning of the last orbit of the Magellanic Clouds about 1.8×10^9 years ago. However, when the test particles were introduced 10^{10} years ago the fit was poor (Fujimoto 1984). This suggests that the Magellanic Clouds have only made one close passage to our Galaxy in their lifetime. It appears possible to still successfully model the Stream with a hyperbolic orbit providing the massive halo of our Galaxy ($10^{12} M_\odot$) used by Murai and Fujimoto is retained.

The concept of a hyperbolic orbit for the Magellanic Clouds is attractive for other reasons as well. They are:

1) The damage to the Magellanic Clouds by our massive Galaxy during seven or more close encounters would have been considerable and much more than observed. (Dynamical friction does not introduce a large enough factor to significantly ease the severity of the tidal forces in the earlier passages, Murai and Fujimoto 1980). Whilst the LMC and SMC have tidally interacted, it is the effect of our Galaxy which would remove significant quantities of material from the system. At the moment there is $5 \times 10^8 M_\odot$ of weakly bound inter-Cloud gas which is about half of the total gas content of the LMC and SMC. It is very likely that this would have been removed long ago if the Magellanic Clouds were bound to the Galaxy. However the gas content of the LMC and SMC is 10% and 30% of their total mass, respectively, which is about normal for Magellanic-type galaxies.

2) The morphology of the LMC is archetypal of its class which would be surprising if it has suffered repeated tidal disruption.

3) The hyperbolic orbit explains why the Magellanic Clouds and the Stream lie near the plane of the Local Group of galaxies because if the galaxies lie in a plane so must their orbits. The Local Group galaxies (including the Magellanic Clouds) show a systematic variation of radial velocity with angle in the plane of the Local Group which is due to the motion of the Galaxy with respect to the Local Group. The fact that the Magellanic Clouds also reflect this motion supports the conclusion that the Magellanic Clouds are not bound to our Galaxy (Mathewson and Schwarz 1976).

4) The proximity of a galaxy as bright as the LMC ($M_V \sim -18.5$) to such a large spiral as our Galaxy is unique which suggests that it is a rare event.

The fact that a group of globular clusters and dwarf spheroidal galaxies with anomalous colour-magnitude diagrams lies close to the orbital plane of the Magellanic Clouds is interpreted by Lynden-Bell (1976) and Kunkel (1979) as relics of tidal interaction between the Magellanic Clouds and the Galaxy on the occasion of an early encounter. However in the context of our present scenario, they would need to be the result of the merger of another galaxy with our Galaxy some 7×10^9 years ago. Schweizer et al. (1983) point out that these near polar rings are favored statistically because of their slow differential precession and consequent longevity; alternatively if the halo of our Galaxy is triaxial, some orbits would tend to migrate toward the poles. Their Figure 2 shows examples of polar rings around galaxies which may be the result of mergers.

REFERENCES

Ardeberg, A., and Maurice, E.: 1979, Astron. Astrophys. 77, 277.
Bajaja, E., and Loiseau, N.: 1982, Astron. Astrophys. Suppl. 48, 71.
Berkhuijsen, E., Haslam, C.G.T., and Salter, C.J.: 1971, Astron. Astrophys. 14, 252.
Briggs, F.H., Wolfe, A.M., Krumm, N., and Salpeter, E.E.: 1980, Astrophys. J., 238, 510.
Brück, M.T.: 1975, Mon. Not. Roy. astr. Soc. 173, 327.
Carozzi, N.: 1974, Astron. Astrophys. Suppl. 16, 277.
Cohen, R.J.: 1981, Mon. Not. Roy. astr. Soc. 196, 835.
Cohen, R.J.: 1982, Mon. Not. Roy. astr. Soc. 199, 281.
Cohen, R.J.: 1982, Mon. Not. Roy. astr. Soc. 200, 391.
Davies, R.D., and Wright, A.E.: 1977, Mon. Not. Roy. astr. Soc. 180, 71.
Dubois, P.: 1975, Astron. Astrophys. 40, 227.
Feast, M.W.: 1968, Mon. Not. Roy. astr. Soc. 140, 345.
Feast, M.W., Thackeray, A.D., and Wesselink, A.J.: 1960, Mon. Not. Roy. astr. Soc. 121, 337.
Feitzinger, J.V.: 1980, Space Sci. Rev. 27, 35.
Feitzinger, J.V., Isserstedt, J., and Schmidt-Kaler, Th.: 1977, Astron. Astrophys. 57, 265.
Florsch, A., Marcout, J., and Fleck, E.: 1981, Astron. Astrophys. 96, 158.
Fujimoto, and Murai, T.: 1984, this volume, p. 115.
Giovanelli, R.: 1981, Astron. J. 86, 1468.
Giovanelli, R., and Haynes, M.P.: 1979, Astrophys. J. 230, 404.
Hart, L., Davies, R.D., and Johnson, S.C.: 1980, Mon. Not. Roy. astr. Soc. 191, 269.
Haynes, M.P.: 1979, Astron. J. 84, 1173.
Hindman, J.V.: 1964, Nature, 202, 377.
Hindman, J.V.: 1967, Aust. J. Phys. 20, 147.
Hulsbosch, A.N.M.: 1975, Astron. Astrophys. 40, 1.
Johnson, P.G., Meaburn, J., and Osman, A.M.I.: 1982, Mon. Not. Roy. astr. Soc. 198, 985.
Kunkel, W.E.: 1979, Astrophys. J., 228, 718.
Large, M.I., Quigley, M.J.S., and Haslam, C.G.T.: 1962, Mon. Not. Roy. astr. Soc. 124, 405.
Lin, D.N.C., and Lynden-Bell, D.: 1977, Mon. Not. Roy. astr. Soc. 181, 37.
Lin, D.N.C., and Lynden-Bell, D.: 1982, Mon. Not. Roy. astr. Soc. 198, 707.
Lynden-Bell, D.: 1976, Mon. Not. Roy. astr. Soc. 174, 695.
McGee, R.X., and Milton, J.A.: 1966, Aust. J. Phys. 19, 343.
McGee, R.X., and Newton, L.M.: 1981, Proceedings of the Astronomical Society of Australia, 4, 189.

Mathewson, D.S.: 1976a, Proceedings of the Astronomical Society of Australia, 3, 20.
Mathewson, D.S.: 1976b, Roy. Greenwich Obs. Bull. No. 182, 217.
Mathewson, D.S., Cleary, M.N., and Murray, J.D.: 1975, Astrophys. J. (Letters), 195, L97.
Mathewson, D.S., Ford, V.L., Dopita, M.A., Tuohy, I.R., Long, K.S., and Helfand, D.J.: 1983, Supernova Remnants and Their X-ray Emission, IAU Symp. #101, ed. J. Danziger and P. Gorenstein, (Dordrecht: Reidel).
Mathewson, D.S., Ford, V.L., and Fisher, J.L.: 1983 (in preparation).
Mathewson, D.S., Ford, V.L., Schwarz, M.P., and Murray, J.D.: 1979, The Large Scale Characteristics of the Galaxy, IAU Symp. #84, ed. W. B. Burton, (Dordrecht: Reidel), p. 547.
Mathewson, D.S., and Schwarz, M.P.: 1976, Mon. Not. Roy. astr. Soc. 176, 47.
Mathewson, D.S., Schwarz, M.P., and Murray, J.D.: 1977, Astrophys. J. (Letters), 217, L5.
Mirabel, I.F.: 1981, Astrophys. J. 250, 528.
Mirabel, I.F.: 1982, Astron. J. 256, 120.
Mirabel, I.F., and Cohen, R.J.: 1979, Mon. Not. Roy. astr. Soc., 188, 219.
Morras, R.: 1983, Astron. J. 88, 62.
Murai, T., and Fujimoto, M.: 1980, Publ. astr. Soc. Japan, 32, 581.
Rogstad, D.H., Lockhart, I.A., and Wright, M.C.H.: 1974, Astrophys. J. 193, 309.
Rohlfs, K., Feitzinger, J.V., and Kreitschmann, J.: 1984, this volume, p. 395.
Schneider, S.E., Helou, G., Salpeter, E.E., and Terzian, Y.: 1983, Astrophys. J. (in press).
Schweizer, F., Whitmore, B.C., and Rubin, V.C.: 1983, Astron. J. 88, 909.
Smith, M.G., and Weedman, D.W.: 1973, Astrophys. J., 179, 461.
Wannier, P., and Wrixon, G.T.: 1972, Astrophys. J. (Letters), 173, L119.
Weaver, H.: 1979, The Large Scale Characteristics of the Galaxy, IAU Symp. No. 84, ed. W. B. Burton, (Dordrecht: Reidel), 295.
Welch, D.L., and Madore, B.F.: 1984, this volume, p. 221.
Westerlund, B.E., and Glaspey, J.: 1971, Astron. Astrophys. 10, 1.

DISCUSSION

Fall: Since your two components of the SMC are seen in projection, the line of sight velocity of 30-40 km/s must be close to the true velocity of separation. At the relative separataion of a few kpc, the pair of SMC components are probably bound i.e. not escaping from each other. Thus they will probably collapse back together and merge in 1 Gyr or less. Should we really think of the SMC as two galaxies in this concept?

Mathewson: The escape velocity of the Mini Magellanic Cloud from the remnant of the Small Magellanic Cloud is about 35 km/s which is about their observed separation velocity. Therefore they may never coalesce.

DISTRIBUTION AND RADIAL VELOCITIES OF LATE SUPERGIANT STARS IN THE LMC
AND THE SMC

N. Martin[1], E. Maurice[2], L. Prévot[1], E. Rebeirot[1], J. Rousseau[3]
[1] Observatoire de Marseille, 2 place Le Verrier,
13248 Marseille Cedex 4, France
[2] European Southern Observatory, Casilla 16317, Santiago 9,
Chile
[3] Observatoire de Lyon, 69230 Saint-Genis-Laval, France

In 1983, our group published two catalogues of KM supergiants, respectively members of the LMC (Rebeirot et al., 1983) and of the SMC (Prévot et al., 1983).

In the LMC, no difference is observed between the general structure of the galaxy, as shown by the O-F supergiants and that shown by the KM supergiants. There are two concentrations of supergiants, one in Shapley III, one in 30 Doradus region, and four small concentrations toward the west.

As noted previously (Martin et al., 1976), there is no correlation between gas and stars; the stellar concentrations are situated in HI poor regions.

Radial velocities of 279 KM stars and 33 F G stars have been measured with Coravel at the 1.5 m Danish telescope at ESO. For 247 stars, correlation peaks are very good, and it is possible to estimate that the radial velocity standard error is 1.5 km s^{-1}.
For 65 stars, the accuracy is not quite as good, but is acceptable. Although the observations and the data reductions are not finished, the present sample is adequate to establish that, within each concentration of stars, the radial velocity dispersion is small - about 6 km s^{-1}.
In the 30 Dor region this dispersion is larger, (14 km s^{-1}) but a more detailed examination reveals that we are dealing with the overlapping of three well-defined velocity fields, one on the western side and two on the eastern side. Within each of these fields, radial velocities have a dispersion of 6 to 8 km s^{-1}.

Comparing these results with those published by McGee and Milton (1965) on the neutral hydrogen complexes, we observe that there is very good agreement between the mean radial velocities of stars and the radial velocities of neighbouring HI complexes.

Study of the 30 Dor field is somewhat more difficult because, on the eastern edge, there are two HI components, one at about 300 km s^{-1}

and the other at about 244 km s^{-1}.

These two components are seen at the southern edge of 30 Dor. We observed only three radial velocities in this region, but it is interesting to note that two of these are high radial velocities (293.5 and 315.3 km s^{-1}), and one is low (221 km s^{-1}).

In the SMC, as in the LMC, the distributions of the KM and the OB supergiants are the same. As noticed by several authors, there is apparently a positive correlation between the surface density of stars and gas, as observed in the 21 cm line (McGee and Newton 1981). However, this correlation disappears when the distributions are compared in various ranges of radial velocity, and the overall correlation might be due to chance. The actual situation looks similar to that in the LMC.

Further radial velocity observations are in progress. We hope to complete this structural and kinematical study of the Magellanic Clouds in the near future.

REFERENCES

McGee, R.X., Milton, J.A.: 1965, *Ap. J.* 191, 317
McGee, R.X., Newton, L.M.: 1981, *Proc. ASA. Contr.* 4 (2)
Martin, N. et al.: 1976, *Astron. Astrophys.* 51, 31
Prévot, L. et al.: 1983, *Astron. Astrophys. Suppl. Ser.* 51, 277
Rebeirot, E. et al.: 1983, *Astron. Astrophys.* 53, 225

A SEARCH FOR LOW-VELOCITY NEUTRAL HYDROGEN IN THE
MAGELLANIC STREAM

N.V. Bystrova
Leningrad Branch of the Special
Astrophysical Observatory
USSR Academy of Sciences

The large velocity range of the HI clouds belonging to the Magellanic Stream, as found by Haynes (1979), makes it reasonable to search for some low LSR velocity neutral hydrogen possibly belonging to the Stream. This gas, if any, would be severely blended by the local galactic gas.

In an attempt to resolve this blending the Pulkovo low velocity HI survey (Bystrova and Rakhimov, 1977) was used. Made with the large Pulkovo radiotelescope (beam 7' x 5°, bandwidth 20 kc/s), this survey contains the sky between declinations -29° and +40°. Therefore more than 1/3 of the Stream's great circle is accessible. The LSR velocity interval is from -21.8 to +25.6 km/sec.

To improve the contrast especially in the middle galactic latitudes the "structureless" component of the HI emission was extracted from the signals (Bystrova, 1980) and the contour maps in the galactic coordinate system were generated and published separately for the two components of HI radio emission. This division in two components was retained, and the contour maps were regenerated in the Magellanic coordinate system.

The position of the pole of this system, as well of its origin, were taken from Wannier and Wrixon (1972). One of the other poles, for example that used by Cohen (1982), puts the Stream rather outside its main plane.

The maps drawn for 10 LSR velocities contain the equal antenna temperature contours and give the regions projected on the Stream without distortions. The full extent of the maps is $\pm$ 70° in "latitude" and 420° in "longitude". For the "structural" component there are two sheets by 210°. The antenna temperature of the lowest contour is 0.75 K or approximately 1.7 K in T_B.

The inspection of the "structural" maps shows how difficult the task is. Three types of details are near the main plain. For positive velocities there are only small spots of several degrees in size, without any extended details. On the maps for negative velocities there are parts of larger structures which extend far beyond the Stream's region and to distinguish there between the far and near gas is very difficult. But also at negative velocities there are some isolated details extending along the plane. They may be suspected to belong to the Stream. A list of such details will be published elsewhere.

As far as the maps for the "structureless" component is concerned it must be mentioned here that the corrections for the seasonal spill-over effects were done only approximately when the data were prepared for the plotter. But just this component of the HI emission is mainly affected by the seasonal false signals. Then the details running almost perpendicular to the plane (along the declination circles) at the velocities from +9.7 to +25.6 km/s may be parts of these uncorrected signals. But for the velocities from +4.5 to -16 km/s there is a distinct area of emission with the coordinates +15 - +95 degrees in longitude and almost $\pm$ 40 or 50 degrees in latitude. Less certainly such broad emission is present for the longitudes 185 - 265°.

The emission for the longitude range 15 - 95° around the Stream's main plane needs special study with the exact correction for the seasonal spill-over effects.

REFERENCES

Bystrova, N.V., Rakhimov, I.A.: 1977, Pulkovo Sky Survey in the interstellar neutral Hydrogen Radio Line, "Nauka", Leningrad
Bystrova, N.V.: 1980, Contour maps to the Pulkovo Sky Survey in the interstellar neutral Hydrogen Radio Line, "Nauka", Leningrad
Cohen, R.J.: 1982, Mon.Not.R.astr.Soc. $\underline{199}$, 281
Haynes, M.P.: 1979, Astron.J. $\underline{84}$, 1173
Wannier, P., Wrixon, G.T.: 1972, Ap.J. $\underline{173}$, L 119

ON THE DISTANCE OF THE MAGELLANIC STREAM

J. Jaaniste
Tartu Astrophysical Observatory
Toravere, Estonia, 202444 USSR

Summary. If the HI clouds in the Magellanic Stream are in pressure equilibrium with the hot gaseous corona of our Galaxy, the distance of each cloud can be estimated from the observed properties of a cloud: position, angular dimensions, surface density of HI and the cloud temperature. Using a dynamical model with massive corona we have found distances of about 30 - 40 kpc for spherical condensations in the northern part of the Magellanic Stream.

1. INTRODUCTION

During the twenty-year history of the study of HI clouds, the determination of their distances has remained one of the central problems. The only physical method used has been the virial one. As we know (see e.g. Hulsbosch 1979), most clouds, except the smallest and coldest ones, have virial distances about a Megaparsec. Evidently some distance standard is needed to check these results.

The Magellanic Stream is the best cloud complex for the calibration of the distance estimates. It is evidently connected to the Magellanic Clouds, and its distance cannot differ much from the distance of the Clouds. The virial distances cannot be used for the Stream. Therefore, the Stream is not gravitationally bound and the observed concentrations of neutral hydrogen must be in pressure equilibrium with the hot gaseous component of the galactic corona. The short thermal expansion time (3×10^7 y for a typical cloud) seems to exclude the possibility of non-equilibrium clouds. In case of pressure equilibrium we can find the distance of a nearly spherical cloud, using its parameters (coordinates, angular dimensions, surface density, and velocity dispersion in 21-cm line) and by making some assumptions on the gaseous corona of our Galaxy.

2. GASEOUS CORONA

Let us suppose that the coronal region of our Galaxy contains hot, fully-ionized, isothermal gas of primeval composition. All these assumptions are not unavoidable, and our method can also be applied to other models, as the adiabatic flow, the galactic fountain (Bregman 1980), etc. As an example the isothermal model serves well enough. We suppose that the coronal gas is in hydrostatic equilibrium:

$$\text{grad } p = -\rho \text{ grad } \phi \tag{1}$$

(p is the pressure of the gas, ρ its density and ϕ the full gravitational potential of our Galaxy that practically does not depend on the density of the coronal gas). The isothermal equation of state leads to a barometric law:

$$p = p_0 \exp\left[-\mu m_H (\phi - \phi_0)/kT_c\right] \tag{2}$$

(μ is the molecular weight and T_c the temperature of the coronal gas). Any dynamical model of our Galaxy (e.g., Einasto et al 1976) can be used to calculate ϕ as a function of galactic coordinates and distance. To find the pressure we must know ϕ_0, p_0 and T_c. If we choose ϕ_0 to be the potential in the solar neighbourhood (given by the model), then p_0 is the pressure of the interstellar gas. It is a poorly determined parameter, the estimates ranging from $p/k = 2000$ K/cm^3 (Field 1965) to 20000 (Shapiro and Field 1966). We choose the latter value and use $T_c = 10^6$ K. Now we have the coronal pressure as a function of the galactic coordinates and distance (with R_{cl} the distance of the cloud):

$$p = p(l, b, R_{cl}) \tag{3}$$

3. METHOD

Consider now a spherical cloud of density

$$n_H = \sigma_H / D_{cl} = \sigma_H / \Theta R_{cl} \tag{4}$$

where σ_H is the central surface density, D_{cl} is the line-of-sight diameter of the cloud, and Θ its mean angular diameter. We suppose that the cloud is homogeneous; this is true if the temperature of the cloud is high enough compared to the virial temperature. The pressure in the cloud

$$p = n_H k T_{cl} \tag{5}$$

must be balanced by the external pressure of the coronal gas:

$$n_H k T_{cl} = p_0 \cdot \exp\left[-\mu m_H (\phi - \phi_0)/kT_c\right], \text{ or} \tag{6}$$

using (4) $\quad \sigma_H T_{cl} / \Theta = R_{cl} p_0 / k \cdot \exp\left[-\mu m_H (\phi - \phi_0)/kT_c\right]. \tag{7}$

Now we have a function of observed cloud parameters at the left-hand side (T_{cl} can be found from the half-velocity width of the 21-cm line profile). The right-hand side is a function of the galactic coordinates and the distance; it varies slowly with l,b and for a single cloud it can be used as a function of distance only. Thus the cloud distance R_{cl} can be found solving the last equation. To illustrate its properties we have drawn the rhs of eq. (7) as a function of R_{cl} in Fig. 1. As every cloud defines a horizontal line in this figure, we see that the number of solutions of eq. (7) ranges from zero to three. In case of a multiple solution, additional information is needed to locate the cloud.

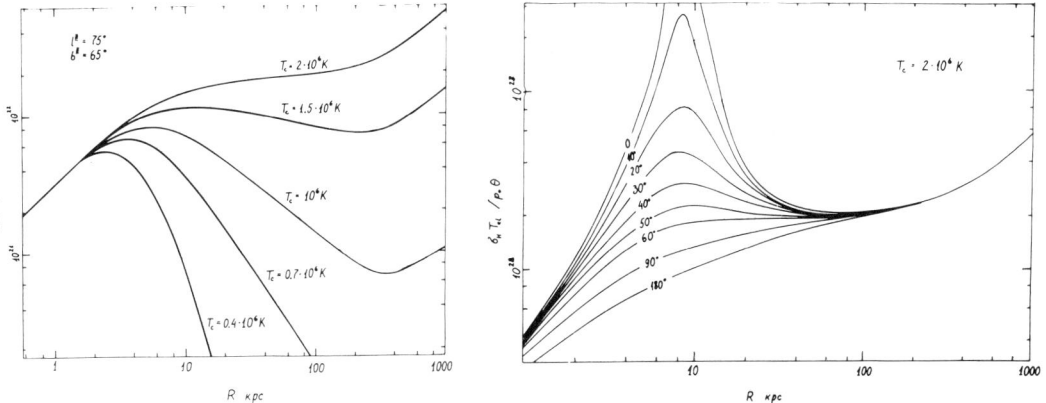

Figure 1. The observational combination $\sigma_H T_{cl}/\theta p_o$ versus cloud distance R_{cl} for different temperatures of the coronal gas (left panel) and for different angular distances from the galactic center for $l = 0, \pi$ (right panel).

4. APPLICATION TO THE MAGELLANIC STREAM

First we must choose suitable clouds in the Stream. The clouds must be spherical and as isolated as possible; the last condition is poorly satisfied in the Stream. The accuracy of our estimates depends strongly on that of the observational parameters, particularly of the half-width of the 21-cm line w (the lhs of (7) is a function of w^2). The best resolution, both angular and in velocity, have the observations of Mirabel (1981) with the Arecibo dish; unfortunately the cloud is not spherical enough. Some clouds in the Stream have been observed with the Mark IA telescope (Mirabel et al. 1979), but their angular resolution is not sufficient. Nevertheless, we chose a cloud from both observations as an example. Of course, there is no guarantee that they are not elongated or compressed along the line of sight; so the distances obtained correspond to the location where the clouds were most symmetrical.

Table. Data and distances for two HI clouds.

No	R.A hms	ζ d/m	σ* deg	n_H** $10^{19}/cm^2$	$w_{1/2}$ km/s	T_3 $10^3 K$	$\sigma_H T_{cl}/\theta$ $10^{25} K/cm^2$	R_{cl} kpc	Ref.
1	23 05 30	12 48	0.09	1.2	19.2***	8.03	6.2	0.97 35	M 1981
2	23 24 40	-9 24	0.26	1.5	31	21.2	7.0	1.1 32	M 1979

* Mean diameter of the last closed (compact) isophote.
** Central surface density, background subtracted.
*** Measured from Fig. 4 by Mirabel et al. 1981.

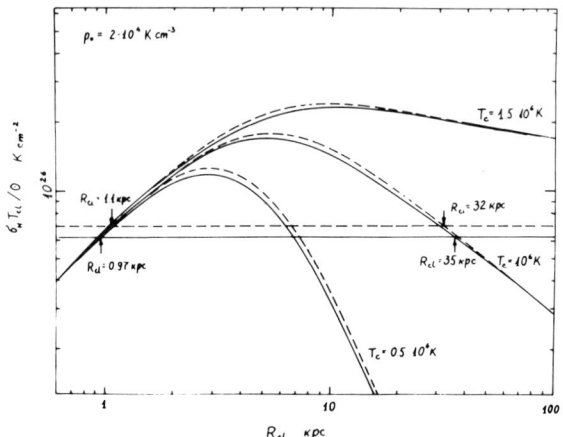

Figure 2. Determination of distances for the HI clouds in the northern part of the Magellanic Stream. Solid lines correspond to cloud 1 from the Table, dashed lines to cloud 2. Horizontal lines are determined by observations.

As you see, in both cases there are two solutions for each cloud, one about 1 kpc, the other 30 - 40 kpc from the Sun. So, our method does not completely solve the problem of the location of the clouds. Of course, for the Magellanic Stream the larger distances are more probable. And, as we have shown above (Fig. 1), the distances found depend strongly on the behaviour of the coronal pressure and especially on its temperature. This ties the problem of external HI clouds together with the problem of the gaseous corona - data on the corona can be used to derive the properties of the clouds and vice versa - cloud data can be used to restrict and verify coronal models.

In order to use this method high-resolution HI data, both in the coordinates and velocities, are badly needed. The two papers cited above show that such data could be obtained now, if the observers were interested in this task. With sufficient angular and velocity resolution it should be easy enough to choose spherical condensations with well-defined temperature as HI distance indicators.

The detailed description of the method will appear in a paper by Dr. E. Saar and me in the Publications of the Tartu Astrophysical Observatory (in Russian). An English version of the paper will be published as a preprint.

REFERENCES

Bregman J.N. 1980, Astrophys. J. **236**, 577
Einasto J., Joeveer M., Kaasik A. 1976, Tartu Astron. Obs. Teated, **54**, 3
Field G.B. 1975, Astrophys. Space Sci. **38** 167
Hulsbosch A.N.M. 1978, Astron. Astrophys. Suppl. **33**, 383
Mirabel I.F. 1981, Astrophys. J. **250**, 528
Mirabel I.F., Cohen R.J., Davies R.D. 1981, M.N.R.A.S. **186**, 433
Shapiro P.R., Field G.B. 1976, Astrophys. J. **205**, 762

THE BRIGHTEST STARS IN THE MAGELLANIC CLOUDS AND OTHER LATE-TYPE GALAXIES

Roberta M. Humphreys
University of Minnesota

The brightest stars always receive considerable attention in observational astronomy, but why are we so interested in these most luminous, and therefore most massive stars? These stars are our first probes for exploring the stellar content of distant galaxies. Admittedly, they are only the tip of the iceberg for the whole stellar population and very interesting processes are occurring among the less massive, older stars, but the most massive stars are our first indicators for studies of stellar evolution in other galaxies. They provide the first hint that stellar evolution may have been different in a particular galaxy because they evolve so quickly. The most luminous stars also highly influence their environments via their strong stellar winds and mass loss and eventually as supernovae.

The brightest stars are also very important as standard candles for the distance scale. We want to know how their luminosities may depend on the morphological type or luminosity of a galaxy and on possible chemical abundance differences. For this purpose we must observe the individual brightest stars in a variety of different galaxies, and we must also understand the evolution of the most massive stars.

In this paper I will be reviewing properties of the brightest stars in the Magellanic Clouds and in other nearby galaxies. The most luminous stars in different spectral type groups will be identified and compared with the population of luminous stars in our region of the Milky Way. This discussion will focus on normal, single stars and will not include peculiar stars, multiple systems and Wolf-Rayet stars. However, later in this paper I will mention the luminous blue variables also known as S Dor or Hubble-Sandage variables. At the end of the paper I will discuss the observed upper boundary to stellar luminosities and its relationship to massive star evolution.

When we study the most luminous stars we want to know how their basic properties, luminosity and mass, may depend on their environment and whether they vary from galaxy to galaxy. To get this information we need spectra and photometry of the individual stars and the distances to

Table 1
Basic Parameters of the Galaxies Used for the Luminosity Calibration[a]

Galaxy	Type	(mag)	App. Dist. Mod.$(m-M)_v$ (mag)	True Dist. Mod.$(m-M)_o$ (mag)	Origin
MW	Sbc (II)	-20.5:	--	-- ±0.25	Clusters and associations[b]
LMC	IR III-IV	-18.5	18.7	18.6±0.1	Cepheids and RR Lyrae[c]
SMC	IR IV/IV-V	-16.8	19.1	19.0±0.1	Cepheids and RR Lyrae[d]
M33	Sc II-III	-18.9	25.0:	24.25±0.2	Cepheids[e]
N6822	IR IV-V	-15.7	24.45	23.2±0.2 -0.0	Cepheids[f]
I1613	IR V	-14.8	24.52	24.3±0.1	Cepheids[g]
N2403	Sc III	-19.0	27.75	26.7±0.2	Cepheids[h]
M101	Sc I	-20.9	29.3 28.9±0.2	28.6-0.1 +0.2	M supergiants[i]

References: a - all distances are on the old Hyades scale; b - Humphreys (1978a); c - Martin, Warren and Feast (1979), Graham (1977); d - Gascoigne (1974), Graham (1975); e - Madore (1983), Sandage (1983); f - Kayser (1967), van den Bergh (1976), Humphreys (1980a); g - Sandage (1971), Humphreys (1980a); h - Tammann and Sandage (1968), Madore (1976); i - Sandage and Tammann (1974), Humphreys and Strom (1983).

the galaxies. The intrinsic visual luminosities are then derived from the photometry corrected for interstellar reddening and the true distance modulus of the galaxy.

Table 1 summarizes the morphological types and adopted distances for the eight galaxies mentioned in this paper. The primary galaxies in this discussion are the Large and Small Magellanic Clouds and the Milky Way. The luminosities for the galactic O-type stars and supergiants are derived from their membership in over 90 stellar associations and young clusters with known distances. The adopted distances to the Magellanic Clouds are based on Cepheids and RR Lyrae stars. The basic data, spectral types and photometry, for the individual stars come from many sources in the literature. For the Magellanic Cloud stars I especially want to mention the observations by Feast, Thackeray and Wesselink (1960) who were the first to identify and classify the brightest stars in the Clouds. Due to the availability of more telescopes and better instruments there has been a great increase in the amount of spectral classification data for the brightest stars; notably the pioneering work by Walborn (1977) on the O-type stars, the extensive observations by Ardeberg and his associates (1972, 1977) and myself (Humphreys (1979a,b) on the red supergiants. These observations rely heavily on the objective prism surveys for blue and red stars (Sanduleak 1968, 1969a,b, 1975; Sanduleak and Philip 1977; Azzopardi and Vigneau 1975; Brunet et al. 1975).

A very efficient way to compare the properties of the luminous star populations in these three galaxies is to look at their HR diagrams. Figures 1 and 2 in Humphreys (1984) compare the M_{Bol} vs. log T_{eff} diagrams for the luminous stars in the Milky Way and LMC.

It is clear from comparison of the HR diagrams that the luminous star populations in both galaxies have similar distributions of luminosities and spectral types. They have several important features in common: (1) a group of intrinsically very luminous hot stars, (2) a lack of supergiants of later spectral type at these high luminosities, and (3) an upper envelope to the luminosities of the late-type supergiants at about $M_{Bol} = -9.5$ mag. The LMC and the solar region of the Milky Way have essentially the same upper envelopes to their stellar luminosities.

Figure 3 in Humphreys (1984) shows the same HR diagram for the SMC, and it is immediately apparent that there are significant differences with the luminous star populations in the LMC and Milky Way. The hottest, most luminous stars in the SMC are fewer in number and are noticeably less luminous than stars of comparable temperatures in the solar region and the Large Cloud, but the large scale features of the HR diagram are similar to the LMC and Milky Way. It is especially important that the upper luminosity boundary for the late-type supergiants is the same in all three galaxies.

The data in Table 2 summarizes the bolometric luminosities for the individual most luminous stars in three broad spectral type groups (early, intermediate and late). The most luminous stars are of course the O-type stars and early B-type supergiants. This table confirms our impressions from the HR diagram that the most luminous stars are comparable in the Large Cloud and Milky Way but significantly fainter in the SMC. Comparison of the luminosities for the intermediate (F-K) type and late (M) type supergiants shows the constancy of their upper luminosity boundary ($M_{Bol} \simeq -9.5$) also mentioned earlier. There are no known high luminosity yellow supergiants in the SMC.

We are also interested in the visually brightest supergiants because of their potential usefulness as distance indicators. The individual visually brightest stars are listed in Table 3 in three spectral type groups. The visually brightest star in each galaxy is an A-type supergiant. The lack of visually bright stars ($M_v \leq -9.0$) in the SMC is very likely due to fewer progenitors evident from its HR diagram. The brightest F-K type supergiants are also comparable in both the LMC and Milky

Table 2

The Most Luminous Stars in the Milky Way and the
Large and Small Magellanic Cloud

Star	Sp. Type	M_{Bol}	Star	Sp. Type	M_{Bol}	Star	Sp. Type	M_{Bol}
HD 93129a	O3 If	−11.0	HDE 268743	O6:nn	−11.0	Sk 8	O9 +neb	−10.4
Cyg OB2 #9	O5 I	−10.9	HDE 269936	O9.5I	−11.0	Sk 80	O7 Iaf	−10.3
+40 4227	O6 Ib	−10.6	HDE 269896	ON9.7Ia+	−10.9	Sk 159	B0 Ia	−10.2
HD 151804	O8 Iaf	−10.5	HDE 269810	O3 If*	−10.6	Sk 157	O9.5 III	−10.1
HD 15570	O4 If	−10.4	HDE 270952	O6 Iaf+	−10.6	Sk 18	O7 +neb	−10.1
HD 93250	O3 V	−10.4	HDE 35517	B0 I	−10.6			
HR 8752	G0–G5 0–Ia	−9.6 to −9.4	HDE 268757	G7 0	−9.4			
RW Cep	K0 0–Ia	−9.6	HDE 269723	G4 0	−9.3			
HR 5171a	G8 0–Ia	−9.5	HDE 269953	G0 0	−9.2			
ρ Cas	F8 pIa	−9.4	HDE 271182	F8 0	−9.0			
HD 96918	G0 Ia+	−9.2	HDE 270046	G0 Ia	−9.0			
μ Cep	M2 Ia	−9.4	MG 46	—	−9.4:	Case 107-1	K5–M0 Ia	−9.0
KY Cyg	M2 Ia	−9.2	Case 46-44	M1 Ia	−9.2	HV 2084	M2 Ia	−8.8
HD 143183	M3 Ia	−9.1	Case 39-33	M4 Ia	−9.1	HV 11423	M0 Ia	−8.7
BD+24°3902	M1 Ia	−8.9	Case 46-32	M0 Ia	−9.0	Case 118-15	M0 I	−8.7
			Case 46-2	M2 Ia	−8.9	Case 106-1A	M0 Ia	−8.7

Table 3

The Visually Brightest Stars in our Galaxy and the Large and Small Magellanic Clouds

Star	Milky Way Sp. Type	M_v	Star	LMC Sp. Type	M_v	Star	SMC Sp. Type	M_v
Cyg OB2 #12	B8 Ia-0	-9.4 to -9.9	HD 33579	A3 Ia-0	-9.8	HD 7583	A0 Ia-0	-9.2
ζ Sco	B1.5 Ia+	-8.7	HD 32034	B9 Iae	-9.2			
HD 134959	B2 Ia+	-8.5	HDE 269647	A0 Ia	-9.0			
α Cyg	A2 Ia	-8.4	HDE 270086	A1 Ia-0	-9.0			
β Ori	B8 Ia	-8.4	HDE 269546	B5 Ia	-9.0			
ρ Cas	F8p Ia	-9.5	HDE 269953	G0 0	-9.2			
HR 8752	G0-G5 0-Ia	-9.4	HDE 271182	F8 0	-9.1			
RW Cep	K0 0-Ia	-9.4	HDE 269723	G4 0	-9.1			
HR 5171	G8 0-Ia	-9.2	HDE 268757	G7 0	-9.1			
HD 96918	G0 Ia+	-9.2	HDE 270046	G0 Ia	-9.0			
μ Cep	M2 Ia	-8.2	Case 46-32	M0 Ia	-8.1	Case 107-1	K5-M0 Ia	-8.2
KY Cyg	M3 Ia	-7.6	Case 46-44	M1 Ia	-8.0	HV 1475	K0-K5 Ia	-8.2
HD 143183	M3 Ia	-7.6	Case 46-39	M1 Ia	-7.7	HV 11423	M0 Ia	-7.7
BD+24°3902	M1 Ia	-7.5	Case 45-38	M1 Ia	-7.7	Case 118-15	M0 I	-7.7
			Case 46-2	M2 Ia	-7.5	Case 106-1A	M0 Ia	-7.6

Way. The brightest M supergiants are very important, because not only are their bolometric luminosities nearly constant, but so are their visual luminosities. The M supergiants have M_v's near -8 mag even in the SMC which lacks the most luminous blue stars and is a much smaller, less massive galaxy.

The visually brightest stars in other members of the Local Group, for which spectra are available (Humphreys 1980a,b), are also all late B or A-type supergiants. Figure 1 in Sandage and Tammann (1982) and in Humphreys (1983b), respectively, show the well known dependence of the brightest star on the luminosity of the parent galaxy. In addition, spectra have been taken of a few candidates in the more distant galaxies NGC2403 and M101, and A-type supergiants have been identified in each (Humphreys 1980c).

As we have already seen for the SMC, the most massive, most luminous stars in the fainter, smaller galaxies are both fainter and fewer in number. As their number decreases in a galaxy, the probability of finding A-type supergiants at a certain luminosity should also decrease. Where there are fewer of the most massive stars the visually brightest stars should be less luminous; consequently, there should be a similar relation between M_{Bol} and galaxy luminosity (Fig. 2 in Humphreys 1983b). This correlation for the most luminous stars suggests that the relation for the visually brightest stars is due to differences in the massive star populations ($>50-60$ $M_\odot$).

In contrast, the situation for the brightest red supergiants is very different. We have already noticed their constancy in both bolometric and visual luminosity in the Milky Way and the Magellanic Clouds. Figure 4 in Humphreys 1983 shows luminosities for the brightest M supergiants in six Local Group galaxies covering a range of nearly six magnitudes in galaxy luminosity. This very tight luminosity calibration is not fortuitous. It is a consequence of massive star evolution which is discussed at the conclusion of this paper.

I have been discussing the basic characteristics of the most luminous known stars with normal spectroscopic and photometric properties. There is also a group of very luminous stars which are distinguished by their emission line spectra and often by their variability which for some stars is often explosive. These stars are known variously as the η Car or S Dor type variables and the Hubble-Sandage variables in M31 and M33. Their spectra are characterized by emission lines of hydrogen, HeI, HeII, FeII and [FeII], many with P Cygni line profiles. Obviously not all of these emission lines are observed in all of the stars. For example, the FeII and [FeII] are most often reported at minimum light. Many of these stars display irregular variability consisting of extended maximum and minimum phases, frequently lasting several years.

In recent years, primarily as a result of ultraviolet (IUE) and infrared observations we have learned considerably more about these peculiar stars and have much greater insight into their important role

in massive star evolution. Information on their temperatures and total
luminosities have been determined by many investigators: η Car, Pagel
(1969), Davidson (1971), Neugebauer and Westphal (1968, 1969), Robinson
et al. (1973), Gehrz et al. (1973) and Hyland et al. (1979); P Cyg,
Cassatella et al. (1979), Underhill (1979), and Lamers et al. (1983);
S Dor, R 71, and R 81 in the LMC, Wolf et al. (1980, 1981a,b) and
Appenzeller and Wolf (1981); and for the Hubble-Sandage variables in
M31 and M33, Humphreys (1975, 1978b), Gallagher et al. (1981), Humphreys
et al. (1984). All of these stars are shown to be hot, very luminous
evolved massive stars with mass loss rates of 10^{-3} to 10^{-5} $M_\odot$/yr. Most
recently, Shore and Sanduleak (1984) have studied spectroscopically the
extreme emission line stars in the LMC and SMC and find that they have
temperatures and luminosities that place them in the same regions of the
HR diagram as their better known counterparts listed above.

Figure 1 is the composite HR diagram for the most luminous known
stars ($L \geq 5 \times 10^5$ $L_\odot$, $M_{Bol} \leq -9.0$ mag) in our galaxy, the Magellanic
Clouds, M33 and NGC6822 and IC1613. The most significant feature of
this HR diagram is the observed upper envelope to the luminosities of
normal stars (Humphreys and Davidson 1979, 1983, Humphreys 1983). This
luminosity boundary declines with decreasing temperature for the hotter
stars, but becomes essentially constant for the cooler supergiants. The
peculiar emission-line luminous stars are shown on this HR diagram and
some of the more famous ones are identified by name. The possible location of R136a, if it is single (Savage et al. 1983), and the proposed
supermassive Wolf-Rayet stars in M33 (Massey and Hutchings 1983) are
also indicated. The evolutionary tracks are from Maeder (1981, 1983).

In the past decade our ideas about the interior structure and evolution of the most massive stars have been radically altered. We now
routinely talk about the effects of mass loss, internal mixing and convective overshooting on the evolution of the most massive stars. The
upper luminosity boundary and the basic features of the observed HR diagram can be explained by models with high mass loss and internal mixing
in which stars >60 $M_\odot$ encounter an instability limit which prevents
further evolution to cooler temperatures. This boundary known as the
'de Jager limit' (de Jager 1980) involves turbulence and has been discussed in some detail by Maeder (1983). It may be the physical explanation for the tight upper limit to the M supergiant luminosities. Stars
that we know to have suffered dramatic instabilities such as η Car and
probably Var A in M33 are near or even above the observed "limit" for
normal stars. They are stars that have already reached the critical
stage in their evolution in which drastic mass loss is the consequence
of the star's approach to the instability limit. It is likely that all
stars above >60 $M_\odot$ pass through a similar critical phase. Very likely,
all of the peculiar emission line stars shown in this HR diagram are
either approaching or have already passed this stage. A star may even
bounce recurrently on the critical line (Davidson 1983, Humphreys and
Davidson 1983). When the star has evolved to this limit, the instability causes an outburst ejecting a fraction of the star's mass. The
star then moves away from the critical limit and temporarily relieves

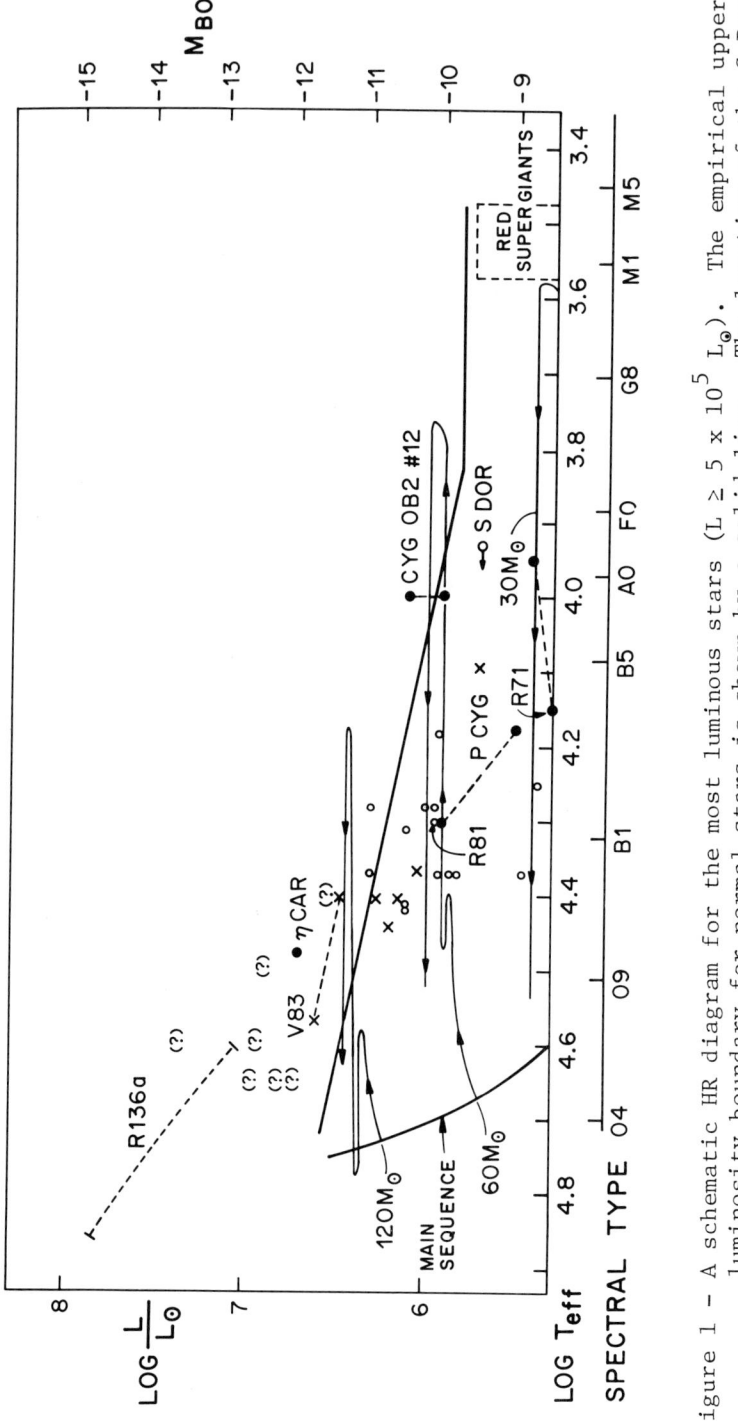

Figure 1 - A schematic HR diagram for the most luminous stars ($L \geq 5 \times 10^5 \, L_\odot$). The empirical upper luminosity boundary for normal stars is shown by a solid line. The location of the S Dor or Hubble-Sandage variables are shown and some of the individual stars are indicated. The possible location of R136a and the suspected supermassive Wolf-Rayet stars in M33 (?) are also shown. The evolutionary tracks for 30, 60 and 120 $M_\odot$ are from Maeder (1981, 1983).

the instability, but then in a few centuries or decades, it evolves back to the limit, and so on, perhaps until the star is reduced to a Wolf-Rayet star, unless it becomes a supernova first.

The most luminous, most massive stars are not merely brilliant beacons in distant galaxies to be observed only as curiosities or as possible distance indicators; they are important astrophysical laboratories where interesting and often unexpected phenomena are occurring.

References
Appenzeller, I. and Wolf, B.: 1981, in "The Most Massive Stars," ed. D'Odorico, Baade and Kjar (Garching: European Southern Obs.), p. 131.
Ardeberg, A., Brunet, J.P., Maurice, E., and Prevot, L.: 1972, Astr. Astrophys. Suppl. 6, p. 24.
Ardeberg, A. and Maurice, E.: 1977, Astron. Astrophys. Suppl. 30, p. 261.
Azzopardi, M. and Vigneau, J.: 1975, Astron. Astrophys. Suppl. 22, p. 285.
Brunet, J.P., Imbert, M. Martin, N., Mianes, P., Prévot, L., Rebeirot, E., and Rousseau, J.: 1975, Astron. Astrophys. Suppl. 21, p. 109.
Cassatella, A., Beeckmans, F., Benvenuti, P., Clavel, J., Heck, A., Lamers, J.H.G.L.M., Machetto, F., Penston, M., and Selvelli, P.L.: 1979, Astron. Astrophys. 79, p. 223.
Davidson, K.: 1971, Mon. Not. R. astr. Soc. 154, p. 415.
Davidson, K.: 1983, private communication.
de Jager, C.: 1980, "The Brightest Stars (Reidel, Dordrecht), pp. 11-14.
Feast, M.W., Thackeray, A.D., and Wesselink, A.J.: 1960, Mon. Not. R. astr. Soc. 121, p. 337.
Gallagher, J.S., Kenyon, S.J., and Hege, E.K.: 1981, Astrophys. J. 249, p. 83.
Gascoigne, S.C.B.: 1974, Mon. Not. R. astr. Soc. 166, p. 25p.
Gehrz, R.D., Ney, E.P., Becklin, E.E., and Neugebauer, G.: 1973, Astrophys. Letters 13, p. 89.
Graham, J.A.: 1975, Publ. Astr. Soc. Pac. 87, p. 641.
Graham, J.A.: 1977, Publ. Astr. Soc. Pac. 89, p. 425.
Humphreys, R.M.: 1975, Astrophys. J. 200, p. 426.
Humphreys, R.M.: 1978a, Astrophys. J. Suppl. 38, p. 389.
Humphreys, R.M.: 1978b, Astrophys. J. 219, p. 445.
Humphreys, R.M.: 1979a, Astrophys. J. Suppl. 39, p. 389.
Humphreys, R.M.: 1979b, Astrophys. J. 231, p. 384.
Humphreys, R.M.: 1980a, Astrophys. J. 238, p. 65.
Humphreys, R.M.: 1980b, Astrophys. J. 241, p. 587.
Humphreys, R.M.: 1980c, Astrophys. J. 241, p. 598.
Humphreys, R.M.: 1983a, Astrophys. J. 265, p. 176.
Humphreys, R.M.: 1983b, Astrophys. J. 269, p. 335.
Humphreys, R.M.: 1984, in Observational Tests of the Stellar Evolution Theory, IAU Symposium No. 105 (Reidel, Dordrecht), in press.
Humphreys, R.M., Blaha, C., D'Odorico, S., Gull, T.R., and Benvenuti, P.: 1984, Astrophys. J., in press.
Humphreys, R.M. and Davidson, K.: 1979, Astrophys. J. 232, p. 409.
Humphreys, R.M. and Davidson, K.: 1983, Science, in press.
Humphreys, R.M. and Strom, S.E.: 1983, Astrophys. J. 264, p. 458.

Hyland, A.R., Robinson, G., Mitchell, R.M., Thomas, J.A., and Becklin, E.E.: 1979, Astrophys. J. 233, p. 145.
Kayser, S.E.: 1967, Astron. J. 162, p. 217.
Lamers, H.J.G.L.M., de Groot, M., and Cassatella, A.: 1983, Astron. Astrophys., in press.
Madore, B.F.: 1976, Mon. Not. R. astr. Soc. 177, p. 157.
Madore, B.F.: 1983, private communication.
Maeder, A.: 1981, Astron. Astrophys. 102, p. 401.
Maeder, A.: 1983, Astron. Astrophys. 120, p. 113.
Martin, W.L., Warren, P.R., and Feast, M.W.: 1979, Mon. Not. R. astr. Soc. 188, p. 139.
Massey, P. and Hutchings, J.B.: 1983, Astrophys. J., in press.
Neugebauer, G. and Westphal, J.A.: 1968, Astrophys. J. Letters 152, p. L89.
Neugebauer, G. and Westphal, J.A.: 1969, Astrophys. J. Letters 154, p. L45.
Pagel, B.E.J.: 1969, Astrophys. Letters 4, p. 221.
Robinson, G., Hyland, A.R., and Thomas, J.A.: 1973, Mon. Not. R. astr. Soc. 161, p. 281.
Sandage, A.: 1971, Astrophys. J. 166, p. 13.
Sandage, A.: 1983, Astron. J., in press.
Sandage, A. and Tammann, G.A.: 1974, Astrophys. J. 194, p. 223.
Sandage, A. and Tammann, G.A.: 1982, Astrophys. J. 256, p. 339.
Sanduleak, N.: 1968, Astron. J. 73, p. 246.
Sanduleak, N.: 1969a, Cerro Tololo Inter-Am. Obs. Contr., No. 89.
Sanduleak, N.: 1969b, Astron. J. 74, p. 877.
Sanduleak, N.: 1975, Astron. Astrophys. 39, p. 461.
Sanduleak, N. and Philip, A.G.D.: 1977, Pub. Warner and Swasey Obs., Vol. 2, No. 5.
Savage, B.D., Fitzpatrick, E.L., Cassinelli, J.P., and Ebbets, D.C.: 1983, Astrophys. J., in press.
Shore, S.N. and Sanduleak, N.: 1984, Astrophys. J., in press.
Tammann, G.A. and Sandage, A.: 1968, Astrophys. J. 151, p. 825.
Underhill, A.B.: 1979, Astrophys. J. 234, p. 528.
van den Bergh, S.: 1976, in "Redshifts and the Expansion of the Universe" (Paris: NCRS), p. 13.
Walborn, N.R.: 1977, Astrophys. J. 215, p. 53.
Wolf, B., Appenzeller, I. and Cassatella, A.: 1980, Astron. Astrophys. 88, p. 15.
Wolf, B., Stahl, O., de Groot, M.J.H., and Sterken, C.: 1981a, Astron. Astrophys. 99, p. 351.
Wolf, B., Appenzeller, I., and Stahl, O." 1981b, Astron. Astrophys. 103, p. 94.

DISCUSSION

Lequeux: I fully agree with the small-number statistics explanation Maeder, Schild, and you give for the apparent lack of extremely bright blue-yellow stars in the SMC. As to the red stars, this effect is compensated by the following effect. Smaller irregular galaxies have lower abundances, hence probably lower stellar mass-loss rates. As discussed by Maeder, Azzopardi, and myself in 1981, this enhances the lifetime of the M supergiant stage and raises the upper mass (thus luminosity limit) where a star can become a M supergiant. Thus we encounter relatively more M supergiants, and brighter ones, in small, low metallicity galaxies and this may explain why you find the upper luminosity of M supergiants independent on that of the galaxy. To be quantitative there are about 6 times less blue-yellow supergiants in the SMC than in the LMC while there are only about 3 times less M supergiants according to the recent Marseille surveys.

Humphreys: Yes, that is correct. The M_{bol} luminosities for the M supergiants are beginning to show a decline between the spiral galaxies and LMC and the irregular galaxies (SMC, NGC6822, and IC1613) very likely due to the smaller population of blue star progenitors. So the lower metallicity in these galaxies does not completely compensate for the smaller sample size in these galaxies. The lower metallicity also produces a shift in the Hayashi track to warmer temperatures resulting in a shift to earlier spectral types for the M supergiants which is observed very dramatically for the SMC red supergiants. This results in a generally smaller bolometric correction so that although the maximum M_{bol} for these stars is slightly lower, the M_V's are still near -8 mag. These two effects are discussed in Humphreys (1983b).

Friedjung: How sure are you that Hubble-Sandage variables belong to the same class? Eta Carinae had a more dramatic behaviour than other stars you mentioned.

Humphreys: No. I do think they are all massive (> 40 M$_\odot$), unstable stars very likely evolving to the Wolf-Rayet stage. Eta Carinae probably represents the upper mass, upper luminosity end of this group of unstable blue stars. Eta Carinae's behaviour was dramatic, but it may not be unique. Var A in M 33 has a light curve which very closely resembles that of Eta Carinae in the 19th century. It also now possesses a circumstellar dust shell.

Hutchings: Please comment on the number count ratios of Blue and Red supergiants and whether this relates to stellar wind differences among galaxies.

Humphreys: The Blue to Red supergiant number ratio does vary with galactocentric distance and from the Galaxy to the LMC and SMC. The B/R ratio gets smaller and this variation is usually presumed to be an indicator of the decreasing metallicity in our galactic disk and the LMC and SMC. One should be careful when using B/R ratios as metallicity indicators, they can be influenced by variations in the IMF. The B/R ratio also depends very much on the luminosity interval being used. We can't just lump all of the blue and red stars together for a meaningful B/R ratio (see Humphreys 1984).

Aaronson: Sandage believes now that the luminosities of the richest M supergiants do depend on parent galaxy luminosity. In some cases he assigns M_V(max) closer to -9, rather than -8. Could you comment on this?

Humphreys: The luminosity calibration depends on the distance used for the galaxy. For M 33 Sandage (1983) now says $M_V \simeq -8.7$ mag for the brightest M supergiants within his new distance modulus of 25.3 mag which assumes little or no reddening. Madore (1983) has a preliminary modulus of 24.6 to 24.7 mag from infrared (H) observations of the Cepheids. I used his distance modulus, corrected to the old Hyades scale, to derive the luminosities of the red supergiants. The difference between these two distance moduli may be due to reddening of at least 0.6 to 0.7 mag in the visual. If this is the case, there is also $\simeq 0.1$ mag of reddening at H. Jones, Sitko and I very recently obtained JHK photometry of the known M supergiants and find that most of them are reddened by $\simeq 1$ mag (A_V). Sandage has also very recently identified bright red supergiant candidates in M 101. He finds his brightest red stars somewhat brighter than those found by Humphreys and Strom (1983) and says that $M_V \simeq -8.9$ mag in M 101. We (Aaronson, Strom, Capps, Lebofsky and Humphreys) have observed 6 of Sandage's 7 brightest candidates in the infrared (JHK). Four are M dwarfs, one is an M supergiant and one is uncertain. The brightest M supergiants therefore are at about $V \simeq 20.6$ to 20.7 mag. Their corresponding visual luminosities will depend on the distance to M 101 which, in my opinion, is very uncertain.

I suspect that the luminosities of the brightest M supergiants do decrease in galaxies less luminous than IC 1613. I doubt if $M_V \simeq -8$ mag is true for galaxies fainter than IC 1613.

MAGELLANIC CLOUD CEPHEIDS: ABUNDANCES, REDDENINGS, P-L AND P-L-C
RELATIONS. A REVIEW.

M.W. Feast
South African Astronomical Observatory

1. INTRODUCTION

Magellanic Cloud cepheids are of special importance for studies of
stellar pulsation, of stellar evolution, of the nature of the Clouds themselves and of the extragalactic distance scale. It is not possible to
cover all aspects of this subject in a short review and fortunately this
is probably unnecessary since a whole symposium on cepheids is planned
for Toronto next year. This paper attempts the much more modest task of
assessing the current status on three main topics; chemical abundances,
reddenings and the P-L and P-L-C relations (including recent infrared
work). Conflicting views on some of these topics have recently appeared
in the literature and a survey of the situation seems rather desirable.

In concentrating on a limited number of topics, discussion has
necessarily been confined to observations which seem of particular
relevance. This has meant that certain important series of photometric
observations of MC cepheids are not explicitly mentioned (though they are
incorporated with other data where appropriate). Besides the pioneering
work by Gascoigne and Kron (1965) and Gascoigne (1969) in BV this group
includes UBVRI observations of long period cepheids by Eggen (1971, 1977)
and BV photometry by Madore (1975) and Martin (1981). In addition to
this photoelectric work there exists a very large body of photographic
photometry. Much of the early photographic work was troubled by photometric scale errors though the material should be very useful for such
matters as period variability. Two recent extensive photographic
investigations have been carried out with considerable attention to the
problem of photometric calibration. The scope of the Dunsink (Butler
1976, 1978, Wayman et al. 1983) and SAAO (Martin et al. 1981) programmes
can be seen from Table 1. Though this material has been partially discussed (Butler 1976, 1978, Martin 1980a) it is very desirable that a
discussion of the complete material be made. These observations should
be particularly valuable for such problems as possible period-amplitude
and period-luminosity-amplitude relations, forms of light curves, period
changes etc. It is however important to remember that the survey of
cepheids in the Clouds is by no means complete. Thus in a 40 x 30

arcmin field around NGC 371 (SMC) where 47 cepheids were previously known, Lloyd Evans and Andrews found 25-30 more variables, almost certainly cepheids and mostly in the 16-18 magnitude range (Lloyd Evans 1977).

TABLE 1

	No. of observations in each colour	No. of cepheids
Dunsink Survey LMC (Fields I + II)	∼ 33	248
SMC	∼ 30	72
SAAO Survey LMC (Fields I + II + III)	∼ 45	213
SMC	∼ 60	180

2. ABUNDANCES

For a good many years a number of observational results have suggested that cepheids in the Clouds are metal deficient compared to those in the solar neighbourhood. These results include differing period-frequency distributions and bluer intrinsic colours. However interpretation is difficult because the period-frequency distribution depends on the (abundance sensitive) evolutionary tracks and also on the mass-function whilst the colours depend on the adopted reddenings. Fortunately the abundance problem has recently been put on a sounder footing by three independent investigations. Harris (1981) measured 45 SMC cepheids in the Washington 4 colour system and found [Fe/H] = -0.54. The calibration is partly from galactic stars of known metallicity and partly from models of Kurucz (1979) and Böhm-Vitense (1972). Pel, van Genderen and Lub (1981) used reddening free indices from the Walraven five colour system to obtain [Fe/H] = -0.70 ± 0.25 for eight SMC cepheids, calibrating their results using Kurucz models. Pel (1981) gives a useful summary of abundance work on cepheids including a preliminary [Fe/H] = -0.3 from Walraven photometry of LMC cepheids. The Washington and Walraven systems are to some extent complementary. There is some inevitable scatter in the Washington results due to a dependence on temperature and the intrinsic width in temperature of the instability strip. This is not a problem in the Walraven system. However the bands of the latter are further to the blue where possible deviations from the adopted reddening law may be more important. Both sets of results depend heavily on the adopted model atmospheres. Laney (1983b) has constructed rough curves of growth for 7 SMC and 5 LMC cepheids (and for 13 Galactic cepheids). He finds [Fe/H] = -0.06 ± 0.10 (LMC), -0.50 ± 0.08 (SMC).

Within the errors these various results are in agreement with those for other young objects (e.g. HII regions cf. the convenient tabulation of Laney). In the following [Fe/H] = -0.15 (i.e. a deficiency of a factor D = 1.4) will be adopted for the LMC and [Fe/H] = -0.60, D = 4 for the SMC.

3. REDDENINGS AND ABSORPTIONS

In the past it has frequently been necessary to adopt some mean value for the reddening of each of the Magellanic Clouds. However for several problems involving cepheids it is essential to have good individual reddenings and for that reason the problem is dealt with at some length here. Attempts have been made to obtain individual reddenings using some form of period-intrinsic colour relation together with a measured colour. However the intrinsic width of the period-colour relation is too great for this method to have any real precision. Attempts have been made to use UBV photometry for reddening determinations of Cloud cepheids but they do not yield satisfactory results, essentially because the reddening and intrinsic lines are nearly parallel (cf. Cogan 1979). At least four multicolour systems do appear capable of handling this problem; the BVI system (Dean, Warren and Cousins 1978), the Walraven system (Pel 1978), the DDO system (Dean 1981a) and the Strömgren system (Feltz and McNamara 1980).

The most extensive work so far has been on the BVI system (Martin and Warren 1979, Martin 1980b, Martin, Warren and Feast (= MWF) 1979, Caldwell and Coulson 1983). The VBI intrinsic line now being used is a slight revision by Dr J Caldwell of that used by MWF. It depends on galactic cepheids of known reddening (in clusters). To determine reddenings the effect of abundance on this intrinsic line has to be estimated. MWF obtained $\overline{E}_{B-V}$ = 0.086 using their uncorrected intrinsic line. They estimated the effect of D = 1.4 on the intrinsic line using Bell and Parson (1974) models and found a corrected $\overline{E}_{B-V}$ = 0.03. Whilst such a value is by no means impossible it is lower than estimates for some other LMC objects. Dr J Caldwell has reinvestigated this problem using later models (Kurucz 1979, Bell and Gustafsson 1978). He finds the BVI intrinsic line to be less sensitive to metallicity than previously thought. These results give

TABLE 2

	D	No.	$\overline{E}_{B-V}$ (Corrected)	Disp
LMC	1.4	33	0.072	0.047
SMC	4.0	46	0.078	0.041

MWF pointed out the small dispersion in reddenings for their sample of LMC cepheids and a similar small dispersion is now found by Caldwell and Coulson for their SMC sample. This scatter is essentially independent of the absolute values of the reddenings which depend on abundance corrections from models. It is important to emphasize this point. An examination of recent papers by Clube and Dawe (1983) and by Stift (1982) shows that their discussions of the MWF P-L-C relation in the LMC could only have some validity if the dispersion in true reddenings for the SAAO sample was much greater (by a factor of ∿ 2) than the observed dispersion in Table 2. Attempts have been made to argue that the observed dispersion will be smaller than the true value due either to observational error or to finite width of the intrinsic line. But obviously, allowance for such effects will decrease the dispersion and the observed dispersion must be an upper limit to the true dispersion in the reddenings.

The extensive work by Pel (1976, 1978) on galactic cepheids in the Walraven system shows that the (B-L), (V-B) diagram is useful for reddening determinations. The method fits observations on the descending branch of the light curve with an intrinsic locus which must be corrected for abundance effects using model atmospheres. It is anticipated that extensive observations of Cloud cepheids in this system will be obtained. Recently van Genderen (1983a, b) has published preliminary observations of a few of the longer period cepheids in each Cloud and has derived reddenings adopting an intrinsic line based on D = 2 (LMC) and D = 5 (SMC). Comparison with BVI reddenings for those of the stars with apparently well determined reddenings (using D = 2 (LMC) and D = 4 (SMC), the nearest available results) gives a mean difference in E_{B-V}, vG - SAAO = + .07 ± .02 (17 stars). In fact the agreement between the two methods is almost certainly better than this. The intrinsic lines for metal deficient cepheids used by van Genderen depends on computations by Lub and Pel (1977) using Kurucz atmospheres. To apply these results to long period cepheids on their descending branches it is necessary to make a considerable linear extrapolation of the Lub-Pel results. If we omit the redder stars ($\overline{V-B}$ > 0.5) for which this effect is most important we find vG - SAAO = 0.03 ± 0.01 (11 stars). The dispersion shows that the standard error of a single comparison is 0.05. Van Genderen finds an average standard error of 0.04 for his reddenings of these stars. This indicates that the standard error of a VBI reddening is ∿ 0.03 which accords with other estimates (e.g. Feast and Balona 1980). Dr J Caldwell has kindly pointed out to me that for 10 cepheids in galactic clusters, the scatter in the BVI reddenings is also 0.03 (as found from a comparison with the cluster OB star reddenings). It is worth noting that the abundance determination for SMC cepheids by Pel, van Genderen and Lub (1981) does not depend on such long extrapolation of the models since the work refers to 10 day cepheids near maximum.

Some DDO photometry of Magellanic Cloud cepheids has been published (Dean 1981a, b) but has not yet been discussed from the point of view of reddening. Similarly McNamara and Feltz (1980) have observed MC cepheids in the Strömgren system. Details are not yet available but apparently they yield low reddenings for LMC cepheids. These workers also find that the galactic foreground absorption is small (E_{B-V} = 0.034 (LMC), 0.019 (SMC)).

Laney (1983a) has made a detailed spectroscopic study of cepheids in both Clouds and has shown that a consistent system of spectral types can be established. This eliminates a previous colour-spectral type anomaly for long-period cepheids. Comparing his results with a $(B-V)_o$ spectral type relation he deduces that the reddenings of his cepheids must be very small (less than ~ 0.05) slightly less than given by the multicolour observations.

Madore (1982) has attempted to derive reddenings for LMC cepheids by a new method. Amongst other things this depends upon the coefficient of the colour term (β) in the P-L-C relation being near 6 rather than near 2.7 as is derived in the next section. This high value of β seems to depend on an attempt to fit a number of low amplitude cepheids, having sinusoidal light curves and log P ~ 0.5, into the P-L-C relation. In view of their distinctive physical characteristics (cf. Connelly 1980) these Cepheids have generally been recognized as overtone pulsators. In that case much too short a period has been adopted by Madore for these stars and in any case his reddenings are incompatible with those derived from multicolour photometry.

The above discussion suggests that the colours of Magellanic Cloud cepheids are naturally interpreted in terms of modest reddenings and metal deficiencies. It does not seem necessary to invoke blueing effects of companions to explain the colours of SMC cepheids as suggested by De Yoreo and Karp (1979). Companions are of course to be expected occasionally and some progress has been made by van Genderen (1977) to sort these out from Walraven photometry. The procedure is not entirely straightforward as can be seen in the extensive literature on the detection of companions to galactic cepheids.

4. P-L AND P-L-C RELATIONS IN B, V

Since Sandage (1958) showed that one would expect cepheids to be more accurately represented by a P-L-C than a P-L relation, several attempts have been made to determine the coefficient of the colour term. A weakness in the early work arises from the fact that for a P-L-C relation of the form,

$$\langle V_o \rangle = \alpha \log P + \beta (\langle B_o \rangle - \langle V_o \rangle) + \gamma \tag{1}$$

theory suggests β will be close to 3. Since this is similar to $R = A_V/E_{B-V}$ it is not possible to distinguish an intrinsic colour term from differential reddening between cepheids unless individual reddenings are available. One of the aims of the BVI photometry as just discussed was to obtain these reddenings and as a result MWF claimed to have established empirically the existence of an intrinsic colour term. The recent literature has shown that there is some misunderstanding of the present status of the P-L-C relation and it is necessary to discuss at least four separate issues. These are: (1) Is the existence of a P-L-C relation empirically established?: (2) How accurately can β be determined?: (3) How well does the P-L-C relation conform to theory?: (4) What are the relative merits of the P-L and P-L-C relations for distance determination?

On the first point the evidence given by MWF consists of 31 LMC cepheids (26 with log P < 1.7) for which individual reddenings were available. Their figure 4 shows that: (a) The $<V>$, log P relation has substantial scatter ($\sigma = 0.30$): (b) The $<V_o>$, log P relation has slightly less scatter ($\sigma = 0.27$): (c) The introduction of a colour term with $\beta = 2.70$ drastically reduces the scatter ($\sigma = 0.14$) except for the five 100 day cepheids. This demonstrates that a colour term exists. Arguments against this reduce to a claim that the real reddening scatter is substantially larger than the observed scatter which as we have seen cannot be the case.

Coulson and Caldwell are at present analysing BVI observations of 52 SMC cepheids. Initial solutions indicate that a significant intrinsic colour term near 3 is present.

It is well known that the application of the method of least squares to an equation such as (1) where there are observational errors in more than one variable, will lead to biased coefficients. In the case of the P-L-C relationship such a bias in determination of β can be demonstrated from actual LMC data (MWF), from an approximate analytical approach (Feast and Balona 1980) or by computer simulation (Brodie and Madore 1980). The method of maximum likelihood was devised to deal with problems of this kind (cf. Kendall and Stuart 1967) and using it, MWF obtained $\beta = 2.70$ for the LMC cepheids. Feast and Balona (1980) showed that this result is not significantly affected by taking into account a reddening induced correlation of errors in $<V_o>$ and $(<B_o> - <V_o>)$. Balona (1983) has set out the method in somewhat greater detail and finds that, for acceptable values of the various observational errors, the uncertainty in β is about 0.2.

The above results use errors estimated in a straightforward way from the observations. If errors in $<V_o>$ and $(<B_o> - <V_o>)$ were entirely dominated by errors in the reddening corrections and if these errors were large enough to entirely dominate the spread in the $<V_o>$, log P and $(<B_o> - <V_o>)$, log P diagrams, it is easily seen that a spurious value of β $(= R = A_V/E_{B-V})$ might be found. Such a suggestion has been made by Clube and Dawe (1983). As already indicated this would require the true spread in the reddenings to be greater than the apparent spread and is generally inconsistent with the numerical values of the errors, intrinsic scatter etc (cf. Feast and Balona 1980). Clube and Dawe also confuse intrinsic spread in the log P, $(<B_o> - <V_o>)$ plane with (possible) intrinsic spread in the $(\overline{V-I})$, $(\overline{B-V})$ plane which is quite a separate issue. Much of their discussion of the spread in the P, C relation is based on a misunderstanding of Feast and Balona (1980) who actually discuss only cepheids with individual reddening corrections.

Some preliminary attempt has been made by Stift (1982) to deal with the P-L-C relation by computer simulation. However amongst other things his published results do not extend to the spread of intrinsic colours shown by MWFs sample of LMC cepheids. The very sweeping claims made by Stift can only be evaluated when full details covering the relevant range of parameters are given.

A detailed comparison with theory will not be given here but an investigation by Cogan (1980) indicates that the empirical P-L-C relation is in good accord with theoretical expectations (LMC). It will be important to study this again when the SMC results become available.

De Vaucouleurs (e.g. 1960) suggested that the LMC was a flattened system at about 27° to the plane of the sky and he attempted to find the sense of the tilt using available photographic data on cepheids. However these results were very uncertain. Gascoigne and Shobbrook (1978) made accurate observations of cepheids on the far East and West side of the LMC and found a significant effect consistent with de Vaucouleurs prediction and showing that the eastern edge of the LMC is nearer to us than the western (a difference in modulus of 0.12 for an angular separation of $6°5$). Apart from this, the scatter about the P-L-C relation is no more than can be anticipated from observational errors (cf. MWF). There can thus be no significant contribution to this scatter from either a large depth of the LMC or from abundance differences between cepheids. First crossing cepheids would be expected to be displaced by $\sim 0^{m}3$ from cepheids on second or subsequent crossings of the instability strip (cf. Becker, Iben and Tuggle 1977). There is no evidence therefore for any of these in the SAAO BVI sample (log P < 1.7) of the LMC. Such cepheids are expected to be a very small percentage of the total population but it might be possible to use the large photographic programmes to select first crossing candidates for detailed photoelectric study.

The scatter of the 100 day cepheids in the P-L and P-L-C diagrams has long been known. A proper understanding of these very luminous stars is essential if for no other reason than their potential importance in extragalactic distance scale problems. Van Genderen (1983) has suggested that the scatter results from a mixture of first and second crossing cepheids and this point obviously deserves detailed consideration. He has also emphasised that the theoretical P-L-C relation contains higher order terms (cf. Iben and Tuggle 1975). It would be important to find these observationally although rough estimates suggest that the effects will be small.

There has been some discussion recently about the relative merits of the P-L and the P-L-C relations as extragalactic distance indicators. It has been urged against the use of P-L-C relation that it is abundance sensitive. However, as Dr J Caldwell has pointed out to me, in the formulation of Iben and Tuggle there is a significant (though lower) abundance effect in the P-L relation. Furthermore whilst the P-L-C relation is relatively insensitive to reddening we require accurate reddening corrections for the P-L relation. If these reddenings are derived from cepheid colours then abundances are needed to estimate intrinsic colours. It is in fact not unreasonable to hope that for any galaxy in which cepheids can be measured it will be possible to obtain abundances for HII regions at least. There remains the major disadvantage that the P-L relation is a strip rather than a line and this can give rise to serious sampling problems.

The zero point of the P-L-C relations is derived from galactic cepheids of known distance (i.e. in clusters) in our Galaxy or from those with radii determinations.

For the LMC P-L-C relation

$$\langle M_V \rangle = -3.80 \log P + 2.70 (\langle B_o \rangle - \langle V_o \rangle) + \phi \qquad (2)$$

MWF found $\phi_1 = -2.39$ based on clusters with ZAMS fitting and scaled to a Hyades modulus of 3.03. Several writers (Stothers 1983, Fernie and McGonegal 1983, Caldwell 1983) have recently reconsidered this problem. Caldwell uses a ZAMS from Schmidt-Kaler (1982) which depends on the Hyades being metal rich and at a modulus of 3.28. He finds $\phi_2 = -2.46 \pm 0.05$ in good agreement with results from radii, $\phi_3 = -2.42 \pm 0.05$ (cf. MWF from Balona 1977). Despite this generally satisfying consistency there has been some concern over distances for calibrating clusters derived from Strömgren Hβ photometry (Schmidt 1980a, b, 1981, 1982a, b, 1983). These average $\sim 0\overset{m}{.}4$ less than those of Caldwell. This discrepancy has been much reduced in recent work by Balona and Shobbrook (1983). They re-calibrate the Hβ index using much new work by Shobbrook and with a tie directly to the Pleiades and ultimately to F stars of known parallaxes. Using their calibration with Schmidt's data and Caldwell's absorptions for the cluster cepheids one finds $\phi_4 = -2.26 \pm .06$. Evidently a mean ϕ_2, ϕ_3 and ϕ_4 differs little from ϕ_1 as adopted by MWF. These results must still be corrected for abundance effects. MWF estimated the correction as $-0\overset{m}{.}11$. Using Bell and Gustafsson models, Caldwell now estimates this to be $-0\overset{m}{.}16$. Thus the best current P-L-C (true) modulus for the LMC is 18.64. A useful estimate for the SMC should follow from the current work of Caldwell and Coulson.

5. P-L AND P-L-C RELATIONS IN THE INFRARED

McGonegal et al. (1982) have shown that important results can be obtained from the study of MC cepheids in the 1 to 2μ region. They find that single (random phase) observations at H (1.6μ) allows them to define a P-L relation with a scatter of only σ = 0.25 (cf. McAlary et al. 1983). Because of this and because infrared observations overcome some of the problems of interstellar absorption,* the method has great promise and is being actively followed up both in the Clouds and other galaxies. Laney (SAAO) and Stobie (ROE) are observing cepheids in both Clouds and in the Galaxy at J, H and K. The aim is to get good light curves for a limited number of cepheids. Matters such as the intrinsic width of the P-L relation can then be examined. They find that some MC cepheids have substantial infrared amplitudes ($\Delta H \sim 0\overset{m}{.}5$). After making small reddening

---

* D Laney has pointed out that the distances to the galactic calibrating clusters depend on $R = A_V/E_{B-V}$ and uncertainties in R can thus significantly affect the calibration of infrared P-L relations.

corrections the SAAO infrared data (log P < 2.0) can be fitted by

$$\langle H_o \rangle = -3.34 \log P + \gamma$$

γ = 16.19 (17 stars LMC) and 16.61 (14 stars SMC). The dispersions, σ, are 0.10 (LMC) and 0.15 (SMC). A considerable reduction over the value for single (random phase) observations. Some evidence for an infrared colour term has also been found. SAAO infrared observations for 8 galactic calibrating cepheids gives a P-L zero point of -2.63. This leads to distance moduli of 18.82 (LMC) and 19.24 (SMC) (Δ mod = 0.42) provided there are no significant abundance effects at H. For the LMC it is probably best to take a straight mean of the above modulus with the P-L-C modulus (18.64) of the last section giving a mean true modulus of 18.73.

This paper depends exceptionally heavily on discussions with and results from Drs L A Balona, J A R Caldwell, I M Coulson, D Laney, R R Shobbrook and R S Stobie.

REFERENCES

Balona, L.A., 1977.   Mon. Not. R. astr. Soc., 178, 231.
Balona, L.A., 1983.   Statistical Methods in Astronomy, Strasbourg, France.   Conference proceedings, in press.
Balona, L.A. and Shobbrook, R.R., 1983.   Private communication.
Becker, S.A., Iben, I. and Tuggle, R.S., 1977.   Astrophys. J., 218, 633.
Bell, R.A. and Gustafsson, B., 1978.   Astr. Astrophys. Suppl., 34, 229.
Bell, R.A. and Parson, S.B., 1974.   Mon. Not. R. astr. Soc., 169, 71.
Böhm-Vitense, E., 1972.   Astr. Astrophys., 17, 335.
Brodie, J.P. and Madore, B.F., 1980.   Mon. Not. R. astr. Soc., 191, 841.
Butler, C.J., 1976.   Astr. Astrophys. Suppl., 24, 299.
Butler, C.J., 1978.   Astr. Astrophys. Suppl., 32, 83.
Caldwell, J.A.R., 1983.   Observatory, in press.
Clube, S.V.M. and Dawe, J.A., 1983.   Astr. Astrophys., 122, 255.
Cogan, B.C., 1979.   Mon. Not. R. astr. Soc., 188, 297.
Cogan, B.C., 1980.   Astrophys. J., 239, 941.
Connolly, L.P., 1980.   Publs. astr. Soc. Pacif., 92, 165.
Coulson, I.M. and Caldwell, J., 1983.   In preparation.
Dean, J.F., 1981a.   Mon. Not. R. astr. Soc., 197, 779.
Dean, J.F., 1981b.   SAAO Circulars, no. 6, 10.
Dean, J.F., Warren, P.R. and Cousins, A.W.J., 1978.   Mon. Not. R. astr. Soc., 183, 569.
de Vaucouleurs, G., 1960.   Astrophys. J., 131, 265.
De Yoreo, J.J. and Karp, A.H., 1979.   Astrophys. J., 232, 205.
Eggen, O.J., 1971.   Astrophys. J., 163, 313.
Eggen, O.J., 1977.   Astrophys. J. Suppl., 34, 1.
Feast, M.W. and Balona, L.A., 1980.   Mon. Not. R. astr. Soc., 192, 439.
Feltz, K.A. and McNamara, D.H., 1980.   Publs. astr. Soc. Pacif., 92, 609.
Fernie, J.D. and McGonegal, R., 1983.   Preprint.
Gascoigne, S.C.B., 1969.   Mon. Not. R. astr. Soc., 146, 1.

Gascoigne, S.C.B. and Kron, G.E., 1965. Mon. Not. R. astr. Soc., 130, 333.
Gascoigne, S.C.B. and Shobbrook, R.R., 1978. Proc. astr. Soc. Aust., 3, 285.
Harris, H.C., 1981. Astr. J., 86, 1192.
Iben, I. and Tuggle, R.S., 1975. Astrophys. J., 197, 39.
Kendall, M.G. and Stuart, A., 1967. The Advanced Theory of Statistics, vol. 2, 2nd ed., Griffin Ltd., London.
Kurucz, R.L., 1979. Astrophys. J. Suppl., 40, 1.
Laney, C.D., 1983a. Publs. astr. Soc. Pacif., submitted.
Laney, C.D., 1983b. Publs. astr. Soc. Pacif., submitted.
Lloyd Evans, T., 1977. Private communication.
Lub, J. and Pel, J.W., 1977. Astr. Astrophys., 54, 137.
Madore, B.F., 1975. Astrophys. J. Suppl., 29, 219.
Madore, B.F., 1982. Astrophys. J., 253, 575.
Martin, W.L., 1980a. Ph.D. Thesis, University of Cape Town.
Martin, W.L., 1980b. SAAO Circulars, no. 5, 172.
Martin, W.L., 1981. SAAO Circulars, no. 6, 96.
Martin, W.L., Thomas, Y., Carter, B.S. and Davies, H.E., 1981. SAAO Circulars, no. 6, 31.
Martin, W.L. and Warren, P.R., 1979. SAAO Circulars, no. 4, 98.
Martin, W.L., Warren, P.R. and Feast, M.W., 1979. Mon. Not. R. astr. Soc., 188, 139 (= MWF).
McAlary, C.W., Madore, B.F., McGonegal, R., McLaren, R.A. and Welsh, D.L., 1983. Preprint.
McGonegal, R., McLaren, R.A., McAlary, C.W. and Madore, B.F., 1982. Astrophys. J., 257, L33.
McNamara, D.H. and Feltz, K.A., 1980. Publs. astr. Soc. Pacif., 92, 587.
Pel, J.W., 1976. Astr. Astrophys. Suppl., 24, 413.
Pel, J.W., 1978. Astr. Astrophys., 62, 75.
Pel, J.W., 1981. 2nd Asian-Pacific Regional Astronomy Meeting, Bandung, Indonesia. Proceedings, in press.
Pel, J.W., van Genderen, A.M. and Lub. J., 1981. Astr. Astrophys., 99, L1.
Sandage, A., 1958. Astrophys. J., 127, 513.
Schmidt, E.G., 1980a. Astr. J., 85, 158.
Schmidt, E.G., 1980b. Astr. J., 85, 695.
Schmidt, E.G., 1981. Astr. J., 86, 242.
Schmidt, E.G., 1982a. Astr. J., 87, 650.
Schmidt, E.G., 1982b. Astr. J., 87, 1197.
Schmidt, E.G., 1983. Astr. J., 88, 104.
Schmidt-Kaler, Th., 1982. Landolt-Börnstein Numerical data and functional relationships in science and technology. Group VI, vol. 2b, p. 19, Springer-Verlag, Berlin.
Stift, M.J., 1982. Astr. Astrophys., 112, 149.
Stothers, R.B., 1983. Preprint.
van Genderen, A.M., 1977. Astr. Astrophys., 54, 737.
van Genderen, A.M., 1983. Astr. Astrophys., 119, 192.
van Genderen, A.M., 1983a. Astr. Astrophys. Suppl., 52, 423.
van Genderen, A.M., 1983b. Preprint.
Wayman, P.A., Stift, M.J. and Butler, C.J., 1983. In preparation.

DISCUSSION

McCarthy: Can you compare for us the distance modulus of the Small Cloud as derived by the IR method and by the P-L-C method? I would be interested to compare the differences between the Clouds.
Feast: Data for the SMC is at present being studied for a P-L-C relation. When this has been done, the comparison you suggest will be of interest.
Stift: I have a list of 11 points of criticism, and like to give them here!
(Editors note: Due to the speed with which these were given, Feast could not comment on each of them at once but later wrote his replies as they appear in the discussion below. Stift expanded slightly some of his questions after that. The references in Feast's answers can be found in his paper. Of the original 11 points 2 have been untracebly lost for eternity.)
Stift: 1. The Cape photographic cepheid surveys exhibit very large photometric errors; they do not constitute an improvement over previous surveys, such as e.g. Butler's (1976, 1978) which keep errors down to 0.04 to 0.05 mag.
Feast: The systematic and random errors in the Cape photographic work have been discussed in detail by Martin et al. (1981) and by Martin (1980a) and have been compared with earlier work. The errors appear to be at the level expected for careful photographic photometry.
Stift: 2. If canonical theory holds -which I do not believe- the Cape photoelectric SMC photometry is of no better precision than the Dunsink photographic LMC photometry, the rms error of a mean V magnitude being of the order of 0.04 to 0.05 for both.
Feast: The Cape photoelectric photometry of SMC Cepheids published earlier (Martin and Warren 1979; Martin 1981) is inadequate for the study of P-L-C relations since it contains BVI observations for very few stars (so individual reddenings cannot be derived) and some stars have very few observations. Discussion was therefore deferred until extensive BVI data became available (Coulson and Caldwell, to be published). The text above contains preliminary results from this latter data set.
Stift: 3. The BVI intrinsic Galactic Cepheid locus has been empirically established using the lowest published reddening values (see Fernie 1967). The intrinsic LMC Cepheid locus has not been shown by Martin, Warren and Feast for good reason, because it turns out that more than 1/3 of the stars exhibit negative or zero reddening.
Feast: The BVI intrinsic line in current use is a slight revision by Caldwell of that given by Dean, Warren and Cousins (1978) which is based on Cepheids with known reddening (i.e. in galactic clusters). Martin, Warren and Feast (1979) adjusted an earlier version of this line for abundance effects (Bell-Parsons models) and found a mean $E(B-V)=0.03$. With such a small mean reddening and with a standard error for a single $E(B-V)$ determination of about 0.03, it is obvious statistically that some apparent negative reddenings will result. As indicated in the text, the newer models show that the BVI diagram is less sensitive to abundance than previously thought and higher mean

reddenings are found. It must be stressed that the mean reddening (which depends on the absolute position of the BVI intrinsic line) is unimportant in establishing the existence of a P-L-C relation. It is the correction for differential reddening between Cepheids that is relevant. With regard to the first sentence it should be noted that Pel's intrinsic colours for Cepheids in the Walraven system (1978 A. Ap. 62, p 75) lead to slightly lower reddenings for galactic Cepheids than the adopted BVI intrinsic colours.

Stift: 4. The Dunsink surveys show that abundance differences within the LMC are probably important (Wayman, Stift, and Butler 1983); this enhances the true scatter in the observed reddenings due to a wider intrinsic Cepheid locus.

Feast: It will be interesting to see the Dunsink results, though it is obviously very difficult to obtain unambiguous evidence of abundance variations from BV photometry alone. As indicated in the text, any intrinsic scatter in the BVI locus will increase the observed reddening scatter which is thus an upper limit to the true scatter. Thus the wider the intrinsic line, the lower the true scatter in differential reddening and the more obvious the need for a P-L-C relation is.

Stift: 5. Compared to all other surveys the Cape photoelectric LMC photometry shows abnormally high scatter about the P-L relation and at the same time abnormally low scatter about the P-L-C relation. It is virtually impossible to rederive the published mean magnitudes from the original observations.

Feast: It is the significant spread (at a given period) in magnitude and colour for Cepheids in the Cape LMC BVI programme that makes the sample well suited for the determination of the P-L-C relation. Whether Cepheids at the extremes of colour and luminosity at a given period (i.e. at the edges of the instability strip) are frequent or rare in the LMC can only be determined when extensive photoelectric multicolour photometry has been carried out. In view of the lack of individual reddenings in other surveys and of the higher errors in photographic work, it is not clear that any higher spread (should it exist) in a P-L-C relation is a matter of significance.

Stift: 6. The initially published Cape SMC Cepheid photometry (Martin et al. 1981) bears no resemblance to the preliminary P-L-C relation presented by Feast. For the SMC the scatter about the original P-L-C relation is about 50% larger than for the LMC.

Feast: The paper mentioned by you contains only photographic (BV) photometry. In view of the lack of individual reddening corrections and the lower accuracy of photographic work compared to photoelectric, a higher scatter would indeed be anticipated. Regarding photoelectric work I refer to what I said after your point 2. **Stift:** The Tables in Martin et al. (1981) clearly say "photoelectric"!

Stift: 7. Stift (1982) and Clube and Dawe (1983) have demonstrated that the maximum likelihood estimator given by Feast and Balona (1980) as applied to MC Cepheids leads to wrong results. Only the assumption of an arbitrary value for the error of a (B-V) determination yields the canonical value of $\beta = 2.70$.

Feast: These statements are incorrect. The approximation adopted by Clube and Dawe is discussed in the text. Essentially they assume the spread in the P-L relation is due to differential reddening. That this is not so for the Cape LMC BVI sample is shown by the small scatter in the observed reddenings (c.f. also the answer at (4) above). From your second sentence it seems that you base your conclusion on Table 3 of Stift (1982). This table is calculated using one possible maximum likelihood estimator (though not the model used by Martin, Warren and Feast, nor that used in the discussion of correlation errors by Feast and Balona 1980 or Balona 1983). In this model a value of $\sigma(B-V)$ is adopted and the remaining parameters are then calculated. Amongst the derived parameters is σV which must evidently be compatible with directly estimated values for σV. The first two colums below are from Stift (1982), Table 3. The values of σV which come directly from these solutions have been added.

$\sigma(B-V)$	β	σV
0.02	2.34	0.10
0.03	2.48	0.08
0.04	2.71	0.03
0.05	3.09	variance negative

From external evidence Martin, Warren, and Feast found $\sigma(B-V)= 0.04$ and $\sigma V= 0.03$. Thus in this model, β cannot deviate from 2.7 by more than 0.2 without σV departing radically from its observed value. (For a detailed discussion of this question in a model which takes into account partial correlation of errors, see Balona 1983).

Stift: 8. Although first crossings can be neglected in the LMC the same is probably not true for the SMC (Stift, this symposium).

Feast: It will be interesting to see if first crossing candidates emerge when the BVI data for SMC Cepheids is fully analysed.

Stift: 9. Why use zero points when it has been demonstrated that chance selection effects are devastating? It should be remembered that the Cape P-L-C relation zero point is extrapolated to log P= 0.0 from a sample of only 26 Cepheids with periods confined between 0.9 and 1.7 in log P!

Feast: This question seems to refer to the discussion as to whether to fix a zero point at log P =0 or at some other value (say log P =0.8). It is clearly necessary, in comparing different cepheid samples to use P-L-C or P-L relations with identical period and colour coefficients (allowing suitably for abundance effects). In that case it is a matter of indifference at which period the zero point is given. Because of the considerable width of the P-L relation, serious statistical selection effects can arise unless care is taken. The P-L-C relation is not subject to this problem. **Stift:** No, see my poster!

Pel: I have only three points...... Firstly, we have now extended our SMC and LMC Cepheid samples, and also done more work on the calibration of the Walraven photometry. My present best numbers for the mean metallicity of Cloud Cepheids are Fe/H =-0.6 for the SMC and -0.2 for the LMC, which is very close to the values you used.

Secondly, I share your hesitation about the rather high reddenings that van Genderen obtained for long-period Cepheids in the Clouds. The Cepheids in our Fe/H programme were observed in a way which is not ideal for reddening determinations, but for these stars with shorter periods I estimate reddenings smaller than 0.10 (including the foreground).

Finally, I became worried also about the galactic zero-point calibration after seeing Schmidt's results for some of the open clusters with Cepheids. I have now analysed quite extensive Walraven photometry for NGC6087, the cluster containing S Nor. By fitting NGC6087 to the Pleiades photometry of van Leeuwen, and using also van Leeuwen's Pleiades distance, which is based on a fit to nearby stars with trigonometric parallaxes, one bypasses the Hyades entirely. The surprising result for NGC6087 is a distance which differs hardly from the one adopted in 1969 by Sandage and Tammann, whereas from the revised Hyades distance one would expect an increase of about 0.25 mag in distance modulus with respect to the old calibration.

Feast: Regarding your last point, this may be another indication that allowance needs to be made for a higher metallicity in the Hyades compared to clusters containing Cepheids. Some allowance for this is made in the Schmidt-Kaler calibration used by Caldwell.

McNamara: The m_1-(b-y) diagram of the LMC Cepheids indicates that the reddening is very small, 0.04 mag, probably due only to galactic foreground extinction.

Frogel: Is it possible to test for a dependence of metallicity of the Cepheids on period (i.e. mass), so that metal enrichment history can be investigated?

Pel: For individual Cepheids the present photometric determinations of metallicity are still very inaccurate, but for sufficiently large samples one can get a reasonably accurate mean Fe/H. One could do this for different period intervals of Cepheids, but this would provide little information on the chemical history of the Clouds, since all Cepheids are very young stars.

Dufour: Nebular abundance studies of LMC and SMC HII regions suggest depletions in the CNO group elements 2 or 3 times greater in eaCh cloud than you used for the Cepheids. How would such significantly larger depletions in reality affect your distance modulus results?

Feast: I don't think that this sort of problem has been investigated in any detail. However, the metallicity effects on the P-L-C relation come mainly through a change in the (B-V)-temperature relation (c.f. Gascoigne, M.N. 166, p 25p, 1974). Since this is primarily due to line blanketing changes, it is the abundances of elements such as iron etc., which are of most importance, I presume.

SPECTROSCOPY OF RED VARIABLES AND OTHER LUMINOUS RED STARS

M. S. Bessell
Mount Stromlo and Siding Spring Observatories
Research School of Physical Sciences
Australian National University

ABSTRACT

Long period variable stars represent the most advanced stage of asymptotic giant branch evolution prior to planetary nebula ejection or catastrophic core collapse. In the Magellanic Clouds through study of the LPVs it has been possible to identify for the first time, stars on the AGB with luminosities right up to the AGB limit ($M_{bol} \sim -7.1$) providing direct evidence that the more massive AGB stars produce supernovae. Because the stars have well defined periods, knowledge of the temperature and luminosity enables the mass to be derived. This review will highlight spectroscopic observations of the LPVs and discuss what information they provide on nucleosynthesis on the AGB in stars of different mass.

1. INTRODUCTION

The spectra of red giant and supergiant stars have always held a fascination because of their bizarre variety and complexity, being dominated by unusual molecular bands. It is believed that nucleosynthesis occurs in red giants during their late evolutionary stages (e.g., Iben and Truran, 1978), but providing observational data to delineate these last stages has proved difficult. This arises for several reasons, firstly the phenomenon is short lived, secondly the masses and luminosities of individual peculiar red giants in the field are practically impossible to determine accurately, and thirdly the atmospheric parameters of such stars are poorly determined. However, spectroscopic and photometric observations made of luminous red giants, (in particular the long period variables) in the Magellanic Clouds have provided a key to the unravelling of the late evolutionary stages of stars of different mass (Wood, Bessell and Fox, 1983: WBF). Many more observations of such stars over a wide wavelength range combined with detailed spectrum synthesis programs such as those of Wehrse (1981), Schmid-Burgk et al. (1981), Tsuji (1978) and Johnson et al. (1980) should permit a detailed understanding of these significant phases in the life of a star.

Long period variables are a special subset of the red giants and supergiants that cannot be considered in isolation from the non-variable or non-regularly variable stars. In this review we will first pay a tribute to the major programs which have identified red stars in the Magellanic Clouds and survey the follow-up spectroscopic or photometric observations. In the second section we will discuss the importance of infrared photometry in deriving atmospheric parameters independently of the spectral peculiarities, and the special place that long period variables (LPVs) have in the understanding of the last stages of asymptotic giant branch evolution. The observed luminosities and derived pulsation masses of a large number of LPVs will next be discussed in some detail. Finally, the spectra of these LPVs of known mass and luminosity will be compared with each other and with some galactic stars, and possible nucleosynthesis scenarios proposed.

2. OBSERVATIONAL OVERVIEW

The discovery of red giants in the Clouds has involved three approaches, objective prism surveys of the field, two color photographic photometry of selected fields and clusters, and long period variable star searches. The value of such time consuming search programs cannot be over-emphasized. Some of these programs were initiated many years ago but have only recently begun to pay spectacular dividends as large telescopes and more sensitive IR and red photometers and electronic imaging devices enable the detailed follow-up work to be done.

Westerlund's original 1N objective prism survey in the LMC (Westerlund, 1961) has yielded the brightest, red stars (M and N) (Westerlund et al. 1978) which have been further investigated by Crabtree et al., (1976), Richer et al., (1978), (1979), Richer and Frogel (1980) and Richer (1983). Sanduleak and Philip (1977) have also made a IIIaJ objective prism survey to discover the hotter C_2 stars.

Blanco et al. (1978; 1980) have conducted grism surveys of the LMC and SMC which have discovered large numbers of M stars and carbon stars, some of which have been subsequently investigated at CTIO by Cohen et al. (1981), Frogel and Blanco (1983), Richer (1983) and Frogel and Richer (1983). These grism surveys will continue to be a major source for red giant studies. Humphreys (1979) used both the Westerlund and the Blanco surveys to select stars for her investigations of the brightest M supergiants in the Clouds.

Many red giants have also been discovered by direct photographic photometry. Originally only B and V plates were used, but more recently V and I plates have proved very successful at identifying red stars. Gascoigne (1963) noted that many of the globular clusters in the Magellanic Clouds contain stars redder than any found in galactic globular clusters (e.g., van den Bergh, 1975) and Feast and Lloyd Evans (1973) showed that some of these were carbon stars. From V, I studies of SMC fields, Lloyd Evans (1978a,b) found many more such red stars and argued that these also were carbon stars. Red stars were similarly searched for and found in many Cloud globular clusters of so called "intermediate-age" by Lloyd Evans (1980a), Aaronson and Mould (1982) and Mould and Aaronson (1982). These discoveries were followed up with spectroscopic and IR photometric observations by Bessell, Wood and Lloyd Evans (1983) (BWL), Lloyd Evans (1980b,c; 1984), Mould and Aaronson (1979; 1980) and Aaronson and Mould (1982) in their investigations of the origin of carbon stars and extended giant branches. We will return to some of these results later.

Finally, the searches for long period variables. Many of these stars, in particular the most luminous ones, were found by Harvard workers blinking Bruce plates. Identification charts are given in Hodge and Wright (1967, 1977) and periods are given by Payne-Gaposhkin (1971), Payne-Gaposhkin and Gaposhkin (1966), and Wright and Hodge (1971). Periods and charts for additional stars in the SMC were published by Dessy (1959) and Lloyd Evans (1978a). More recently Moore 1983 (private communication), Paltoglou et al. (1984) and Lloyd Evans (1983b) have identified red variables using I plates taken over several years with the SRC Schmidt. These latter programs used automatic measuring machines and promise a steady yield of new red variables as more areas are searched. Infrared photometry of some of these LPVs has been published over the last few years by Glass (1979), Feast et al. (1980) Catchpole and Feast (1981), Glass and Feast (1982) and W.B.F.

3. INFRARED PHOTOMETRY AND LPVS

Although this review ostensibly concerns the spectra of red variables, the IR magnitudes and colors of the stars are of fundamental importance in permitting the bolometric magnitude and effective temperatures to be derived essentially independently of the spectral appearance. Much of the confusion surrounding the properties of the carbon stars and the S stars in the galaxy arises from the adoption of temperature composition and luminosity indicators which are not independent of atmospheric composition. BWL and WBF adopt (J-K)-temperature scales based on occultation measurements of non-variable M giants, Mira variables and carbon stars, (there is some disagreement in the literature as to the correctness of these adopted temperatures and it is hoped that model atmosphere colors can be computed to explore these color-temperature relationships) and use these temperatures for comparing theoretical predictions, such as evolutionary tracks and pulsation properties, with observed magnitude and colors.

The importance of long period variables in stellar evolution studies lies in the fact that these stars are passing through the most luminous phase of their lives, the penultimate short-lived phase on the asymptotic giant branch prior to disruption through envelope expulsion, and that nucleosynthesised elements reaching the surface of the star during this time will be expelled at disruption and enrich the interstellar medium. Study of long period variables should tell us a great deal about these catastrophic events and the processes of nucleosynthesis. But most importantly, the fact that the star has a period means that a pulsation mass can be derived, given knowledge of the luminosity, temperature and value of the pulsation constant Q.

Unlike the galactic field stars, stars in the Magellanic Clouds suffer little interstellar obscuration and are at known distances or at least relatively the same distances in each Cloud, thus the absolute magnitude can be simply derived from the apparent magnitude. As noted above, the (J-K) color can be calibrated in terms of temperature while the bolometric magnitude correction is a function of (J-K) and atmosphere composition (Frogel et al., 1980; Bessell and Wood, 1983b). Finally, the values of pulsation constant Q appropriate for LPVs have been investigated by Fox and Wood (1982) and are further discussed in WBF. Therefore we can derive masses for LPVs in the Magellanic Clouds on the basis of their periods and observed IR magnitudes.

4. THE MASSES, LUMINOSITIES AND SPECTRA OF MAGELLANIC CLOUD LPVS.

In Figure 1 is plotted the absolute bolometric magnitude versus period diagram for the red variables of known period in the Clouds. This data is taken from WBF. The stars seem to fall into two sequences separated by approximately 1 magnitude, and the line of separation appears to coincide with the theoretical luminosity limit for AGB stars. WBF identif the upper group as the young massive core burning supergiants and the lower group as the shell burning (asymptotic giant branch) AGB stars.

Lee (1970) states that Ia, Iab, Ib and II supergiant luminosity classifications correspond approximately to M_{bol} = -7.8, -7.0, -5.7 and -3.6 respectively for M supergiants; thus all the core burning supergiants in fig. 1 should be class Ia or Iab, and the brightest AGB stars should be class Iab or Ib. There are 10 core burning supergiants in fig. 1 which are common to the list of Humphreys (1979). The bolometric magnitudes from the IR colors are on the average 0.3 mag fainter than given by Humphreys, and from the 4 stars with luminosity types, it appears that Iab stars are fainter than -7.5, in good agreement with Lee. We must be aware when interpreting the spectra of stars, that stars classified spectroscopically as supergiants could be either shell burning AGB stars of intermediate mass or core burning supergiants of large mass. Humphreys observed many HV stars which are brighter than any in the WBF sample but this is not significant because WBF only observed stars with known periods and the brightest supergiants could have periods of over 1000 days and small amplitudes, which makes period determinations very difficult. An additional, complication in the interpretation of the spectra of LPVs is the effect the pulsation has on the atmosphere. One difference between LPVs and class III M giants is indicated by the occultation measurements which show that LPVs have a different (J-K) temperature relation than do non-variable stars. Another noted by Wing (1967), was that TiO band strength versus continuum-temperature relations differ between phases on the rising and falling parts of the light curve in some LPVs. Yet another difference is the existence of strong H_2O bands in

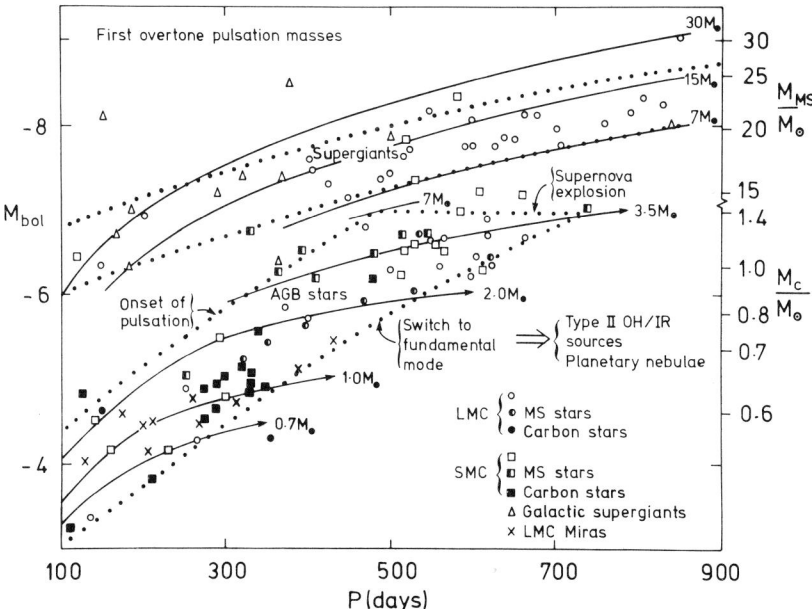

Figure 1. M_{bol} plotted against P for LPVs in the SMC and LMC. Some galactic supergiant variables are also shown. The dotted lines delineate the regions occupied by supergiants and AGB stars and the continuous lines are lines of constant mass assuming that the LPVs are first overtone pulsators. At each luminosity, the core masses (M_c) of AGB stars and the main-sequence masses (M_{MS}) of red supergiants are indicated.

the IR in Miras (LPVs) compared to supergiants (Hyland, 1974). Theoretical exploration of these differences is justified and it seems likely that the degree of "extension" of the atmosphere (Scholz and Wehrse, 1982) is the parameter.

Low dispersion spectra from $\lambda = 0.60\mu$ to 0.85μ have been obtained by WBF for almost all the LPVs of fig. 1. Most of the spectra were obtained with the IDS on the AAT and a sample are shown in fig. 2 together with some galactic stars for comparison. A few SMC stars and comparison stars were obtained with the CCD on the AAT in order to investigate the region of the infrared system bands of ZrO; the spectra cover the range 0.70μ or 0.85μ to 1μ and are shown in fig. 3. Spectral types on the MK system were assigned to all the spectra and details are given in WBF. Many of the stars were observed at different phases and large difference in spectral type were seen. The sample of core-burning supergiant LPVs had spectral types from M0 to M5.5; no unusual heavy elements were evident in their spectra and the CN bands were strong only at some phases for the bluest stars (see eg., HV2255 fig. 2). The spectra of the AGB supergiants

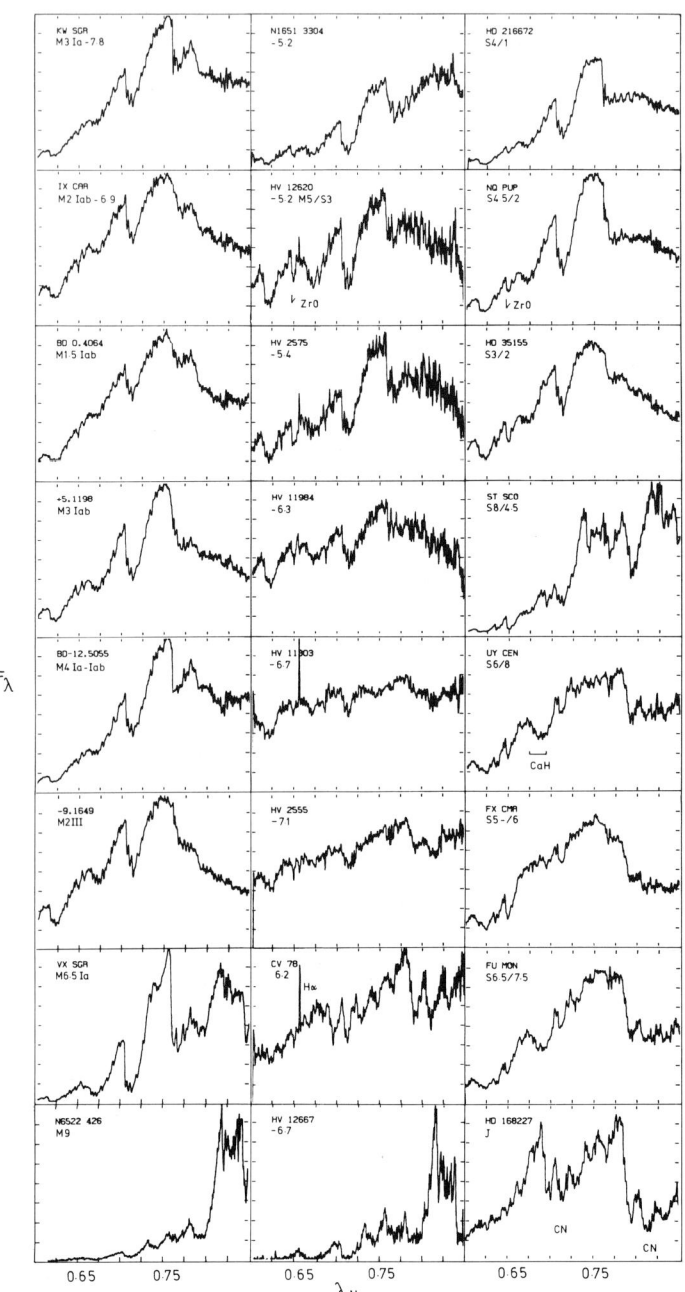

Figure 2. F_λ plotted against λ for a sample of LPVs in the Magellanic Clouds, and a variety of galactic comparison stars. Absolute bolometric magnitudes are given for the LPVs. Data were obtained with the IDS.

and giants were more interesting. No carbon stars were found brighter than M_{bol} = -6.2 (CV78), which is close to the known upper luminosity limit for cloud carbon stars (Cohen et al., 1981; and Richer et al., 1979) but more luminous O-rich stars were found, many of which showed strong ZrO bands near λ = 0.65μ. Comparison of these S stars with the galactic standards show clearly that none of them are pure S stars with C/O = 1; they all have very strong TiO. In searching for enhanced ZrO, the γ-system of ZrO bands around 0.65μ can be seen reasonably well up to spectral types as late as M5. Later than this, the infrared system bands can be used. Unfortunately there is some confusion due to possible H_2O absorption at these wavelengths, which occurs in LPVs at some phases. It is also not clear in late M stars how much TiO bands with heads at 0.9224μ and 0.9284 contribute to the 0.93μ feature, or how strong the IR ZrO bands are in "normal" M stars. Clearly more CCD spectra of standard M stars are needed. With regard to normality it is interesting that two of the random galactic supergiant stars observed and shown in fig. 2 (i.e., -9°1629 and +5°1198) have quite strong ZrO bands. These stars could be galactic counterparts of these intermediate age Cloud S stars. Many of the upper AGB stars had spectral types later than M7, and one star, shown in fig. 2, HD 12667 (M_{bol} = -6.7) had a spectral type of M9+. There was also another interesting carbon star HV 2379 which lay beyond the maximum period line indicated by the other stars. Bessell and Wood (1983a) have discussed the properties of this star in detail. It appears to have ejected a shell during November 1981.

What then do the spectra of these LPVs indicate about nucleosynthesis on the asymptotic giant branch? Let us first consider what investigations of non-variable red stars have shown. The intermediate-age cluster investigations of BWL and Aaronson and Mould and the field carbon star investigations of Richer et al. show that stars in the Magellanic Clouds with masses as low as ~1 $M_\odot$ will all become carbon stars, which is not the case in our galaxy. Using the results of dredge-up calculations, BWL show that there is a minimum stellar mass for which dredge-up and carbon star formation can occur once the luminosity exceeds a critical value and that this minimum mass depends on the envelope metallicity. The metallicity difference between the Clouds and the Galaxy is sufficient to restrict carbon star formation in the galaxy to more massive stars (M ≳ 1.4 $M_\odot$). In a most significant discovery BWL found that many M stars in these intermediate-age clusters with luminosities close to the transition (M→C) luminosity showed strong ZrO bands in their spectra indicating enhanced s-process elements. This was clear evidence that the S stars occur during asymptotic giant branch evolution and that the sequence is M→S→C. Lloyd Evans (1984) and Mould (private communication) have verified this phenomenon in many other clusters. Now theoretical stellar evolution calculations show that helium shell flashes on the AGB produce ^{12}C via the 3-α process and also s-process elements due to neutron capture on heavy nuclei. Furthermore, the carbon and s-process elements can be dredged up to the stellar surface during helium shell flashes (e.g., Iben 1975). The existence of the three spectral types M, S and C among the AGB stars is direct evidence for the above-mentioned nucleosynthetic processes. The carbon stars have dredged up sufficient carbon during shell flashes to produce C/O > 1 as required in a carbon star atmosphere. The S stars which lie on the AGB between the pure M and C stars, have begun dredging up ^{12}C and s-process elements so that the bands of ZrO are enhanced; however, insufficient ^{12}C has been dredged up to produce a ratio C/O > 1. Since the envelopes of AGB stars are lost to the interstellar medium via stellar winds, planetary nebulae ejection and supernova explosions, these stars make a major contribution to the enrichment of the interstellar medium in ^{12}C and s-process elements (Iben and Truran 1978; Renzini and Voli 1981).

However, the O-rich spectra of the luminous AGB LPVs, and the upper luminosity limit to the field carbon stars indicates that there is an upper mass limit, above which C stars are no longer produced. WBF discuss two possible reasons for this observation that C/O > 1 in the upper AGB stars.

1. the envelope mass is so large that the cumulative amount of ^{12}C dredged up by successive shell flashes on the AGB is not sufficient to produce a ratio C/O > 1 before the termination of AGB evolution, or

2. the ^{12}C dredged up is converted to ^{14}N during quiescent evolution between shell flashes (Iben 1975; Renzini and Voli 1981). The direct way of distinguishing between these two cases would be to measure ^{14}N abundances in the upper AGB stars.

There is some observational evidence which favors explanation (2). In the sample of carbon stars given by Cohen et al. (1981), the J stars (i.e., those carbon stars with a high ^{13}C/^{12}C ratio) all tend to lie near the upper limit ($M_{bol} \simeq -6$) of carbon star luminosity. One of the results of CNO cycling is an increase in the ^{13}C/^{12}C ratio from ~1/90 to ~1/4. Hence, we interpret the existence of J stars at $M_{bol} \simeq -6$ as evidence for some CNO cycling occurring in the envelopes of these stars during quiescent evolution. Theoretical calculations indicate that the amount of envelope CNO cycling increases with both mass and luminosity (Scalo, Despain and Ulrich 1975) so that it seems reasonable to assume that the AGB stars of type S that we have found and which have luminositie above the most luminous carbon stars have indeed converted their dredged up ^{12}C to ^{14}N. From Figure 1 we would estimate that the critical initial mass above which nitrogen rich stars rather than carbon stars are formed is ~3.5 $M_\odot$ (in the Magellanic Clouds). Another piece of observational evidence that appears to favor the existence of nitrogen rich AGB stars is the existence of nitrogen enhancements in planetary nebulae of Type 1 and in some supernova remnants.

The nitrogen produced in these massive AGB stars is primary in origin as it is synthesized from the hydrogen and helium in the star at its birth. Thus the AGB stars may be the source of the primary nitrogen that some observations require (Alloin et al. 1979; Bessell and Norris, 1982; Tomkin and Lambert, 1983). Furthermore Tomkin and Lambert find that [N/C] $\simeq$ 0 and [N/Fe] $\simeq$ 0 is the rule for the disk dwarfs and the majority of halo dwarfs, suggesting that these three elements C, N and Fe are formed in the same stars or in stars of similar evolution i.e., the carbon core asymptotic giant branch stars.

We believe that spectroscopic observations and IR photometric observations of LPVs in the Magellanic Clouds can be used to follow the entire story of AGB evolution, but it is essential that quantitative analysis of the data be pursued and that a larger sample of stars be observed.

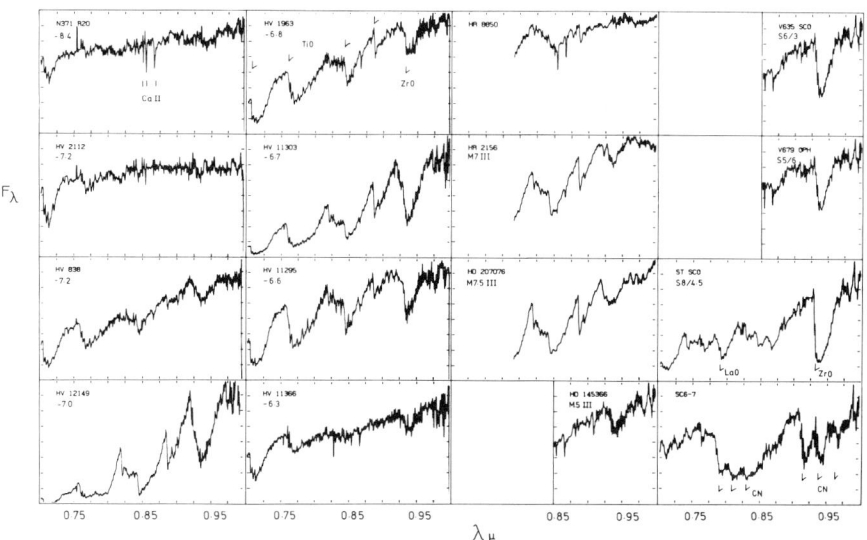

Figure 3. F_λ plotted against λ for selected LPVs in the SMC and some galactic comparison stars. Spectra were obtained with the CCD in order to show the region of the infrared system bands of ZrO near 0.93μ. Absolute bolometric magnitudes are given for the LPVs.

REFERENCES

Aaronson M. and Mould J.R., 1982, Astrophys. J. Suppl., 48, p161.
Alloin, D., Colin-Souffrin S., Joly M. and Vegroux L.: 1979, Astron. Astrophys., 78, p200.
Bessell M.S. and Norris J.E.: 1982, Astrophys. J. (Letters), 263, L101.
Bessell M.S. and Wood P.R.: 1983a, Mon. Not. Roy. Astr. Soc., 202, p316.
Bessell M.S. and Wood P.R.: 1983b, Publ. Asp. submitted.
Bessell M.S., Wood P.R., Lloyd Evans T.: 1983, Mon. Not. Roy. Astr. Soc., 202, p59.
Blanco B.M., Blanco V.M. and McCarthy M.F.: 1978, Nature 271, p638.
Blanco B.M., McCarthy M.F. and Blanco.: 1980, Astrophys. J., 242, p938.
Cohen J.G., Frogel J.A., Persson S.E., Elias J.H.: 1981, Astrophys. J., 249, p481.

Crabtree D.R., Richer H.B., Westerlund B.E.: 1976, Astrophys. J., 203, L81.
Dessy J.L.: 1959, Bol. Inst. Mat. Astr. Fis., Cordoba 1, no 2.
Feast M.W. and Lloyd Evans T.: 1973, Mon. Not. Roy. Astr. Soc., 193, p377.
Feast M.W. and Lloyd Evans T.: 1973, Mon. Not. Roy. Astr. Soc., 164, p15.
Fox M.W. and Wood P.R.: 1982, Astrophys. J., 259, p158.
Frogel J.A. and Blanco V.M.: 1983, preprint.
Frogel J.A., Persson S.E. and Cohen J.G.: 1980, Astrophys. J., 239, p495.
Frogel J.A. and Richer H.B.: 1983, preprint.
Glass I.S. and Feast M.W.: 1979, Mon. Not. Roy. Astr. Soc., 186, p317.
Glass I.S. and Feast M.W.: 1982, Mon. Not. Roy. Astr. Soc., 199, p245.
Hodge P.W. and Wright F.W.: 1967, "The Large Magellanic Cloud" (Smithsonian Press: Washington).
Hodge P.W. and Wright F.W.: 1977, "The Small Magellanic Cloud" (Univ. of Washington Press: Seattle).
Humphreys R.: 1979, Astrophys. J. Suppl., 39, p389.
Hyland A.R.: 1974, Galactic Radio Astronomy (ed Kim and Simonson) p439.
Iben I.: 1975, Astrophys. J., 196, p525.
Iben I. and Truran J.W.: 1978, Astrophys. J., 220, p980.
Johnson H.R., Bernat A.P. and Krupp B.M.: 1980, Astrophys. J. Suppl., 42, p501.
Lee T.A.: 1970, Astrophys. J. 162, p217.
Lloyd Evans T.: 1978a, Mon. Not. Roy. Astr. Soc., 183, p305.
Lloyd Evans T.: 1978b, Mon. Not. Roy. Astr. Soc., 183, p319.
Lloyd Evans T.: 1980a, Mon. Not. Roy. Astr. Soc., 193, p87.
Lloyd Evans T.: 1980b, Mon. Not. Roy. Astr. Soc., 193, p97.
Lloyd Evans T.: 1980c, Mon. Not. Roy. Astr. Soc., 193, p333.
Lloyd Evans T.: 1983a, preprint.
Lloyd Evans T.: 1984, this volume, p. 217.
Mould J.R. and Aaronson M.: 1979, Astrophys. J., 232, p421.
Mould J.R. and Aaronson M.: 1980, Astrophys. J., 240, p464.
Mould J.R. and Aaronson M.: 1982, Astrophys. J., 263, p629.
Paltoglou G., Wood P.R., Bessell M.S., Ratnatunga K.,: 1984, this volume, p. 219.
Payne-Gaposhkin C.H.: 1971, Smithsonian Contributions to Astrophysics, 131, p1.
Payne-Gaposhkin C.H. and Gaposhkin S.: 1966, Smithsonian Contributions to Astrophysics, 9, p1.
Renzini A. and Voli M.: 1981, Astrophys. J., 94, p175.
Richer H.B.: 1983, preprint.
Richer H.B. and Frogel J.A.: 1981, Astrophys. J., 249, p481.
Richer H.B., Olander N. and Westerlund B.E.: 1979, Astrophys. J., 230, p724.
Sandaleak N. and Philip A.G. Davies,: 1977, Pub. Warner Swasey Obs., Vol 2, no 5, p105.
Scalo J.M. and Despain K.H. and Ulrich R.K.: 1975, Astrophys. J., 196, p805.
Schmid-Burgk J., Scholz M. and Wehrse R.: 1981, Mon. Not. Roy. Astr. Soc., 194, p383.

Scholz M. and Wehrse R.: 1982, Mon. Not. Roy. Astr. Soc., 200, p41.
Tomkin J. and Lambert D.L.: 1983, preprint.
Tsuji T.: 1978, Astr. Ap., 62, p29.
van den Bergh S.: 1975, Ann. Rev. Astr. Ap. 13, p217.
Westerlund B.E.: 1964, IAU Sym. 20 The Galaxy and Magellanic Clouds, ed Rodgers.
Westerlund B.E., Olander N., Richer H.B. and Crabtree D.R.: 1978, Astronomy and Astrohpysics Suppl., 31, p61.
Wing R.F., Dissertation, University of California, Berkeley.
Wood P.R., Bessell M.S. and Fox M.W.: 1983, Astrophys. J. submitted.
Wood P.R., Bessell M.S.: 1983, in preparation.
Wehrse R.: 1981, Mon. Not. Roy. Astr. Soc., 195, 553.
Wright F.W., Hodge P.W.: 1971, Astr. J. 76, p1003.

DISCUSSION

Mould: What lower limit on the supernova rate in the LMC, from this source alone, can you put from the counts of stars at the AGB limit?
Wood: The number of pulsating supernova precursors in Wood, Bessell and Fox indicates one type I supernova in 50.000 years. This is a very minimum supernova rate as the search of LPVs is very incomplete, and periods are known for only a small fraction of them.
Shull: Some SNRs are rich in N compared to their ambient ISM. How much N is produced, expressed as a percentage of C mass or stellar mass?
Bessell: We have not been able to measure N abundance in these AGB stars, but one might expect that the amount of N produced would be from one to several times the initial envelope carbon abundance.
Frogel: Two remarks on temperature scales:
a) my comments don't affect your fundamental conclusions;
b) my comments don't imply your scale is incorrect, just that there is an alternative interpretation. For C stars I think J-K is strongly affected by CN blanketing. Cool M stars, LPV's included, have strong water absorptionbands which will affect JHK colors. Also hot dust (800 to 1200 K) will affect JHKL colors. This latter effect is probably not important for hotter M's in LMC.
Bessell: We chose the $(J-K) - T_e$ relations found for M stars, Miras and carbon stars based on lunar occultation measurements. Admittedly these are few in number but we cannot ignore them. With regard to the blanketing possible in the JHK bands of these late stars I don't think that it seriously affects the temperatures of most of our stars, but it is a very important problem that we should attack both observationally and via model atmospheres.
Pel: What is known about the presence of OH/IR stars in the Clouds?
Bessell: There have been no serious OH surveys of the Clouds because it is anticipated that they would be too faint to observe. In view of the very luminous AGB stars that we have seen, an OH survey of the LMC Bar would be a very worthwhile exercise.
Gardner: No detections have been reported to date. With present sensitivity detections would not be expected.

DETECTION AND PHOTOMETRY OF RED GIANTS IN THE MAGELLANIC CLOUDS

Marc Aaronson
Steward Observatory, University of Arizona

ABSTRACT

This review will focus on photometric observations of evolved stars in the Magellanic Clouds. Emphasis is placed on red and near-infrared measurements, as these allow reasonable estimates to be made of bolometric magnitude and temperature for reliable placement in the physical HR diagram. The review is divided into three parts: field stars, cluster stars, and red variables; a summary of the surveys for objects in each of these areas is also given. Particular attention is drawn to the intermediate-age stellar population, as this component appears to be the primary tracer of the star formation rate and chemical enrichment history of the Clouds.

1. INTRODUCTION

Studies of the red stellar content in the Magellanic Clouds over the last ten years have led to new insight concerning both the star formation history in the Clouds and the evolutionary nature of the red giants themselves. A special incentive on both of these fronts has been provided by the discovery and subsequent photometry of large numbers of carbon stars in both the field and clusters of the Clouds. These stars appear to be signposts for a metal-poor, intermediate-age population which may in fact be the dominant stellar component in both the Large and Small Cloud.

The review that follows is divided into three sections on field stars, cluster stars, and red variables. Each part begins with a brief summary of the various survey results, followed by a discussion of the key photometric observations. For the latter, measurements at near-infrared wavelengths are emphasized. Major advances in instrumentation have occurred since the last symposium on the Clouds, and photometry to the few percent level at JHK wavelengths is now routine on large telescopes for evolved Cloud stars near or brighter than the first giant branch tip. Such photometry permits accurate and straightforward estimation of bolometric magnitude, a quantity providing the critial link with stellar evolution theory. In this regard, working in the Clouds offers the well-known advantage of eliminating the (relative) distance uncertainties among objects so that, for instance, reliable luminosity functions can be determined. The IR also provides a better temperature estimator than can be obtained from optical photometry alone, although as mentioned further below the precise correspondence with effective temperature still remains a somewhat thorny issue.

Finally, it should be noted that the discussion here is confined largely to observational aspects of the red giants. The reader is referred to the reviews elsewhere in this volume by Mould (1983) for an extended treatment of theoretical comparisons, and by Humphreys (1984) for a discussion of the red supergiants.

2. FIELD RED GIANTS

In order to study the red giants we must first find them. With regard to the field, while two-color photometry has been employed to isolate red stars (e.g. Walker et al. 1969, Tifft and Snell 1971, Mendoza and Gomez 1973), it has been of limited use for detailed study of the stars themselves. Of more value have been objective prism surveys carried out with Schmidt-type telescopes and "grism" surveys with 4-m class telescopes, the latter reaching considerably fainter than the former, but at the cost of wide area coverage.

The most important objective prism surveys for the LMC have been those of Sanduleak and Philip (1977), who report ~ 600 M supergiants and ~ 470 carbon stars; Westerlund et al. (1978), who report ~ 300 C stars (most of which do not overlap with the Sanduleak and Philip sample, possibly owing to the use of red CN bands rather than C_2 bands as the discriminating feature); and Westerlund et al. (1981), who report ~ 500 M supergiants and ~ 600 M giants. These surveys cover nearly the entire LMC area, but apparently sample only the more luminous carbon and late-M giants. An additional list of ~ 1000 red stars in selected LMC regions obtained from ultra-low dispersion spectra is given by Bappu et al. (1978 and references therein). For the SMC, objective prism plates have been primarily used to identify red supergiants, of which ~ 500 are listed by Sanduleak (1975 and references therein) and Azzopardi and Vineau (1975). More recent objective prism surveys of Rebeirot et al. (1983) for the LMC and Prévot et al. (1983) for the SMC have led to detection of ~ 200 additional red supergiants in each Cloud.

Field grism surveys have been conducted by Blanco et al. (1978 and references therein), Blanco et al. (1980), and Richer and Westerlund (1983). A new Cloud grism survey is also underway by Westerlund (1983) and collaborators. Although some warmer-temperature, weak-banded carbon stars may be missed, these efforts nevertheless appear to reach well below the peak in the carbon star and late-M giant luminosity functions, providing good samples for statistical investigation. The most ambitious effort to date and the one whose results we shall briefly summarize is that of Blanco and McCarthy (1983). These authors have obtained grism plates of 37 SMC and 52 LMC selected areas, reaching from the center to the periphery of these systems. In the SMC, they identify some ~ 860 C stars, ~ 60 stars of type M6 or later, ~ 130 M5 stars, and ~ 1060 M2-M4 stars; while in the LMC, the corresponding numbers are ~ 1040, ~ 480, ~ 820, and ~ 4440.

Blanco and McCarthy (1983) use their remarkable sample to examine the surface brightness distribution of the carbon stars, which they find correlates reasonably well with the red isophotes of de Vaucouleurs (1957). They estimate the total number of C stars to be ~ 11,000 and ~ 2,900 in the LMC and SMC, respectively. These numbers are quite interesting because their ratio is comparable to that of the Clouds' integrated luminosities, suggesting a basic similarity in star formation history. For example, the work of Butcher (1977) and Stryker and Butcher (1981) implies that a major and possibly dominant star formation event took place in the LMC ~ 2 - 4 Gyr ago. Perhaps the similarity in the carbon star number (and luminosity function as discussed below) between the LMC and SMC is reflective of a similar event having occurred in the Small Cloud.

Blanco and McCarthy (1983) also find the ratio of C to M6 and latter types to be ~ 2 for the LMC and ~ 14 for the SMC. These ratios become ~ 0.2 and ~ 0.6 when stars as early as type M2 are considered. The difference in C/M ratio is in accord with what is known about the relative metallicities of the Clouds. In particular, a decrease in abundance is expected to (a) diminish the number of late M stars via decreasing TiO absorption and warming of the giant branch; and (b) increase the number of C stars through the need to dredge up less carbon (e.g. Iben 1982) and the possible increase in efficiency of the dredge-up process itself (Wood 1981). Given that the relative numbers of Large and Small Cloud C stars are the same, the former mechanism would seem to be more important in explaining the change in C/M ratio, although this conclusion rests on the plausible yet unproven assumption that the mean carbon star ages in the Small and Large Cloud are similar. A final result of Blanco and McCarthy worth mentioning is the radial dependence of the C/M6+ ratio, which is constant over the face of the LMC, but apparently changes from ~ 19 to ~ 5 in going from the center of the edge of the SMC. While small number statistics may account for part of the latter effect, it nevertheless appears to be real. The explanation is unclear, but a naive interpretation would be that metallicity of the intermediate-age population increases from the center outwards, contrary to expectations.

The general completeness of the Blanco and McCarthy survey has been verified by Frogel and Richer (1983), who scanned an area of the LMC Bar West field at 2.2 and 3.5 μm in an unsuccessful search for "missing" AGB stars. Elias (1982) has conducted a similar mapping over selected Cloud areas at 1.6 and 10 μm, and the results of this work are awaited with interest, as are those from the orbiting telescope IRAS. One final comment on the grism effort is that, except for a few notable exceptions (Blanco et al. 1981; Richer and Frogel 1980), it has not lead to the discovery of many S stars. This appears to reflect the difficulty of distinguishing early S from M stars at low dispersion, rather than a real absence of the former, as shown by the recent spectroscopic work of Lloyd-Evans (1983a).

Extensive VRI photometry for field carbon stars in the Clouds has been published by Richer et al. (1979) and Richer (1981; see also Crabtree et al. 1976). The major JHK study has been that of Cohen et al. (1981), who use their data to convert the Blanco et al. (1980) I luminosity functions to true bolometric ones. The results are virtually identical in the LMC and SMC: both luminosity functions show a roughly Gaussian shape with a peak at $M_{bol} \sim -5$ mag, and a range from ~ -4 to ~ -6 mag. The two-fold discrepancy of these results with theoretical expectations (e.g. the absence of high-luminosity C stars with $M_{bol} < -6$, and the presence of so many low luminosity ones with $M_{bol} > -5$) has been discussed extensively in the literature.

The similarity in the LMC and SMC carbon star luminosity functions is perhaps surprising in view of the Clouds' known present day abundance differences. All else being equal, dredge-up theory would predict a lower mean luminosity for SMC carbon stars relative to LMC carbon stars (e.g. Iben 1982). The absence of this result suggests three possibilities: (a) the ease of conversion to a C star may bottom out at low metallicities when only one thermal pulse is required for the conversion to occur; (b) errors in relative distance moduli and/or differences in star formation history have conspired to mask the dredge-up effect and produce identical luminosity functions; or (c) the progenitor stars had similar metallicities. Evidence against the last suggestion comes from Cohen et al.'s (1981) study of the JHK two-color diagram. These authors show that the loci of carbon stars are displaced such that J-H is bluer at fixed H-K as one progresses from the Milky Way to the LMC to the SMC. Cohen et al. cite blanketing changes in accord with present day metal differences as explanation for the effect. If so, it would seem that estimation of carbon star effective temperatures from J-K colors is severely hampered for the Clouds, although Aaronson and Mould (1982) have shown that such color temperatures for galactic C stars do provide a reliable match with the Ridgway et al. (1980) occultation results.

The field M giants have not yet received quite the attention of the carbon stars. Recently, however, Frogel and Blanco (1983a) obtained IR photometry for a complete sample of known M giants in the LMC Bar West field. These authors find evidence for a bi-modal distribution in the (K, J-K) color-magnitude diagram which they suggest indicates the presence of two separate star formation bursts: the conventional 3 - 5 Gyr old event and a more recent 10^8 year old one. Additional supporting statistics for this conclusion via study of other Cloud fields would be welcome.

3. CLUSTER RED GIANTS

The populous clusters provide the key for understanding of the chemical enrichment history in Clouds. The clusters also are the ultimate testing ground for theories of advanced stellar evolution, as both luminosities and ages of the red giants can (in principal) be directly determined. In this regard two questions are particularly relevant today. The first concerns the luminosity of the transition from M to C type and its dependence on age and metallicity. The second concerns the dependence of the maximum extent of the asymptotic giant branch (AGB) on these same quantities. This latter question is especially important for assessing the mechanism responsible for the so-called "missing" luminous AGB field stars (cf. Reid and Mould 1983), the primary choices here being variation in the star formation rate or dependence of stellar mass loss rate on age and abundance.

That the Cloud clusters contain very red stars has been known since the work of Arp and others in the late fifties (e.g. Arp 1958). However, it was sometime before the first of these were identified as carbon stars by Feast and Lloyd-Evans (1973); and only within the last few years have general surveys of cluster red stellar content been undertaken. Two-color photographic surveys at V and I have been published now by Lloyd-Evans (1980), Aaronson and Mould (1982), and Mould and Aaronson (1982). These efforts have largely concentrated on the red Cloud clusters, i.e., these having integrated B-V color $\gtrsim 0.3$ mag. A new grism survey by Frogel and Blanco (1983b) is extending these results to bluer Cloud clusters.

The grism and two-color techniques are somewhat complementary for cluster investigations. In rich clusters the grism method is limited by crowding, an example being Blanco and Richer's (1979) work on NGC 419, which identified only 6 of the 15 or so carbon stars in this cluster found by Aaronson and Mould (1982). On the other hand, while the two-color method enables a complete exploration of the upper AGB, additional photometry or spectroscopy is required for stellar classification. Such observations have now been obtained for red stars in about a dozen SMC clusters and ~ 35 LMC clusters, where ~ 70 carbon stars, ~ 20 M stars, and ~ 20 S stars have at present been located (Feast and Lloyd-Evans 1973; Mould and Aaronson 1979, 1980, 1982; Aaronson and Mould 1982, Lloyd-Evans 1980, 1983a,b; Bessell et al. 1981, 1983; Blanco and Richer 1979; Danks 1982). IR photometry has been secured for most of these stars and can be found in the cited references and additionally in Frogel et al. (1980) and Frogel and Cohen (1982).

A major photometric result to come out of the above efforts has been the demonstration that most of the red Cloud globulars have extended giant branches reaching well above the first giant branch tips of galactic globulars. This can be taken as prima facie evidence that the majority of the red clusters are of intermediate age. A similar conclusion has been reached by Rabin (1982) from spectrophotometric data, and by Searle et al. (1980) from integrated cluster photometry on the Gunn four-color system. A question of current interest is just how many of the Cloud clusters are really as old as galactic globulars? In particular, five out of six Searle et al. (1980) SWB type VII clusters appear to have extended giant branches, including such famous "old" objects as NGC 121, 1841, and 2257 (Mould and Aaronson 1982; Frogel and Cohen 1982). Deep color-magnitude diagrams and accurate metallicities will ultimately settle this issue, but in the meantime, it would be nice to have spectroscopic confirmation of membership for the one or two key stars in the older clusters.

Carbon stars populate the upper AGB and define the tip luminosity in most of the intermediate-age globulars. Their luminosity function appears similar to that of the field C stars, with a range in M_{bol} from ~ -4.2 to ~ -5.6. Except for the peculiar variable V8 in NGC 121, C stars have not been located in SWB type VII or types I-III clusters, being apparently confined between types III-IV to types VI. This conclusion for the early SWB types rests at present primarily on the integrated infrared colors (see below) of Persson et al. (1983), and detailed exploration of the red stellar content in these younger clusters is awaited.

The JHK two-color diagram has proven to be an effective way of selecting carbon from non-carbon stars (cf. Figures 2 and 3 in Aaronson and Mould 1982), owing to the generally much redder colors of the former. There is, however, a gray area in which the photometric colors and luminosities overlap. It is precisely this region where Bessell et al. (1983) and Lloyd-Evans (1983a,b) have located a number of S stars, dramatically confirming the association of these objects with the transition region between M and C types. Lloyd-Evans (1983a,b) and Frogel and Blanco (1983b) have explored this transition zone in detail. Their principle conclusions can be summarized as follows: (a) Most of the non-carbon cluster stars above an M_{bol} of -4.3 are S stars. (b) The ratio of S to C stars is greater in the younger groups. (c) The C stars are generally not found less luminous than ~ 0.2 mag from the brightest non-carbon stars. (d) The transition from M→S and S→C appears brighter in the younger groups, ranging from $M_{bol} \sim -4.5$ mag at SWB type VI to $M_{bol} \sim -5$ mag at SWB type IV. These results are in qualitative agreement with dredge-up theory; in particular, younger and more massive stars are expected to have to undergo more thermal pulses and thus climb higher on the AGB before turning themselves into carbon stars. Whether the results are in detailed quantitative agreement with theory must await accurate ages and metallicities for the clusters and fuller explorations of the transition region.

The tip luminosity has been used by Mould and Aaronson (1982 and references therein) as a cluster age estimator. For sparse clusters only an upper limit can be obtained, but for rich clusters with well-populated AGB's a specific age is deducible. Figure 1 illustrates the correspondence between SWB type and observed AGB extent for the sample from Mould and Aaronson. A rough correlation is present, but there is considerable spread at a given SWB type, with an apparent turnover between types IV and VI. The appearance of Figure 1 can be understod in part because (a) $M_{bol,max}$ provides only an upper age limit; (b) several lower luminosity clusters populate the lower border of the type V and VI lines, and stochastic effects may result in the complete absence of upper AGB stars in them; and (c) the SWB types themselves may have a wide spread in age at fixed type. Primarily because of the last reason, the results in Figure 1 cannot be used to meaningfully gauge $M_{bol,max}$ as an age estimator, as some authors have attempted. The true test will come when more main sequence turn-off ages become known. The presently available turn-off ages appear to correspond well to $M_{bol,max}$ for clusters older than 1 Gyr, but poorly for clusters younger than this (Mould 1984); the latter result if upheld may ultimately require modification of the adopted mass-loss rates. In any event, the addition of earlier SWB types to Figure 1 will clearly be of importance.

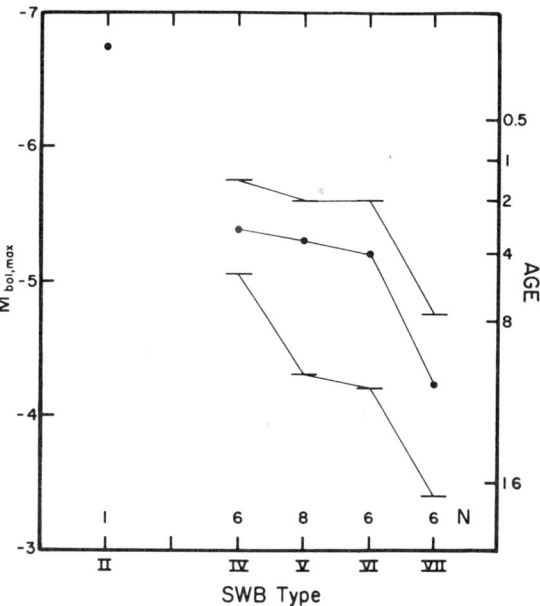

Figure 1. Maximum cluster AGB extent and corresponding age calibration from Mould and Aaronson (1982) are plotted against SWB type from Searle et al. (1980). The points are median values, while the brackets define the extremes in the cluster sample. SWB types III-IV and V-VI have been absorbed into types IV and VI, respectively.

The cluster carbon stars in the SMC appear to have in the mean bluer JHK colors than these in the LMC, similar to the above-mentioned results found by Cohen et al. (1981) for the field C stars. This may again reflect a lower mean abundance in the SMC clusters as compared with those in the LMC. Aaronson and Mould (1982) have also noted a peculiarity of the non-carbon cluster stars involving their tendency to deviate substantially from the (J-K, V-K) two-color relation characterizing galactic field and cluster giants. The explanation for this effect remains unclear, but its presence obviously complicates the derivation of believable effective temperatures.

In closing this section, we mention the dramatic influence that the upper AGB and particularly carbon stars have on the integrated infrared colors of the red Cloud clusters (Persson et al. 1983). This effect may be of use for locating intermediate-age clusters in other nearby galaxies, and may also be of importance for understanding the IR colors of high-z systems. However, it should be noted that Blanco and McCarthy (1983) have shown the total field bolometric contribution from carbon stars to be only ~ 3% for the LMC and ~ 4% for the SMC. Because of their extreme redness, the C stars have much less effect on integrated cluster colors either in the optical (e.g. van den Berg 1981 and references therein) or UV (de Boer 1981).

4. RED VARIABLES

Luminous long period variables in the Clouds are rare, only 24 having been identified by Payne-Gaposchkin and Gaposchkin (1966) in the SMC, and 46 having been identified by Payne-Gaposchkin (1971) in the LMC, although these authors have located a number of red irregular and semi-regular variables as well. The frequency of lower luminosity red variables is less clear, as the only published results remain those of Lloyd-Evans (1978a,b) for selected regions near the SMC clusters NGC 371 and 419. The data base for investigation of these objects is thus sparse, and additional searches for them such as those of Lloyd-Evans (1984), Paltoglou et al. (1984), and Reid (1983) are most welcome. Earlier reviews of the variables are given by Lloyd-Evans (1975), Feast (1973), and references therein.

Infrared photometry of the red Cloud variables can be particularly valauable for studying their pulsational properties and determining pulsational masses, and possibly as well for application to the extragalactic distance scale, since such stars are measurable out to quite large moduli. Until recently, work in these areas has largely been carried out by the South African Group (Catchpole and Feast 1981, Feast et al. 1980, Glass 1979, Glass and Feast 1982, and Glass and Lloyd-Evans 1981). Two principal results that have come from these efforts are (a) the period-luminosity relation for supergiants has been shown to be consistent with the theoretical relation for massive stars undergoing core helium burning; and (b) the Mira variables are favored as being first overtone pulsators. Unfortunately, this latter conclusion rests with the still considerably uncertain effective temperature calibration. (Photometry of individual variables of interest has also been presented by Elias et al. 1980, Bessell and Wood 1983, and Frogel 1983.)

Recently, studies by Wood et al. (1981, 1983) appear to have somewhat clarified the situation with regard to the red Cloud variables. Figure 6 in the last reference shows a clear segregation of the variables into two groups; these are identified as luminous small-amplitude ($\Delta K \sim 0.25$ mag) supergiant core helium burners, and lower-luminosity large amplitude ($\Delta K \sim 0.5 - 1$ mag) AGB stars. The latter extend up to the AGB limit at $M_{bol} \sim -7$. S stars are found up to this limit and indicate that dredge-up is occurring at these luminosities, but the absence of carbon variables above $M_{bol} \sim -6$ lead Wood et al. to conclude that dredged-up carbon is being eliminated through CN cycling.

It is interesting that in their Figure 6 SMC variables are seen with the same P-L properties as the LMC Miras (which in turn appear similar to the 47 Tuc variables). Hence, although SMC Miras have not been specifically identified as such, stars which may correspond to them do appear to exist.

The results of Wood et al. (1983) do not on the face of it bode well for distance scale measurements to other galaxies. In particular, the supergiant and AGB tracks are well separated by ~ 1 mag at the long period end and ~ 2 mag at the short period end. P-L relations alone cannot disentangle the two tracks. P-L-A relations involving amplitude may help, though, and need to be explored. Even so, the spread in magnitude at fixed period on the AGB track is quite large at ~ 1 mag. It does not seem that two P-L relations would be sufficient to characterize this spread, as has been suggested by Feast (1981), but further investigation of this question via P-L-A and P-L-C relations is clearly required.

Preparation of this article was partially funded by NSF grant AST 81-17365. The author would also like to thank the organizers of IAU Symposium No. 108 for travel support.

REFERENCES

Aaronson, M., and Mould, J.: 1982, Ap. J. Suppl., 48, 161.
Arp, H. C.: 1958, A. J., 63, 273.
Azzopardi, M., and Vineau, J.: 1975, Astr. Ap. Suppl., 22, 285.
Bappu, M. K. V., Parthasarathy, M., and Searia, K. K.: 1978, Kodaikanal Obs. Bull., Ser. A., 2, 184.
Bessell, M. S., and Wood, P. R.: 1983, M.N.R.A.S., 202, 31p.
Bessell, M. S., Wood, P. R., and Lloyd-Evans, T.: 1981, Proc. A.S.A., 4, 201.
_____ : 1983, M.N.R.A.S., 202, 59.
Blanco, B. M., Blanco, V. M., and McCarthy, M. F.: 1978, NATURE, 271, 638.
Blanco, V. M., Frogel, J. A., and McCarthy, M. F.: 1981, Pub. A.S.P., 93, 532.
Blanco, V. M., and McCarthy, M. F.: 1983, preprint.
Blanco, V. M., McCarthy, M. F., and Blanco, B. M.: 1980, Ap. J., 242, 938.
Blanco, V. M., and Richer, H. B.: 1979, P.A.S.P., 91, 659.
Butcher, H.: 1977, Ap. J., 216, 372.
Catchpole, R. M., and Feast, M. W.: 1981, M.N.R.A.S., 197, 385.
Cohen, J. G., Frogel, J. A., Persson, S. E., and Elias, J. H.: 1981, Ap. J., 249, 481.
Crabtree, P. R., Richer, H. B., and Westerlund, B. E.: 1976, Ap. J. Letters), 203, L81.
Danks, A. C.: 1982, Astr. Ap., 106, 4.
de Boer, K. S.: 1981, in IAU Coll. No. 68, "Astrophysical Parameters for Globular Clusters", eds. A. G. D. Philip and D. S. Hayes (Schenectady: L. Davis Press), p. 3.

de Vaucouleurs, G.: 1957, A. J., 62, 69.
Elias, J. H.: 1982, private communication.
Elias, J. H., Frogel, J. A., and Humphreys, R. M.: 1980, Ap. J. (Letters), 242, L13.
Feast, M. W.: 1973, in IAU Symp. No. 59, "Stellar Instability and Evolution", eds. P. Ledoux, A. Noels, and A. W. Rodgers (Dordrecht: Reidel), p. 93.
_____: 1981, in "The Most Massive Stars", eds. S. D'odorico, D. Baade, and K. Kjar (Garching: ESO), p. 217.
Feast, M. W., Catchpole, R. M., Carter, B. S., and Roberts, G.: 1980, M.N.R.A.S., 193, 377.
Feast, M. W., and Lloyd-Evans, T.: 1973, M.N.R.A.S., 164, 15p.
Frogel, J. A.: 1983, Ap. J., 272, 116.
Frogel, J. A., and Blanco, V. W.: 1983a, preprint.
_____: 1983b, presented at IAU Symp. No. 105, "Observational Tests of Stellar Evolution Theory".
Frogel, J. A., and Cohen, J. G.: 1982, Ap. J., 253, 580.
Frogel, J. A., Persson, S. E., and Cohen, J. G.: 1980, Ap. J., 239, 495.
Frogel, J. A., and Richer, H. B.: 1983, Ap. J., in press.
Glass, I. S.: 1979, M.N.R.A.S., 186, 317.
Glass, I. S., and Feast, M. W.: 1982, M.N.R.A.S., 199, 245.
Glass, I. S., and Lloyd-Evans, T.: 1981, NATURE, 291, 303.
Humphreys, R. M.: 1984, in IAU Symp. No. 108, "Structure and Evolution of the Magellanic Clouds", this volume, p. 145.
Iben, I.: 1982, Ap. J., 260, 821.
Lloyd-Evans, T.: 1975, in "Variable Stars and Stellar Evolution", eds. V. Sherwood and L. D. Plaut (Dordrecht: Reidel), p. 531.
_____: 1978a, M.N.R.A.S., 183, 305.
_____: 1978b, M.N.R.A.S., 183, 319.
_____: 1980, M.N.R.A.S., 193, 87.
_____: 1983a, M.N.R.A.S., in press.
_____: 1983b, M.N.R.A.S., submitted.
_____: 1984, presented at IAU Symp. No. 108, "Structure and Evolution of the Magellanic Clouds", this volume, p. 217.
Mendoza, E. E., and Gomez, T.: 1973, Pub. A.S.P., 85, 439.
Mould, J. R.: 1984 in IAU Symp. No. 108, "Structure and Evolution of the Magellanic Clouds", this volume, p. 195.
Mould, J., and Aaronson, M.: 1979, Ap. J., 232, 421.
_____: 1980, Ap. J., 240, 464.
_____: 1982, Ap. J., 263, 629.
Paltoglou, G., Wood, P. R., and Bessell, M. S.: 1984, presented at IAU Symp. No. 108, "Structure and Evolution of the Magellanic Clouds", this volume, p. 219.
Payne-Gaposchkin, C. H.: 1971, Smith Contr. Ap., No. 13.
Payne-Gaposchkin, C.H., and Gaposchkin, S.: 1966, Smith. Contr. Ap., No. 9.
Persson, S. E., Aaronson, M., Cohen, J. G., Frogel, J. A., and Matthews, K.: 1983, Ap. J., 266, 105.
Prévot, L., Martin, N., Maurice, E., Rebeirot, E., and Rousseau, J.: 1983, Astr. Ap. Suppl., 53, 255.

Rabin, D.: 1982, Ap. J., 261, 85.
Rebeirot, E., Martin, N., Miares, P., Prévot, L., Robin, A., Rousseau, J., and Peyrin, Y.: 1983, Astr. Ap. Suppl., 51, 277.
Reid, N.: 1983, privately communicated by I. S. Glass.
Reid, N., and Mould, J. R.: 1983, in preparation.
Richer, H. B.: 1981, Ap. J., 243, 744.
Richer, H. B., and Frogel, J. A.: 1980, Ap. J. (Letters), 242, L9.
Richer, H. B., Olander, N., and Westerlund, B. E.: 1979, Ap. J., 230, 724.
Richer, H. B., and Westerlund, B. E.: 1983, Ap. J., 264, 114.
Ridgway, S. T., Jacoby, G. H., Joyce, R. R., and Wells, D. C.: 1980, in IAU Coll. No. 59, "Physical Processes in Red Giants", eds. I. Iben and A. Renzini (Dordrecht: Reidel), p. 47.
Sanduleak, N.: 1975, Astr. Ap., 39, 461.
Sanduleak, N., and Philip, A. G. D.: 1977, Pub. Warner Swasey Obs., 2, No. 5.
Searle, L., Wilkinson, A., and Bagnuolo, N.: 1980, Ap. J., 239, 803.
Stryker, L., and Butcher, H.: 1981, in IAU Coll. No. 68, "Astrophysical Parameters for Globular Clusters", eds. A. G. D. Philip and D. S. Hayes (Schnectady: L. Davis Press), p. 255.
Tifft, W. G., and Snell, C. M.: 1971, M.N.R.A.S., 151, 365.
van den Bergh, S.: 1981, Astr. Ap. Suppl., 46, 79.
Walker, M. F., Blanco, V. M., and Kunkel, W. E.: 1969, A. J., 74, 44.
Westerlund, B. E.: 1983, private communication.
Westerlund, B. E., Olander, N., and Hedin, B.: 1981, Astr. Ap. Suppl., 43, 267.
Westerlund, B. E., Olander, N., Richer, H. B., and Crabtree, D. R.: 1978, Astr. Ap. Suppl., 31, 61.
Wood, P. R.: 1981, in IAU Coll. No. 59, "Physical Processes in Red Giants", eds. I. Iben and A. Renzini (Dordrecht: Reidel), p. 135.
Wood, P. R., Bessell, M. S., and Fox, M. W.: 1981, Proc. A.S.A., 4, 203.
_____: 1983, Ap. J., 272, 99.

DISCUSSION

Hesser: Is NGC2257 one of the SWB VII clusters that you suspect has an extended giant branch? I ask because there is a poster paper by Hesser, McClure, and Harris (see this volume) reporting a new CMD based on CCD data reaching the upper $\simeq$1.5 mags of the main sequence. The CMD is very similar to that of M92, as previously suggested by Stryker and as we also found from SIT vidicon photometry. NGC2257 also has a substantial RR Lyrae star population. Consequently I would be surprised if NGC2257 is younger than the Galactic globular clusters.

Aaronson: That's why membership of the key one or two stars needs to be checked. If these can be demonstrated to be members, then there is clearly something different going on in the giant branch evolution between the LMC and the Galaxy.

Frogel: In at least one SWB VII cluster the brightest star, which is apparently more luminous than the brightest stars in Galactic globulars, is a close (about 1") double. Both stars are of nearly equal magnitude and color. Thus in at least this cluster (I think it is NGC1841), the problem you mentioned with luminous stars goes away.

Aaronson: Good !

RED GIANT STARS: COMPARISON OF OBSERVATIONS AND THEORY

Jeremy Mould
Palomar Observatory
California Institute of Technology

ABSTRACT

Recent observations in both the field and the clusters of the Magellanic Clouds suggest a higher mass loss rate during or at the end of the asymptotic giant branch phase than previously supposed. Recent theoretical investigations offer an explanation for the frequency of carbon stars in the Clouds, but a rich parameter space remains to be explored, before detailed agreement can be expected.

1. INTRODUCTION

The study of the red giants of the Magellanic Clouds, as opposed to the more accessible supergiants, is substantially confined to the period since the last symposium on the Clouds. Since that time, deeper spectroscopic surveys, the development of infrared techniques and analysis of Magellanic Cloud clusters on a broad front have presented us with a wealth of information on the red giants. By even greater good fortune, this period has coincided with the investigation by theorists, principally from Illinois, Mt. Stromlo and Bologna, of the evolution of intermediate mass double-shell-source stars. The result has been exactly what one might hope for, an interplay between theory and observation and a stimulus to our understanding of both the physical processes in red giants and the evolutionary history of the Magellanic Clouds.

Although this review, constrained by time limits, will skip directly into the present to take a current perspective on the issues, it would be wrong not to indicate briefly the milestones of the past few years. The spectroscopic surveys begin with the I-N plate survey at the Uppsala Schmidt by Westerlund (1965) and Westerlund, Olander, Richer and Crabtree (1978). Other surveys followed, in the visible region by Sanduleak and Philip (1977) and in the infrared, using the Curtis Schmidt at Tololo, by Blanco and McCarthy (1975). These have culminated in the most complete survey, in terms of limiting magnitude, with IV-N plates at the

CTIO 4-m prime focus by Blanco and McCarthy (1983).

Work on the clusters of the Magellanic Clouds has been summarized by Hodge (1984). It is worth recalling that all of 25 years ago Arp (1958) noted the existence of very red stars at the giant branch tips of Cloud clusters. He and van den Bergh (1968) suggested they were carbon stars, but this was not confirmed until Feast and Lloyd-Evans (1973) obtained image tube spectra. The link with the "third dredge-up" mechanism of Iben (1975) was made by Crabtree, Richer and Westerlund (1976) in their field sample and by Mould and Aaronson (1979) in the red globular clusters. Subsequent spectroscopic work is reviewed here by Bessell (1984).

The most significant milestone in the asymptotic giant branch (AGB) theory was the discovery by Iben (1975) that during the thermal pulse power-down phase in model stars of intermediate mass, freshly produced carbon and s-process isotopes are dredged up into the convective envelope. The lure of predicting AGB properties and the consequences for nucleosynthesis with little extra work then became too great. A parameterized dredge-up law was introduced by Iben and Truran (1978), as a substitute for full exploration of the parameter space (X,Y,Z,1/H,M). A parameterized mass-loss law was introduced on to the AGB by Renzini (1977), and a parameterized planetary nebula ejection law by Renzini and Voli (1980). These "fudges" have permitted theoretical predictions to be made, but, not surprisingly, some of them have compared unfavorably with observations, as we shall see in the following sections. An excellent review of theoretical developments is that of Iben and Renzini (1983).

2. THE AGB LUMINOSITY FUNCTION

The prediction (Paczynski 1970) of a core-mass/luminosity relation for AGB stars has as an immediate consequence (since $dL/dt \propto L$) a flat luminosity function : ϕ (stars/mag. interval) = constant (see Renzini 1977). The most direct test is to construct an AGB luminosity function for the most populous intermediate age cluster NGC 419. According to the combined photometry of Mould and Aaronson 1982, Bessell et al. 1982, and Aaronson and Mould 1982, NGC 419 passes this test with 3 or 4 stars per 0.25 mag for $-5.5 < M_{bol} < -4.5$. But the sample is not complete to $M_{bol} = -4$, and the statistics are poor. Another test of the core-mass / luminosity relation is reviewed in § 6.

A test of theory which has no problem from the statistical point of view is the field luminosity function in the Magellanic Clouds. But in order to predict the AGB luminosity function for the field, we need to know 1) the mass-loss law, 2) the star formation history, and 3) (less importantly) the initial mass function. Hence it is less clear what is being tested. The largest observational sample available is that of Reid and Mould (1983), chosen photometrically from a 16 sq. deg field and corrected for foreground stars (a 10% effect). Figure 1 shows the star

counts in 0.25 mag bins, together with a prediction (solid line) for a
constant star formation rate (SFR) over the last 7 Gyrs (SFR = 0 before
that time). An additional contribution from stars with initial mass
greater than $4M_\odot$ is given by the lower solid curve. Other assumptions
which contribute to this prediction are the Reimers (1975) mass-loss
law (with an efficiency parameter η_R = 0.45) and the Renzini and Voli
(1981) planetary ejection law (with b = 1).

It is clear that the data do not fit the theory. Yet the data only
confirm the picture from a spectroscopic plus infrared survey done at
CTIO. The histogram in Figure 1 is a scaling (times 20) of the M(light)
and C(heavy) star luminosity function for the Bar West field of Blanco,
McCarthy and Blanco (1980). Additional stars found by Frogel and Richer
(1983) have been added, scaled appropriately (open histogram).

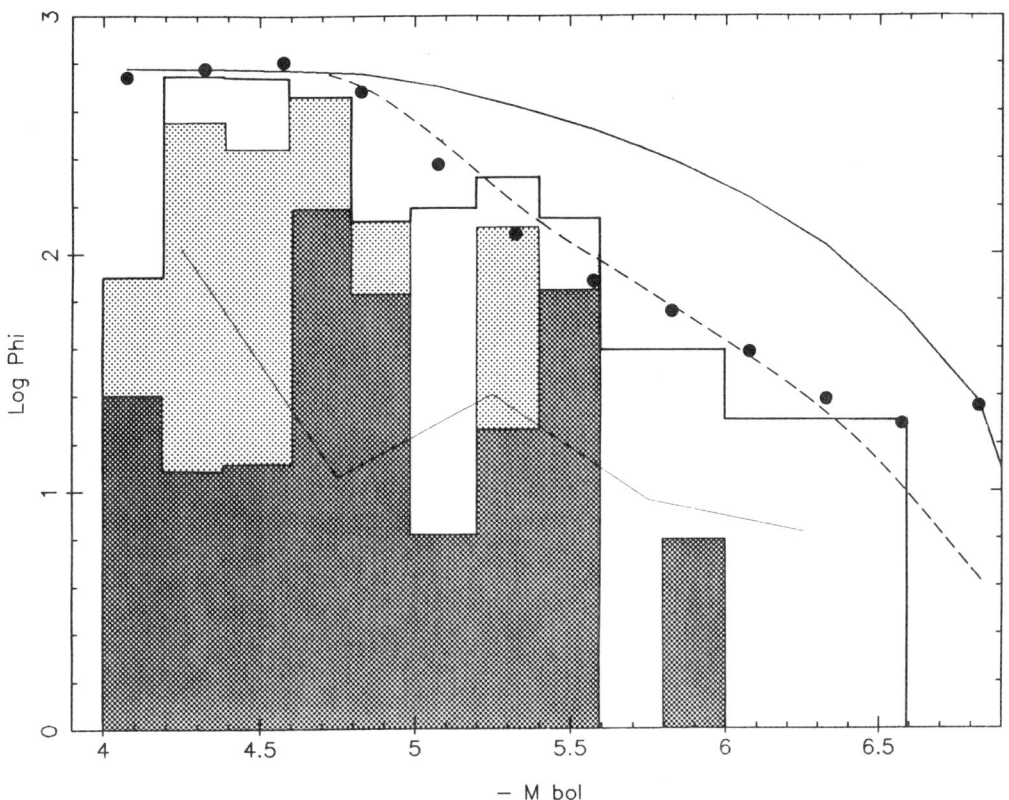

Figure 1
Bolometric Luminosity Functions for AGB Stars in the Large Cloud

Iben (1981) was the first to notice this discrepancy. It represents one facet of his carbon star `mystery´, "where have all the high mass ones gone?" At that time Iben offered three possible explanations: 1) that envelope burning of carbon to nitrogen in AGB stars was converting carbon stars to (undetected) M stars. This hypothesis is contradicted by the complete samples now available. 2) It was suggested that a pause in star formation in the LMC might be responsible for the missing mass range. This was subsequently ruled out by the identification of cepheids in the Bar West field by Becker (1982). 3) It is possible that massive carbon stars bury themselves in an optically thick dust shell after M_{bol} < -5.5. An infrared survey by Frogel and Richer (1983) has excluded objects in half the Bar West Field with M_{bol} < -6.4 and blackbody temperature greater than 700 K.

The current consensus in this matter (Frogel and Richer 1983; Reid and Mould 1983) is that none of these explanations is really complete. Envelope burning (C to N) may, or may not, take place: spectroscopy of larger samples is required to determine this. A star formation history could be contrived to reproduce the distribution in Figure 1. For example, the upper dashed curve in Figure 1 supposes SFR = 0 in the LMC until 7 Gyrs ago, followed by an exponential decline with an e-folding time of 2 Gyrs. Such an extreme decline does not seem to fit in with the data available from other sources. The IRAS satellite is expected to put more severe limits on the dust star hypothesis. More probably, the faulty component of the theory is the adopted mass-loss law. We shall return to scrutinize this more closely in § 4. We note in passing that the star formation history in the field does modulate the AGB luminosity function, as shown in the spatial variation detected by Reid and Mould (1983).

3. WHEN DOES AN AGB STAR BECOME A CARBON STAR?

Observationally, this is a spectroscopic question, to be reviewed more fully by Bessell (1983). The question is also a vexed one, in which models and data have been in conflict. So we begin by highlighting the qualitative agreement between theory and observation.

1. Of the Magellanic Cloud clusters surveyed (and published) to date, carbon stars are confined to types III-VI of Searle, Wilkinson and Bagnuolo (1980). These correspond to large enough ages (Rabin 1982) that only in AGB evolution could stars attain the observed luminosities.

2. Mixing on the AGB is predicted by theory to be episodic, i.e. discrete amounts of carbon rich material are added to the convective envelope. This leads naturally to a situation in which M stars evolve into S stars which evolve into C stars, as observed (Bessell, Wood and Lloyd-Evans 1982).

Theory can also be made to predict, however, the luminosity at which this M to C transition takes place. And that is where the trouble began, the other half of Iben's (1981) carbon star `mystery´, "why do the low mass ones become such?"

Initially the problem was that, according to the interpolated dredge-up law, only large core mass stars ($M_{bol} < -5.2$ from Iben 1981) were predicted to become carbon stars (c.f. Figure 1). Since that time the conflict has been softened by the following advances.

1. Carbon star formation can be likened to a titration experiment, in which carbon is added to an oxygen rich envelope (forming CO). If the envelope is metal poor, it takes less carbon to achieve `neutrality´. In this way Renzini and Voli (1981) were able to reduce the transition luminosity to $M_{bol} \sim -4.9$.

2. A refined treatment of the carbon opacity in the vicinity of the helium burning shell allowed Iben and Renzini (1982) to reach a minimum luminosity of $M_{bol} \sim -4$.

3. A much fuller exploration of the parameter space by Wood (1981), Wood and Zarro (1981) and Iben (1983) has led to the discovery that mixing results are critically dependent on the mixing-length assumed for convection and on Y and Z, which control the strength of the instigating thermal pulse.

This last result indicates that firm a priori predictions are not an expected product of the theory. Rather, the observations can be expected to lead the theory into a reasonable accommodation with real AGB evolution. Critical in this process will be observations of the transition luminosity in Magellanic Cloud clusters. One wants to know the two dimensional dependence of transition luminosity on age and metallicity (c.f. Lloyd-Evans 1983, Frogel and Blanco 1984).

Two points should be made in passing. First, according to theory (Iben and Renzini 1983) AGB stars spend 20% of their time approximately 0.5 mag below their quiescent luminosity. So in the vicinity of the C-M transition some C stars should be fainter than K, M or S stars. It would be useful to verify this effect, and it is important to make the distinction between minimum and quiescent luminosity. Second, according to Aaronson and Mould (1983) very metal poor dwarf spheroidal galaxies have M_{bol} (transition) ~ -4.2. The distribution of carbon stars suggests that this represents M_{bol} (quiescent).

4. THE LUMINOUS EXTENT OF THE AGB AS A FUNCTION OF AGE

Renzini (1977) was first to point out that in the absence of mass loss Galactic globular cluster stars would climb 2 magnitudes above the helium flash luminosity in the course of their AGB evolution, before the hydrogen burning shell reached the surface of the star. This is not observed, and mass loss is credited with curtailing the evolution by removing approximately 0.3 $M_\odot$ from the star during or at some point in its lifetime.

The rich intermediate age globular clusters of the Magellanic Clouds offer us the opportunity of learning how this net mass loss varies with increasing initial mass. Figure 2 shows the final luminosity on the AGB (from Mould and Aaronson 1983) for clusters whose ages are known by observation of the main sequence turnoff. The solid symbols are preliminary results from new observations with the prime focus CCD camera at the CTIO 4-m telescope. These color magnitude diagrams are of high quality and go 2 mag fainter than the best published 4-m photographic data. An example (Kron 3) is presented by Rich, Mould and Da Costa (1984). Open symbols are from the compilation by Hodge (1983b) (with the exception that NGC 416 and 419 are omitted for lack of really adequate CM diagrams). Note that only for three clusters were Mould and Aaronson (1983) able to estimate the location of the AGB tip. Other points plotted are lower limits on the luminosity, because of the small numbers of AGB stars available. The expected relation for a Reimers (1975) mass loss law (with $\eta_R = 0.45$) and a Renzini and Voli (1981) planetary ejection law (with b = 1) is also shown (from Mould and Aaronson 1983, Table 3).

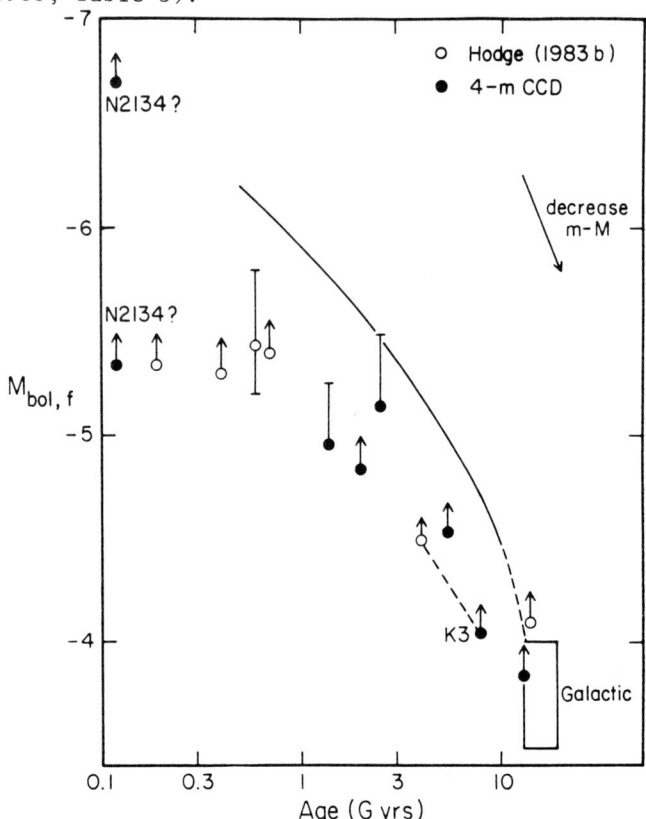

Figure 2
The Peak Luminosity on the AGB as a Function of Cluster Age

A number of points deserve to be made in respect of Figure 2.

1. The data do not fit the expected relation, even if one is reasonably generous in drawing an upper envelope. A similar point was made by Hodge (1983b). The deviation is in the sense that more mass loss is taking place, either during ascent of the AGB or in the nebula ejection (either the "wind" or the "superwind") than predicted.

2. This conclusion is reasonably independent of the distance moduli adopted for the Magellanic Clouds. The solid symbols are based on $(m-M)_0 = 18.3$ (LMC) and 18.8 (SMC), which is the "short" distance scale for the Clouds (de Vaucouleurs 1978, Eggen 1977). This was done, because a noticeably better fit resulted to theoretical main sequence isochrones. Data from Hodge (1983b) is mostly on the "long" distance scale: $(m-M)_0 = 18.7$ (LMC) and 19.3 ± 0.1 (SMC) (Gascoigne 1972, Martin et al. 1979, Sandage and Tammann 1981). A shift indicated by the arrow in Figure 2 will put the latter data on the former scale.

3. The major uncertainty in Figure 2 is in the interval $0.1 <$ age < 1 Gyrs. The cluster NGC 2134 has been plotted twice, because it is unclear whether star 3 of Mould and Aaronson (1983) is a member or not. With a little work other clusters could readily be added to tie down the high mass end. The trend in the data towards an upper limit to $M_{bol,f}$ at young ages should not be given too much weight at present. For there is evidence of AGB stars in the LMC field at $M_{bol} = -7$ (see § 6).

4. A revised mass loss law based on Figure 2 (or preferably more complete data) would undoubtedly produce a better fit to the field luminosity function (Figure 1) also.

5. ANOMALOUS OR SUPERLUMINOUS GIANTS

According to Flower et al. (1980) and Hodge (1981) several Magellanic Cloud clusters younger than 1 Gyr contain numbers of unaccountable giant stars spread over a range of color and averagely 2.7 mag above the main sequence tip. A number of possibilities have been discussed in respect of these stars:

1. They are field stars. Control fields analysed close to the cluster seem to rule this out. Radial velocities (Olszewski 1982) are comparable with that of the Magellanic Clouds for most of these stars.

2. They are coalesced stars. This suggestion (Flower 1980) does not seem to be realistic.

3. They are post-AGB stars. Given the problem (§ 2) of the missing AGB stars, this might seem like a welcome solution. Post-AGB stars, however, would be expected to cross from red to blue on a short timescale and at a luminosity $M_{bol} \leq -5$. Most of the anomalous giants

are much fainter.

4. A further possibility, which seems to have received less attention, is that mass transfer in binary systems may be responsible. After the primary had evolved to a red giant, a binary system with a mass ratio of order 1 would concentrate most of its mass on the secondary, which would appear as a superluminous giant in its core helium burning (blue loop) stage. The long lifetime of the blue loop would not require an excessive fraction of binaries even in the archetypal anomalous cluster, NGC 1868.

Further work is required to resolve the problem of the anomalous giants. High dispersion spectroscopy would seem particularly valuable.

6. PERIOD LUMINOSITY RELATIONS

We conclude this review by pointing out a remarkable confirmation of theory in observations discussed by Wood, Bessell and Fox (1981). These authors constructed a period-luminosity diagram for long period variables in the LMC, using infrared magnitudes, which relate readily to bolometric luminosities. They discovered a bifurcation of the general period luminosity correlation. Stars which they interpret as massive supergiants continue on to the highest luminosities, while a second sequence with the characteristics of AGB stars ceases abruptly at $M_{bol} = -7.0$. According to the Paczynski (1970) relation this corresponds to a core mass of 1.4 $M_\odot$ i.e. the Chandrasekhar mass. Confirmation that such an AGB limit exists and occurs at the predicted luminosity validates the core-mass/luminosity relation observationally.

Other observations of LMC Miras (Glass and Lloyd-Evans 1981, Glass and Feast 1982) have assisted materially in determining the luminosities of these stars. The question of the mode of pulsation of long period variables remains in debate, because of uncertainties in their radii/temperatures.

7. CONCLUSION

It is clear that the study of red giants in the Magellanic Clouds is a rich area just opening up for fuller investigation. Key issues at present are the amount of mass loss as a function of initial stellar mass and the details of observed and predicted nucleosynthesis.

I would like to thank the many respondents to a request for reprints and preprints. Preparation of this review was aided by NSF grant AST 8306139. Travel support from the IAU, the American Astronomical Society and the Deutsche Forschungsgemeinschaft is gratefully acknowledged.

REFERENCES

Aaronson, M. and Mould, J.R.: 1982, Astrophys. J. Suppl., 48, pp. 161-184.
Aaronson, M. and Mould, J.R.: 1983, (in prep.).
Arp, H.C.: 1958, Astron. J., 63, pp. 487-491.
Becker, S.A.: 1982, Astrophys. J., 260, pp. 695-706.
Bessell, M.S.: 1984, this volume, p. 171.
Bessell, M.S., Wood, P.R., and Lloyd-Evans, T.: 1982, Monthly Notices Roy. Astron. Soc., 202, pp. 59-76.
Blanco, V.M. and McCarthy, M.F.: 1975, Nature, 258, pp. 407-408.
Blanco, V.M.: 1983, (preprint).
Blanco, V.M., McCarthy, M.F., and Blanco, B.M.: 1980, Astrophys. J., 242, pp. 938-964.
Crabtree, D.R., Richer, H.B., and Westerlund, B.E.: 1979, Astrophys. J. Letters, 203, pp. L81-85.
de Vaucouleurs, G.: 1978, Astrophys. J., 223, pp. 730-739.
Eggen, O.J.: 1977, Astrophys. J. Suppl., 34, pp. 1-34.
Feast, M.W. and Lloyd-Evans, T.: 1973, Monthly Notices Roy. Astron. Soc., 164, pp. 15P-18P.
Flower, P.J.: 1980, Astrophys. J. Letters, 249, pp. L11-14.
Flower, P.J., Geisler, D., Hodge, P.W., Olszewski, E.W.: 1980, Astrophys. J., 235, pp. 769-782.
Frogel, J.A. and Blanco, V.M.: 1984, IAU Symp. No. 105, (in press).
Frogel, J.A. and Richer, H.B.: 1983, (preprint).
Gascoigne, S.C.B.: 1972, Quart. J. Roy. Astron. Soc., 13, p. 274.
Glass, I.S. and Feast, M.W.: 1982, Monthly Notices Roy. Astron. Soc., 199, pp. 245-253.
Glass, I.S. and Lloyd-Evans, T.: 1981, Nature 291, pp. 303-304.
Hodge, P.W.: 1981, J. astrophys. Astron., 2, pp. 161-164.
Hodge, P.W.: 1984, this volume, p. 7.
Hodge, P.W.: 1983b, Astrophys. J., 264, pp. 470-475.
Iben, I.: 1983, Astrophys. J. Letters, (in press).
Iben, I.: 1975, Astrophys. J., 196, pp. 525-547.
Iben, I.: 1981, Astrophys. J., 246, pp. 278-291.
Iben, I. and Truran, J.W.: 1978, Astrophys. J., 220, pp. 980-995.
Iben, I. and Renzini, A.: 1983, Ann. Rev. Astron. Astrophys., 21, p.271.
Iben, I. and Renzini, A.: 1982, Astrophys. J. Letters, 263, pp. L23-27.
Lloyd-Evans, T.: 1983, Monthly Notices Roy. Astron. Soc., (in press).
Martin, W.L., Warren, P.R., and Feast, M.W.: 1979, Monthly Notices Roy. Astron. Soc., 188, pp. 139-157.
Mould, J.R. and Aaronson, M.: 1979, Astrophys. J. 232, pp. 421-427.
Mould, J.R. and Aaronson, M.: 1982, Astrophys. J., 263, pp. 629-638.
Olszewski, E.W.: 1982, Ph.D. Thesis, Univ. of Washington.
Paczynski, B.: 1970, Acta astron. 20, pp. 47-58.
Rabin, D.: 1982, Astrophys. J., 261, pp. 85-101.
Reid, N. and Mould, J.R.: 1983, (preprint).
Reimers, D.: 1975, "Problems in Stellar Atmospheres and Envelopes", ed. B. Baschek, W. H. Kegel, and G. Traving (Berlin: Springer-Verlag), p. 229.
Renzini, A.: 1977, "Advanced Stages in Stellar Evolution", eds. P.

Bouvier and A. Maeder (Sauverny: Geneva Obs.), pp. 151-267.
Renzini, A. and Voli, M.: 1981, Astron. Astrophys., 94, pp. 175-193.
Rich, R.M., Mould, J.R., and Da Costa, G.S.: 1984, this volume, p. 45.
Sandage, A. and Tammann, G.A.: 1981, "A Revised Shapley-Ames Catalog of Bright Galaxies", (Washington: Carnegie Institution).
Sanduleak, N. and Philip, A.G.D.: 1977, Pub. Warner and Swasey Obs., 2, No. 5.
Searle, L., Wilkinson, A. and Bagnuolo, W.B.: 1980, Astrophys. J., 239, pp. 803-814.
van den Bergh, S.: 1968, J. Roy. Astron. Soc. Can., 62, pp. 1-73.
Westerlund, B.E.: 1964, in IAU Symp. No. 20, eds. F.J. Kerr and A.W. Rodgers (Canberra: Australian Academy of Sciences), p. 239.
Westerlund, B.E., Olander, N., Richer, H.B. and Crabtree, D.R.: 1978, Astrophys. Astron. Suppl., 31, pp. 61-82.
Wood, P.R.: 1981, Proc. A.S.A., 4, pp. 145-148.
Wood, P.R., Bessell, M.S., and Fox, M.W.: 1981, Proc. A.S.A., 4, pp. 203-205.
Wood, P.R. and Zarro, D.: 1981, Astrophys. J., 247, pp. 247-256.

DISCUSSION

Wood: The results of Wood, Bessell and Fox (1983 Ap.J., in press) show that AGB stars do exist at luminosities brighter than $M_{bol} \simeq -6$ and, in fact, that the stars extend up to the AGB limit $M_{bol} \simeq -7.1$. We therefore disagree with the "consensus" view that mass loss terminates the AGB for all initial stellar masses at M fainter than $\simeq -6.5$. Our pulsation masses indicate that only stars with ages less than about 100 million years will reach luminosities near the AGB limit.

Mould: That is not the "consensus" view! Figure 2 suggests that a higher rate of mass loss occurs on the AGB than is normally assumed. The luminosity function in Figure 1 is in agreement with your result that AGB stars with $M_{bol} = -7$ exist in the LMC in small numbers. We need to examine more clusters to find out what initial mass stars are their progenitors. The pulsation masses, as you know, are highly contentious.

Frogel: The problem is that luminous AGB stars do not exist in anywhere the numbers required by theory. There may be a few of them, but mass loss seems to operate in such a way that nearly all luminous stars are evaporated.

Aaronson: Might not a star burst model be able to produce better agreement with the observed luminosity function?

Mould: Yes.

Demarque: In connection with your comments on the "superluminous" giants reported by Flower et al., I wish to draw attention to a paper by Alan Hirshfeld (published in 1981 in the Ap.J.) which treats the problem of mass transfer binaries in the core helium burning phase. Hirshfeld's calculations were made in the context of anomalous Cepheids, but they might also be applicable to the superluminous giants.

Westerlund: I do not believe that the surveys for carbon stars in the Magellanic Clouds are as complete as suggested here. Already the Schmidt telescope surveys by Sanduleak and Philip (IIIa-J plates) and Westerlund et al (I-N plates) showed that the former survey contained a large number of hotter carbon stars than the latter. A IIIa-J GRISM survey of SMC by Azzopardi, Breysacher, Lequeux and myself gives about 15 percent carbon stars not found by Blanco et al. in their field in the Bar. GRISM surveys by us of the Fornax dwarf galaxy show similar results. Possibly these carbon stars, which may be among the more luminous, fill the gap in Mould's Figure 1.
Also the number of luminous ZrO-rich M giants in the LMC is higher than previously known, as found recently by Lundgren and myself. They may contribute to filling the gap, too.

Blanco: Remarks: 1) In regard to incompleteness, as far as the carbon and very late M stars are concerned, the surveys by Westerlund based on near-infrared spectra and the ones in the blue-green spectral region (Swan bands) by Sanduleak and Philip are found by McCarthy and myself to be very incomplete. This explains the lack of overlaps when the Westerlund and Sanduleak-Philip surveys are intercompared and has no bearing on the incompleteness problem discussed by Mould. In our CTIO GRISM surveys with the 4 m Telescope one can be very sure of completeness because the carbon and very late M giants are extremely bright. Also please notice that the missing stars are supposed to be the bright ones which are the easiest ones to find in any survey.
2) Another myth that should be cleared up is that in the Swan band surveys hot carbon stars were discovered that do not show up in the near-infrared surveys based on CN-bands. In numerous examples examined by us an extremely high percentage of carbon stars showing Swan bands also show CN-bands in the near-infrared. I know of only one exeption to this general rule and that star showed CN-bands and no or very weak Swan bands. Thus we cannot say either that the near-infrared surveys missed hot luminous carbon stars. Nevertheless, in Westerlund's survey the brighter carbon stars found were the brighter ones that could be found in the near-infrared and his star list even though very incomplete does include the more luminous carbon stars except in the bars of the Clouds where spectral overlaps caused him to miss many carbon stars. It follows from all this that the missing luminous AGB stars must be something other than carbon stars for sure and it is also unlikely that they are late M giants.

Renzini: The missing AGB stars are not necessarily carbon stars, they can be something else, for example, even luminous K stars; this is expected from theory, and supported by the existence of the long-period variables (which are not carbon stars).
My impression is that current theory with a mass loss rate calibrated on Galactic globulars is OK for MC clusters older than $\simeq 1$ Gyr, which is to say for stars less massive than $\simeq 2 \, M_\odot$. The same applies to the comparison with field MC stars. So the difficulty seems to arise only for stars more massive than $\simeq 2 \, M_\odot$. Maybe that it is not by chance that this occurs just at $\simeq 2 \, M_\odot$, as less massive stars develop a degenerate He core during the first red giant branch phase, while more massive stars do not. Perhaps the <u>rotational</u> history of the core of

stars more massive than 2 M⊙ is correspondingly different from that of lower mass stars, and this may provide a hint to understand the apparent lack of bright AGB stars. However, still I'm not totally convinced of the completeness of existing surveys, particularly as far as late K and early M type stars are concerned.

Mould: A final word on this subject of survey completeness. I believe that the order of magnitude agreement between spectroscopic surveys and photometric surveys, such as the one I have reported here, implies that there is a real deficiency of luminous AGB stars in the LMC.

RR LYRAE STARS AND NOVAE IN THE MAGELLANIC CLOUDS

J.A. Graham
Cerro Tololo Inter-American Observatory
Casilla 603, La Serena, Chile

ABSTRACT

RR Lyrae variable stars and classical novae can be used very effectively for locating and studying the old stellar populations in the Magellanic Clouds. RR Lyrae stars are found in large numbers in both Clouds at about 19^m. Apart from the concentrated searches within Cloud clusters, intensive surveys for field variables have so far been made in four areas, each about 1° square. Novae have the advantage of being as bright as 10^m at maximum light. They can be detected with small telescopes and surveys have covered almost the entire area of each Cloud. Needs and prospects for future surveys are discussed. Both types of object are suitable for investigating the early chemical composition of the Clouds and its subsequent enrichment with heavy elements. Studies of the old populations in the Clouds are reinforcing the view that, while some stars in the Magellanic Clouds are as old as any in our Galaxy, the major bursts of star formation came along comparatively recently.

INTRODUCTION

The Magellanic Clouds are the closest independently evolving galaxies in which star formation is going on today. Stellar birth, evolution and decay have been proceeding for at least 10^{10} years with little or no interchange of matter with the Galaxy and it is through the study of the oldest stars that we can hope to obtain insight into the early history of these two systems. The two types of variable star which I am discussing here are by no means the only old stars in the Magellanic Clouds but they do have the advantage of being easy to identify and isolate for detailed study. They are both suitable for investigating the early chemical composition of the Clouds and the subsequent enrichment with heavy elements from evolving stars. Studies of the space distribution give us information about the early star formation in each system. Distance checks can also be made which are useful in that they derive from an older population than that most commonly used for calibration purposes.

RR LYRAE STARS

i) Surveys

RR Lyrae stars were first detected in the Magellanic Clouds by Thackeray (1951, 1958). As Thackeray (1974) pointed out in a later review "the impact of these discoveries was twofold. Firstly, it proved that the Clouds contain old populations as well as the conspicuous young Population I... Secondly, the fact that the variables appeared at 19th magnitude instead of the expected 17.5 confirmed Baade's 1952 revision of the distance scale." In the initial discovery in the Small Magellanic Cloud (SMC), 3 variables were found in the cluster NGC 121 and 3 in the surrounding field outside the cluster boundaries. RR Lyraes are faint, but, as highly evolved stars of low mass, they can be confidently assigned ages of the order of 10^{10} years. Their presence marks out the location of a very old stellar population.

Almost all searches to date have been centered on Magellanic Cloud clusters. This has been largely because sequences of known stellar magnitudes are preferentially found in such places but also because of the interest in studying variable stars within the clusters themselves. Following the initial discovery, Thackeray found variables in the Large Magellanic Cloud (LMC) cluster NGC 2257 (Alexander 1960). A large number of clusters in both Clouds have now been searched for RR Lyrae stars. The present situation is discussed by Graham and Nemec (this volume).

Surveys for field variables had to wait for the availability of wide photographic fields with large telescopes. (Graham 1975, 1977) took advantage of the $1° \times 1°.5$ field of the CTIO 1.5m telescope to search intensively two fields, one around NGC 121 (SMC) and another around NGC 1783 (LMC). A total of 143 variables were found showing that RR Lyrae stars are indeed abundant in the Clouds. Photographic plates also exist for two similar fields, one between NGC 361 and NGC 362 (SMC) and the other, about NGC 1835 (LMC). So far these have been examined in only a cursory manner.

The main conclusions of the Graham studies were as follows:

1) The mean apparent magnitudes seem to cluster closely about a single value. In the LMC, the time averaged $<\bar{V}>$ and $<\bar{B}>$ are $19^m.2$ and $19^m.6$ while in the SMC $<\bar{V}>$ and $<\bar{B}>$ are $19^m.6$ and $20^m.0$.

2) ab type variables with large amplitudes and periods less than $0^d.45$ are rare or absent in both Clouds although such stars are relatively common in the solar neighborhood.

3) The space density in each field is about 15 RR Lyrae stars $(kpc)^{-3}$. It is greater than that found in the Galactic halo but of the same order as that found in the solar neighborhood.

4) A preliminary survey of the two additional fields revealed that there does not appear to be a strong concentration of RR Lyrae variables towards the central region of either Cloud. In the SMC, the RR Lyraes concentrate preferentially to the west of the main body of bright stars. A similar conclusion was drawn by Brück (1980) from faint stars counts of the old population in general.

5) There are differences between the RR Lyrae variables in each Cloud. The period distribution in the LMC resembles that found in high latitudes in the Galaxy. The period distribution in the SMC is more characteristic of the anomalous distributions found among dwarf spheroidal galaxies. The SMC also contains a small group of RR Lyraes which appear normal in all respects except that they have a mean apparent magnitude about $1\overset{m}{.}5$ brighter than the average value. In a similar LMC sample, no such variables are found.

Apart from the cluster searches described in the Graham-Nemec paper at this Symposium, intensive searches have been made in only two other fields to date, one around NGC 2257 (Hesser, Nemec & Ugarte 1980) and one around NGC 1841 (Kinman, Stryker and Hesser 1976). The latter was especially interesting as it turned up 3 normal RR Lyrae variables which lie in the general field beyond the tidal radii of both the LMC and SMC.

ii) Chemical Abundance Studies

An important study of some of the Cloud variables has been carried out by Butler, Demarque and Smith (1982). They found that RR Lyrae stars of ordinary luminosity in the LMC have [Fe/H] = -1.4 while those in the SMC have [Fe/H] = -1.8. The difference is only marginally significant but clearly demonstrates the presence of a metal poor component among the field stars of the Magellanic Clouds. Metal abundances were also determined for the overluminous SMC variables. These were found to have [Fe/H] = -0.4 and are much richer in metals than their fainter counterparts. Pulsation masses and evolutionary ages were also determined. Ages range from 2.5×10^8 years for the overluminous stars to 14×10^9 years for the ordinary RR Lyrae stars.

iii) Future Directions

RR Lyrae stars may be useful for mapping in depth, as well as in area, the distribution of the oldest populations in the Magellanic Clouds. Sandage (1981) has shown that, with high quality color and amplitude data, the absolute magnitude uncertainty may be reduced to the $0\overset{m}{.}1$ level thus making RR Lyraes a more precise distance calibrator than formerly thought. The advent of efficient, linear CCD detectors makes the determination of precise magnitudes and colors a far more straightforward process than with the older photographic techniques. Several observatories now have these operating on telescopes of small or moderate size.

To date, an unfortunate restriction of the surveys for RR Lyrae stars is that they have been limited to small areas of approximately

one square degree. There is a great need for surveys to cover much
larger areas in each Cloud and I am wondering if the time is perhaps
right to be doing this with one on other of the big southern Schmidts
in conjuction with automatic measuring machines. Certainly large
numbers of measurements would need to be made and the key to success
would be in processing the data to isolate the variable stars with RR
Lyrae characteristics. Relative photometry only should be attempted
at first with accurate photometry following later on carefully chosen
samples.

There is hope for improvement of the basic absolute magnitude
calibration for RR Lyrae stars. We depend at present on statistical
parallaxes and are slowly accumulating data from the application of
the Baade-Wesselink method to Galactic stars. Space astrometry should
finally provide the definite zero points. Stothers (1983 preprint) has
shown that a mean visual absolute magnitude $M_{<V>} = +0\overset{m}{.}6$ seems the best to
use for now. With this value, the Graham photographic photometry gives
apparent distance moduli of $18\overset{m}{.}6$ and $19\overset{m}{.}0$ for the LMC and the SMC. A
correction for foreground interstellar absorption of $0\overset{m}{.}1 - 0\overset{m}{.}2$ should be
taken into account to convert these values to distances. An overall
uncertainty of $\pm 0\overset{m}{.}3$ is probably applicable. Let us hope that our
efforts in the next 10 years can reduce this to no more than $\pm 0\overset{m}{.}1$.

NOVAE

i) Surveys

Novae have been recognized in the Magellanic Clouds since 1927
(Luyten 1927) but systematic searches have been carried out only
sporadically. During their brief appearance, novae become the brightest
members of the old stellar population of a galaxy. They are thought to
result from post-main sequence evolution of close binary stars with
components of a solar mass or less and are thus representative of a
stellar population whose age is greater than 5×10^9 years. Like RR
Lyrae stars, they are good tracers of the distribution of old stars
within a galaxy. The main disadvantage is that they are infrequent.
(Graham 1979a) estimated a frequency of 2 or 3 novae per year in the
LMC and 1 nova every 3 or 4 years in the SMC. A list of known novae is
given in Table 1. 8 were discovered during a survey which was carried
out between 1970 and 1979 at CTIO. Details of the observing technique
were given by Graham and Araya (1971). The survey was discontinued in
1979 but we now plan to reinitiate it in a limited way during the 1983-
1984 observing season. A watch is kept on the Magellanic Clouds by the
University of Chile group which is conducting a systematic search for
extragalactic supernovae. They found one nova in 1981 which was widely
observed.

TABLE I

Known Novae in the Magellanic Clouds

	Year	RA 1975		Dec		Centroids (1975)
SMC						
	1897	0^h	$59^m.1$	$-70°$	$23'$	
	1927	0	33.2	-73	24	
	1951	0	34.3	-73	06	$0^h\ 40^m.7\ -72°\ 58'$
	1952	0	47.6	-73	39	
	1974	0	25.0	-74	10	
LMC						
	1926	5	14.9	-66	51	
	1936	5	07.4	-66	41	
	1937	5	57.2	-68	55	
	1948	5	38.6	-70	22	
	1951	5	13.0	-70	00	
	1968	5	10.4	-71	42	
	1970a	5	33.5	-70	36	
	1970b	5	36.0	-70	48	$5^h\ 25^m.0\ -69°\ 05'$
	1971a	4	58.4	-68	08	
	1971b	5	40.6	-66	41	
	1972	5	28.8	-68	50	
	1973	5	15.5	-69	41	
	1977a	6	05.9	-68	38	
	1977b	5	05.4	-70	11	
	1978	5	05.8	-65	55	
	1981	5	32.4	-70	23	

ii) Novae as Distance Indicators

For many years it has been known that there is a tight relation between absolute visual magnitude at maximum and the rate of decline of novae. They can thus be used as good indicators of distance. The Galactic calibration is based on a comparison between the observed angular expansion of the nova shell and its radial velocity. For useful distances, good photometric coverage of a nova through maximum is essential. In the Magellanic Clouds, data of this quality is available for only two novae, both in the LMC (Ardeberg and de Groot 1973, Canterna and Schwartz 1977). Absorption corrected distance moduli of $18^m.5$ and $18^m.6$ were derived from these two studies with uncertainties of about $\pm 0^m.2$.

iii) Distributions

Figure 1 shows the surface distribution of the 5 novae known in the SMC. The numbers are small but it may be significant that, like other old populations, the known novae group preferentially on the west side of the SMC outside the main bar. In the LMC, with 16 novae now known, the statistics are better. Figure 2 shows the distribution. The most significant feature is the lack of concentration towards the crowded bar. This is not a selection effect. With the overlaps in the survey, the bar region was searched 4 times more frequently than the rest of the LMC. Figure 2 shows that the centroid of the LMC novae is significantly north of the bar and close to the center of rotation determined by Feitzinger (1980).

iv) Spectroscopy

Detailed spectroscopic observations of two novae by Canterna and Thompson (1981) and by Andrillat and Dennefeld (1983) confirm that LMC novae are basically similar to those known in our Galaxy. Subtle differences may exist. Andrillat and Dennefeld remark that, the weakness of the [O I] and [O II] lines in the 1981 LMC nova may indicate low oxygen abundance in the LMC. They point out that this interpretation may not necessarily be correct because of the wide variety of spectra one sees in novae and there is clearly a need for more examples to be studied. With systematic surveys, occasionally low dispersion spectra are obtained before maximum light. This happened with the 1978 LMC nova for which a pre-discovery objective prism plate was taken when the nova was about 5 magnitudes below maximum. (Graham 1979b). The spectrum of LMC nova 1970b is described by Havlen, West and Westerlund (1971).

v) Novae as a Source of Element Enrichment

Williams (1982) has pointed out that novae may be important contributors to the nitrogen content of galaxies. Chemical abundance analyses of old nova shells in the Galaxy have demonstrated marked enrichments of nitrogen in their envelopes. This, when combined with the known frequency of nova outbursts, suggests that novae may exceed planetary nebulae and Wolf-Rayet stars in enriching the interstellar medium with nitrogen. He suggests the possibility of checking this scheme by making abundance studies of novae and old novae shells in the Magellanic Clouds. It is a difficult job but just possible with large telescopes and currently available detectors.

EARLY STELLAR EVOLUTION IN THE MAGELLANIC CLOUDS

RR Lyrae stars and novae share the characteristic of only slight increases in surface density towards the crowded central regions of the Large and Small Magellanic Clouds. In this way, they differ from their Milky Way analogs which show strong concentrations towards the center of the Galaxy. These distributions imply that the numerous

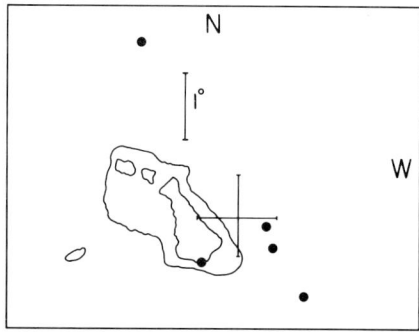

Fig. 1: Distribution of Novae in the SMC. The vertical cross marks the centroid and its extent indicates the standard error of the mean value.

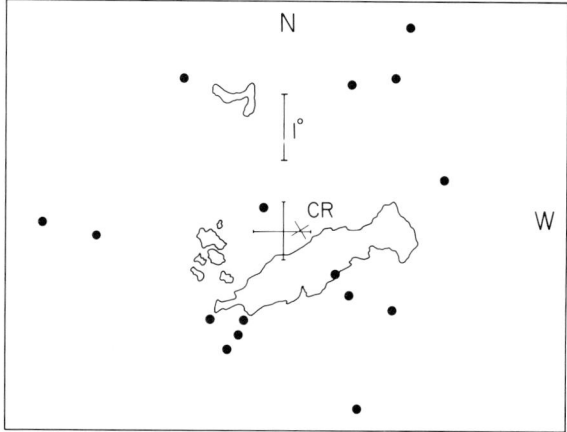

Fig. 2: Distribution of Novae in the LMC. The vertical cross marks the centroid and its extent indicates the standard error of the mean value. The diagonal cross shows the center of rotation used by Feitzinger (1980).

low mass stars in the two Magellanic Cloud bars are not sufficiently old yet to become RR Lyrae stars and novae. We cannot give an answer to the question as to whether the oldest stars in the Magellanic Clouds structurally form disk or halo-type systems. With increased samples (especially for the RR Lyrae stars) and improved calibrations, the answer could be obtained in a straightforward manner. Use of globular clusters (Freeman, Illingworth and Oemler 1983) has shown promise but is basically limited by the small number of globular clusters and the difficulty of isolating the few very old globular clusters from the large number with intermediate ages.

Oort (1971) drew attention to the fact that galaxies like the Clouds have no strong, dominating central mass concentration. The distribution of the old stellar populations seems to confirm that no major gravitational collapse with a contemporaneous major burst of star formation occurred in the early history of either Cloud. Star formation has been, in the long term, a more gradual affair. Butler et al. (1982) concluded that the SMC only now is reaching the degree of chemical enrichment which the Galaxy attained 10^{10} years ago. R.E. Williams (private communication) has pointed out another constraint. By making an account of probable abundance enhancement rates from novae, planetary nebulae and Wolf-Rayet stars, and then comparing this with the present day abundances of the interstellar medium (Cloud H II regions), he notes that the present rates cannot have occurred for longer than about 3×10^9 years. In several ways, studies of the old populations in the Clouds are leading us to accept that, while some stars in the Magellanic Clouds are as old as any in our Galaxy, the major bursts of star formation came along comparatively recently.

REFERENCES

Ardeberg, A. and de Groot, M.: 1973, A. & A. 26, p. 53.
Alexander, J.B.: 1960, MNRAS 121, p. 97.
Andrillat, Y. and Dennefeld, M.: 1983, A. & A. (in press).
Brück, M.T.: 1980, A. & A. 87, p. 92.
Butler, D., Demarque, P. and Smith, H.A.: 1982, Ap. J. 257, p. 592.
Canterna, R. and Schwartz, R.D.: 1977, Ap. J. Letters 216, p. L91.
Canterna, R. and Thompson, L.F.: 1981, PASP 93, p. 581.
Feitzinger, J.V.: 1980, Space Science Rev. 27, p. 35.
Freeman, K.C., Illingworth, G. and Oemler, A.: 1983, Ap. J. 272 (in press).
Graham, J.A.: 1975, PASP 87, p. 641.
Graham, J.A.: 1977, PASP 89, p. 425.
Graham, J.A.: 1979a in "Changing Trends in Variable Star Research" IAU Colloquium 46, eds. F.M. Bateson, J. Smak, and I.H. Urch (University of Waikato, Hamilton, N.Z.), p. 96.
Graham, J.A.: 1979b, PASP 91, p. 79.
Graham, J.A. and Araya, G.: 1971, Astr. J. 76, p. 768.
Havlen, R.J., West, R.M. and Westerlund, B.E.: 1971, A. & A. 16, 404.
Hesser, J.E., Nemec, J.M. and Ugarte, P.: 1980, in "Star Clusters" IAU Symposium 85, ed. J.E. Hesser (D. Reidel Dordrecht), p. 347.
Kinman, T.D., Stryker, L.L. and Hesser, J.E.: 1976, PASP 88, p. 393.
Luyten, W.J.: 1927, Harvard Bull. No. 847, p. 8.
Oort, J.H.: 1971, in "The Magellanic Clouds", ed. A.B. Muller (D. Reidel Dordrecht), p. 184.
Sandage, A.: 1981, Ap. J. Letters 244, p. L23.
Thackeray, A.D.: 1951, Observatory 71, p. 219.
Thackeray, A.D.: 1958, MNRAS 118, p. 117.
Thackeray, A.D.: 1974, MN Astr. Soc. So. Africa 33, p. 66.
Williams, R.E.: 1982, Ap. J. Letters 261, p. L77.

DISCUSSION

Mould: After correction for an extinction of $A_V = 0.2$ from the Cepheid studies and $M_V = +0.6$, which may be an underestimate for Galactic RR Lyraes, one obtains from your $<V>$ a true distance modulus of 18.4 for the LMC, which seems to support the short distance scale of Eggen and de Vaucouleurs, rather than the 18.73 we heard earlier today. Globular cluster main sequences also seem better fitted with such a short modulus. Is there a discrepancy between the Cepheids and the RR Lyraes?

Graham: I think $A_V = 0.2$ may be a bit high, but probably the greatest uncertainty is the photometry itself. New photometry for a small sample with a CCD detector would be a very useful check. The overall uncertainty of the MC distance moduli is still about 0.2 mag.

Fall: The kinematic study of globular clusters by Freeman, Illingworth, and Oemler suggests that there may never have been a "halo" phase in the history of the LMC. In this context, the kinematics of the RR Lyrae variables would be of great interest. If they are in a disk, the velocity dispersion should be about 20 km/s or less, whereas, if they are in a spheroid, the velocity dispersion should be about 40 km/s or more. Thus, it may be possible to decide between these alternatives with only a few radial velocities.

Graham: Yes, indeed, and this should now be possible with currently available techniques.

Renzini: Let me warn the observers that one should not take for granted that if a MC globular contains RR Lyraes then its age should be virtually identical to that of Galactic globulars. Indeed, a cluster can maintain a population of RR Lyraes for some 3 to 5 Gyr, and, even more important, there is no reason to believe that the [CNO/Fe] ratio in old MC globulars is the same as in Galactic globulars. For instance, if [CNO/Fe] $\simeq 0$ in MC old globulars, while [CNO/Fe] $\simeq 0.5$ to 0.7 in the Galactic globulars, then the former can be several billion years younger than the latter ones, in spite of them containing RR Lyraes.

Graham: On the whole I think this unlikely for field stars in view of the radically different distributions of the RR Lyraes and intermediate age stars like carbon stars. Presumably we will be able to check it for the MC globular clusters when good color-magnitude diagrams become available.

A SURVEY FOR RED VARIABLES IN THE MAGELLANIC CLOUDS

T. Lloyd Evans
South African Astronomical Observatory

ABSTRACT. One LMC and two SMC fields of 0.3 sq. deg. have been searched for red variables. Carbon stars of V $\sim$ 16-17 are common and are usually of small amplitude, while the LMC alone contains numerous faint M type variables of small amplitude. M giants of small amplitude generally have much shorter periods than carbon stars. The LMC contains numerous Miras with a P-L relation similar to that of galactic Miras, while the SMC has few Miras but many bright red variables of large amplitude which have a steeper P-L relation.

Lloyd Evans (1971a) made a study of Magellanic Cloud fields with the 1.9m reflector from 1966-71. New and published results are discussed in the light of present ideas. The initial priority was to detect Mira variables and a sufficient number were found in a field in the Bar of the LMC to show that they obey a P-M_{bol} relation similar to that of galactic Miras (Glass & Lloyd Evans 1981). Subsequent spectroscopic work shows that this relation holds for M, MS and carbon stars. Very few Miras have been found in the SMC; the distance is not so great as to be responsible. However the SMC contains much brighter red variables of large amplitude, including some of the Harvard variables as well as stars which are fainter in blue light because they are carbon stars (Lloyd Evans 1971b, 1980a, in preparation). The surface density of these stars is comparable to that of normal Miras in the LMC and the masses are estimated to be $\leq$ 7 $M_\odot$ (Lloyd Evans 1971b). Wood, Bessell & Fox (1981, preprint: WBF) deduce from extensive infrared photometry that all the large amplitude red variables are AGB stars with a range of mass and chemical composition which occupy a broad instability strip of slope in the P-M_{bol} diagram intermediate between the relations which would be fitted to LMC Miras and the more luminous variables separately. Removal of stars of small amplitude from this diagram shows two well-defined loci, linear in M_{bol} - log P, which represent the Mira and luminous Mira-like stars respectively and meet near P = 280 days. Representatives of both Clouds fall on each locus and it remains to be seen whether a larger sample would fill the instability strip proposed by WBF.

WBF and Frogel (preprint) show that luminous red variables of small amplitude, all of which are K or M stars, are massive stars in the core He - or C - burning stage of evolution. These are present in moderate numbers in our fields but most small amplitude red variables have $V \sim 16-17$. Most of those in the SMC are carbon stars while there are also many M stars in the LMC. The distribution of periods is quite different: carbon stars occupy a wide range from 170-360 days, with a few at longer or shorter periods (Lloyd Evans 1978), while nearly all M stars have $100 \leq P \leq 150$ days.

REFERENCES

Glass, I.S. and Lloyd Evans, T., 1981. Nature, 291, pp 303-304.
Lloyd Evans, T., 1971a. The Magellanic Clouds, pp 74-78, ed. Muller, A.B., Reidel, Dordrecht, Holland.
Lloyd Evans, T., 1971b. Observatory 91, pp 118-120.
Lloyd Evans, T., 1978. Mon. Not. R. astr. Soc., 183, pp 305-317.
Lloyd Evans, T., 1980. Mon. Not. R. astr. Soc., 193, pp 333-336.
Wood, P.R., Bessell, M.F. and Fox, M.W., 1981. Proc. astr. Soc. Australia, 4, pp 203-205.

A SEARCH FOR RED VARIABLE STARS IN THE LMC

G. Paltoglou, P. R. Wood, M. S. Bessell and K. Ratnatunga
Mount Stromlo & Siding Spring Observatories,
Australian National University, Canberra.

1. OBSERVATIONS AND REDUCTION

The observational material for this study consists of a series of 19 I plates (IVN+RG715) of the southern half of the LMC taken by the UK Schmidt Telescope over a six year period from 1977-1983. A small region of size 28'x56' centred on $\alpha(1950) = 5\ 28\ 50.7$, $\delta(1950) = -69\ 31\ 56$ was scanned on all plates with a PDS microdensitometer and magnitudes were derived for typically 14000 stars per scan area. All magnitudes were converted to a common system by comparison of magnitudes to a standard plate; this comparison showed the individual rms measurement error to be 0.13 mag. No photoelectric standard sequence exists in the region studied but the instrumental magnitudes have been converted to preliminary I_K magnitudes by using the I_K magnitudes given by Blanco, McCarthy and Blanco (1980) for some of the red stars in the field.

Variables were defined initially as those stars whose standard deviation in magnitude about their mean I from all plates was > 0.14 mag. Light curves were then plotted for all (233) of these stars and 80 stars were selected with obvious amplitudes in I of > 0.5 mag. Periods were then searched for using the technique of Stellingwerf (1978). This search yielded 44 stars with a distinct period, 20 with $\Delta I > 1.0$ mag. and the remainder with $0.5 < \Delta I < 1.0$ mag.

In the field studied, Gaposhkin (1970) lists 46 Cepheids, 10 irregular variables, 7 red variables with periods, and 3 eclipsing variables. This survey found 6 of the Cepheids, 4 irregulars, and 5 of the red variables with periods (although only one, HV12048, was found by us to have a regular period). The present survey technique has therefore greatly increased the number of known red variables in the region studied. It is not, however, an efficient method for locating Cepheid variables.

2. DISCUSSION

The period distribution for the present sample of stars is shown in Figure 1 and compared with the period distribution for local Mira variables from Wood and Cahn (1977). The LMC red variables found in the

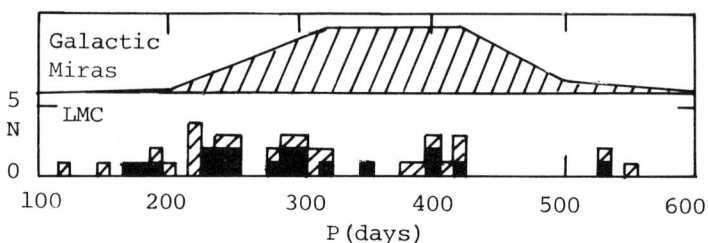

Fig. 1 Period distributions for LMC red variables with amplitude $\Delta I \gtrsim 1.0$ mag. (solid) and with $1.0 > \Delta I \gtrsim 0.5$ mag. (hatched). The period distribution of Galactic Miras is also shown.

present survey tend to be of shorter period than the local Mira variables (note that there will be a bias in our sample against periods of ~ 1 year). This result might be expected since the LMC is more metal poor than the Galaxy and Galactic Mira variables show a decrease in mean period with decrease in metal abundance. The 45% of our sample with I amplitudes > 1 mag. (similar to the amplitudes of ~1.5 mag. common in Galactic Mira variables - Eggen 1975) has a period distribution similar to that of the smaller amplitude variables (Fig. 1) so it is unlikely that the shorter period distribution the LMC objects can be accounted for by their being semi-regular variables rather than genuine Mira variables. Typical magnitudes at maximum are I_{max} ~ 13.4, with HV12048 (P=526d) being brightest at $I_{max} \approx 12.7$ mag. The magnitude $I \approx 13.4$ ($M_I \approx -5.3$) is in reasonable agreement with I_{max} for Galactic Miras ($M_I \approx -6.0$, Eggen 1975) and SMC red variables of Lloyd Evans (1978) ($M_I \approx -5.0$). The bolometric corrections for red variables given by Bessell and Wood (1983) indicate a mean bolometric magnitude for the bulk of the LMC variables in the present sample of ≈ -5 mag.

In summary, we have identified 80 red variables with amplitude $\Delta I > 0.5$ mag. in a 28'x56' region for the LMC and have obtained periods for 44 of these. We suggest that most of the stars are similar to the metal poor Mira variables in the Galaxy. The few longer period objects may be relatively massive AGB stars ($2 < M/M_\odot < 9$) (Wood, Bessell and Fox (1983).

We are extremely grateful to the UK Schmidt Telescope Unit for taking, and giving us access to, the plate material used in the present study.

REFERENCES

Bessell, M.S., and Wood, P.R.: 1983, submitted to Publ. A.S.P.
Blanco, V.M., McCarthy, M.F., and Blanco, B.M.: 1980, Astrophys. J. 242, pp. 938-964.
Eggen, O.J.: 1975, Astrophys. J., 195, pp. 661-678.
Gaposhkin, S.: 1970, Smithsonian Astrophys. Obs. Special Report, No. 310.
Lloyd Evans, T.: 1978, Mon. Not. Roy. astron. Soc., 183, pp. 305-317.
Stellingwerf, R.F.: 1978, Astrophys. J., 224, pp. 953-960.
Wood, P.R., Bessell, M.S., and Fox, M.W.: 1983, Astrophys. J. in press.
Wood, P.R., and Cahn, J.H.: 1977, Astrophys. J., 211, pp. 499-508.

JHK OBSERVATIONS OF MAGELLANIC CLOUD CEPHEIDS

D.L. Welch, B.F. Madore
Department of Astronomy, David Dunlap Observatory
University of Toronto, Toronto, Ontario, Canada M5S 1A1

ABSTRACT

The pioneering work of McGonegal et al. (1982), originally employing random-phase data for the LMC near-infrared period-luminosity (P-L) relation, is refined to mean light and extended to the SMC using new observations. We use a procedure for reducing single-phase infrared observations to mean light to obtain JHK magnitudes for Magellanic Cloud Cepheids. The new <H> P-L relations are presented and discussed.

INTRODUCTION

New JHK observations of Cepheids in the LMC and SMC obtained on the 2.5m Dupont reflector at Las Campanas in Jan. 1983 have been combined with previous observations in an effort to determine intensity-weighted mean magnitudes at $H(1.6 \mu m)$. As the amplitude at H is only one-third that at V, and metallicity, duplicity, and extinction effects are very small in the infrared, we expect the dispersion of the points about the <H> P-L relation to reflect only intrinsic width and depth.

DISCUSSION

Full H lightcurves obtained for 23 galactic Cepheids from Welch et al (1984) have been used to reduce random-phase observations of LMC and SMC Cepheids to mean light by a procedure described in that paper. We have done this for 40 LMC and 27 SMC Cepheids and obtain the following mean-light P-L relations:

$$<H> = 15^m\!.89 - 3.06 \log P \qquad \sigma = 0^m\!.22 \qquad (LMC)$$

$$<H> = 16.48 - 3.25 \log P \qquad \sigma = 0.18 \qquad (SMC)$$

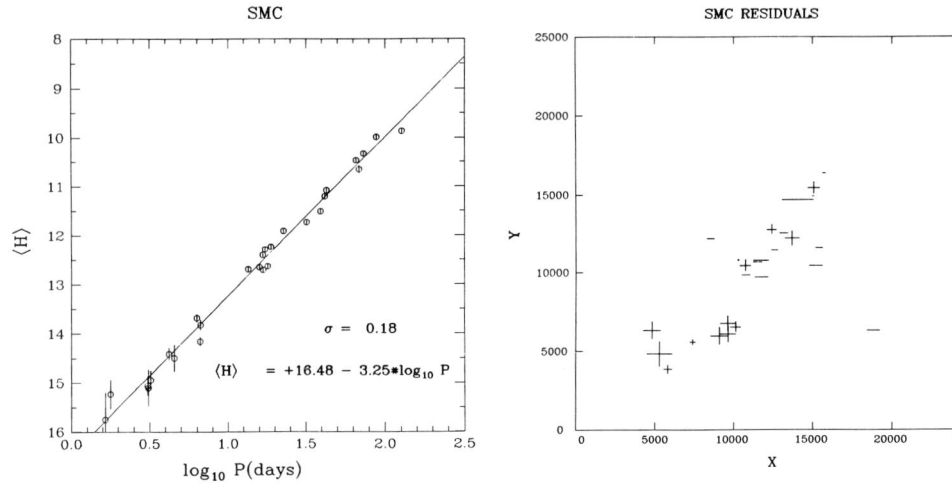

Figure 1. The <H> P-L relation for 27 Cepheids in the SMC. Observations were obtained at CTIO and MWLCO.

Figure 2. (O-C) residuals from the SMC <H> P-L relation. X and Y are in arcseconds from Payne-Gaposchkin and Gaposchkin (1966).

The SMC <H> P-L relation is displayed in Fig. 1. The small dispersion of the points about the least-squares regression line illustrates the advantages of the infrared approach. This small dispersion is even more remarkable when it is found that the residuals from the mean P-L relation correlate strongly with position on the sky (see Fig. 2). One is forced to conclude that depth effects are important and that the intrinsic dispersion of the <H> P-L is even smaller than that reported above. These points will be discussed more fully in a future paper.

REFERENCES

McGonegal, R., McLaren, R.A., McAlary, C.W., and Madore, B.F. 1982, Astrophys. J. (Letters), L33.
Payne-Gaposchkin, C. and Gaposchkin, S. 1966, Smithson. Contr. Astrophys. 9.
Welch, D.L., Wieland, F., McAlary, C.W., McGonegal, R., Madore, B.F., McLaren, R.A., and Neugebauer, G. 1984, Astrophys. J. Suppl. (submitted).

SPECTROSCOPY OF OVERLUMINOUS CEPHEIDS IN THE MAGELLANIC CLOUDS

Horace A. Smith
Michigan State University

Leo Connolly
Southeast Missouri State University

The Small Magellanic Cloud is known to contain types of short period Cepheid variable stars not yet discovered in either the Large Magellanic Cloud or, with the exception of a single star, in the Galaxy. These variables can be divided into two categories: anomalous Cepheids and Wesselink-Shuttleworth (WS) stars. The former, which have also been found in dwarf spheroidal systems and in the globular cluster NGC 5466, have periods of 0.4-3 days, but average 0.7-1.0 mag. brighter than RR Lyrae and BL Her stars of equal period. The stars we call WS stars have periods less than about 1.1 day and, at M_v = -1 to -2, are brighter than anomalous Cepheids of equal period.

Connolly (1980, 1982) has recently proposed that some variables in the direction of the LMC, hitherto thought to be foreground RR Lyrae stars, are in fact LMC members, perhaps related to the WS stars of the SMC. This suggestion was based on the apparent existence of a period-luminosity relation for these stars, even though they had periods of under a day, and the determination that several of these variables have radial velocities consistent with LMC membership. However, in the direction of the LMC, radial velocity measures cannot always distinguish between stars in the LMC and stars in the halo of the Galaxy.

Butler, Demarque, and Smith (1982 -- BDS) reported results of low resolution spectroscopy of 3 anomalous Cepheids and 3 WS stars in the SMC. We have obtained new spectra of Wesselink and Shuttleworth (1965) variables 2, 4, and 5. Spectra have also been obtained for 5 of the suspected LMC variables of Connolly (1982). All spectra were obtained with the cassegrain Reticon spectrograph on the du Pont 2.5 m telescope of the Las Campanas Observatory.

Metal abundances for the variables have been determined by the ΔS method (Preston 1959, Butler 1975). These measures confirm the result of BDS that the SMC WS are, as a group, only mildly metal deficient, at [Fe/H] = -0.6. BDS estimated an age of 2.5×10^8 years for the WS stars in the SMC, and an age of 3×10^9 years for the more metal poor anomalous Cepheids ([Fe/H] = -1.3). If the stars examined so far are representative, then the SMC interstellar medium may have been substantially enriched in heavy elements during the past few Gyr.

Four of the five Connolly variables are confirmed to have radial velocities consistent with LMC membership. Nonetheless, these stars range from [Fe/H] = -0.7 to -1.7. They thus display no common metal abundance which might have served to establish their identity as a distinct group of variable stars. Neither can they be distinguished from galactic halo RR Lyrae stars on the basis of metal abundance alone.

This work has been supported in part by the U.S. National Science Foundation.

REFERENCES

Butler, D. 1975. Astrophys. J. 200, pp. 68.
Butler, D., Demarque, P., and Smith, H. A. 1982. Astrophys. J. <u>257</u>, pp. 592.
Connolly, L. P. 1980. Sp. Sci. Rev. <u>27</u>, pp. 443.
Connolly, L. P. 1982. Pulsations in Classical and Cataclysmic Variable Stars, ed. by J. P. Cox and C. J. Hansen, pp. 188.
Preston, G. W. 1959. Astrophys. J. <u>130</u>, pp. 507.
Wesselink, A. J. and Shuttleworth, M. 1965. M.N.R.A.S. <u>130</u>, pp. 433.

SPECTRAL TYPES, COLOR INDICES, AND ABUNDANCES FOR CEPHEIDS IN THE MAGELLANIC CLOUDS AND THE GALAXY

C.D. Laney and D.H. McNamara
Department of Physics and Astronomy, Brigham Young University
Provo UT 84602, USA

Abstract: Accurate MK temperature and luminosity classes have been obtained for 58 Cepheids in the Galaxy and in the Magellanic Clouds from 121 spectrograms. Classification criteria for the spectral region 3700X - 6000X in the spectra of F-K supergiants are described. The spectra of galactic and LMC Cepheids are found to be similar, and Cepheids in both systems appear to obey the same relation between $(B-V)_o$ and spectral type. The low $E(B-V)$ values of Parsons (1970) and certain other recent authors are confirmed. The spectra of SMC Cepheids show slightly weaker metal lines relative to galactic Cepheid spectra, and SMC Cepheids average about 1.3 subclasses earlier in spectral type than LMC and galactic Cepheids at the same value of $(B-V)_o$. Reddenings for both Clouds are found to be less than 0.05. Among supergiants ($M_v = -4$ to -8), stars of the same spectral type have the same $(B-V)_o$ regardless of luminosity class. Spectral types at minimum light are found to be later than those reported by Feast (1974) when luminosity effects are allowed for. Spectroscopic surface gravities near minimum light appear to be lower than those near maximum and slightly fainter M_v values are found at each luminosity class than those usually cited. SMC Cepheids appear to have lower surface gravities than LMC Cepheids at the same absolute magnitude, as one would expect from the lower metal abundances.

A full account is being prepared for the Pub. Astron. Soc. Pacific.

PERIOD CHANGES IN MAGELLANIC CLOUD CEPHEIDS

H. Deasy and P.A. Wayman
Dunsink Observatory, Dublin, Ireland.

It has been found possible to obtain information on period change in data on 115 cepheid variable stars in the Magellanic Clouds (84 LMC cepheids and 31 SMC cepheids). Harvard Observatory data of the period 1910 to 1950 (collated by Payne-Gaposchkin and Gaposchkin) are combined with Dunsink Observatory observations carried out by C.J. Butler in 1966/67 and with South African Astronomical Observatory observations covering the years 1975-1977 by Martin, Thomas, Carter and Davies to derive mean periods for the intervals between the various data sets. Using these new periods in conjunction with the very accurate Harvard periods, seperate estimates of the time averaged fractional change of period per day, $d/dt(\ln P)$, with corresponding estimated errors, could be evaluated for two epochs, one around 1950 and the other around 1971. It was found that 70 stars give rates of change of period that are not significantly different from zero, that 20 stars have two values of rate of change of period that are in agreement at the two epochs (indicative of secular period change), while 22 stars give two disparate values of rate of change of period (indicative of irregular period changes).

While neither positive nor negative period changes dominate, it was found that "large" ($>10^{-8}$ days^{-1}) significant fractional period changes were largely confined to cepheids with periods greater than about 10 days, which is probably due to the greater instability of the extended envelopes of these stars. The fraction of significant period change estimates which indicate irregular period changes is less for the LMC sample than for the SMC sample, and according to Szabados (1977, 1980, 1981) this fraction is still less for galactic cepheids. Van Genderen (1983) also finds a large fraction of irregular period changes for Magellanic Cloud cepheids. This indicates a possible link between composition and period changes.

The size of the period changes increases with period in good agreement with the trend predicted by evolutionary theory, with most of the data falling around the values associated with the second and third crossings of the instability strip. Indeed, even the irregular period changes, whose time averaged rates are greater than the corresponding errors, follow this trend, although it seems likely that these period fluctuations are non-evolutionary in character. Although the binary light-time effect may account for some of the observed period changes, natural constraints on the extent of this effect exclude it as an explanation for the larger period changes. For instance the possibly cyclic variation in the period of HV 900 cannot be due to orbital motion.

References

L. Szabados (1977) Mitt. Sternw. Ungar. Akad. Wiss., No.70
L. Szabados (1980) Mitt. Sternw. Ungar. Akad. Wiss., No.76
L. Szabados (1981) Mitt. Sternw. Ungar. Akad. Wiss., No.77
A.M. Van Genderen (1983) Astron. Astroph. Supp. Ser. 52,432-442

MAGELLANIC CLOUD CEPHEIDS: INTRINSIC AND EXTRINSIC PROPERTIES AS INFERRED FROM NUMERICAL SIMULATIONS OF TWO-COLOUR PHOTOMETRY

M.J. Stift
Institut für Astronomie
Türkenschanzstr. 17
A-1180 Wien, Austria

Controversy is raging over the question which of the relations linking periods, luminosities and colours of classical cepheids should be used to the greater advantage of cosmologists in desperate need of a reliable primary distance indicator. Adopting a cepheid evolutionary scenario of utmost simplicity, Sandage and Tammann (1969) (=ST) have introduced the concept of an universal period-luminosity-colour (PLC) relation which is expected to minimize the Malmquist bias by virtue of its small intrinsic spread; the PLC relation constitutes the only primary pillar of their controversial distance scale. More recently, Stift (1982) and Clube and Dawe (1983) have shown that there are good reasons to consider a more realistic and hence more complex picture of cepheid evolution and photometric behaviour. The PLC based distance scale advocated by ST will be seriously affected by a number of non-canonical effects such as mass loss, abundance differences and varying relative crossing times. Cepheids no longer obey a single universal PLC relation as in canonical theory but a different PLC relation with a period-dependent colour coefficient applies for every crossing, mass loss rate and chemical composition (see also van Genderen 1983b). It is possible to model cepheid photometric behaviour in a way similar to that proposed by Stift (1982) and to analyze existing LMC and SMC cepheid surveys in order to put constraints on intrinsic and extrinsic properties. The <B> and <V> magnitudes are given by

$$(\langle B \rangle - \langle V \rangle)_0 = a \log P_0 + b + SBV \cdot U\{-0.5, +0.5\} \quad \text{canonical} \quad (1)$$
$$(\langle B \rangle - \langle V \rangle)_0 = a \log P_0 + b + G\{SBV, CBV\} \quad \text{iconoclastic} \quad (2)$$
$$\langle V \rangle_0 = \alpha \log P + \beta (\langle B \rangle - \langle V \rangle)_0 + \gamma \quad (3)$$

where
$$\log P = \log P_0 + CR \quad \text{crossing difference} \quad (4)$$
$$\log P = \log P_0 \cdot (1 + SML \cdot U\{0,1\}) \quad \text{differential mass loss} \quad (5)$$
$$\log P = \log P_0 + G\{SAB, CAB\} \quad \text{abundance differences} \quad (6)$$

and finally
$$E(B-V) = T\{\nu, \lambda\} \quad (7)$$
$$\langle V \rangle = \langle V \rangle_0 + G\{SV, CV\} + R \cdot E(B-V) \quad (8)$$
$$\langle B \rangle = \langle V \rangle_0 + (\langle B \rangle - \langle V \rangle)_0 + G\{SV, CV\} + (R+1) \cdot E(B-V) \quad (9)$$

$U\{m,n\}$ stands for a deviate from an uniform distribution between m and n; $G\{\sigma,c\}$ for a deviate from a Gaussian distribution with standard deviation σ and cutoff at $c \cdot \sigma$. $T\{\nu, \lambda\}$ for a deviate from a reddening distribution resulting from absorption in a mean of ν clouds, $1/\lambda$ being the mean amount of absorption per cloud - ν follows a Poisson distribution, $1/\lambda$

an exponential distribution (Tscharnuter and Stift 1983) (=TS). R denotes the ratio of total visual to selective absorption, SBV the strip width in B-V at constant period and SV the standard error of an observation. SML is taken to be ≈0.02-0.03, CR≈0.10 and SAB≈0.02. Cutoffs range between 2σ and 3σ. Comparison of the predicted with the observed scatter about the mean PL, PC and Wesenheit relations - TS give the numerical values for the 5 surveys used in this investigation - leads to the following conclusions (see also Stift 1982):

(1) If the canonical view holds true both photoelectric and photographic LMC cepheid photometry must be credited with much higher accuracy than the corresponding SMC photometry. Given the small differential distance modulus of $\Delta\mu < 0\overset{m}{.}4$ and given the fact that the surveys have been carried out with comparable or even identical equipment this conclusion appears paradoxical.

(2) The paradox can be resolved by assuming that up to 10-15% of the SMC cepheids are observed in the 1st, 4th and 5th crossings, as compared to only a few percent of the LMC cepheids. Some support for this explanation comes from theory which suggests a correlation between enhanced relative importance of 1st crossings and lower metallicity (Becker et al. 1977).

(3) Another way of resolving the paradox consists in the adoption for the SMC of a value of R substantially in excess of R=3.2. It can be shown that Isserstedt's (1980) claim for R=2 has to be rejected; on the other hand my estimate of 3<R<5 is well in line with Harris' (1981) findings.

(4) A reliable estimate of total LMC and SMC extinction turns out to be impossible. It appears however that to a very high degree of probability, SMC and LMC suffer reddening of a comparable order of magnitude, viz. $E(B-V) \approx 0\overset{m}{.}11 - 0\overset{m}{.}14$ (van Genderen 1983a); the extremely low values of $E(B-V) < 0\overset{m}{.}04$ sometimes found in the literature can safely be excluded. Reddening estimates are somewhat higher on the average in the canonical case than in the iconoclastic case.

(5) By no means is the zero-point of the PLC relation insensitive to chance selection effects especially in view of an empirical determination of this relation. The uncertainty may attain a few $0\overset{m}{.}1$ in the canonical case and is aggravated by a period-dependence of the colour-coefficient as well as by the above-mentioned non-canonical effects. Minimization of chance selection effects can best be achieved by the use of mean magnitudes at standard period following de Vaucouleurs (1978).

References

Becker, S.A., Iben, I., Tuggle, R.S.: 1977, Astrophys. J. 218, 633
Clube, S.V.M., Dawe, J.A.: 1983, Astron. Astrophys. 122, 255
van Genderen, A.M.: 1983a, Astron. Astrophys. 119, 192
van Genderen, A.M.: 1983b, Astron. Astrophys. 124, 223
Harris, H.C.: 1981, Astron. J. 86, 1192
Isserstedt, J.: 1980, Astron. Astrophys. 83, 322
Sandage, A., Tammann, G.A.: 1969, Astrophys. J. 157, 683
Stift, M.J.: 1982, Astron. Astrophys. 112, 149
Tscharnuter, W.M., Stift, M.J.: 1983, in preparation
de Vaucouleurs, G.: 1978, Astrophys. J. 223, 351

A MORE COMPLETE LISTING OF LMC PLANETARY NEBULAE

N. Sanduleak
Warner and Swasey Observatory
Case Western Reserve University, Cleveland, Ohio

In an earlier paper by Sanduleak et al. (1978) a listing was given of 102 confirmed and probable planetary nebulae in the Large Magellanic Cloud detected on objective-prism plates taken with the Curtis Schmidt telescope at Cerro Tololo. Subsequently, deeper coverage was obtained on nitrogen-baked Kodak IIIa-J plates plus GG 455 filter exposed for 90 minutes. The thin prism was again used to provide a dispersion of about 1500 Å mm^{-1} at Hβ and the spectra were unwidened. An additional 25 planetary nebula candidates were found on this new plate material to show the requisite characteristics, i.e. they display (a) a stellar appearance, (b) [OIII] λλ5007, 4959 strongly in emission, and (c) no evidence of a continuum.

These additional candidates are given in Table I along with an outlying planetary recently reported by Savage et al. (1982) which is positioned just outside of the region covered by our plate material. Figure 1 is the updated version of Figure 3 in Sanduleak et al. (1978) and shows the surface distribution of 127 spectroscopically detected planetaries (No. 103 is not included). The centroid of the distribution and the earlier conclusions concerning the radial dependence of the surface density of the planetaries are essentially unchanged by the addition of these 26 objects.

REFERENCES

Jocoby, G.H.: 1980, Astrophys. J. Suppl. 42, 1.
Lindsay, E.M., and Mullan, D.J.: 1963, Irish Astron. J. 6, 51.
Sanduleak, N., MacConnell, D.J., and Philip, A.G.D.: 1978, Pub. A.S.P. 90, 621.
Savage, A., Murdin, P.G., and Clark, D.H.: 1982, Observatory 102, 229.

Table I. Additional LMC Planetary Nebulae

No.	R.A.	1975	Dec.	Remarks or Name(s)
103	$3^h 31^m.7$	$-69°$	30'	Or faint emission-line galaxy.
104	4 24.7	-69	43	LM1-1
104A	4 25.4	-66	51	Savage et al. (1982)
105	5 02.7	-69	28	
106	5 03.2	-68	35	
107	5 06.9	-69	17	LM2-6
108	5 07.5	-69	22	
109	5 12.0	-69	25	LM2-13, J5
110	5 12.3	-68	31	
111	5 15.6	-69	30	J 15
112	5 18.9	-69	12	J 20
113	5 20.2	-69	27	J 26
114	5 21.2	-70	06	J 31
115	5 21.5	-69	44	J 33
116	5 24.8	-69	07	LM2-24, J 38
117	5 25.2	-69	17	
118	5 26.4	-69	02	J 41
119	5 29.6	-70	21	
120	5 29.8	-70	18	
121	5 30.7	-71	15	
122	5 34.6	-69	35	
123	5 34.7	-70	29	LM2-36
124	5 40.8	-67	18	
125	5 42.5	-67	38	
126	5 53.5	-70	25	
127	5 55.0	-71	08	

LM1 = Lindsay and Mullan (1963), Table 1.
LM2 = ibid, Table 2.
 J = Jacoby (1980).

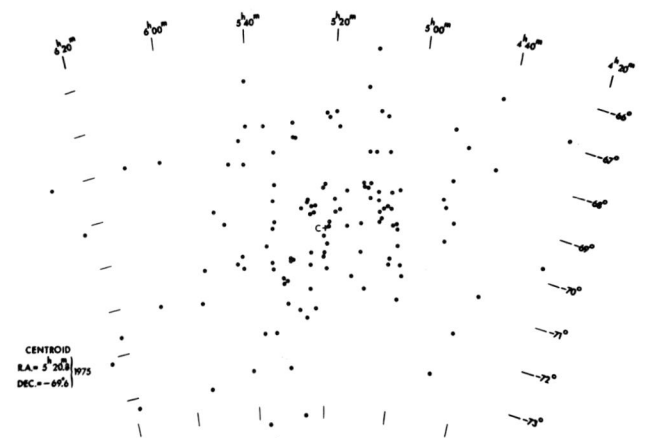

Figure 1. Surface distribution of 127 LMC planetary nebulae

SPECTROPHOTOMETRY OF EARLY-TYPE SUPERGIANTS IN THE LMC

K. Nandy, G.I. Thompson, D.H. Morgan
Royal Observatory, Edinburgh, Scotland

L. Houziaux
Institut d'Astrophysique, Liège, Belgium

ABSTRACT. Measures of the equivalent widths of major stellar features in the visible and UV regions of a sample of the LMC supergiants whose types range from O5 to A1 have been derived. Their low resolution UV spectra were obtained with the IUE and complemented in the visual with spectra from ESO 3.6 m and SAAO 1.95 m telescopes. The methods used are entirely numerical and as far as possible free from subjective estimates. The data for the LMC members have been compared with those obtained for their Galactic counterparts. The variation of CIV 1550 Å and Si IV 1400 Å absorptions with spectral type has been investigated. It may be that the intensity of HeI lines is smaller for the middle and late B supergiant members of the LMC as compared with those Galactic standards, otherwise there is no systematic difference within the observational errors between the Galactic and LMC supergiants in the strengths of the principal stellar features observed in the visible as a function of spectral type. Galactic O and early B supergiants have greater equivalent widths of Si IV λ 1400 and C IV λ 1550 absorptions than the LMC supergiants of similar spectral types, confirming earlier observations of Hutchings.

As a part of a programme oriented to study the interstellar medium, ultraviolet and visible spectra of thirteen O and B supergiants have been studied and the spectroscopic features have been measured.
The programme stars include SK-67-111 (O5f), -67-108 (O7IIIf) -68-135 (O9,5), -68-107 (B0), -69-253 (B0), -67-110 (B1(e)), -65-91 (B1), -69-213 (B1), -68-177 (B2), -69-274 (B2,5), -69-108 (B3), -65-11 (B5), -69-247 (B8). For fourteen other objects we have either UV or visible spectra. Comparison stars were the white dwarfs L 870-2E and E 99. They have been observed either with the ESO 3.6 m telescope (cassegrain spectrograph equipped with IDS) or with the SAAO 1.9 m telescope and the IPCS.

The ultraviolet wavelength range has been observed from the I.U.E. in the usual ranges and with the low resolution mode. ESO observations cover the region 400-480 nm at 2.9 nm/mm reciprocal dispersion and the region 480-550 nm at 11nm/mm, while the SAAO spectra are taken at 6 nm/mm and extend from 395 to 510 nm.

The 29 nm/mm dispersion spectra were reduced to give data points every 0.3 Å, and subsequently averaged over five data points. The accuracy of the wavelength scale was found to be ± 0.05 nm. Lower dispersion spectra were averaged over three pixels leading to a resolution of 0.5 nm.

Spectral features observed are mostly blends. HeI lines include 4388, 4471, 4713, and 4920 Å. HeII lines at 4542, 4686 and 5412 Å are present in the spectra of stars hotter than B1. In order to improve the statistical accuracy, we have made sums of HeI equivalent widths (sum HeI) and compared to corresponding sums in galactic supergiants of the same spectral subclasses taken from previous work published at ROE. While there is a close agreement between LMC and galactic OB stars in both HeI and HeII, the HeI line strength for LMC stars later than B2 are consistently weaker in the LMC stars than in their galactic counterpart. There is also agreement between the observed sum (HeI and equivalent widths computed in non-LTE by Auer and Mihalas using our previous temperature determinations), except for SK-68-107 where the observed Sum (HeI) is almost twice the computed value. In the ultraviolet, CIV and SIV doublets have P Cygni profiles. They are weaker than galactic stars. Terminal velocities ranging from 1000 to 2500 km/s have been determined.

The only metal line represented throughout the sequence is Mg II 4481 Å. This line is stronger in our LMC stars than in the galactic standards, especially around B0-B1. Other features observed include Hβ (in emission in SK-68-135 and SK-67-110) Hγ, blends of NIII at 4511-4515 Å, SiIII at 4568-4575 Å, FeII 4584 Å in SK-65-11, -69-247, -69-250, and NII/CIII/OII blend at 4642-4650 Å, and finally one feature at 4662-4676 Å.

As a conclusion, we can state that no significant difference is found in the strengths of the major stellar features for LMC and galactic O and early B supergiants. There is however a suggestion that HeI lines may be weaker in middle and late B LMC supergiants.

References

Auer L., and Mihalas D : 1972 Astrophys. J. Suppl. Ser. 24,19:
———————————————— : 1973 Astrophys. J. Suppl. Ser. 25,43:
Thompson G.I., Nandy K., Morgan D.H., Willis A.J., Wilson R., and Houziaux L. : 1982 Monthly Notices R. Astron. Soc. 200,551

FROM WHAT RADIUS DO THE WINDS OF MAGELLANIC CLOUD SUPERGIANTS
PRODUCE Fe II EMISSION LINES?

M. Friedjung and C. Muratorio
Institut d'Astrophysique, Paris, France and
Observatoire de Marseille, Marseille, France

Abstract
Maximum values for the inner radii at which Fe II emission line formation can take place in the winds of three Magellanic Cloud supergiants have been determined. The lie between 3×10^{13} and 3×10^{14} for R50, R82, and R126.

We have studied the formation of emission lines of singly ionized iron in stellar winds in which hydrogen is almost completely ionized, using the theory of Viotti (1976). The wind was assumed to have a constant velocity in the line forming region, so that the electron density varied as $1/r^2$; r being the distance from the centre of the wind. Two cases of line formation were considered:
  (c) The local random (thermal plus "turbulent") velocities are much less than the wind velocity (Sobolev case of line formation in a moving envelope).
  (b) The local random velocities are much greater than the wind velocity. This case is suggested when P Cygni profiles are not observed; it is, however, much less realistic physically.
In each of these cases theoretical self-absorption curves were calculated. Such a curve is defined as being the graph of $\log F_c + \log W_\lambda - \log(gf) + 3 \log \lambda$ against $\log(gf) + \log \lambda$ for lines of the same multiplet. Here F_c is the relative continuum flux per angstrom, W_λ the equivalent width, λ the wavelength, and gf the oscillator strength multiplied by the statistical weight of the lower level. When all lines are optically thin, as is usually the case for the forbidden lines, the self absorption curve is replaced by a horizontal line.

In order to compare observations with theory, the values of F_c were first dereddened. This was done by supposing that the forbidden lines of multiplets 6F and 20F, having the same upper term, lay on the same optical thin self-absorption curve (horizontal line); any observed difference being due to differential reddening. After dereddening the observed self-absorption curves of permitted multiplets were shifted vertically and horizontally so as to give a curve which agreed best with the theoretical one. In this method multiplets having the same upper term must not have a relative vertical shift, while those having the same lower term must not have a relative horizontal shift of the observed curves (Friedjung and Muratorio 1980). The shifts gave relative populations of terms of different multiplets. The comparison of theoretical and observed self-absorption curves is shown in Figs. 1, 2, and 3.

Maximum inner radii of the FeII line formation region were found by comparing the self-absorption curve of multiplet 38 with that of multiplets 6F and 20F, whose upper term is the lower term of multiplet 38. The method is similar to that used by Friedjung and Muratorio (1980) to determine the radius of the FeII emitting region of S22.

R50 gave, with an E(B-V) of 0.10 and a wind velocity of 70 km s^{-1}, a maximum inner radius of 10 10^{13}cm. A wind velocity of 100 km s^{-1} gives a radius of 8   10^{13}cm. In the case of R82, with a zero E(B-V) and a wind velocity of 150 km s^{-1}, a maximum inner radius of 3 10^{13}cm is obtained. In both these cases an (b) type (see with figure) theoretical self-absorption curve was used for the permitted lines. R126 was analyzed using a (c) type self-absorption curve. E(B-V) was taken as 0.09 and a mean deconvolved line half-intensity width of 0.54Å led to a radius of 2 to 3 10^{13}cm.

Friedjung, M., Muratorio, A. 1980, Astron. Astrophys. **85** 233
Viotti, R.   1976, Astrophys. J. **204** 293

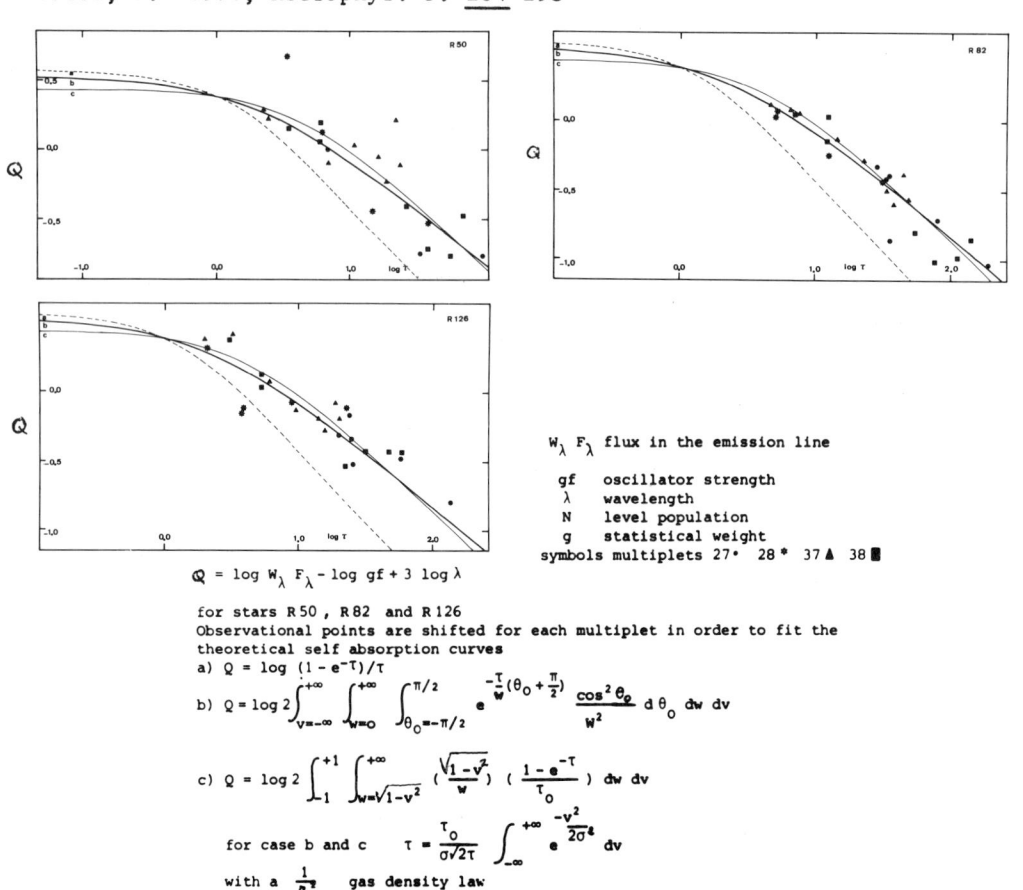

$Q = \log W_\lambda F_\lambda - \log gf + 3 \log \lambda$

for stars R50, R82 and R126
Observational points are shifted for each multiplet in order to fit the theoretical self absorption curves

a) $Q = \log (1 - e^{-\tau})/\tau$

b) $Q = \log 2 \int_{v=-\infty}^{+\infty} \int_{w=0}^{+\infty} \int_{\theta_0=-\pi/2}^{\pi/2} e^{-\frac{\tau}{w}(\theta_0 + \frac{\pi}{2})} \frac{\cos^2 \theta_0}{w^2} d\theta_0 \, dw \, dv$

c) $Q = \log 2 \int_{-1}^{+1} \int_{w=\sqrt{1-v^2}}^{+\infty} (\frac{\sqrt{1-v^2}}{w}) (\frac{1-e^{-\tau}}{\tau_0}) \, dw \, dv$

for case b and c $\tau = \frac{\tau_0}{\sigma\sqrt{2\pi}} \int_{-\infty}^{+\infty} e^{\frac{-v^2}{2\sigma^2}} dv$

with a $\frac{1}{r^2}$ gas density law

$W_\lambda F_\lambda$ flux in the emission line
gf   oscillator strength
λ   wavelength
N   level population
g   statistical weight
symbols multiplets 27 • 28 * 37 ▲ 38 ■

R127: AN S DOR TYPE VARIABLE INTERMEDIATE BETWEEN Of AND WN

O. Stahl, B. Wolf
Landessternwarte, Königstuhl, 6900 Heidelberg 1

On 1982 Jan. 9 a brightening of 0.75 mag of the OIafpe star R127 (=HDE 269858) was detected. Subsequently extensive photometric (optical and infrared) and high resolution spectroscopic (ground-based and IUE) observations covering a period of more than one year have been carried out. These observations have shown R127 to be a new S Dor variable which is hotter than any previously detected and studied variable of this class. For its maximum phase we derived the following basic parameters: $M_{bol} = -10.6$, $T_{eff} = 16000$ K, $R_* = 150\ R_\odot$. The stellar wind is characterized by a small mean velocity (v = 110 km s^{-1}), a high mass loss rate ($\dot{M} = 6 \cdot 10^{-5}\ M_\odot yr^{-1}$), and a decelerated velocity field. The optical spectra show strong Balmer lines with P Cygni profiles. The He I lines changed within the one year period from strong P Cygni type lines to complex line profiles with extremely wide (FWZI = 3000 km s^{-1}) shallow emission line wings. The IUE spectra are dominated by crowded absorption lines of singly ionized metals. A very complex shell phenomenon is indicated by the multiple substructure of these ultraviolet lines. We suggest that R127 is a massive (M > 60 $M_\odot$) Of star evolving via a short-lived S Doradus phase to a late WN star.

A detailed paper by O. Stahl, B. Wolf, G. Klare, A. Cassatella, J. Krautter, P. Persi and M. Ferrari-Toniolo is forthcoming in Astron. Astrophys.

AN UNSTABLE Ofpe STAR IN THE LMC

Nolan R. Walborn*
Laboratory for Astronomy and Solar Physics
NASA/Goddard Space Flight Center

The extreme Ofpe star HDE 269858 = Radcliffe 127 in the Large Magellanic Cloud (Walborn 1977), which has ejected nitrogen-rich material (Walborn 1982), was observed to be in a state of outburst on December 20, 1982. Spectroscopic observations at 2.5Å resolution with the Cerro Tololo Inter-American Observatory 1.5 meter SIT vidicon system showed a blue magnitude of 10.2, more than 1 magnitude brighter than previously, and a qualitatively changed spectrum. The Of emission features and the strong He I P Cygni profiles (especially $\lambda\lambda 3889$ and 4471) had disappeared, while the hydrogen Balmer emission remained very intense. IUE observations by S. Shore and N. Sanduleak provide evidence that the star has been in this "high" state since at least March 1981. Hence, a phenomenon related to those of P Cygni and/or Eta Carinae may be indicated. A current preprint by O. Stahl, B. Wolf, G. Klare, A. Cassatella, J. Krautter, P. Persi, and M. Ferrari-Toniolo announces the prior discovery of the outburst at the European Southern Observatory in January 1982 and discusses extensive observations of its development, making a convincing case for interpretation of HDE 269858 as the hottest known Hubble-Sandage variable and a prime candidate for an O star becoming a WR via episodic enhanced mass loss.

REFERENCES

Walborn, N. R. 1977, Ap. J. 215, 53.
_____. 1982, Ap. J. 256, 452.

*National Research Council Senior Associate

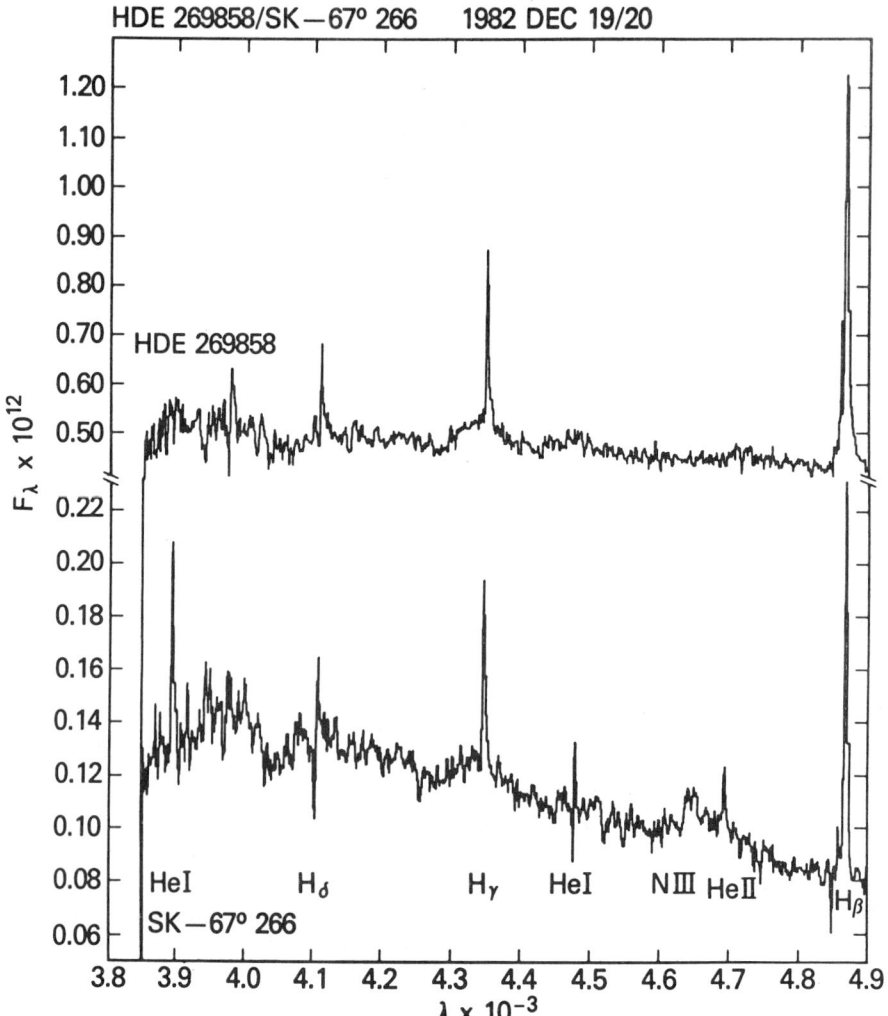

The spectra of HDE 269858 and Sanduleak −67° 266, as recorded with the SIT vidicon in December 1982. Note the absence of He I λλ3889 and 4771 as well as of the N III and He II Of emission features in HDE 269858; in 1975 these features were stronger than in Sk −67° 266 (see Figure 2 of Walborn 1977). Also at that time HDE 269858 was only a few tenths of a magnitude brighter than Sk −67° 266. The broad depression between λλ4000 and 4100 and the hump near Hγ in the SIT data are due to imperfections of the flux calibration.

EVIDENCE FOR A BLACK HOLE IN LMC X-3

A. P. Cowley, D. Crampton, J. B. Hutchings, and R. Remillard
Arizona State Univ., Dominion Astrophysical Obs., and M.I.T.

The best X-ray position (Einstein Observatory HRI - Giacconi et al 1979) for LMC X-3 confirms its identification with the early type star first suggested by Warren and Penfold (1975). Our spectroscopic observations obtained with the CTIO 4-m telescope show the WP star is a slightly reddened B3 V star with $m_v \approx 16.9$. Large radial velocity variations ($\Delta v \approx 500$ km s^{-1}) reveal an orbital period of 1.7049 days. From the orbital elements (Table 1) one can determine the mass function $f(M) = (M_x \sin i)^3/(M_{opt} + M_x)^2 = 2.3\ M_\odot$, which shows without <u>any</u> assumptions about the mass of the optical star, the orbital inclination, or the mass ratio the unseen X-ray object has a mass $> 2.3\ M_\odot$. Detailed analysis of the HEAO-1 scanning modulation collimator X-ray data shows that the system does <u>not</u> eclipse, implying that the orbital inclination is $\leq 65°$. Assuming the B star mass lies between 4 and 8 $M_\odot$ (an average mass for a normal B3 V star would be about 6 - 7 $M_\odot$), the mass of the unseen companion must lie between 7 and 13 $M_\odot$ (see Fig. 4a - Hutchings, this volume). Smaller inclinations of course give even higher masses. An important point is that the unseen star must have a mass <u>larger</u> than that of the B star, and thus if it were any kind of normal <u>star it</u> should be easily seen in the spectrum. Thus the X-ray emitting object is a very good candidate for a black hole.

It is important to establish the absolute magnitude of the optical star, as together with the temperature it can be used to determine a radius which can be equated with the Roche radius of the star. This in turn yields the mass ratio in the system and hence another way of deriving the minimum stellar masses. Unfortunately, at present, the M_v of the optical star is uncertain by more than ≈ 0.5 magnitudes for a variety of reasons: The reddening, and hence the extinction, is poorly known. Our spectroscopic observations show all of the absorption lines are about 50% weaker than in standard stars, indicating the probable presence of a luminous accretion disk in the system (as is seen in SMC X-1 and LMC X-4). Finally the overall variability (≈ 0.6 mag) of the system introduces uncertainty into the absolute magnitude. Once M_v is determined, masses can be read from Fig. 4 of Cowley et al (1983). A similar discussion about the mass determination has been made by Paczynski (1983), although he fails to consider the large range in uncertainty of the absolute magnitude.

Some confirming evidence has been presented to establish further that the orbital period is correctly determined and that a black hole may be present. Van der Klis and van Paradijs (1983) have found photometric variations (ellipsoidal) modulated on the 1.7 day period. The amplitude of these variations is consistent with the mass ratio and orbital inclination range suggested here. Also, White and Marshall (1983) have shown that the exceptionally soft X-ray spectrum of LMC X-3 is similar to that of another black hole candidate, Cyg X-1 in its high state.

Although both the mean radial velocity of the optical star and its luminosity are consistent with LMC X-3 being a member of the Large Cloud, its distance has almost no bearing on the dynamical arguments which imply the probable presence of a black hole.

TABLE 1. Orbital Elements for LMC X-3

$P = 1.70491 \pm 0.00007$ days
$K = 235 \pm 11$ km s^{-1}
$V_o = 310 \pm 7$ km s^{-1}

$T_o = $ HJD 2445278.45 ± 0.01
$e = 0$
$\omega = 0$ (assumed)

REFERENCES

Cowley, A.P., Crampton, D., Hutchings, J.B., Remillard, R., and Penfold, J. E. 1983 Ap. J. 272, in press.
Giacconi, R. et al, 1979 Ap. J. 230, 540.
Paczynski, B. 1983 preprint.
Warren, P. R. and Penfold, J. E. 1975 M.N.R.A.S. 172, 41p.
White, N.E., and Marshall, F.E. 1983 preprint.
van der Klis, M., and van Paradijs, J. 1983 I.A.U. Circ. #3765.

THE STELLAR CONTENT OF 30 DORADUS

Nolan R. Walborn*
Laboratory for Astronomy and Solar Physics
NASA/Goddard Space Flight Center

ABSTRACT

The supergiant H II region 30 Doradus is placed in context as the optically most spectacular component in a much larger region of recent and current star formation in the Large Magellanic Cloud, as shown by deep Hα photographs and the new IRAS results. The current state of knowledge concerning the concentrated central cluster in 30 Dor is summarized. Spectroscopic information exists for only 24 of the brightest members, most of which are WR stars; however, photometry shows over 100 probable members earlier than B0. The spectral classification of these stars is a difficult observational problem currently being addressed; in the meantime their hypothetical ionizing luminosity is calculated from the photometry and compared with that suggested for the superluminous central object R136a alone, and with the H II region luminosity. With reference to related regions in the Galaxy, the likelihood that many of the brightest objects in 30 Dor are multiple systems is emphasized. An interpretation of R136a as a system containing a few very massive stars (as opposed to a single supermassive object) is in good accord with the observations, including the visual micrometer results. The study of 30 Dor and its central cluster is vital for an understanding of the numerous apparently similar regions now being discovered in more distant galaxies.

I. INTRODUCTION: THE 30 DORADUS REGION

The 30 Doradus (Tarantula) Nebula, with a diameter of 200 pc and nearly 10^6 $M_\odot$ of ionized hydrogen, might properly be categorized as a supergiant H II region, with the Great Carina Nebula, smaller by a factor of five in diameter and by two orders of magnitude in mass, relegated by comparison to mere giant status. Yet despite its spectacular attributes, in another sense the Tarantula is only a hot spot, the optically most prominent component of a much larger region of

*National Research Council Senior Associate

recent and current star formation in the Large Magellanic Cloud. This aspect of the 30 Doradus region is well shown by the beautiful Hα photographs from the UK Schmidt (Elliott et al. 1977; Meaburn 1979), which reveal apparently related filamentary structure on a scale of 2 kpc. To the south are the fascinating N158/N159/N160 complexes (Henize 1956), which contain LMC X-1 (Hutchings and Crampton 1983), a young SNR (Mathewson et al. 1980), a molecular maser (Scalise and Braz 1981; Caswell and Haynes 1981), an infrared protostar (Gatley et al. 1981), and other strange objects (Heydari-Malayeri and Testor 1982). There is also strong CO emission from this area (Israel et al. 1982). Indeed, some of the first observations with the IRAS satellite have evidently discovered a hotbed of current star formation in this area south of the Tarantula (Sky and Telescope 1983).

In contrast, the 30 Doradus Nebula itself appears to be a well evolved H II region. No compact far IR sources were found there by Werner et al. (1978), and only some weak CO emission was detected by Israel et al. (1982). On the other hand, it harbors large internal motions seen in both the nebular emission lines (Smith and Weedman 1972; Cantó et al. 1980; Meaburn 1981; Cox and Deharveng 1983) and the interstellar absorption lines (Blades 1980; Walborn 1980; Blades and Meaburn 1980; de Boer, Koornneef, and Savage 1980), as well as an extended X-ray source (No. 72) discovered by the Einstein Observatory (Long, Helfand, and Grabelsky 1981). The latter was originally classed as an SNR, but it may well represent instead the effects of the energetic stellar winds within the Nebula, as also observed in the Carina Nebula (Seward et al. 1979; Seward and Chlebowski 1982; Walborn 1982a; Walborn and Hesser 1982).

II. FIELD STARS

All of this impressive radiative and dynamical activity in the 30 Doradus Nebula is of course ultimately due to the massive stars it contains, so it is appropriate to focus our attention on what is currently known about them. I shall begin by specifying some objects which probably do not belong to the stellar content of 30 Doradus; in this connection the status of the Nebula as a component within a larger young region containing a range of evolutionary states is relevant. From interesting near-IR surveys, Hyland, Thomas, and Robinson (1978) and McGregor and Hyland (1981) discovered 13 M supergiants in the field of the Tarantula, and they suggested a causal relationship to the younger exciting stars of the Nebula in terms of successive star formation events. However, as they also pointed out, the late-type supergiants likely belong to the population of the larger region which contains 30 Doradus. Seven of the M stars are associated with two small clusters 3' northwest and 8' southeast of R136, which appear unrelated to the Nebula, and only one (No. 18) is near the central exciting cluster. In fact, three of the original Radcliffe stars (Feast, Thackeray, and Wesselink 1960) are probably related to the M supergiants and not to the Nebula, on the basis of both their

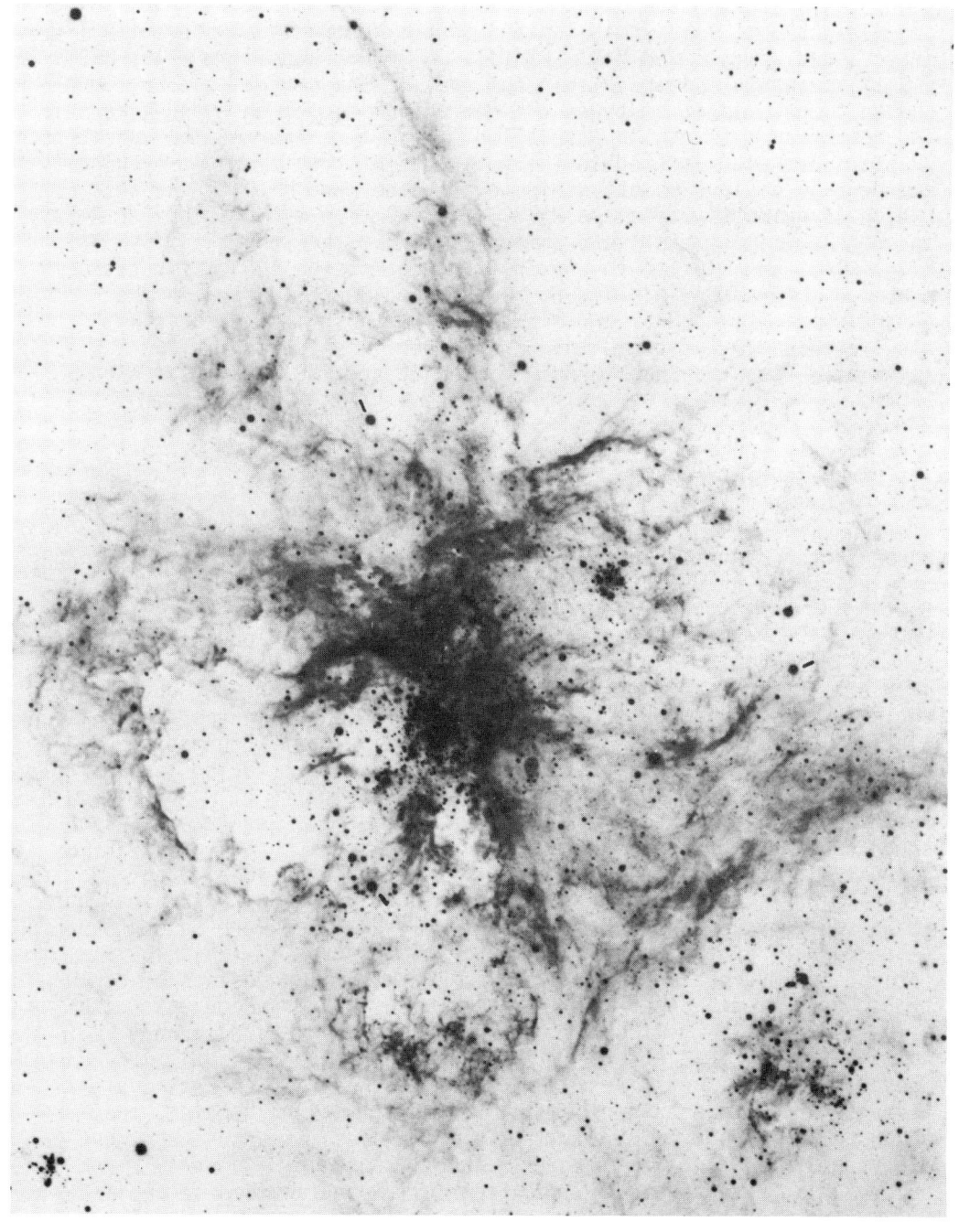

The 30 Doradus Nebula in the light of [S II] $\lambda\lambda 6716,6731$; 30 min. exposure at the CTIO 4m prime focus, through a 100Å interference filter on 098-04 emulsion. North is at the top and east to the left. Tick marks identify the peripheral stars R130 to the west, R144 to the north, and R143 to the south. The latter two are separated by 6.1 in declination, corresponding to 93 pc in projection. N157B is at the SW corner.

relatively late spectral types and their peripheral locations: they are R131 (B9 I), R132 (B-A, located in the northwest cluster with several red stars), and R143 (F7 Ia). A fourth Radcliffe star, R138 (A0 I), probably appears within the area of the central cluster by chance. These spectral types are characteristic of an h and χ Persei and not of an H II region population. Moreover, four of the Radcliffe WR stars are also located in the outskirts of the Tarantula nebulosity and bear no obvious relationship to it: they are R130 to the west (which is evidently of mixed WN/WC or of WC type and possibly an X-ray source; P. S. Conti, private communication; Mathewson et al. 1983); and R144, R146, and R147 to the north. Indeed, of the 100 LMC WR stars listed by Breysacher (1981), no fewer than 40 are located in the general vicinity of N157 and N158. One of these is the recently discovered AB 10 (Azzopardi and Breysacher 1979), which may be associated with the SNR N157B (Danziger et al. 1981; Gilmozzi et al. 1983).

III. THE CENTRAL CLUSTER

Thus, I take the stellar content of the 30 Doradus Nebula to be its concentrated central cluster, which clearly interacts with the nebulosity and evidently provides its energy source. I dearly wish that I had definitive spectral classifications for all of its members to present to you - perhaps at the next Magellanic Cloud symposium in about five years! They constitute a very difficult observational problem, not only because of their faintness and crowding, but also due to the bright, inhomogeneous nebulosity which contaminates the He I lines. A sky-subtracting detector and detailed data processing will be essential. At the present time, spectroscopic information is available for only 23 likely members of this cluster exclusive of R136 - a small fraction of its brightest stars, as we shall see. Several new WR spectra have recently been discovered and discussed (Melnick 1978, 1982, 1983; Azzopardi and Breysacher 1979; Phillips 1982; Conti 1982). A significant new result is that two of them are of type WC; previously it was believed and often stated that only WN types were to be found in this region. Spatially resolved observations by M. M. Phillips (private communication) have established that R140 consists of two WR stars, one of each type. I have observed a few non-WR objects with the CTIO 1.5-meter SIT vidicon system in December 1982; preliminary spectral classifications for five of them are given in Table 1. These results are not definitive due to the need for increased signal to noise and more detailed nebular line subtraction; however, they are adequate to confirm the original Radcliffe classifications and most importantly to establish that R137 and R142 are in fact early B supergiants (recently they have mistakenly been reported to be of type WN). Three additional stars of the central cluster are probably early B supergiants as well. The presence of these and the WC types raises a basic question about the 30 Doradus stellar population. It seems unlikely that all of them could be chance alignments on the central cluster. I often refer to the following three distinctly different

Table 1. New Spectral Classifications

R133	O7:
R137	B0.7 – 1.5
R138	A0 I
R139	O6 – 7 Iaf
R142	B0.5 – 0.7 I

galactic regions as evolutionary paradigms for massive young clusters: (1) the Eta Carinae Association, the brightest members of which are of types O3 and WN, containing no OB supergiants (except for the peculiar variable QZ Car), but associated with substantial H II and dust; (2) Scorpius OB1, brightest members late O–early B supergiants, contains one WN and one WC star, but no late B/AF/M supergiants, and no significant H II or dust; and (3) h/χ Persei, visually brightest members A supergiants, many M supergiants. Hence, at least part of the 30 Doradus population corresponds to the Sco OB1 stage, but associated with strong H II and dust as in Eta Car. Either evolutionary differences from the galactic regions (due, e.g., to the greater masses or lower metal content), or a mixture of stellar ages in 30 Doradus, may be indicated, but a conclusive interpretation of this interesting cluster must await spectroscopy of its abundant fainter members.

In the meantime, a data set which includes most members of the 30 Doradus central cluster and can provide an indication of its possible nature is the photographic narrow-band photometry by B. E. Westerlund (1964; private communication). The numbers of early-type stars present in the central region (exclusive of R136) are listed by half-magnitude intervals of M_V in Table 2, together with the possible alternative corresponding spectral types. The absolute visual magnitudes have been computed by simply adopting for all stars $A_V = 1^m.2$ as determined by Savage et al. (1983) for R136, and an LMC true distance modulus of $18^m.6$; a more elaborate procedure is not warranted by the available information or the present purpose. Table 2 contains 109 stars with M_V between –4.0 and –8.3; spectroscopic data are available only for 20 of those brighter than –6.0 and for one at –5.4 (O3 V, Melnick 1983 No. 4). The potential ionizing radiation from these stars has been calculated on the assumption that the earlier spectral types apply; if they are predominantly later-type giants the ionizing luminosity will of course be far less. For this purpose, incompleteness factors of 20% at O3-6 V, 50% at O7-9 V, and 0 for $M_V \leq$ –6.0 have been estimated from Westerlund's charts and lists; this estimate adds 25 stars with M_V between –4.0 and –6.0, and it is conservative in view of the probable incidence of unresolved multiple systems and spectroscopic binaries. Two new WN stars from Melnick (1983) have also been added, giving a total of 136 stars. The ionizing luminosities given by

Table 2. Content of the 30 Doradus Central Cluster

M_V Range	No. of Stars	Possible Spectral Types	
-4.0/-4.4	20	O8-9 V	B1 III
-4.5/-4.9	14	O7 V	B0 III
-5.0/-5.4	26	O6 V	O9.5 III
-5.5/-5.9	15	O3-5 V	O6-9 III, II
-6.0/-6.4	13	O3-5 III, O3 I	Ib
-6.5/-6.9	10	WR	
-7.0/-7.4	4	and/or	Ia
-7.5/-7.9	6	Multiple	
-8.0/-8.4	1	systems	Ia^+

Note: Stars with photometry by Westerlund, excluding R136.

Cruz-González et al. (1974) have been adopted, but scaled to the absolute visual magnitude calibration of Walborn (1973a). The value for type O9 I has been used for all WR stars. These results are presented in Table 3, where they are also compared with the ionizing luminosity of R136a implied by the spectral mix matched to its UV flux by Savage et al. (1983), and with the H II region luminosity within a diameter of 140 pc determined at 6 cm by Mills, Turtle, and Watkinson (1978). The ionizing luminosity of the 22 spectroscopically classified stars is consistent with that found by Feitzinger et al. (1980), but the potential contribution from the remaining cluster stars is seen to be much greater, and altogether comparable to that from R136a alone. The total ionizing radiation from the cluster is in good agreement with the H II region luminosity, considering the uncertainties involved in the stellar computations. The parallel discussion by Melnick (1983) leads to essentially the same conclusions reached here.

IV. R136

With reference to some related galactic regions, I would now like to emphasize further the high probability that many of the brightest stellar images in the 30 Doradus cluster correspond to multiple systems. In the Carina Nebula, for instance, the systems HD 93128/93129AB, 93160/93161AB, and 93204/93205AB, each containing at least three massive stars, have projected dimensions of about 0.25 pc, corresponding to 1" at the LMC. There one also finds the double spectroscopic binary HD 93206 (QZ Carinae; Leung, Moffat, and Seggewiss 1979; Morrison and Conti 1980) – four massive stars along the same line of sight at a distance of only 2.8 kpc. Even more germane is the spectacular HD 97950 system at the center of the galactic supergiant H II region NGC 3603 (Walborn 1973b; Moffat and Seggewiss 1983), which

Table 3. Ionizing Luminosities in 30 Doradus

Spectral Types	No. of Stars	$L_H(10^{49} s^{-1})$	Total $L_H(s^{-1})$
Photometrically observed stars			
O3 V	6	36	
O4 V	6	25	
O5 V	6	19	
O6 V	31	47	
O7 V	21	13	
O8 V	15	5	
O9 V	15	3	1.5×10^{51}
O3 III, I	4	51	
O4-5 III	10	74	1.2×10^{51}
Spectroscopically observed stars (excluding R136)			
WN	12[a]	} 21	
WC	2[b]		
O6-7 I	3[c]	9	
B I	5[d]	---	0.3×10^{51}
Total			3×10^{51}
R136a (Savage et al. 1983)			
O3 III	15	193	
WN3	15	22	2×10^{51}
H II luminosity (Mills et al. 1978)			8.7×10^{51}

[a] R134, 135, 140, 145; Melnick A, C, G = 42, H = 39, J, 30, 35N; Azzopardi-Breysacher 11
[b] R140, Melnick E
[c] R133, 139; Westerlund 6
[d] R137, 141, 142; Westerlund 10, 29

I submit as a closeup view of R136! Here a volume of diameter 0.25 pc contains HD 97950ABCDEF, each of which may be at least double from the appearance of the stellar images. From published photometry of nearby stars (Moffat 1974, 1983; van den Bergh 1978) and my plates, I have estimated V ~ 9.6 for the 0".6 double AB, which with a distance of 7.5 kpc and $A_V = 3^m9$ implies $M_V = -7.9$ for each component. Moreover, through a 20" aperture (0.75 pc) V = 9.2, so that AB contributes 0.7 of the visual light from that volume. The integrated spectral type of HD 97950 is O5-6(n) + WN6-A(B) (Walborn 1982b).

Well, what about R136? Feitzinger et al. (1980) isolated the relatively faint components b and c at about 2" and 3".5, respectively, to the southeast of a, while Schmidt-Kaler and Feitzinger (1981) derived V = 10.77 for the latter. I wish to emphasize that there are now four independent and concordant observations of the $R136a_1$-a_2 structure with a separation of 0".48, position angle 220°, and magnitude difference $\sim 1^m$: visual micrometer measurements by Innes (1927) and Worley (1983), photography by Chu and Wolfire (1983), and speckle interferometry by G. Weigelt (quoted by Moffat and Seggewiss 1983). Moreover, Weigelt has reportedly now observed an a_3 component at 0".1 and $\Delta m \sim 1^m$, and fainter components a_4 - a_5. With $A_V = 1\overset{m}{.}2$ from Savage et al. (1983), these numbers imply M_V = -8.3, -7.7, and -7.3 for $R136a_1$, a_2, and a_3, respectively. Furthermore, through a 3" aperture (0.75 pc) V ~ 10.1 (Y.-H. Chu, private communication), so that R136a contributes 0.5 of the visual light from that volume. The integrated spectral type of R136 (HD 38268) is OB(n) + WN5-A(B) (Walborn 1977). Hence, the virtually identical structures of R136 and HD 97950 should be clear, in agreement with the more detailed discussion of Moffat and Seggewiss (1983).

Both UV (Savage et al. 1983) and near-IR (Vreux, Dennefeld, and Andrillat 1982) spectroscopic observations show that one or more components of R136 emit on O3-type spectrum. If one assumes that these are the visually brightest components, they can be compared directly to HD 93129A in the Carina Nebula, which has M_V = -6.6 and ~100 $M_\odot$ (Walborn 1982c and references therein). $R136a_1$ is $1\overset{m}{.}7$ brighter, which with a linear mass-luminosity relation leads to 480 $M_\odot$. The mass-luminosity relations of Maeder (1983) would reduce this value to 220 or 280 $M_\odot$. This is a weak upper limit for the most massive object in R136, because of the likely possibilities that $R136a_1$ is itself multiple, and that the visually brightest components are not the hottest.

Recent high signal-to-noise spectroscopic observations of supergiant H II regions in more distant galaxies have frequently revealed the presence of WR emission bands in their central regions; more than two dozen cases are now known (D'Odorico and Rosa 1982; Rosa 1983; Massey and Hutchings 1983). It is evident that the LMC will play its canonical role as stepping-stone to the galaxies in the context of this problem: an understanding of R136 and the 30 Doradus central cluster will be essential to the interpretation of the more distant objects.

REFERENCES

Azzopardi, M. and Breysacher, J. 1979, Astron. Astrophys. 75, 243.
Blades, J. C. 1980, M.N.R.A.S. 190, 33.
Blades, J. C. and Meaburn, J. 1980, M.N.R.A.S. 190, 59P.
Breysacher, J. 1981, Astron. Astrophys. Suppl. 43, 203.
Cantó, J., Elliott, K. H., Goudis, C., Johnson, P. G., Mason, D., and
    Meaburn, J. 1980, Astron. Astrophys. 84, 167.

Caswell, J. L. and Haynes, R. F. 1981, M.N.R.A.S. 194, 33P.
Chu, Y.-H. and Wolfire, M. 1983, Bull. American Astron. Soc. 15, 644.
Conti, P. S. 1982, in IAU Symp. 99, Wolf-Rayet Stars: Observations, Physics, Evolution, ed. C. W. H. de Loore and A. J. Willis (Dordrecht: Reidel), p. 551.
Cox, P. and Deharveng, L. 1983, Astron. Astrophys. 117, 265.
Cruz-González, C., Recillas-Cruz, E., Costero, R., Peimbert, M., and Torres-Peimbert, S. 1974, Rev. Mexicana Astron. Astrof. 1, 211.
Danziger, I. J., Goss, W. M., Murdin, P., Clark, D. H., and Boksenberg, A. 1981, M.N.R.A.S. 195, 33P.
de Boer, K. S., Koornneef, J., and Savage, B. D. 1980, Ap. J. 236, 769.
D'Odorico, S. and Rosa, M. 1982, in IAU Symp. 99, Wolf-Rayet Stars: Observations, Physics, Evolution, ed. C. W. H. de Loore and A. J. Willis (Dordrecht: Reidel), p. 557.
Elliott, K. H., Goudis, C., Meaburn, J., and Tebbutt, N. J. 1977, Astron. Astrophys. 55, 187.
Feast, M. W., Thackeray, A. D., and Wesselink, A. J. 1960, M.N.R.A.S. 121, 337.
Feitzinger, J. V., Schlosser, W., Schmidt-Kaler, Th., and Winkler, Chr. 1980, Astron. Astrophys. 84, 50.
Gatley, I., Becklin, E. E., Hyland, A. R., and Jones, T. J. 1981, M.N.R.A.S. 197, 17P.
Gilmozzi, R., Murdin, P., Clark, D. H., and Malin, D. 1983, M.N.R.A.S. 202, 927.
Henize, K. G. 1956, Ap. J. Suppl. 2, 315.
Heydari-Malayeri, M. and Testor, G. 1982, Astron. Astrophys. 111, L11.
Hutchings, J. B. and Crampton, D. 1983, I.A.U. Circ. No. 3791.
Hyland, A. R., Thomas, J. A., and Robinson, G. 1978, A. J. 83, 20.
Innes, R. T. A. 1927, Southern Double Star Catalogue (Johannesburg: Union Obs.).
Israel, F. P., de Graauw, T., Lidholm, S., van de Stadt, H., and de Vries, C. 1982, Ap. J. 262, 100.
Leung, K.-C., Moffat, A. F. J., and Seggewiss, W. 1979, Ap. J. 231, 742.
Long, K. S., Helfand, D. J., and Grabelsky, D. A. 1981, Ap. J. 248, 925.
Maeder, A. 1983, Astron. Astrophys. 120, 113.
Massey, P. and Hutchings, J. B. 1983, preprint.
Mathewson, D. S., Dopita, M. A., Tuohy, I. R., and Ford, V. L. 1980, Ap. J. Letters 242, L73.
Mathewson, D. S., Ford, V. L., Dopita, M. A., Tuohy, I. R., Long, K. S., and Helfand, D. J. 1983, Ap. J. Suppl. 51, 345.
McGregor, P. J. and Hyland, A. R. 1981, Ap. J. 250, 116.
Meaburn, J. 1979, Astron. Astrophys. 75, 127.
_____. 1981, M.N.R.A.S. 196, 19P.
Melnick, J. 1978, Astron. Astrophys. Suppl. 34, 383.
_____. 1982, in IAU Symp. 99, Wolf-Rayet Stars: Observations, Physics, Evolution, ed. C. W. H. de Loore and A. J. Willis (Dordrecht:Reidel), p. 545.
_____. 1983, The Messenger, ESO, No. 32, p. 11.
Mills, B. Y., Turtle, A. J., and Watkinson, A. 1978, M.N.R.A.S. 185, 263.
Moffat, A. F. J. 1974, Astron. Astrophys. 35, 315.
_____. 1983, Astron. Astrophys. 124, 273.

Moffat, A. F. J. and Seggewiss, W. 1983, preprint.
Morrison, N. D. and Conti, P. S. 1980, Ap. J. 239, 212.
Phillips, M. M. 1982, M.N.R.A.S. 198, 1053.
Rosa, M. 1983, Highlights Astron. 6, 625.
Savage, B. D., Fitzpatrick, E. L., Cassinelli, J. P., and Ebbets, D. C., 1983, Ap. J., in press.
Scalise, E., Jr. and Braz, M. A. 1981, Nature 290, 36.
Schmidt-Kaler, Th. and Feitzinger, J. V. 1981, in The Most Massive Stars, ed. S. D'Odorico, D. Baade, and K. Kjär (Garching: ESO), p. 105.
Seward, F. D. and Chlebowski, T. 1982, Ap. J. 256, 530.
Seward, F. D., Forman, W. R., Giacconi, R., Griffiths, R. E., Harnden, F. R., Jr., Jones, C., and Pye, J. P. 1979, Ap. J. Letters, 234, L55.
Sky and Telescope 1983, 65, 322.
Smith, M. G. and Weedman, D. W. 1972, Ap. J. 172, 307.
van den Bergh, S. 1978, Astron. Astrophys. 63, 275.
Vreux, J. M., Dennefeld, M., and Andrillat, Y. 1982, Astron. Astrophys. 113, L10.
Walborn, N. R. 1973a, A. J. 78, 1067.
_____. 1973b, Ap. J. Letters 182, L21.
_____. 1977, Ap. J. 215, 53.
_____. 1980, Ap. J. Letters 235, L101.
_____. 1982a, Ap. J. Suppl 48, 145.
_____. 1982b, A. J. 87, 1300.
_____. 1982c, Ap. J. Letters 254, L15.
Walborn, N. R. and Hesser, J. E. 1982, Ap. J. 252, 156.
Werner, M. W., Becklin, E. E., Gatley, I., Ellis, M. J., Hyland, A. R., Robinson, G., and Thomas, J. A. 1978, M.N.R.A.S. 184, 365.
Westerlund, B. E. 1964, in IAU Symp. 20, The Galaxy and the Magellanic Clouds, ed. F. J. Kerr and A. W. Rodgers (Canberra: Australian Academy of Science), p. 316.
Worley, C. E. 1983, Washington Neighborhood Meeting, Univ. of Maryland.

DISCUSSION

Appenzeller: Is it possible to derive an age for the stars in the 30 Doradus region from the fact that we seem to see so many evolved massive stars (WR stars, supergiants, ...)?

Walborn: The age of the Eta Carinae association has been estimated at 3 million years, and Scorpius OB1 would appear to be 5 to 7 million years old by interpolation between the former and h/χ Persei. The HII and O3 stars in 30 Dor indicate an Eta Carinae stage, while the B supergiants and WC stars correspond to Sco OB1. This mixture could indicate either evolutionary differences from the galactic regions, or a range of ages, in the 30 Dor central region.

Humphreys: Is anything known about circumstellar dust shells around any of the hot stars in the 30 Dor region or in NGC3603? We know that evolved very massive stars (60 to 100 $M_\odot$) such as η Car often eject processed material. Have any stars in these two very massive star complexes evolved to that stage?

Walborn: McGregor and Hyland (1981) found near-IR excesses in several of the brightest hot stars in 30 Dor, which they interpreted in terms of free-free emission and mass-loss rates. No η Car like objects are known in 30 Dor or NGC3603.

Geyer: Concerning the age of the 30 Dor association it might perhaps be of interest that the young globular cluster NGC2100, which is boardering on the east side of the complex, has an age of about 8-10 million years.

Walborn: 30 Doradus is located within a larger young region containing both earlier and more advanced evolutionary states.

UV EXTINCTION IN THE 30 DORADUS NEBULA AND THE UV ENERGY DISTRIBUTION
OF R136a.

Blair D. Savage and Edward L. Fitzpatrick
Washburn Observatory, University of Wisconsin-Madison

The properties of ultraviolet interstellar extinction in and near the
core of the 30 Doradus Nebula are studied. The pair method is employed
using nine reddened stars from within 5' (80pc) of the core and nine
unreddened stars from a variety of locations in the large Magellanic
Cloud. All of the 30 Doradus stars examined appear to be reddened by
$E(B-V) \simeq 0.12$ with an extinction law similar in wavelength dependence to
those derived for the LMC by Koornneef and Code (1981) and Nandy et al.
(1981). Several of the stars, including R136a, R145, and R147, are found
to be additionally reddened by $E(B-V) \simeq 0.18$ with an extinction law
qualitatively similar in wavelength dependence to the law found in the
Orion region. A two-component model, featuring a layer of "LMC foreground
dust" which affects all of the stars and a deeper layer of "nebular dust"
which affects some of the stars, provides the simplest explanation of the
extinction properties. The 2200 Å extinction bump is present in both
curves. The wavelength positions of the bump and the bump profiles, when
normalized to a linear "background extinction", are indistinguishable
from the average Galactic bump. The strengths of the bumps, relative
to $E(B-V)$, are 20-30% weaker than for the Milky Way Curve.

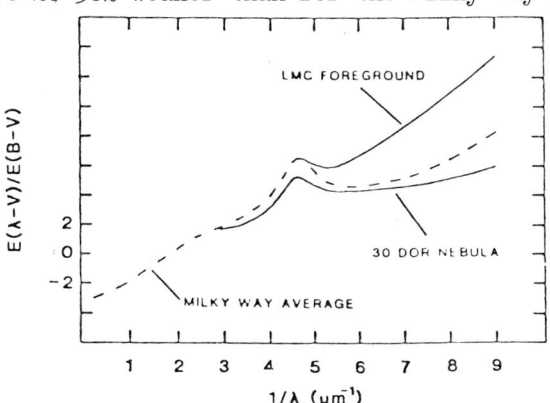

Figure 1. Extinction derived for "LMC foreground dust" and 30 Doradus
"nebular dust" are shown along with the average Milky Way extinction.

The extinction curves derived in this study have been applied to the ultraviolet energy distribution of R136a to determine its intrinsic continuum shape. The resultant UV continuum is similar in shape to that found for the LMC O3 stars R122 and Sk22-67. However, R136a is brighter than these stars by a factor of 13 and 43, respectively. This result implies an extremely large luminosity for R136a, because R122 is considered the most luminous "normal" star in the LMC, with $L \simeq 3 \times 10^6 L_\odot$. The dereddened energy distributions for R136a and R122 are shown in figure 2. Complete results of this investigation are contained in Savage et al. (1983) and Fitzpatrick and Savage (1984).

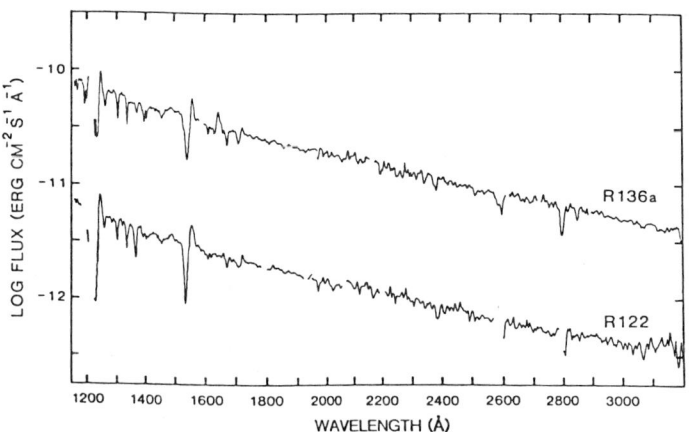

Figure 2. Dereddened IUE spectrum for R136a and R122. R136a was dereddened with $E(B-V) = 0.06$, 0.11, and 0.17 to account for Galactic foreground reddening, LMC foreground reddening and 30 Doradus Nebular reddening. The R136a fluxes have been scaled upwards by a factor of 2 in order to account for lightloss in the IUE small aperture. This correction assumes the ultraviolet radiation from R136a is dominated by radiation from component a1. A correction factor of 1.8 is required to correct for the small aperture light loss experienced by the extended emission around component a1 detected by Chu (this volume) at visible wavelengths. The comments by Melnick (1983) relating to the application of this correction factor are incorrect since the extended component has an extension comparable to the size of the IUE 3" aperture. R122 (O3III(f*)) is the brighter of the 2 LMC O3 stars, with $M_V \simeq -6.5$, and has been dereddened with $E(B-V)=0.07$ for Galactic foreground extinction.

REFERENCES

Fitzpatrick, E.L., and Savage, B.D.: 1984, Astrophys.J., in press.
Koornneef, J., and Code, A.D.: 1981, Astrophys.J., 247, pp. 860.
Melnick, J.: 1983, The Messenger, 32, pp. 11.
Nandy, K., Morgan, D.H., Willis, A.J., Wilson, R., and Gondhalekar, P.M.: 1981, M.N.R.A.S., 196, pp. 955.
Savage, B.D., Fitzpatrick, E.L., Cassinelli, J.P., and Ebbets, D.C.: 1983, Astrophys.J., (October 15).

ON THE NATURE OF R 136, THE CENTRAL OBJECT OF 30 DOR. A COMPARISON WITH
THE GALACTIC CLUSTER NGC 3603.

Michael Rosa [1], Jorge Melnick [1,2], Preben Grosbol [1]
1 European Southern Observatory, Garching/Munich, FRG
2 Universidad de Chile, Santiago, Chile

ABSTRACT
    The massive H II region NGC 3603 is the closest galactic counterpart
to the giant LMC nebula 30 Dor. Walborn (1973) first compared the ionizing
OB/WR clusters of the two H II regions and suggested that R 136, the unresolved luminous WR + O type central object of 30 Dor, might be a multiple
system like the core region of NGC 3603. Suggestions that the dominant
component of R 136, i.e. R 136A, might be either a single or a very few
supermassive and superluminous stars (Schmidt-Kaler and Feitzinger 1982,
Savage et al. 1983) have recently been disputed by Moffat and Seggewiss
(1983) and Melnick (1983), who have presented spectroscopic and photometric
evidence to support the hypothesis of an unresolved cluster of stars. We
have extended Walborn's original comparison of the apparent morphology of
the two clusters by digital treatment of the images to simulate how the
galactic cluster would look like if it were located in the LMC

    Electronographs of the H II regions were secured in the same night
with the 40 mm McMullan camera at the Danish 1.5m telescope at La Silla
(Melnick and Grosbol 1981). The V band films have a seeing of 1.8 arcsec
FWHM and were digitized with a resolution of 0.2 arcsec. We have adopted
a distance ratio of 7 between the 30 Dor complex (52 kpc) and NGC 3603
(7 kpc). To reproduce the dimming of NGC 3603 when moved away to the LMC
we have corrected the film densities by the ratio in exposure times and
additional factors of 1/49 due to the increase in distance and 14 due to
the different foreground extinction, 4.5 mag to NGC 3603 and about 1.7 mag
to R 136 in 30 Dor. The "moved" image of NGC 3603 has been convolved with
a seeing of 1.8 arcsec and artificial noise with a gaussian distribution
similar to that in the background of the 30 Dor image has been added to
make the "observations" comparable.

    Here we have to limit our presentation to the contour plots produced
from the final images. We note however that the apparent similarities in
the morphology of the two clusters, as seen at the same scale on the sky, are
striking (see figures in Walborn 1973) but misleading. In fact, the image
of NGC 3603 (3 arcmin overall diameter) as seen at the distance of the LMC
(Fig.1) covers only the innermost 25 arcsec of 30 Dor (Fig.2). The blurred

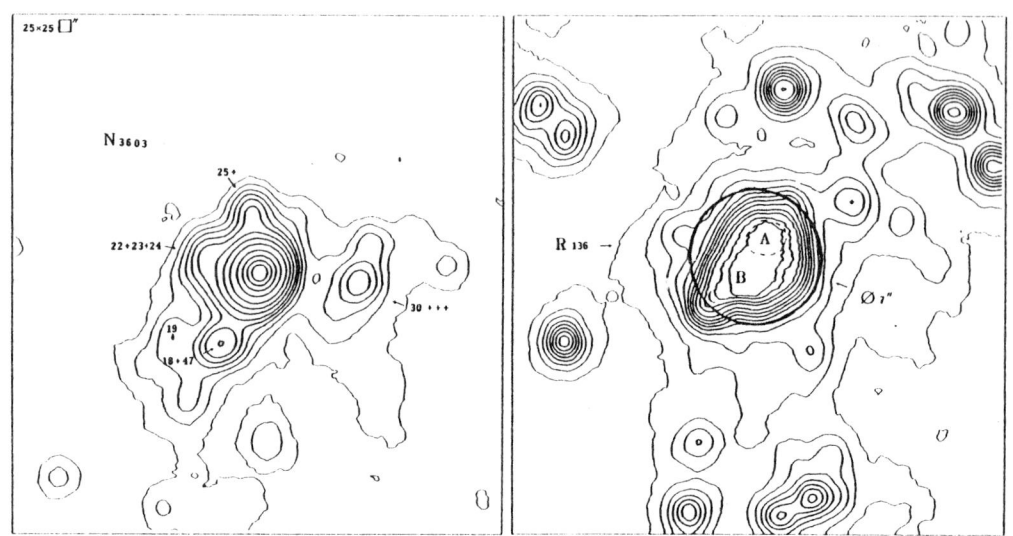

Fig.1　NGC 3603 at the LMC distance　　Fig.2　The vicinity of R 136

image of the whole NGC 3603 cluster is very similar to R 136 and its nearby environment. The merging of the individual stellar images in NGC 3603 into unresolved stellar like blobs at the distance of the LMC is indicated in Fig. 1 by identification of the dominant objects. The compact source R 136A seems to be comparable to the unresolved core of a populous bright cluster. This conclusion is even more enforced by the analysis of the central part of NGC 3603 that is going to make up the unresolved bright stellar like peak in Fig. 1. Within an area of 1.5 arcsec in diameter (10" at galactic distance) are located 10 to 20 stars of types O5 and earlier, including at least one WN 6 (Moffat 1983). The quantitative comparison of this cluster of O stars in NGC 3603 with R 136A has been presented in detail by Moffat and Seggewiss (1983). The intriguing result is an absolute V magnitude of -9.3 for NGC 3603 as compared to -8.3 for R 136A within 1.5 arcsec.

With our *a priori* knowledge of the "true" nature of NGC 3603 we have to support the earlier conclusions that R 136 is the almost unresolved core of a dense cluster, consisting of dozens of very massive but normal OB and WR stars, and that its components, namely R 136A, are still unresolved groups of theoretically wellknown objects below the 140 $M_\odot$ threshold.

REFERENCES

Melnick,J.,Grosbol,P.: 1981, Astron.Astrophys. 107, p 23
Melnick,J.: 1983, The Messenger, No 32, p 11
Moffat,A.F.J.: 1983, Astron.Astrophys. 124, p 273
Moffat,A.F.J.,Seggewiss,P.: 1983, Astron.Astrophys. 125, p 83
Savage,B.D.,Fitzpatrick,E.L.,Mathis,J.S.,Ebbets,D.C.: 1983, Ap J, in press
Schmidt-Kaler,Th.,Feitzinger,J.V.: 1982, in "The Most Massive Stars",
　　　Proc. of ESO Workshop, S.D'Odorico,D.Baade,K.Kjaer (eds), p 105
Walborn,N.R.: 1973, Ap J Lett. 182, p L21

NEW OPTICAL OBSERVATIONS OF R136

You-Hua Chu
Washburn Observatory, University of Wisconsin at Madison

Introduction

R136 is the luminous central object of the 30 Doradus nebula in the LMC. Its bluest and brightest component R136a has been interpreted as a supermassive star with a mass of approximately 2000 $M_\odot$, based on its unusual UV spectral properties and the assumption that it is responsible for most of the ionization of the 30 Dor nebula (Feitzinger et al. 1980; Cassinelli et al. 1981; Savage et al. 1983). On the other hand, R136a has also been argued to be the core of a dense cluster, since the integrated light distribution from 0.2 to 22 pc radius centered on R136a is similar to that of a globular cluster (Moffat and Seggewiss 1983), and its optical spectrum can be synthesized by its nearby early-type stars (Melnick 1983).

Observations

We obtained a series of prime focus plates of R136 on the 4m telescope at Cerro Tololo Inter-American Observatory during two observing runs in January and February 1983. These new plates are characterized by their short exposure time and the use of narrow interference filters. The exposure time is typically 30 sec to 4 min for the single exposure plates, and 2 to 10 sec for each exposure on the multiple exposure plates. The filters used are centered on 1) blue continuum, 4765 Å, 2) HeII line, 4686 Å, 3) red continuum, 6485 Å, and 4) Hα line, 6563 Å. These filters were chosen to allow us to detect line emission objects and determine the colors of the stellar components.

The star R131 (B9I, V=10^m24) was used to calibrate both the flux and the seeing for each exposure.

Results

1) Light Distribution in R136 - R136 is clearly resolved into a, b, and c components. R136a apparently has a bright component and several fainter components superposed on an extended background (3"x4"). The shapes of the innermost isophotes are consistent with the micrometer measurements of a_1 and a_2, a_2 being at (0".5, 220°) from a_1. Farther away from a_1, there are two components about 2-3 mag fainter at (1".2, 80°) and (1".5, 160°), a much fainter component at (1".6, 10°), and

a still fainter component at (2″.6, 320°). Since these components and the extended background are detected in passbands that exclude nebular emission lines, the light sources must be stellar, not nebular.
2) Color and Brightness of R136a$_1$ and the Background - Assuming that R136a has two unresolved components a$_1$ and a$_2$, and an extended background, we can use the 2-D brightness profile of R131 as point spread function to decompose R136a. In this analysis, the ratio of a$_2$:a$_1$ may be uncertain by 50%, depending on the estimated positions of a$_1$ and a$_2$. However, the brightness of a$_1$ can be relatively accurately determined to 10%. We have also integrated the total light within a 3″ diameter area. The brightness of R136a$_1$ and 3″-aperture in the four passbands is summarized in the following:

	BC	HeII	RC	Hα
R136a$_1$:R131	0.42:1	0.54:1	0.33:1	0.44:1
(3″) :R131	1.14:1	1.34:1	0.92:1	1.12:1
R136a$_2$:R136a$_1$*	0.3:1	0.3:1	--	0.4:1

* This ratio may be uncertain by 50%.
Note that the brightest unresolved component a$_1$ only contributes about 37% of the light from a 3″ area.

We have interpolated between the BC and RC passbands to obtain V magnitudes. Assuming an A$_V$ of $1^m.2$ for R136, we derive
$$V(R136a_1) = 11^m.22, \quad V(3″) = 10^m.13 \quad \text{and}$$
$$M_V(R136a_1) = -8^m.58, \quad M_V(3″) = -9^m.67$$
The flux ratio $F_\lambda(4700Å)/F_\lambda(6500Å)$ is essentially a color indicator. We find this ratio to be 3.5 and 3.4 for R136a$_1$ and (3″), respectively. This flux ratio for a B9 star or a WN5 star is about 2.5, while for hot stars with T$_{eff}$=35,000 to 45,000 the flux ratio is about 3.6.

Conclusion

The new data of R136a presented here do not exclude either a single supermassive star hypothesis or a core of a cluster hypothesis; however they do set tighter constraints on the properties of R136a.
a) Single Supermassive Star - So far this hypothesis is still consistent with all observations. If the stellar components of R136a all have similar spectral energy distribution, then R136a$_1$ only contributes 37% of the UV flux, and its mass would be reduced accordingly to $\approx$ 750 M$_\odot$.
b) Core of a Dense Cluster - If R136a$_1$ consists of "normal" stars, then about 6-8 luminous O3 stars like R122 or about 15-20 "normal" O3 stars like HD 93250 are needed to provide the UV spectral features. So many O3 stars in a volume of diameter $\approx$ 0.03 pc would be more remarkable than any known clusters.

References

Cassinelli, J.P., Mathis, J.S., and Savage, B.D. 1981, Science, 212, 1497.
Feitzinger, J.V., Schlosser, W., Schmidt-Kaler, Th., and Winkler, C. 1980, Astr. Ap., 84, 50.
Melnick, J. 1983, The Messenger, No. 32, p. 11.
Moffat, A.F.J. and Seggewiss, W. 1983, Astr. Ap., 125, 83
Savage, B.D., Fitzpatrick, E.L., Cassinelli, J.P., and Ebbets, D.C. 1983, Ap. J., in press.

THE OBJECT R136 IN THE CORE OF THE 30 DORADUS NEBULA

W. Seggewiss
Observatorium Hoher List, Universität Bonn, F.R. Germany

A.F.J. Moffat
Observatorium Hoher List, Universität Bonn, F.R. Germany
and Dépt. de Physique, Université de Montréal, Canada

R136 = HD 38268 is the luminous and *diffuse* central object ($V \sim 10^m$ in a region of 7 arc sec diameter) of the young populous cluster NGC 2070 at the centre of the giant H II region 30 Doradus.

New spectroscopic and photometric observations confirm the interpretation of R136 as the dense stellar core of the cluster (Walborn 1973).

We point out with special emphasis:

(a) The visual light distribution is definitely not that of a single star. It can be fitted to cluster profiles with R136 as the core and NGC 2070 as the halo. R136 is not variable and emits $\sim 1/4$ of the total cluster light (visual and UV). As a cluster core it has a relaxation time of $\sim 10^7$ years.

(b) Visual and speckle observations resolve R136 in at least eight stars above an unresolved background of mainly *stellar* light. The spectrum changes across the face of R136. Also, the sum of the spectra of resolved stars in NGC 2070 resembles closely that of the centre of R136 (Melnick 1983, who also gives a CMD of the cluster).

(c) Comparison with the central cluster of the massive galactic H II region NGC 3603 shows
   - the striking similarity between the optical and the UV spectrum of R136 and HD 97950, the partially resolved core of NGC 3603 and
   - that the stellar core of NGC 3603 is even brighter and denser than R136.

An extended version of the paper was published in Astron. Astrophys.

REFERENCES

Melnick, J.   1983, ESO Messenger 32, 11
Moffat, A.F.J., Seggewiss, W.   1983, Astron. Astrophys. 125, 83
Walborn, N.R.   1973, Astrophys. J. 182, L21

PANEL DISCUSSION ON R136

The second panel of this symposium discussed the nature of R136 in 30 Doradus. The panel consisted of Y.-H. Chu, J.V. Feitzinger, B.D. Savage, W. Seggewiss, N.R. Walborn and T. Schmidt-Kaler (chairman). The panel members presented their views of R136, followed by replies, comments, and some discussion, while a few remarks were made from the audience. We reproduce here the opening statements of the panelists. All references have been collected at the end.

Th. Schmidt-Kaler
Ruhr University, Bochum

1. <u>Introduction</u>. Conventional theory of stellar structure places an upper limit for stellar masses at 60-120$M_\odot$. At the 1979 IAU Assembly in Montreal I announced the discovery of R136a as the first supermassive star (Schmidt-Kaler 1980), its mass exceeding 500 $M_\odot$. Cassinelli et al. (1980,1981) confirmed the discovery at the 1981 AAS Spring meeting by means of IUE observations and gave a model of the star with mass of the order of 2500 $M_\odot$. Subsequently R136a became the object of many investigations. Furthermore, quite a few examples of supergiant HII regions in other galaxies have been identified which apparently must be excited by similar stars (Conti and Massey 1981, Osterbrock and Cohen 1982, D'Odorico and Rosa 1982, Massey end Hutchings 1983, Grothus and Schmidt-Kaler 1983). We have new results from optical and ultraviolet spectra, and from surface photometry of the surrounding nebula confirming the nature of R136a and showing that R136a is responsable for energizing and structuring the whole supergiant HII region 30 Doradus.

2. <u>Basic observational parameters</u>. The various observers seem to reach agreement on important parameters:
a) The visual magnitude of R136a_1. The best estimate of the magnitude difference between a_1 and a_2 seems to be 1.2 mag. This yields

Schmidt-Kaler,Feitz.(1981)	(2".2)	V(a_1+a_2)= 10.77	V(a_1)= 11.08
Chu (1984)	(3")	–	11.22
Schlosser (1984)	(1".3)	10.62	<u>10.93</u>
			11.08±0.08

b) The reddening E(B-V)

Schmidt-Kaler,Feitzinger (1981)	0.46	various methods
Walborn (1982)	0.44	nebula (central object)
Melnick (1983)	0.46	MK and B-V of O-stars
Fitzpatrick,Savage (1983)	0.34	UV spectra in LMC.

The difference may be due to the fact that the last authors intercompare only LMC stars. Foreground dust common to all stars would not be noticed.

c) The ratio $R = A_V/E_{B-V}$. For the galactic foreground and LMC general dust $R = 3.25$, for 30 Dor nebular dust it is

    Le Marne (1968)   7
    Melnick (1978)   5.0
    Israel, Koornneef (1979)   6.5
    McGregor, Hyland (1981)   5.0
    Clayton et al. (1983)   4.4
    Savage et al. (1983)   3.7, 5.0
                                       5.5

d) The extinction is therefore at least $A_V = 1\overset{m}{.}5$, but more probably $2\overset{m}{.}0$. It depends on the relative contributions of galactic, and LMC foreground, and 30 Dor dust.

e) The optically determined luminosity then is $M_{bol} = -15.0$ or $L = 7 \cdot 10^7 L_\odot$ (using B.C. = -5.5 for $T = 62000$ K). The UV liminosity according to Savage et al. (1983) is $6 \cdot 10^7 L_\odot$.

f) The spectrum. Recent optical spectral classifications are

    Ebbets, Conti (1982)   abs (O)+WN4.5
    Vreux et al. (1982)   O3f
    Moffat, Seggewiss (1983)   WN6+O3-4
    Schmidt-Kaler (1983)   O2f (+WN4.5)

The photographic spectra of Moffat and Seggewiss are contaminated by stray light from the central object in conjunction with atmospheric dispersion. Our 1.5m-IDS sky-corrected spectrum is in excellent agreement with Melnick's 1983 4m digital spectrum, while in Moffat's spectra many important features are missing or unclear. R136 shows no SiIV and HeI in absorption, HeII $\lambda 4542$ is in wide absorption (± 1000 km/s) plus central emission, H_β like H_α with wide emission wings (± 3000 km/s).

3. The physical properties of R136a are listed in the following table.

	Multiple System or Cluster Core	Single Star R136a1		
		Schmidt-Kaler and Feitzinger 1981 Bochum	Panagia et al. 1983 Bologna	Savage et al. 1983 Wisconsin
Luminosity	$10^7 L_\odot$	$7 \cdot 10^7 L_\odot$	$6 \cdot 10^7 L_\odot$	$6 \cdot 10^7 L_\odot$
Temperature	---	62000 K	52000 K	75000 K
Radius	---	65 $R_\odot$	84 $R_\odot$	50 $R_\odot$
Mass	several stars with $M \leq$ 220 to 280 $M_\odot$	2500 $M_\odot$	>2000 $M_\odot$	$\approx 2100 M_\odot$ (>800 $M_\odot$)
Mass-loss	----	$4 \cdot 10^{-4} M_\odot/a$	$5.2 \cdot 10^{-4} M_\odot/a$	$5 \cdot 10^{-4} M_\odot/a$

R136a shows apparently normal chemical abundances from its spectrum (Feitzinger et al. 1983).

4. Formation of Supergiant HII-regions. From our survey and Hodge's (1983) size distribution function of HII regions of galaxies the supergiant HII regions represent a seperate class of objects. The same appears to be true for the mass function of the exciting stars. The supermassive stars are probably formed by a new seperate mechanism, when four (or more) adjacent, strongly expanding HII regions powered by (nearly simultaneously formed) massive O stars squeeze a molecular cloud in between then into collapse. In this way stars of extremely high masses may be formed.

PANEL DISCUSSION ON R136

N.R. Walborn
Goddard Space Flight Center

I wish to emphasize the following three points, which are discussed at greater length in my review (Walborn 1984).

1). The potential contribution to the ionization of the 30 Doradus Nebula by the cluster stars other than R136 has been substantially underestimated in some recent discussions. More than half of the ionizing luminosity probably is provided by stars other than R136.

2). Detailed comparisons between R136 and HD97950 (NGC3603) show that they are virtually identical multiple systems (see Seggewiss and Moffat 1984, and Rosa, Melnick and Grosbol 1984).

3). The previous conflict between two speckle interferometry observations of R136a, one showing an unresolved and the other a multiple object, has been definitely decided in favor of the latter by independent visual micrometer and photographic observations (see also Chu 1983).

The following spectroscopic points are also relevant. The wind terminal velocity of 3600 km/s seen in R136a is, in fact, typical for O3-O5 stars; indeed the terminal velocity of the O3 comparison star R122 is 4000 to 5000 km/s. The recent discovery of "intermediate" O3/WN objects such as Sk-67°22 and Melnick 30, 35N and 39 (Melnick 1983), is also important in the interpretation of R136; these objects may provide an O3 UV flux and relatively weak WR bands simultaneously. The Melnick stars are adjacent to R136, with M_V= -6.8 for No 39. Melnick 42 appears to be an O3If* object with M_V= -7.5.

B.D. Savage
Washburn Observatory

This discussion will concentrate on the UV measurements of R136a contained in Cassinelli, Mathis, and Savage (1981), Savage et al. (1983), and Fitzpatrick and Savage (1984). The second of these papers represents a reasonably complete analysis of all the stellar UV data relating to R136a that we have obtained with the IUE satellite. R136a is a difficult object to study because it lies in a crowded field. Observations in different spectral regions experience differing amounts of contamination from stars near R136a depending on aperture size and shape, and on the effects of atmospheric seeing. In Savage et al. (1983) this contamination problem was carefully addressed and it was found highly desirable to restrict the analysis to the IUE small (3") aperture data. However, the contamination problem doesn't go away with the use of the small aperture. The 3" diameter aperture accepts radiation from components R136a1, R136a2, and the extended emission which is approximately centered on component a1 (see Chu 1984).

The de-reddened small aperture UV spectrum of R136a is illustrated in Figure 2 of Savage and Fitzpatrick (1984). To produce that spectrum the IUE small aperture fluxes were increased by a factor of 2 to correct for the small aperture light losses expected for a point source. A similar correction factor is required to correct for the aperture light loss experienced by the extended component detected by Chu.

The line spectrum of R136a resembles to that of O3 stars. Strong P Cygni lines of N V, N IV, O V, and C IV are present while the Si IV P Cygni feature is absent. In addition the He II 1640Å line is P Cygni in appearance with a strong broad emission component. He II emission this strong is not normally seen in O3 spectra. If the spectrum of R136a is produced by a group of normal stars the requirement is $\simeq$ 13 R122's to produce the O3 characteristics and $\simeq$ 15 normal luminosity early WN stars to produce the He II emission. R122$_6$(O3 III(f*)) is the most luminous normal star in the LMC with $L \simeq 3\ 10^6 L_\odot$ and a mass estimated to be about 200 $M_\odot$. If less luminous O3 stars, such as Sk67-22, are producing the spectrum of R136a, then the number requirement for O3 stars increases to about 40. The C IV line in R136a has a blue edge which rises rapidly between 3000 and 3600 km/s. If a group of stars is producing the R136a spectrum, the dominant UV emitters in that group must have C IV P Cygni lines with edge velocities between 3000 and 3600 km/s. In the LMC, only the O3 stars have wind velocities that large. It was primarily on the basis of the C IV line profile that Cassinelli, Mathis, and Savage (1981) concluded that the UV light from R136a is more likely dominated by the emission from a single object. On the basis of Chu's visible imagery, it now appears that R136a may be that object. With $M_V \simeq -8.5$ and an O3 spectrum implying $T \gtrsim 50,000$ K, R136a1 would have a luminosity of $\gtrsim 1.5\ 10^7 L_\odot$ and a mass of $\gtrsim 600\ M_\odot$. The speckle measurements of Weigelt (private communication) imply that component a1 has a diameter of less than 0.02 pc. We can't rule out the possibility that component a1 is a multiple system of perhaps ten of the most luminous O3 stars. In this case the He II emission might be provided by component a2 if it is a superluminous Wolf-Rayet star. The formation of such a multiple system of O3 stars seems as implausible as the formation of a single supermassive object.

W. Seggewiss
Observatorium Hoher List

HD38268 = R136a is a most remarkable diffuse object (V= 10.0 mag in a diaphragm of 7" diameter) in the centre of the young populous cluster NGC2070 in the 30 Dor nebula. Already in 1982 we found (Moffat and Seggewiss 1983) that:
a) the spectrum changes across the face of R136a; the WN spectrum peaks in the centre whereas the early O-type spectrum is approximately constant over R136a (spectral type of the centre is WN6 + O3-4);
b) the visual light distribution is definitely not that of a single star.

In early 1983 we "rediscovered" the visual observations by van den Bos (Innes 1927) which resolve R136a into four components A, B, C and D. This is confirmed by photographic observations by Chu (1984) who found in addition two more faint components. Weigelt (1981) announced that speckle interferometry has revealed more stars in the main component

R136A (=R136a1). Thus, the apparent visual magnitude of R136A can not be much brighter than V = 12 mag. This compares to the magnitude of the brightest WN6 star in the surrounding cluster NGC20270: V = 11.9 mag. In contrast, early O-type stars (see above (a); V = 14 for O3V stars) contribute to the main unresolved stellar background light and to the UV flux.

The recognition, however, of the huge population of early-type stars in the central region of the 30 Dor nebula by Walborn (1984) and the construction of the colour-magnitude diagram of NGC2070 by Melnick (1983; 55 stars brighter than M_V = -5.2 mag, the absolute magnitude of an O4 ZAMS star) demonstrate that R136 is not the main contributor to the ionizing Lyman continuum flux of the nebula.

Note: New determinations revise downward not only the interstellar absorption A_V to 1.2 (Savage et al. 1983; Melnick 1983) but also the effective temperature for both the ionizing source of the nebula to T(eff) = 38000 K (Lequeux et al. 1981) and the O3-type stars to T(eff) = 45000 K (Simon et al. 1983).

The comparison between the galactic cluster NGC3603 and R136a was made by Walborn (1973, 1984). We like to point out the drastic difference between the open clusters of the Galaxy and young populous clusters of the LMC. The main body of the cluster NGC3603 can be placed inside R136a, the core of the populous cluster NGC2070, i.e. R136a, is not such an unusually compact object. Calculations of the relaxation time of R136a (as a cluster core), using a realistic initial mass function, lead to values of the order of 10^7 years. The central stellar density is below densities observed in globular clusters.

While in the visual spectral range the "problem" R136a seems to be solved, the enormous UV flux, as derived by Savage et al. (1983) and Savage and Fitzpatrick (1984), deserves further attention. They conclude that, within a circle of 3" (the IUE small aperture), R136a is brighter in the UV than 13 O3f-type stars like R122.

> J.V.Feitzinger
> Ruhr University, Bochum

New optical and UV observations favour the interpretation of R136a1 as a supermassive object.

1. <u>Surface Photometry</u>: R136a1 lies in the centre of the extremely young cluster NGC2070, with the most recent star forming event having taken place less than 10^6 years ago. The age of the giant star population from IR measurements confirms this (McGregor and Hyland 1981). Fitting of King's luminosity profile of isothermal spherical clusters and their extrapolation to the innermost core (Moffat and Seggewiss 1983) is, in the present case, inadequate. The extrapolation used our seeing-limited (FWHM ≃ 1".2) surface photometry (Feitzinger et al. 1980). To see if the

innermost luminosity structure (< 1") has a spike or not requires resolution better than $\simeq$ 0".3. The situation is similar to that in the core of M82, where conventional seeing-limited data do not reveal the central luminosity spike; a simple King profile fit being clearly inadequate (Young et al. 1978). Chu (1984, Chu and Wolfire 1983) with the best resolved photographs ($\simeq$ 0".2) of R136a showed that the "light distribution in R136a is strongly peaked to the center and disagrees with Moffat and Seggewiss (1983) interpretation". Speckle observations (Weigelt 1981, Meaburn et al. 1982) are discordant regarding the faint background, but concordant in revealing the existence of a small bright component, with an upper limit of the diameter of 0".08 (or even 0".02) = 0.02 pc. This is also found by Chu (1984). For the brightness of the central object R136a1 we obtain V = 11.08 mag, carefully taking into account contributions from the companion R136a2 (0".5 apart, ca. 1.5 mag fainter) and from the nebula.

2. The Spectrum: The IUE spectrum (Feitzinger et al. 1983) and the optical spectrum, together with UBV photometry (Feitzinger and Isserstedt 1983) allows temperature determinations from emission lines and by continuum fitting. From the emission lines we deduce 60000 K $\leq$ T $\leq$ 90000 K. By continuum fitting we derive for a blackbody 55000 K, for a model approximation 63000 K. This high temperature means that at least 50 % - 70 % of the 30 Dor nebula is ionized by the central object. Walburn's stellar population (this Symposium) can account for at most 68 % of the excitation (assuming the earliest spectral classifications), and at least 8 % (assuming the latest types), with a median of 20 %.

3. The environment of R136a: The stellar wind power of 4 10^{39} erg/s of R136a is sufficient to energize the major part of the 30 Dor shells and to structure the whole nebula. The expansion velocity of the inner shell is 25 km/s. The deduced age is 4 10^5 yrs. The central object cannot be much older than 1 10^6 yrs.

The Bochum group therefore favours the interpretation of R136a1 as a single object, equivalent to approximately 30 O3 stars in an area of less than 0.02 pc. At the moment the possibility of three or four stars of 300 $M_\odot$ - 400 $M_\odot$ each (or two of 700 $M_\odot$ and 1600 $M_\odot$) can not be ruled out completely. This might be the case if the visually brightest component should not be responsible for the UV flux. In any case the core region would be extraordinary in luminosity and density (a factor 100 greater than the globular cluster centers). Whatever the nature of R136a is, this object plays the key role in understanding other supergiant HII-regions.

Y.-H. Chu
Washburn Observatory

Of course, R136a is the core of a dense cluster. But, does it contain a supermassive star? And, how supermassive is it? The assumption of a dominating single source within a 3" aperture would simply predetermine the nature of this source. Comparison between NGC2070 and NGC3603 or other systems cannot place any real constraints on the nature of R136a, since similar appearance is a necessary but not sufficient condition for identical nature. Before high spatial resolution (better than 0".1) UV photometry and spectra of the stars in R136a are available, I don't think anybody can convincingly answer the questions.

R136a is definitely worth further investigations.

REFERENCES

Cassinelli, J.P., Mathis, J.S., Savage, B.D. 1980, B.A.A.S. **12**, 796
Cassinelli, J.P., Mathis, J.S., Savage, B.D. 1981, Science **212**, 1497
Chu, Y.-H. 1984, this volume, p. 259.
Chu, Y.-H., Wolfire, M. 1983, B.A.A.S. **15**, 644
Clayton, G.C., Martin, P.G., Thompson, I. 1983, Astrophys. J. **265**, 194
Conti, P.S., Massey, P. 1981, Astrophys. J. **249**, 471
D'Odorico, S., Rosa, M. 1982, Astron. Astrophys. **108**, 339
Ebbets, D.C., Conti, P.S. 1982, Astrophys. J. **263**, 108
Feitzinger, J.V., Hanuschik, R.W., Schmidt-Kaler, Th. 1983,
    Astron. Astrophys. **120**, 269
Feitzinger, J.V., Isserstedt, J. 1983, Astron. Astrophys. Supp. **51**, 505
Feitzinger, J.V., Schlosser, W., Schmidt-Kaler, Th., Winkler, Ch. 1980,
    Astron. Astrophys. **84**, 50
Fitzpatrick, E.L., Savage, B.D. 1984, Astrophys. J. in press
Grothus, H.-G., Schmidt-Kaler, Th. 1983, in preparation
Hodge, P.W. 1983, Astron. J. **88**, 1323
Innes, R.T.A. 1927, Southern Double Star Cat., Union Obs., Johannesburg
Israel, F.P., Koornneef, J. 1979, Astrophys. J. **230**, 390
Le Marne, A.E. 1968, Proc. Astron. Soc. Australia, **1**, 97
Lequeux, J., Maucherat-Joubert, M., Deharveng, J.M., Kunth, D. 1981
    Astron. Astrophys. **103**, 305
Massey, P., Hutchings, J.B. 1983, preprint
McGregor, P.J., Hyland, A.R. 1981, Astrophys. J. **250**, 116
Meaburn, J., Hebden, J.C., Morgan, B.L., Vine, H. 1982, Mon. Not. Royal
    Astr. Soc. **200**, 1 p
Melnick, J. 1978, Ph.D.
Melnick, J. 1983, ESO Messenger **32**, 11
Moffat, A.F.J., Seggewiss, W. 1983, Astron. Astrophys. **125**, 83
Osterbrock, D.E., Cohen, R.D. 1982, Astrophys. J. **261**, 64
Panagia, N., Tanzi, E.G., Tarenghi, M. 1983, Astrophys. J. **272**, 123
Rosa, M., Melnick, J., Grosbol, P. 1984, this volume, p. 257.
Savage, B.D., Fitzpatrick, E.L. 1984 this volume, p. 255.
Savage, B.D., Fitzpatrick, E.L., Cassinelli, J.P., Ebbets, D.C. 1983
    Astrophys. J. **273**, in press
Schlosser, W. 1983, priv. commun. to Schmidt-Kaler
Schmidt-Kaler, Th. 1980, Transactions IAU **XVIIB**, 208
Schmidt-Kaler, Th. 1983, in preparation
Schmidt-Kaler, Th., Feitzinger, J.V. 1981, in "The Most Massive Stars",
    p 105, Ed. S. D'Odorico, D. Baade, K. Kjaer; ESO Munich

Seggewiss, W., Moffat, A.F.J. 1984, this volume, p. 261
Simon, K.P., Jonas, G., Kudritzki, R.P., Rahe, J. 1983,
    Astron. Astrophys. **125**, 34
Vreux, J.M., Dennefeld, M., Andrillat, Y. 1982, Astron. Astrophys.
    114, L 10
Walborn, N.R. 1973, Astrophys. J. **182**, L 21
Walborn, N.R. 1982, Astrophys. J. **256**, 452
Walborn, N.R. 1984, this volume, p. 243
Weigelt, G. 1981, in "Scientific Importance of High Angular Resolution
    at Infrared and Optical Wavelengths", p 93, ESO Conference; and
    priv. commun. to Seggewiss
Weigelt, G. 1981, see in Schmidt-Kaler and Feitzinger, 1981, op. cit.
Young, P.J., Westphal, J.A., Kristian, J., Wilson, P., Landauer, F.P.
    1978, Astrophys. J. **221**, 721

30 Doradus

Doradus, your stars are shining bright above
Doradus, your central object is my love
I press you, caress you,
    and bless the day you taught me to care
To always remember
    the supermassive cluster you bear
Doradus, when day is done you'll hear my call
Doradus, I'd like to solve your puzzles all
I dread the dawn when I awake to find you gone
Doradus, I need you, my own

Post-Panel Statement, Monastery Bebenhausen, between 7 and 10 p.m.

THE MAGELLANIC CLOUD SUPERNOVA REMNANTS

Michael A. Dopita
Mt. Stromlo & Siding Spring Observatories
Research School of Physical Sciences
Australian National University, Canberra

1. SURVEYS FOR SNR IN THE MAGELLANIC CLOUDS

It was some twenty years ago, at the previous IAU symposium devoted to the Magellanic Clouds, that Mathewson and Healey (1964) announced the discovery of the first supernova remnant (SNR) in the LMC. The method used was to identify a non-thermal radio source with a filamentary optical emission object and this confirmed N49 in the catalogue of Henize (1956) as an SNR and suggested N63 and N132 D as likely candidates.

Westerlund and Mathewson (1966) and Mathewson and Clarke (1972, 1973a,b,c) developed and exploited a technique whereby collisionally ionised gas could be distinguished from photoionised gas. This relies on the fact that the [S II] lines are much stronger with respect to Hα in plasma excited by collisions. Thus nebulosities that appear equally bright on [S II] and Hα plates are likely to be SNRs, those that are much brighter at Hα are H II regions and objects that are about equally bright at V, Hα and [S II] are probably stars or distant galaxies. By searching in the vicinity of known non-thermal radio sources, they identified twelve SNR in the LMC and two in the SMC. (see also Lasker 1976).

Such a selection procedure, although very successful, can now be recognized with the benefit of hindsight to carry with it a bias towards the detection of only evolved SNR. Young oxygen-rich SNR, although strong sources of non-thermal radio emission, must be found in the [O III] lines (Mathewson et al. 1980). Furthermore, several SNR have been discovered to be of the collisionless-shock type, showing only Balmer lines in their spectra (Tuohy et al. 1982) and relatively weak non-thermal radio emission. We thus now have as optical selection criteria for SNR (a) strong [S II] lines c.f. Hα (b) Hα and Hβ emission only and (c) strong [O III]; Hα, Hβ absent!

The launching of the Einstein Observatory led to the most recent SNR discoveries in the Clouds. In the LMC, Long, Helfand and Grabelsky

(1981) listed a total of twenty-six SNR of which five were optical nebulosities in the catalog of Davies, Elliot and Meaburn (1976) and six were found to be extended in follow-up HRI observations. In the SMC, Seward and Mitchell (1981) concluded that one source was an SNR; confirmed by Dopita, Tuohy and Mathewson (1981), and that four others were probably SNRs. A complete catalogue of optical identifications and X-ray maps is given by Mathewson et al. (1983a).

In the SMC, the X-ray survey of Tanaka (1983), by going deeper than the Seward and Mitchell (1981) survey, resulted in several tentative SNR identifications (Mills et al. 1982). This survey together with the deep radio survey of Mills et al. (1983) in both the SMC and LMC has allowed the additional optical identification of five SNR in the SMC and two in the LMC (Mathewson et al. 1983b).

Detailed study of these SNR has proven to be very rewarding in its Astrophysical yield. The chemical abundances in both the Interstellar Medium (ISM) and the supernova ejecta can be studied, information on the mass of precursor stars can be derived in some cases and a great deal can be inferred about the physical structure of the ISM. These aspects will be the subject of the remainder of this review, with an emphasis on the younger SNR.

2. THE OXYGEN-RICH SNR

For many years only the fast moving knots of Cas-A were known to display the characteristic spectrum of this type of remnant, namely very strong forbidden lines of oxygen and a few heavier elemental species, but a total absence of hydrogen and helium recombination lines. (Chevalier and Kirshner 1979). Since then one other SNR of this type has been discovered in each of our Galaxy (Goss et al. 1979) and NGC 4449 (Balick and Heckman 1978). However, as many examples again are now known in the Magellanic Clouds; two in the LMC (Danziger and Dennefeld 1976; Mathewson et al. 1980) and one in the SMC (Dopita, Tuohy and Mathewson 1981).

Astrophysically, the Cas-A type SNRs are very exciting, since their peculiar spectra strongly suggest that we see directly matter ejected from deep within the core of a massive progenitor star from which vital clues about the nucleosynthesis processes and processes of ejection during the supernova event may be gathered.

From a dynamical viewpoint there is little doubt that the matter which is at present radiating in the visible has undergone little or no deceleration since ejection. In Cas-A, for example Kamper and van den Bergh (1976) show that the knots have a space velocity which is directly proportional to their distance from the center of expansion and that no deceleration of individual knots can be detected. If the knots are excited at the time they run into the reverse shock generated by interaction of the ejecta with the ISM, then they must map the dynamics of a particular zone in the ejecta. What evidence there is

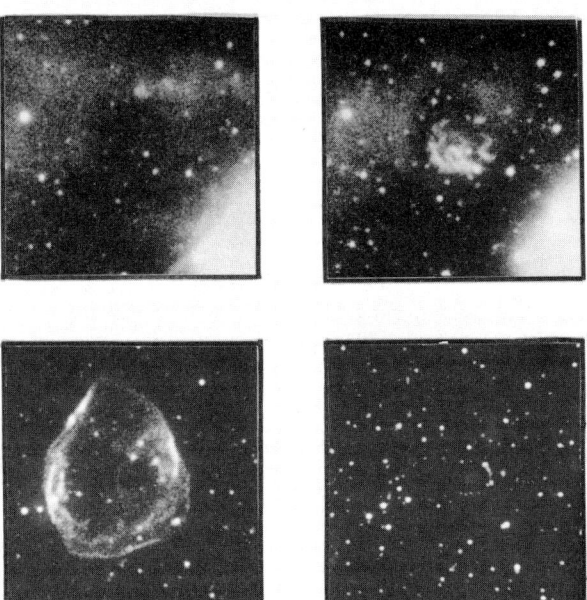

Fig. 1. Two SNR in the Magellanic Clouds, left Hβ right [O III]. The top remnant, 1E0102-7219, is a young SNR in the SMC with oxygen-rich filaments and diffuse highly ionised halo. The lower object is DEM 71, in the LMC and is an example of the non-radiative shock SNR.

suggests that this zone is not spherical. Lasker (1978) showed that, rather, the high-velocity material is N132 D maps into an annular feature apparently expanding at 2250 km s^{-1} in a plane inclined at 45° to the line of sight. The existence of similar high velocity rings of ejecta has since been inferred in Cas-A from X-ray dynamical evidence (Winkler et al. 1982), and in the Galactic SNR G292.0+1.8, from high-resolution X-ray mapping (Tuohy, Clark and Burton 1982). However, perhaps the most curious ring is seen in the SMC remnant 1E0102.2-7219 (see Fig. 1). The dynamics of this remnant (Tuohy and Dopita 1983) show a perturbation in the polar direction which oscillates as $\cos 3\theta$ in azimuthal angle and with an amplitude that varies with azimuthal angle. Even with such a complex structure, the underlying expanding ring motion appears to be preserved. Such a complex structure argues strongly for a global instability of the ejecta either caused by plasma instability during the ejection or else during the collapse phase.

The fact that ring ejection appears to be a common feature of oxygen-rich (Type II) supernova events is an important clue in the physics of the collapse phase. At the present time only the model of Bodenheimer and Woosley (1982) appear to generate such a feature naturally. This model (for a 25 $M_\odot$ precursor star) generates a thick shocked disk during collapse in which explosive nuclear burning coupled with a rotational bounce near the core drives out high velocity equatorial ejection. Such a model is capable of producing a large degree of core overturn and mixing and this may be observable by spectrophotometric studies of the knots. If temperatures rise high enough, explosive burning of neon may occur. The 'ashes' of this are very sensitive to temperature (Woosley and Weaver 1980) and traces of this process may also be found in the ejecta.

Although the knots and filaments of these remnants offer the potential to study the processes of nucleosynthesis in massive stars, this potential has not been realised for two reasons; the lack of an adequate number of diagnostic line ratios in the visible, and failing this, the lack of a believable model for the excitation of the knots.

It has generally been assumed that the optical emission in the fast-moving material is the result of a radiative shock passing through it (Peimbert and van den Bergh 1971, Lasker 1978, Chevalier and Kirshner 1978, 79, Goss et al. 1979, Murdin and Clarke 1979, Kirshner and Blair 1980, Blair, Kirshner and Winkler 1983). This would naturally result from the interaction between cloudy ejecta and a hot reverse-shocked intercloud medium. The cloudy ejecta could be generated by operation of a Rayleigh-Taylor instability at the core envelope interface in the star (Falk and Arnett 1973, Chevalier 1975, 76) or at the ejecta/ISM interface (Gull 1973) or else by early thermal instability in the oxygen-rich ejecta (Dopita, Binette and Tuohy 1983).

Shock-wave models for the filaments have been developed by Itoh (1981a,b) for the pure-oxygen case and by Dopita, Binette and Tuohy (1983) for a reasonable mix of elements. The models display some extraordinary properties resulting from the very efficient cooling of oxygen which include a large separation between ion and electron temperatures, a 'freezing' of states of high ionisation to very low temperature, and a very extensive region of pre-ionisation. Entertaining as these effects may be, the computed spectra bear little resemblance to those observed (Lasker 1978; Dopita and Tuohy 1983), and Dopita, Binette and Tuohy (1983) conclude that the shock excitation theory is no longer tenable. If not shocks, then what? Three possible mechanisms were suggested, photoionisation by X-rays, electron conduction and direct heating by suprathermal ions (which will occur when the cloud is smaller than the stopping distance for these ions).

Whatever the mechanism of excitation, it is common to all SNR of this class. From the spectra of Table 1, it is clear that although there are obvious differences resulting from different chemical compositions (e.g. presence or absence of lines of neon, calcium and sulphur) the physical conditions are very similar. For example, the electrom temperature given by the [O III] $\lambda 4363/\lambda 5007$ Å ratio is very little different from $T_e \sim 2000°K$ and the degree of excitation given by [O I]: [O II]: [O III] line ratios varies within very narrow limits. We can therefore confidently expect that a successful model will yield abundances that can be believed.

TABLE 1. Spectrophotometry of Young SNR (Dopita + Tuohy 1983)

Ion	λ (Å)	G292	0540	N132 D*	1E0102
[Ne V]	3426	–	–	–	3.5
[O II]	3726, 9	79	54	165	130
[Ne III]	3869	7	–	9	14
	3967	2	–	3	5
[S II]	4068, 76	–	4	–	–
[O III]	4363	3	3	5	4
	4959	32	30	31	34
	5007	100	100	100	100
[O I]	6300	5	3	11	3
	6363	–	1	4	1
Ca I]	6573	–	17	–	–
[S II]	6717, 31	–	63	–	–
[O II]	7319, 30	–	(2)	10	–

*Below 5007, global spectrum; above 5007 single filament spectrum.

Spectral similarities apart, each of the Magellanic Cloud oxygen-rich remnants have individual peculiarities. N132 D has, apart from the oxygen-rich ring, a few very bright knots of more normal composition which were discovered by Mathewson and Clarke (1973b) and a 26 pc diameter halo first described by Lasker (1978). This halo appears to be normal shock excited gas and is visible in X-rays over the same or slightly greater diameter (Mathewson et al. 1983a). Given the very rapid expansion of the oxygen-rich ring (Lasker 1980), it is difficult to avoid the conclusion that N132 D is a double SNR event resulting possible from the closely sequential explosion of the components of a binary. The other LMC remnant, SNR 0540-69.3 (in N158 A) shows broad [S II] lines and, probably, a Ca I] line in its spectrum. These have very similar velocity dispersion to the [O III] lines. (Dopita and Tuohy 1983). There is some evidence that these other elements are confined in a smaller annular ring than the [O III] emission (Mathewson et al. 1980), this may be caused by the limited bandpass of the filter used to image the remnant, but would be consistent with their zone of nucleosynthesis in the parent star.

The SMC remnant 1E0102.2-7219 shows a remarkable halo of high excitation which largely surrounds it. This appears to be excited by the remnant itself and has very strong He II lines in its spectrum (Dopita, Tuohy and Mathewson 1981, Tuohy and Dopita 1983). It is separated from the remnant by a dark annulus which in its outer boundary appears to delineate the position of the blast-wave (see Fig. 1). This spatial separation between oxygen-rich filaments and blast-wave is expected in the reverse-shock hypothesis.

3. THE NON-RADIATIVE SHOCK SNR

Amongst the SNR identified by Long, Helfand and Grabelsky (1981) purely on their X-ray properties, four were found to emit in the visible only in the Balmer lines (Tuohy et al. 1982) (see Fig. 2). Three of these were not previously known to have an associated optical nebulosity. A further example of transitional class was later found in the SMC (Mathewson et al. 1983b). These discoveries more than doubled the SNR of this type, since only Tycho (Kirshner and Chevalier 1978) and the remnant of SN 1006 (Schweizer and Lasker 1978) are similar; although both the Cygnus Loop (Raymond et al. 1980) and IC 443 (Lozinskaya 1979) show some regions with similar properties.

The theory of such spectra has been developed by Chevalier and Raymond (1978), Chevalier, Kirshner and Raymond (1980) and Raymond et al. (1983) as due to a high-velocity non radiative shock passing through a partially ionised ISM. In this case the neutral hydrogen is collisionally excited by the high temperature electrons before it is finally ionised and joins the high-velocity post-shock plasma. Since the neutral gas is not initially coupled to the bright velocity stream by Coulomb interactions, the Balmer lines emitted by it are very narrow. However, a broad component to these lines may arise by resonant charge-ex change with the ionised stream. Where this is seen, as in SNR 0519-69.0 this can be used to estimate the current blast-wave velocity directly.

All the non-radiative shock SNR are young (or youthful) and there is fairly strong circumstantial evident to comment them with Type I supernova events. For SNR 0519-69.0, the following parameters are derived for the mass and energy of the supernova event, respectively:

$$1.8 \times 10^{51} < E_o < 2.2 \times 10^{51} \text{ ergs}$$

$$1.2 < M < 4.0 \ M_\odot$$

These figures are similar to those derived by Chevalier, Kirshner and Raymond (1980) for Tycho's SN except that the mass range was brighter, $0.9 - 2.8 \ M_\odot$. This event was certainly of Type I from its high curve (Clark and Stephenson 1977). Furthermore, Wu et al. (1983) recently observed an OB star projected against SNR 1006 in the UV and found strong absorption lines of iron with a velocity dispersion of about 5000 km s^{-1}. This would be expected from the theory of Type I supernova events which are thought to result from carbon detonation or deflagration of a white dwarf or the degenerate core of a moderate mass star (up to 9 $M_\odot$) (Wheeler 1982), with consequent ejection of iron-peak elements.

Although there is little remarkable about the X-ray properties of these remnants, both the Magellanic Cloud and Galactic examples of this class of SNR appear to be underluminous at radio frequencies by factors of order ten (Tuohy et al. 1982). This implies that either different physical conditions or different physical processes moderate the radio emission from these remnants.

4. THE EVOLVED (RADIATIVE) SNR

As McKee (1983) points out, SNRs are thought to generate a hot intercloud component of the interstellar medium (ISM) and thus govern the inhomogeneity of the ISM to a significant extent. In turn, the cloudy structure of the ISM strongly influences the evolution and appearance of the SNRs, and evolved remnants may therefore be used to probe this structure.

The simplest demonstration that the ISM is not homogeneous comes from a cumulative number/diameter relationship. This was first plotted for the LMC remnants by Mathewson and Clark (1973b) and has been updated several times since (Dopita 1982, Mills 1983, Mathewson et al. 1983) and we present in Figure 2 the latest version of this diagram (which excludes the non-radiative SNR and so refers to Type II supernovae) According to Sedov theory for blast waves in a uniform medium, this should have a slope of 5/2. In fact, it is nowhere near as steep as this and can be best fitted by a linear relationship $N(D) = 0.4$ D pc. This implies that, even out to diameters of 50 or 60 parsecs, the LMC remnants appear to be expanding with a uniform velocity.

For a uniform ISM, the free expansion phase is terminated when the mass of ISM swept up about equals the mass ejected. A reverse shock then heats the ejecta and the remnant enters the Sedov phase in which the majority of the energy is thermal. The LMC cumulative number/diameter relationship therefore either implies free expansion or the storage of an appreciable fraction of energy in the kinetic form until quite late in the evolution. This was suggested by Clarke (1976) and was confirmed observationally by Dopita (1979) who measured directly the pressure in the optical filaments in a number of LMC remnants. Since this is very nearly constant, a thermal energy content increasingly almost as fast as R^3 is inferred. For constant driving pressure, the blast-wave would, of course, move at constant velocity; which would explain the cumulative numbers/diameter relationship.

The inescapable conclusion of these observations is that both the interstellar medium and the ejecta are cloudy. The theory of this geometry has been developed by McKee (1983). In this type of model, the optically radiative shocks are moving into dense pre-existing interstellar clouds which are pressurised by the external hot gas. The position of the blast-wave will be delineated by non- radiative shocks (as in the case of IC 443 and Cygnus mentioned in the previous section), by coronal emission of species such as [Fe XIV] and by the outer boundary of X-ray shell. A relatively smooth [Fe XIV] has been mapped for the brightest of the LMC remnants, N49, by Dopita and

Mathewson (1979). This did indeed lie for most of its length outside the brightest of the optical filaments. Little has so far been said about X-ray observations.

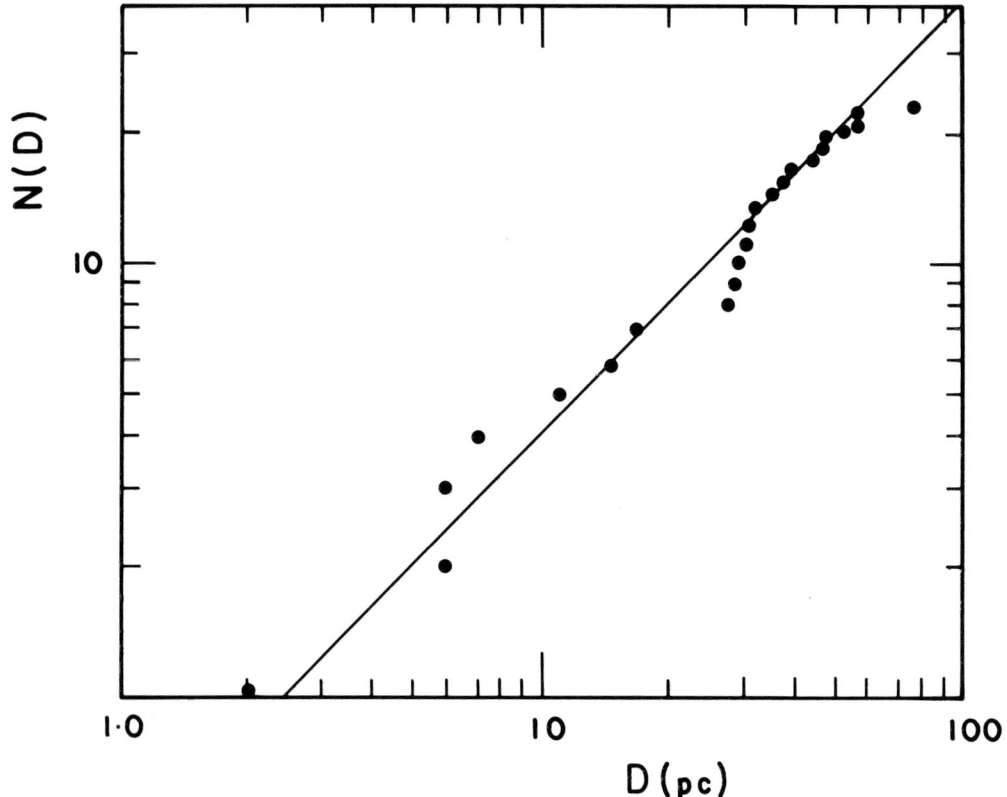

Fig. 2. The cumulative number, N(D), diameter relationship for the Type II SNR in the LMC.

The X-ray surface brightness declines rapidly with radius for the LMC remnants, as first shown by Long and Helfand (1979). The X-ray surface brightness between 0.15 and 4.5 KeV, L_X, can be fitted best by the relationship $L_X = 1.1\ Dpc^{-3.3}$ (erg cm s^{-2} s^{-1} sr^{-1}). In this respect it has similar behaviour to the radio surface brightness at 408 MHz, L_R, which is fitted by $L_R = 1.5 \times 10^{-16}\ Dpc^{-2.6}$ (W.M.$^{-2}$ Hz^{-1} sr^{-1}) (Mathewson et al. 1983a). A variety of different effects can determine this relationship. Density variations in the ejecta and ISM will lead to large pressure variations which in turn strongly influence the X-ray morphology. In particular, McKee (1983) points out the the impact of a blob of ejecta on an interstellar cloud will result in rapid thermalisation of the kinetic energy and the production of a secondary blast-wave and X-ray hot spot. This process may be responsible for the scolloped appearance of some SNR. Another effect that cannot be ignored is the high abundances of important X-ray coolants in the ejecta which will enhance the emissivity of this material in the early phases of the evolution (Long, Dopita and Tuohy 1982). A third effect is the production of an X-ray synchrotron continuum by a pulsar. The SSS of the Einstein Observatory was used to measure the spectra of six bright SNR in the LMC. Four of these showed a normal thermal spectrum, but two young remnants N157 B and 0540-69.3 had featureless continuum a consistent with a power law (Clark et al. 1982).

In conclusion, much observational data now exist to produce a self consistent model of the ISM in the Magellanic Clouds and the evolution of the SNR in this medium using data from the optical, X-ray and radio domains. It is clear that this will involve a multiphase ISM (e.g. McKee and Ostriker 1977) but must also properly account for the structure and abundances in the ejecta. Further work on the young SNRs is required both observationally and theoretically to provide this data and to examine the physics of the supernova event in its own right.

ACKNOWLEDGEMENTS

I would like to thank Cerro Tololo Observatory for their financial and secretarial support under their Visiting Scientist Program and for providing a convivial working atmosphere which greatly assisted the writing of this paper.

REFERENCES

Balick, B. and Heckman, T.: 1978, Ap. J. (Letters) 226, L7.
Blair, W.P., Kirshner, R.P. and Winkler, P.F.: 1983, Ap. J. (in press).
Bodenheimer, P.B. and Woosley, S.E.: 1982, B.A.A.S. 12, 833.
Chevalier, R.A.: 1975, Ap. J. 200, 698.
_____.: 1976, Ap. J. 207, 872.
Chevalier, R.A. and Kirshner, R.P.: 1978, Ap. J. 219, 931.
_____.: 1979, Ap. J. 233, 154.
Chevalier, R.A. and Raymond, J.C.: 1978, Ap. J. (Letters) 225, L27.
Chevalier, R.A., Kirshner, R.P. and Raymond, J.C.: 1980, Ap. J. 235, 186.
Clark, D.H. and Stephenson, F.R.: 1977, The Historial Supernovae Pergamon Press, Oxford.
Clark, D.H., Tuohy, I.R., Long, K.S., Szymkowiak, A.E., Dopita, M.A., Mathewson, D.S. and Culhane, J.L.: 1982, Ap. J. 255, 440.
Clarke, J.N.: 1976, M.N.R.A.S. 174, 393.
Danziger, I.J. and Dennefeld, M.: 1976, Ap. J. 207, 394.
Davies, R.D., Elliott, K.H. and Meaburn, J.: 1976, Mem. R.A.S. 81, 89.
Dopita, M.A.: 1979, Ap. J. Suppl. 40, 455.
_____.: 1982, Supernovae: A Survey of Current Research. Ed. M.J. Rees and R.J. Stoneham, D. Reidel, p. 483.
Dopita, M.A. and Mathewson, D.S.: 1979, Ap. J. (Letters) 231, L147.
Dopita, M.A., Tuohy, I.R. and Mathewson, D.S.: 1981, Ap. J. Letters 248, L105.
Dopita, M.A. and Tuohy, I.R.: 1983, Ap. J. (in press).

Dopita, M.A., Binette, L. and Tuohy, I.R.: 1983, Ap. J. (in press).
Falk, S.W. and Arnett, W.D.: 1973, Ap. J. (Letters) 180, L65.
Goss, W.M., Shaver, P.A., Zealey, W.J., Murdin, P. and Clark, D.H.: 1979, M.N.R.A.S. 188, 357.
Gull, S.F.: 1973a, M.N.R.A.S. 161, 47.
Henize, K.G.: 1956, Ap. J. Suppl. 2, 315.
Itoh, H.: 1981a, P.A.S. Japan 33, 121
_____.: 1981b, P.A.S. Japan 33, 521.
Kamper, K. and van den Bergh, S.: 1976, Ap. J. Suppl. 32, 361.
Kirshner, R.P. and Blair, W.P.: 1980, Ap. J. 236, 135.
Kirshner, R.P. and Chevalier, R.A.: 1977, Ap. J. 218, 142.
Lasker, B.M.: 1976, RGO Bull. #182, 185.
_____.: 1978, Ap. J. 223, 109.
_____.: 1980, Ap. J. 237, 769.
Long, K.S. and Helfand, D.J.: 1979, Ap. J. (Letters) 234, L77.
Long, K.S., Helfand, D.J. and Grabelsky, D.A.: 1981, Ap. J. 248, 925.
Long, K.S., Dopita, M.A. and Tuohy, I.R.: 1982, Ap. J. 260, 202.
Lozinskaya, T.A.: 1979, Australian J. Phys. 32, 113.
McKee, C.F.: 1983, Paper presented at IAU Symposium #101 Supernova Remnants and their X-Ray Emission, p. 87.
McKee, C.F. and Ostriker, J.P.: 1977, Ap. J. 218, 148.
Mathewson, D.S. and Clarke, J.N.: 1972, Ap. J. (Letters) 178, L105.
_____.: 1973a, Ap. J. 179, 89.
_____.: 1973b, Ap. J. 180, 725.
_____.: 1973c, Ap. J. 182, 697.
Mathewson, D.S., Dopita, M.A., Tuohy, I.R. and Ford, V.S.: 1980, Ap. J. (Letters) 242, L73.
Mathewson, D.S. and Healey, J.R.: 1964, The Galaxy and the Magellanic Clouds, Eds. F.J. Kerr and A.W. Rodgers. Aust. Acad. of Science, p. 283.
Mathewson, D.S., Ford, V.L., Dopita, M.A., Tuohy, I.R., Long, K.S. and Helfand, D.J.: 1983, Ap. J. Suppl. Ser. 51, 345.
Mathewson, D.S., Ford, V.L., Dopita, M.A., Tuohy, I.R., Mills, B.Y., Turtle, A.J., Little, A.G. and Durdin, J.M.: 1983, Ap. J. (in press)
Mills, B.Y.: 1983, IAU Symp. #101: Supernova Remnants and their X-Ray Emission, p. 551
Mills, B.Y., Little, A.G., Durdin, J.M. and Kesteven, M.J.: 1982, M.N.R.A.S. 200, 1007.
Mills, B.Y., Turtle, A.J., Little, A.G. and Durdin, J.M.: 1983 (in prep.).
Murdin, P. and Clarke, D.H.: 1979, M.N.R.A.S. 189, 505.
Peimbert, M. and van den Bergh, S.: 1971, Ap. J. 167, 223.
Raymond, J.C., Davis, M., Gull, T.R. and Parker, R.A.R.: 1980, Ap. J. (Letters) 238, L21.
Schweizer, F. and Lasker, B.M.: 1978, Ap. J. 226, 167.
Seward, F.D. and Mitchell, M.: 1981, Ap. J. 243, 736.
Tanaka, Y.: 1983, IAU Symp. #101: Supernova Remnants and their X-Ray Emission, Venice, 30 Aug-2Sept.
Tuohy, I.R., Clark, D.H. and Burton, W.M.: 1982, Ap. J. (Letters) 260, L65.
Tuohy, I.R. and Dopita, M.A.: 1983, Ap. J. (in press).

Tuohy, I.R., Dopita, M.A., Mathewson, D.S., Log, K.S. and Helfand, D.J.: 1982, Ap. J. 261, 473.
Westerlund, B.E. and Mathewson, D.S.: 1966, M.N.R.A.S. 131, 371.
Wheeler, J.C.: 1982, Supernovae: A Survey of Current Research, eds. M.J. Rees and R.J. Stoneham, D. Reidel, p. 167.
Winkler, P.F., Canizares, C.R., Markent, T.H. and Szymkowiak, A.E.: 1982, ibid p. 501.
Woosley, S.E. and Weaver, T.A.: 1980, Ap. J. 238, 1017.
Wu, Chi-Chao, Leventhal, M., Sarazin, C.L. and Gull, T.R.: 1983, Ap. J.(Letters) 269, L5.

DISCUSSION

Dufour: Do you observe any evidence of slow-moving pre-SN ejected material enriched with nitrogen (or other) within, or in front of, the O-rich material in young oxygen-rich Cloud SNRs?
Dopita: Only in the case of N132D are dense shocked cloudlets seen.
Dufour: So the young O-rich SNRs in the Clouds are not like Cas A with substantial pre-SN mass loss?
Dopita: No, there is probably nothing corresponding to the quasi-stationary flocculi of Cas A.
Lasker: The detection of [ArIII]$\lambda 7132$ in the young O-rich SNRs still appears marginal. Please elaborate on this and on the other trace elements that are suggested by the stellar interior models.
Dopita: I agree that the fainter lines are indeed very faint. The spectra I have shown are the summation of many spatial pixels shifted to zero velocity. This has extended the sensitivity limit to very faint lines by a factor of order five. It appears that, except for N158A, very faint lines of other species are almost entirely absent.

A RADIO CONTINUUM SURVEY OF THE MAGELLANIC CLOUDS

B.Y. Mills & A.J. Turtle
School of Physics, University of Sydney, N.S.W. 2006, Australia

Continuum radio emission from the Large Magellanic Cloud was first detected just 30 years ago (Mills & Little, 1953). Subsequently, surveys of the Clouds were made after each new advance in southern hemisphere instrumentation and the principal surveys are listed and briefly described in Table I. They cover a range of frequencies from 19.7 MHz to 8.4 GHz. The early surveys at low frequencies were chiefly concerned with the overall synchrotron emission from the Clouds but, as resolution improved, emphasis has shifted to the individual sources, both emission nebulae and supernova remnants, which can be recognized as Cloud members. Of recent years the unique position of the Clouds for studying radio sources in external galaxies has been undermined by the development of powerful synthesis telescopes in the northern hemisphere; these have provided equivalent sensitivity and better spatial resolution on M31, and several other northern galaxies can also be studied effectively. However, with the commissioning of the Molonglo Observatory Synthesis Telescope (MOST) in 1981 the Clouds have been restored to their rightful place befitting the closest galaxies.

The MOST is a unique rotational synthesis telescope which, in 12 hours observation, can synthesise a map in real time from a comb of fan beams (Mills 1981). The operating frequency is 843 MHz and the synthesised beam has a half-power width of 43 x 43 cosecδ arcsec. The basic field size defined by the comb of fan beams is 23 x 23 cosecδ arcmin (11 x 11 cosecδ arcmin for early observations). The field may be expanded by a factor of either two or three using a multiplexing system with a corresponding reduction of sensitivity. For the observations described here the rms noise varied between about 0.4 mJy and 0.7 mJy depending on the field size chosen. The sensitivity has now been significantly improved by the installation of low noise FET preamplifiers.

Observations of the Clouds with the MOST began in September 1981 with a study of supernova remnant candidates. These candidates were chosen from the X-ray catalogues of Seward & Mitchell (1981), for the SMC, and Long et al. (1981) for the LMC. Previously suggested SNR identifications from earlier optical and radio observations were also

Table 1: Principal radio surveys of the Magellanic Clouds

Frequency MHz	Beam Size	Reference	Notes
19.7	85'x104'	Shain (1959)	Contour maps of both Clouds
85.5	48'x 59'	Mills (1959)	Contour maps of both Clouds
408	2!8x3!5	Clarke et al. (1976)	A definitive catalogue of radio sources in both Clouds with maps of individual sources
408 1410	48' 14'	Mathewson & Healey (1964)	Contour maps of both Clouds
2700	8'	Broten (1972)	Contour maps of both Clouds
5000	4!1	McGee et al. (1972)	A catalogue of radio sources in the LMC and maps
5000 8400	4!1 2!6	McGee et al. (1976)	A catalogue of radio sources in the SMC and maps
8400	2!6	McGee et al. (1978)	Measurements on selected sources in the LMC and maps

included. Only a small area of the Clouds was actually mapped in these programs but 6 SNRs were confirmed in the SMC (Mills et al. 1982) and 23 plus 6 possible SNRs found in the LMC (Mills, 1983). There was good reason for believing that many large diameter SNRs were missed by this procedure because the X-ray surveys discriminated against them. Accordingly in the 1982 season the SMC was surveyed with the MOST, essentially completely.

THE SMC SURVEY

Fourteen overlapping fields were mapped, one with a size of 70 x 74 arcmin and the remainder 46 x 48 arcmin, covering about 4 square degrees of the SMC including the 'bar' and the 'wing'. An exceptionally rich area from one of these fields is shown in Figure 1. It is an 'uncleaned' map of the type available immediately after the observation. Three categories of radio source are found.

(i) The most common are background, usually unresolved, sources which are either radio galaxies or quasars. In general their identification is not possible because of the crowded star field. In the rich field of Figure 1, these background sources are outnumbered by the SMC sources.

A RADIO CONTINUUM SURVEY OF THE MAGELLANIC CLOUDS

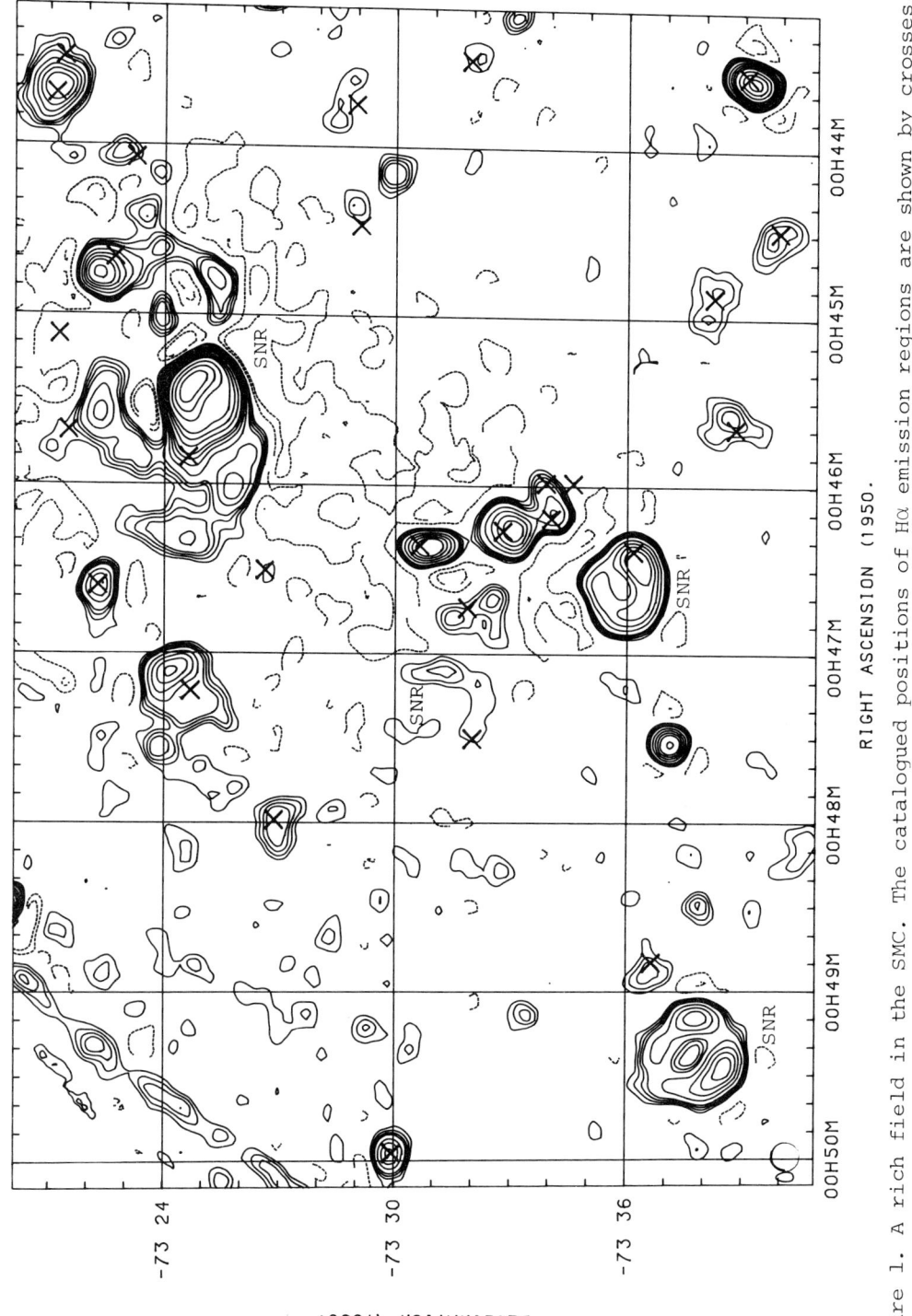

Figure 1. A rich field in the SMC. The catalogued positions of Hα emission regions are shown by crosses. Contour levels (mJy/beam) : -4, -2, 2, 3, 4, 5. 7.5, 10, 12.5, 15, 25, 30, 40, 50, 60, 70.

(ii) Emission nebulae: The detectability of H II regions appears to be better than in the Hα survey by Henize (1956) but generally inferior to the Hα survey of Davies et al. (1976). For a nebula of size between about 30 arcsec and 5 arcmin the minimum emission measure for reliable detection is in the range 300-500. The positions of the nebulosities catalogued by Davies et al. (1976) are indicated by crosses in Figure 1. A comprehensive catalogue of the radio sources is planned in which the integrated flux density and a map of each will be provided. This catalogue will yield information about mass and excitation parameters and, if used in conjunction with Hβ photometry, a direct measure of the interstellar extinction.

(iii) Supernova remnants: These stand out as resolved radio sources, usually with an associated catalogued nebulosity, in which the radio brightness is significantly greater than expected from the Hα brightness classification given by Davies et al. (1976). The larger remnants show a shell or arc-like structure. Confirmation may be obtained by data at other radio frequencies yielding a non-thermal but flattish spectrum, by an apparently associated X-ray source in the catalogues of Seward & Mitchell (1981) or Tanaka (1983) and by optical data. Optical data are often essential to eliminate the possibility of misidentification of a background radio and X-ray galaxy. In collaboration with Mt. Stromlo astronomers, five radio sources from the survey were confirmed as new supernova remnants (Matthewson et al., submitted); three of these are included in the X-ray catalogue of Tanaka (1983). Eleven remnants have now been found in the SMC viz: 0045-739, 0046-735, 0047-735, 0049-736, 0050-728, 0056-725, 0058-718, 0101-724, 0102-722, 0103-726, 0104-723.

THE 30 DORADUS REGION

A systematic survey of the LMC is planned during the next two years using a field size of 70 x 75 arcmin. Apart from the areas mapped around SNR candidates very little data are yet available. It is obvious, however, that the LMC contains much stronger radio sources, both thermal and non-thermal than the SMC. As an example, the region around the 30 Doradus nebula and the interesting emission nebulae to the south are shown in Figure 2; this map has been constructed from the overlap of two 46 arcmin fields and, because of the complexity and brightness of the emission, both fields have been 'cleaned' using a standard algorithm. The 'cleaning' process effectively removes the first negative sidelobe of -8% and has no other significant effects.

The peak Hα and radio brightness are much greater in this region than in the SMC field mapped in Figure 1 and as a result the lowest contour level has been set some five times higher. Also, the catalogue of Davies et al. (1976) is not suitable for very bright complex regions so we have given the Henize (1956) catalogue numbers to the main peaks. However, well away from the 30 Doradus nebula itself, low levels of emission are easily detected and the sensitivity is comparable to that

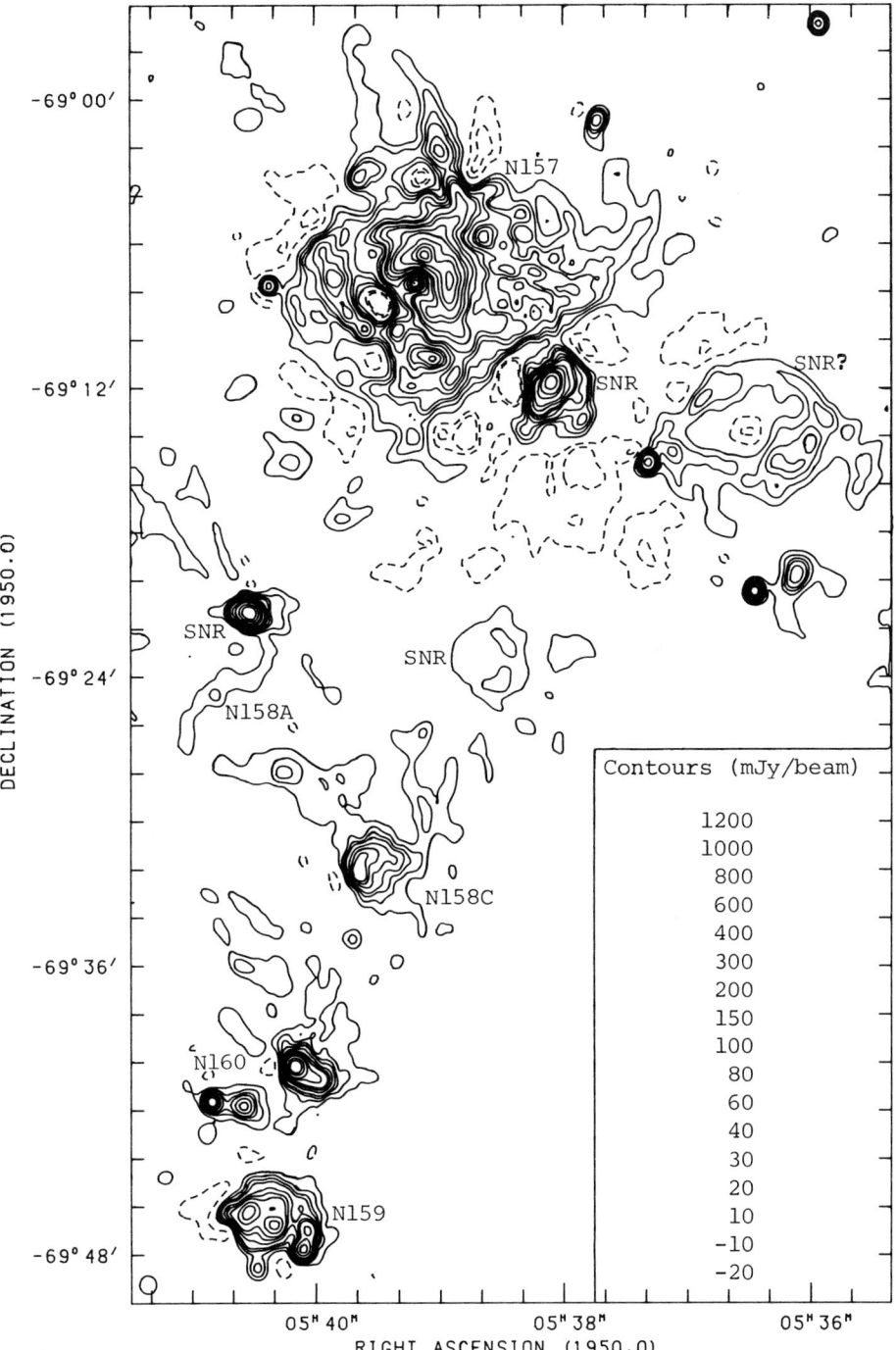

Figure 2. A 'cleaned' map of the 30 Doradus region.

shown in Figure I. The whole region has been discussed by Mills et al. (1978) and it was shown that the radio emission is predominantly thermal, arising from the numerous H II regions. The present results confirm the main features of the earlier 1415 MHz map obtained with the Fleurs Synthesis Telescope but the greater sensitivity reveals many new correspondences with fainter Hα emission features.

Three SNRs have been identified in the region, 0538-691, 0538-693 and 0540-693. The large 6 arcmin diameter ring source 0536-692 is regarded as a possible old SNR; although its radio spectrum is predominantly thermal there is some evidence for non-thermal emission on the eastern side (Clarke et al. 1976) and a weak X-ray source is also catalogued in the region (Long et al., 1981). Two of the SNRs, 0538-691 and 0540-693, appear to have angular sizes much greater than the optical and X-ray remnant (Mills et al., submitted). However in both cases this could be the result of unassociated background emission and better resolution will be needed to decide.

STATISTICS OF THE SNRs

The SNR candidate survey of the LMC has led to the confirmation of 27 remnants (Mills et al., submitted). Four of the six 'possible' remnants previously listed by Mills (1983) have now been confirmed. It is believed that a reasonably complete sample of SNRs has been obtained with diameters less than about 40 pc but comparisons with the SMC suggest that many large diameter remnants remain to be found when the LMC has been surveyed in its entirety. Mills (1983) described the preliminary statistics based on 29 remnants in both Clouds and showed that the results could not be reconciled with the conventional picture of their properties and evolution. Using the present catalogues containing 38 remnants these preliminary results have been examined more deeply and generally confirmed. There is great diversity in the radio properties and morphologies of the SNRs and no significant correlation between the luminosity and diameter of a remnant is apparent unless selections based on size or luminosity are imposed. No evolutionary track can be recognized. A direct application of the maximum likelihood method to the 24 remnants with diameters between the resolution limit of the radio-telescope, 7 pc, and the maximum diameter for reasonable completeness, 40 pc, yields the number-diameter relation:

$$N(< D) = \frac{1}{3} D^{1.2 \pm 0.35}$$

This result cannot be reconciled with the common assumption that all radio detectable SNRs spend most of their lifetime in the adiabatic or 'Sedov' expansion phase which requires a relation of the form $N \propto D^{2.5}$. The majority of remnants in the Clouds must have expanded to their present sizes with little deceleration of the supernova ejecta. Furthermore there is evidence that the radio remnant does not persist for very long as a recognizable object after significant deceleration has occurred; it appears that the synchrotron-emitting phase of a supernova remnant is probably a transient phenomenon lasting perhaps a few thousand years instead of some tens of thousands.

Comparison with the catalogue of Galactic SNRs compiled by Clark & Caswell (1976) produces no convincing evidence that the Galactic SNRs are any different. Because the great majority of Galactic remnants have no independent measure of their distances, a directly observed N-D relation cannot be derived but properties, when they can be compared, are similar. If the simplest assumption is made that the populations are identical and have an N-D relation of slope 1.2, the occurrence rate of SNRs in both Galaxy and Clouds may be estimated using the historical supernovae in the Galaxy as calibrators. The rate is found to be about four per century in the Galaxy and one per century in the two Magellanic Clouds combined. From the distribution of SNR diameters in the Clouds it seems likely that there was a burst of supernova activity in the LMC about four or five hundred years ago; the X-ray and optical data suggests that one remnant may be much younger (0540-693).

SUPERNOVA REMNANTS AS DISTANCE CALIBRATORS

In the course of comparisons between the Galactic and Cloud SNRs, Mills et al. (submitted) derive a distance scale from the Cloud remnants and apply it to the Galactic calibrators. This scale is an improvement on the simple luminosity scale of Mills (1983), used for a similar purpose. It makes use of the regressions of L, the luminosity, and D, the diameter, on Σ, the brightness, of the SNRs. In deriving the scale, distances of 55 kpc and 63 kpc were assumed for the LMC and SMC respectively. The scale is given by

$$d_{kpc} = 46\, S_{843}^{-0.275}\, \theta^{-0.45}$$

where d_{kpc} is the distance to the SNR in kiloparsecs, S is the measured flux density in jansky at 843 MHz and θ is the effective angular diameter in minutes of arc. Application of the scale to the remnants in the Clouds results in a dispersion of about 40 percent for individual SNR distances although the mean distances are, of course, close to those assumed. When applied to the group of Galactic calibrators tabulated by Clark & Caswell (1976) good agreement is found with, however, two minor differences. The calibrators with distances based on optical considerations, on average, have scale distances slightly greater than the optically derived distances. The calibrators using kinematic distances from radio absorption lines are found to have scale distances significantly less, by about 20%. However, these kinematic distances have been based on a distance to the Galactic centre of 10 kpc and this is now believed to be too large.

These comparisons can be used to provide an independent measure of the distances to the Magellanic Clouds as well as the distance to the Galactic centre. Taking the optically derived distances of relatively close SNRs as basic calibrators, distances to the LMC, SMC and Galactic centre are 49 kpc, 60 kpc and 7.2 kpc respectively. The statistical uncertainties are less than 10% but there are further uncertainties in the assumption that the Cloud distance scale may be directly applied to the group of Galactic calibrators and that the optically derived distances have no systematic bias.

REFERENCES

Broten, N.W.: 1972, Aust.J.Phys., 25, 599.
Clark, D.H. & Caswell, J.L.: 1976, M.N.R.A.S., 174, 267.
Clarke, J.N., Little, A.G. & Mills, B.Y.: 1976, Aust.J.Phys.Astrophys. Suppl., No. 40, 1.
Davies, R.O., Elliott, K.H. & Meaburn, J.: 1976, Mem.R.A.S., 81, 89.
Henize, K.G.: 1956, Astrophys.J.Suppl., 2, 315.
Long, K.S., Helfand, D.J. & Grebalsky, D.A.: 1981, Astrophys.J., 248, 925.
McGee, R.X., Brooks, J.W. & Batchelor, R.A.: 1972, Aust.J.Phys., 25, 581.
McGee, R.X., Newton, L.M. & Brooks, J.W.: 1974, Aust.J.Phys., 27, 79.
McGee, R.X., Newton, L.M. & Butler, P.W.: 1978, M.N.R.A.S., 183, 799.
Mathewson, D.S. & Healey, J.R.: 1964, "The Galaxy and the Magellanic Clouds", (eds) F.J. Kerr and A.W. Rodgers (Canberra; Australian Academy of Science), p.283.
Mathewson, D.S., Ford, V.L., Dopita, M.A., Tuohy, I.R., Mills, B.Y. & Turtle, A.J: (submitted), Astrophys.J.
Mills, B.Y.: 1959, Handbuck der Physik, Ed. S. Flugge, Springer-Verlag, 239.
Mills, B.Y.: 1981, Proc.astr.Soc.Aust., 4, 156.
Mills, B.Y.: 1983, IAU Symposium 101, Supernova Remnants and their X-ray Emission, p. 551.
Mills, B.Y. & Little, A.G.: 1953, Aust.J.Phys., 6, 272.
Mills, B.Y., Little, A.G., Durdin, J.M. & Kesteven, M.J.L.: 1982, M.N.R.A.S., 200, 1007.
Mills, B.Y., Turtle, A.J. & Watkinson, A.: 1978, M.N.R.A.S., 185, 263.
Mills, B.Y., Turtle, A.J., Little, A.G. & Durdin, J.M., (submitted), Aust.J.Phys.
Seward, F.D. & Mitchell, M.: 1981, Astrophys.J., 243, 736.
Shain, C.A.: 1959, Paris Symposium on Radio Astronomy, Ed. R.N.Bracewell, Stanford University Press, 328.
Tanaka, Y.: 1983, IAU Symposium 101, Supernova Remnants and their X-ray Emission, p. 535.

DISCUSSION

Dopita: I worry about the use of SNR radio emission for distance estimation, since VLA work by D'Odorico and his co-workers (including myself) showed that the M31 and M33 relations were different between each other and from that of the LMC. Thus it seems that intrinsic efects (structure of the ISM, IMF, etc.) influence a L-d relationship more than is acceptable for this method to be a useful distance indicator.

Mills: I agree that one must be cautious. There are several reasons for thinking that the Cloud and Galactic SNRs are similar, but this may not be so for other galaxies. In any event, because of selection effects I do not believe that the method could usefully be employed on other galaxies at present.

van den Bergh: How would your interpretation of the V versus D relation be affected if there were a very large range of interstellar magnetic field strengths in the Magellanic Clouds?

Mills: Provided the sensitivity limit of the radio telescopes is well below the SNR emissions, as we seem to have here, I cannot see any significant effect.

Preite-Martinez: I'd like to draw your attention to our work presented at this symposium (Fusco-Femiano and Preite Martinez), in which we show that the N-D relation in the LMC can be explained by a population of SNRs in the adiabatic phase, expanding in a non-homogeneous interstellar medium.

Mills: Our interpretations may not be so very different. I consider that SNRs are only likely to emit at radiofrequencies after significant slowing of the ejecta and that the emission is quenched shortly thereafter. That is, we think it likely that most of the observed SNRs are rather close to the adiabatic phase.

X-RAY SURVEYS OF THE MAGELLANIC CLOUDS

David J. Helfand
Columbia Astrophysics Laboratory, Columbia University

The proximity, well-determined distance, and low foreground obscuration of the Magellanic Clouds makes them an ideal target for the study of galactic X-ray source populations. Steady progress since the initial detection of X-rays from the LMC in 1968 culminated in the recent Einstein Observatory surveys from which over three dozen supernova remnants and ~10 compact binaries have been identified. In this review, we record the 15-year history of Magellanic Cloud X-ray research, summarize our current knowledge, and offer a brief prospectus of what the next 15 years may hold.

1. INTRODUCTION

The last IAU Symposium dedicated to the Magellanic Clouds was held just one year after the discovery of the first extrasolar X-ray source in 1963. The detection of any radiation above 10 eV from the Clouds in those days would have required a trip with a rocket to Australia and a target source with a luminosity in excess of 10^{41} ergs s^{-1}. Twenty years later, we have entered (and perhaps exited) a new era in X-ray astronomy. Instrumental sensitivities have increased a millionfold and angular resolution approaching that of optical astronomy is now attainable, although, at the present time, opportunities for studying X-rays from the Clouds are extremely limited. In this paper we will review the progress that has been made in studying the X-ray source populations in our neighbor galaxies from the first detection of X-rays in 1968 through the most recent optical and radio work on the ~150 sources catalogued in surveys carried out the Einstein Observatory. A brief prospectus outlining the future of Magellanic Cloud X-ray research, including a discussion of the contribution which could be made by an AXAF-class facility, concludes this review.

2. THE EARLY YEARS (1968-1971)

A two-page Astrophysical Journal Letter published in early 1969 marks the beginning of the X-ray study of the Magellanic Clouds (Mark et al. 1969). It reported on the results of a five minute rocket flight from Johnston Atoll in the South Pacific on October 29, 1968; the proportional counter detectors on board had a 5° field of view and a limiting sensitivity in the 1.5-10.5 keV band of ~10^{-9} ergs cm^{-2} s^{-1}. In an 80° scan through the positions of the Clouds, the LMC was detected as a ~4 sigma excess in two adjacent 5° bins with a flux of ~1.5×10^{-9} ergs cm^{-2} s^{-1} and a spectrum slightly softer than that of the diffuse X-ray background; the SMC was not detected. The authors estimated that the LMC flux corresponded to an X-ray luminosity of ~$4 \pm 1 \times 10^{38}$ ergs s^{-1} and, using an estimate of the Milky Way's luminosity in the same band (which was rather uncertain at the time), concluded that "the relative populations of X-ray sources are apparently the same in our Galaxy and the LMC." While some subtle differences in these populations are beginning to emerge from the ensuing fifteen years of X-ray research, we now know that the numbers of bright X-ray binaries and supernova remnants do indeed roughly scale with the mass ratio of these two glaxies.

A similar rocket flight two years later by the same group from Lawrence Livermore Laboratory obtained further information on the spatial and spectral distribution of the X-rays from the Large Cloud and reported the first evidence of emission from the SMC (Price et al. 1971). A few months later, Leong et al. (1971) published the first Uhuru results on the Cloud sources, resolving the LMC emission into three steady and one possible highly variable source (designated LMC X-1, X-2, X-3, and X-4) plus a possible diffuse component extending over much of the galaxy. The three SNRs known at the time (N49, N63a, and N132d) were not detected, and upper limits of 2×10^{-11} ergs cm^{-2} s^{-1} were placed on their emission in the 2-7 keV Uhuru band. The SMC flux was found to be consistent with a single, rapidly variable source in the wing of the galaxy. Fluctuations on timescales as short as a few minutes were observed, leading the authors to claim SMC X-1 as the first confirmed example of a stellar X-ray source in an external galaxy. No diffuse emission from the Small Cloud was detected and a limit of ~1×10^{-11} ergs cm^{-2} s^{-1} was set on any additional individual sources in either galaxy.

3. THE 2-10 KEV SURVEYS (1971-1978)

Over the ensuing seven years, it became clear that, in fact, all five of these Magellanic Cloud X-ray emitters were binary systems consisting of a normal star and a collapsed companion. Nonperiodic variability on a variety of timescales was established for all four LMC sources through observations made with Uhuru, SAS-3, and Ariel V (Griffiths and Seward 1977 and references therein). More recently 1.4 day X-ray eclipses (Li, Rappaport, and Epstein 1978), 13.5 s X-ray pulsations (Kelly et al. 1982) and a 30.5 day flux modulation reminiscent of that in Her X-1 (Lang et al. 1981) have all been seen in LMC X-4. X-ray pulses (0.7 s) and binary eclipses (3.9 d) are also present in SMC X-1 (Lucke et al. 1976; Schreier et al. 1972). Optical counterparts to all five sources have been identified (see Hutchings, this volume, and references therein); four have massive, early type components similar to the Population I X-ray sources in the Milky Way, while LMC X-2 is associated with a faint blue object in which the visible light may well be dominated by an accretion disk as in the Sco X-1 type Population II sources in the Galaxy. Clark et al. (1978) have noted that the X-ray luminosities of the Cloud binaries are systematically higher than those of the comparable galactic systems and suggest that the lower metal abundance (and, thus, lower opacity) of the accreting material may be responsible. We discuss the comparison of the Clouds' X-ray binary populations with those of the Galaxy further below.

The only other sources detected in the direction of the Clouds during this period were a number of transients, only one of which has been seen since. Two new sources (SMC X-2 and X-3) were reportd in the Small Cloud by Clark et al. (1978). They were detected by SAS-3 and had X-ray fluxes a factor of ~5 less than SMC X-1. An SMC-member O star was found in each ~1' error box (Crampton, Hutchings and Cowley 1978) although these identifications remain most uncertain. Both sources had disappeared six months later (Clark, Li, and van Paradijs 1979) and neither was detectd in the Einstein Observatory SMC survey of 1979-80 to a level at least 10^2 times lower than the original detections. Three similarly elusive sources have been reported in the direction of the LMC. LMC X-5 was first seen by Markert and Clark (1975) and was apparently confirmed in Ariel V data (Griffiths and Seward 1977); however, it did not appear in several Einstein pointings at a flux level 150 times below its reported value. LMC X-6 was observed only once (Griffiths and Seward 1977); 0544-665, first reported on the basis of a 1977/78 HEAO-1 modulation collimator survey (Johnston, Bradt, and Doxsey 1979), had dropped by a factor >500 during a single Einstein observation two years later.

The one transient Cloud X-ray source, concerning whose existence there is no doubt, is the remarkable source A0538-66. First reported by Ariel V (White and Carpenter 1978), a number of outbursts reaching a peak luminosity in excess of 10^{39} ergs s^{-1} and lasting from a few hours to several days were seen by HEAO-1 (Johnston et al. 1979; Johnston, Griffiths, and Ward 1980); an outburst recurrence time of 16.7 days was indicated, although X-rays are not seen at every predicted maximum. The source was not seen in the initial Einstein survey of the LMC, which set a minimum on- to off-state luminosity ratio of $\sim 10^4$:1. However, subsequent observations did detect 0538-66, confirming the suggested B star optical counterpart (Skinner et al. 1982) and revealing an X-ray pulsation period of 67 milliseconds. Although the underlying system of a B-star/neutron star binary is typical of Pop I X-ray systems, the extreme pulse-period, luminosity, and variability sets this source apart from all known galactic sources, marking it as an important object for further study.

All of the observations reviewed above were conducted in the ~2-10 keV X-ray band. This is the energy range in which the all-sky surveys carried out during the 1970s revealed accreting binary systems as the dominant galactic source population. However, other source classes were also discovered during this period, including SNR, white dwarfs, novae and star formation regions for which the dominant flux appears in the soft X-ray (0.1-2 keV) band (see Gorenstein and Tucker 1976 for a review). At these energies, interstellar absorption becomes important: X-rays below 1 keV can penetrate only a few kiloparsecs in the galactic plane (to $N_H \sim 10^{22}$ cm^{-2}). The line of sight to the Magellanic Clouds, however, has a foreground column density of only $\langle N_H \rangle \sim 3 \times 10^{20}$ cm^{-2}; column densities through the Clouds themselves range from 0.1 to 3×10^{20} cm^{-2} for the LMC and 1 to 10×10^{21} cm^{-2} for the SMC, rendering sources in the Clouds good targets for observation in this band.

Soft X-ray emission from the LMC was first detected in a series of rocket flights in the mid-seventies, although its interpretation remained confused (Rappaport et al. 1975; Long, Agrawal, and Garmire 1976; Borken 1976). The only detection of a discrete source of ~1 keV X-rays prior to the launch of Einstein (the SNR N132D) was reported by McKee et al. (1980; the other two sources suggested by their data were not confirmed by the Einstein results). Observations of the SMC in this band were directed primarily at attempts to determine the extragalactic component of the soft X-ray background through the diminution in its intensity by the interstellar medium of the Cloud (McCammon et al. 1971; Bunner, Sanders, and Nousek 1979). The Einstein results discussed below confirm the conclusion of McCammon et al. (1976) that the failure to detect such a diminution resulted not from a bright population of SMC soft X-ray sources which compensated for it, but from the fact that most of the diffuse background at these energies is local in origin. No discrete soft sources were detected toward the SMC during these observations.

In summary, as of early 1979, seven discrete sources of X-ray emission had been detected in the direction of the Magellanic Clouds: six X-ray binaries and one SNR. Another five or so transient sources had also been reported. At the detection threshold characteristic of this work (~5×10^{37} ergs s^{-1}), the results were not inconsistent with a simple extrapolation from Galactic source populations scaled by the mass ratios of the galaxies. The characteristics of these sources, both individually and collectively, were interesting - short pulse periods, high luminosities, extreme variability - and the extension of these surveys to flux levels 1,000 times lower held great promise for comparative studies of celestial X-ray emitters.

4. THE EINSTEIN SURVEYS (1979-1981)

The Einstein Observatory, the first satellite-borne imaging X-ray telescope, was launched in November of 1978. It carried a complement of four focal plane instruments for soft X-ray imaging and spectroscopy and 2-30 keV monitor proportional counter. The imaging proportional counter (IPC) provided 1' spatial resolution over a 1 square degree field of view and modest spectral resolution ($\Delta E/E \sim 1$) in the 0.15 to 4.0 keV band. Typical sensitivity in a 3,000 second observation was $\sim 10^{-13}$ ergs cm^{-2} s^{-1}. The IPC was the prime instrument for survey work and was used extensively in the Magellanic Cloud surveys discussed below. The count rate-to-flux conversion for a typical (kT $\sim$ 1 keV) spectrum with low interstellar absorption is $\sim 2.5 \times 10^{-11}$ ergs cm^{-2} s^{-1} (IPC ct)$^{-1}$. The high resolution imager (HRI) complemented the IPC by offering $\sim$5" spatial resolution over a 25' field of view in the slightly softer, 0.1 to 3.0 keV band. With this instrument, absolute positions could be obtained to $\sim$3".5; the HRI was used extensively in followup observations to determine accurate source locations and to map diffuse sources such as SNR. The count rate-to-flux conversion for this instrument is $\sim 10^{-10}$ ergs cm^{-2} s^{-1} (HRI ct)$^{-1}$. The two spectrometers on board included the solid state spectrometer (SSS) used for obtaining moderate resolution spectra ($\Delta E \sim$ 250 eV) between 0.5 and 4.0 keV, and the focal plane crystal spectrometer which achieved high resolution ($\Delta E/E \sim 10^3$) over narrow bandpasses near lines of astrophysical interest. Both of these instruments were used to observe the brighter SNR detected in the LMC (Clark et al. 1982; Winkler, private communication). Further details concerning the Observatory and its instrumentation may be found in Giacconi et al. (1979).

The LMC and SMC surveys were carried out as part of the Einstein programs of the Columbia Astrophysics Laboratory and the Smithsonian Astrophysical Observatory, respectively, with extensive follow-up observations by a number of participants in the Guest Observer Program. The observations were conducted between 1979 February and 1981 March; a total of 1.0×10^6 s of good data were accumulated, representing $\sim$25 days of telescope time or $\sim$3% of the satellite's life. The initial IPC surveys were designed to provide complete coverage over the bulk of both Clouds to a sensitivity $\sim 10^2$ times that previously available; these results are described in Long, Helfand, and Grabelsky (1981) for the LMC and Seward and Mitchell (1981) for the SMC. Deeper IPC pointings were then undertaken in optically prominent regions such as the Bar regions of both galaxies and around 30 Doradus; these data have been (or will be) reported by Inoue, Koyama, and Tanaka (1983; 1984) and Gull and Bruhweiler (1984) for the SMC, and by Long, Helfand, and Grabelsky (1981) and Hamilton and Helfand (1984) for the LMC. The survey statistics are summarized in Table 1; the number of known sources in the direction of the Clouds has increased 30-fold and the flux of the weakest source is a factor of $\sim$1,000 below previous survey limits.

Extensive optical and radio follow-up observations have been in progress over the past few years with the aim of identifying as many Cloud members as possible and separating out the numerous foreground and background X-ray sources which appear superposed against both galaxies. Many new SNRs in both the LMC (Mathewson et al. 1980; Tuohy et al. 1982) and SMC (Dopita, Tuohy, and Mathewson 1981; Mills et al. 1982) have been discovered as a result of this work. In addition,

TABLE 1: The __Einstein__ Observatory Magellanic Cloud Surveys.

Survey		LMC	SMC
Area covered (sq. deg.)		37	40
Strongest source detected	(IPC ct s^{-1})	11	5
Weakest source detected	"	0.006	0.003
Completeness limit	"	~0.03	~0.05
	(ergs s^{-1})	~3 × 10^{35}	~10^{36}
Sources detected		103	45
Above completeness limit		52	9
Cloud members		~40	~5

several new binary source candidates have been located and nearly twenty interlopers have been identified (Cowley et al. 1983; Hutchings, this volume; Pakull, this volume). At the present time, then, we have a rather complete picture of the X-ray source populations above luminosities of 3×10^{35} ergs s^{-1} for the LMC and 10^{36} ergs s^{-1} for the SMC, and considerable information on sources a factor of ~5 fainter. These results are summarized in Table 2 and discussed by source class below.

4.1 Interlopers

Roughly half of the ~150 X-ray sources now known in the direction of the Clouds are interlopers - either background galaxy clusters and active galactic nuclei (AGN) or foreground galactic stars. The numbers of expected and identified sources in each category are summarized in Table 2. The extragalactic objects are consistent with expectations. In the LMC, however, the number of foreground stars seen is considerably higher than expected from other surveys of stellar X-ray emission (see Cowley et al. 1983 for a detailed discussion). We have investigated the possibility that a young open cluster lies in the direction of the LMC and contributes a number of sources to the foreground population. Although one such cluster is known (Sanduleak and Philip 1968), it contributes none of the detected stars (nor would we expect it to, based on its distance and richness). Thus, we conclude either that there is an unrecognized young ($\lesssim 10^8$ yr) group of stars at ~100 pc in the direction of the LMC, or our detection of 11 X-ray emitting stars in the flux-limited portion of our survey (~3 were expected) is a statistical fluke. While few other foreground stars remain to be identified (since they are all brighter than $m_v \sim 15$ and have already been found) the number of remaining background interlopers among the weaker portions of both samples is still fairly large. The difficulty of detecting these faint AGN and clusters in the crowded MC star fields is likely to be a permanent limitation on the completeness of identification programs for the fainter sources.

TABLE 2: X-Ray Source Populations of the Magellanic Clouds

Galaxy:	LMC		SMC	
Sample:	X-ray Complete (>0.032 IPC ct s^{-1})	Total	X-Ray Complete (>0.05 IPC ct s^{-1})	Total
Number of sources:	52	102	8	45
INTERLOPERS:				
AGN				
Identified:	2	3	0	0
Candidates:	1	3	0	3
Addt'l no. expected	~3	(~12)*	~2	(~4)
GALAXY CLUSTERS				
Identified:	0	0	0	0
Candidates:	0	0	0	0
Addt'l no. expected:	~2	(~7)	~1	(~2)
STARS				
Identified:	10	12	2	4
Candidates:	1	7	0	2
Addt'l no. expected:	~0	(~1)	~0	(~1)
TOTAL				
ID plus candidates	14	25	2	9
Addt'l no. expected:	~5	(~20)	3	(~5)
SNRs				
Identified:	17	24	2	9
Candidates:	1	4	0	0
POP I BINARIES:				
Identified:	4	4		1
Candidates:	2	3	0	0
POP II BINARIES:				
Identified:	2	2	0	0
Candidates:	0	0	0	0
UNIDENTIFIED:				
Expected interlopers:	~5	(~20)	3	(~5)
Expected Cloud members:	~7	(~20)	0	(~20)

*Estimates in parenthesis are uncertain by a factor ~2 owing to uncertainty in the completeness level of the surveys.

4.2 Supernova Remnants

In view of their accurately determined distances (and, thus, luminosities and diameters), the sample of Magellanic Cloud SNR will prove uniquely important in future studies of remnant evolution. Twenty-four LMC and nine SMC objects are now detected in all three (radio, optical, and X-ray) bands, with five more seen in two of the three regimes; half of the total were discovered first as X-ray sources. The statistics of these samples are becoming particularly well defined with planned or recently finished complete, flux-limited surveys in both the radio and X-ray bands. For example, an analysis of all data for the flux-limited portion of the LMC X-ray survey shows that there are 19 ± 1 remnants with $L_x > 3 \times 10^{35}$ ergs s^{-1} in that galaxy. The only remaining sources of incompleteness are the lack of observations in the outermost regions of the Cloud and the possibility that we have missed a few of the largest diameter remnants of low surface brightness (this latter defect will be rectified in future analysis, but we are stuck with the former for ~5 years).

The full implications of these results for the Clouds' SN rate and the evolution of remnants in general is only just now beginning to be explored. Several authors have noted that the cumulative number-diameter relation for the LMC remnants (which exhibits a slope of order unity) is in conflict with the conventional assumption of adiabatic remnant expansion and suggests a relatively long-lived phase of free expansion for the SN shock (Clarke 1976; Mills 1983, 1984; Mills et al. 1983; Mathewson et al. 1983). More recently, however, Fusco-Femiano and Preite-Martinez (this volume) and Hughes, Helfand, and Kahn (1983) have pointed out that, for the X-ray sample at least, the effect of a finite flux threshold (in this case, equivalent to a luminosity threshold) is to introduce a diameter cutoff which is dependent on the initial explosion energy and the density of the surrounding medium. This fact can flatten the N(D) curve from the slope of ~2.5 expected for the adiabatic case to a slope of ~1 as is observed, even when most remnants are actually seen during their adiabatic phase. Further work on this problem, coupled with an analysis of the luminosity function (and L(D) relation) for Magellanic Cloud remnants should prove useful in problems ranging from a determination of the supernova rate, to the effects of nonequilibrium ionizaton and a cloudy, multiphase interstellar medium on SNR evolution.

4.3 Binary X-Ray Sources

There are now six confirmed and two candidate binary X-ray sources in the flux-limited portion of the LMC sample; SMC X-1 remains the only confirmed nontransient such source in the Small Cloud. Six of the LMC systems contain a bright early type stellar companion, analogous to the Pop I binaries of the Galaxy. There are ~50 such Galactic systems implying that this population, representative of recent ($\lesssim 10^7$ yr) star formation history, scales simply with the galaxian mass ratios of ~10:1. Our optical identification program is relatively complete for objects with $M_v \lesssim 0$ (Cowley et al. 1983) implying few if any new identifications will be made to this scaling. The single, bright SMC binary is also consistent with this rule.

Only two of the known binaries have faint companions similar to the Pop II X-ray binaries of the Galaxy where ~75 exist above our X-ray luminosity limit. Here, however, the optical identifications are far from complete. We estimate above (see Table 2) that there remain ~7 unidentified LMC members in the complete portion of the sample; all but 1 or 2 are excluded from being SNR as a result of observed variability or anomalously high IPC/HRI flux ratios. Thus, we expect that most of these will ultimately be identified as Pop II binaries (faint blue candidates exist for many) bringing the total to ~8. For this much older population ($\gtrsim 10^9$ yr) then, we again see simple scaling by mass.

While the number of MC binaries per unit mass appears to be very similar to that for the Galaxy, there are important differences in the populations. The two fastest of the 14 known Pop I X-ray pulsars (and three of the four fastest) are found amongst the Cloud sources. Also, the luminosities of the Cloud binaries are substantially higher than the mean of their Galactic counterparts (Clark et al. 1978; Hutchings, this volume); the peak luminosity of A0538-66 is higher than that of any other known binary X-ray source. Both of these facts are most likely related to the accretion process, possibly, as first suggested by Clark et al. (1978), the lower metallicity of the accreting gas in the Cloud sources. Also, Hutchings (this volume) has pointed to an apparent higher frequency of black hole companions in the Cloud sources. Further observational and theoretical work on all of these questions may provide important new insights into a number of problems in galaxy evolution.

5. THE NEAR FUTURE (1983-1988)

While few opportunities for observing the Clouds at X-ray wavelengths will exist during the next five years, continuing optical and radio followup work, as well as a more complete analysis of the <u>Einstein</u> X-ray database, will allow for significant progress during this interval. The complete radio survey of the LMC planned by Mills and his collaborators (Mills, this volume) may well identify a number of the remaining X-ray sources as SNR, in addition to answering important questions about the selection effects in the present sample and the coevolution of remnant X-ray and radio emission. More detailed comparisons of the optical and X-ray maps of the brighter SNR are also planned. Considerable optical work on the new identified and candidate binaries in the Large Cloud will be needed to define these orbital parameters, testing, in the process, the intriguing suggestion that the Cloud sources harbor more black hole companions than their galactic counterparts. Should better X-ray positions become available (see below) more optical identifications may be forthcoming. The background AGN we have discovered could be exploited as useful probes of the Clouds' ISM through 21 cm and optical absorption line studies.

We are currently in the process of constructing a single X-ray map of the LMC from the newly reprocessed IPC data. Overlapping and redundant pointings, as well as better background rejection and spectral selectivity, will provide an improvement in our point source (D ≲ 2') sensitivity by a factor of ≳ 2 throughout the survey regions with up to a factor of ~4 enhancement in heavily observed areas; the result will be another ~50 new sources in the luminosity range 3×10^{34} to 3×10^{35} ergs s^{-1} (for the Cloud members). More importantly, however, this new map will allow us to search for larger scale features – both the large diameter SNR (40 ≲ D ≲ 100 pc) so important to remnant evolution studies, and features tracing the hot regions of the Cloud's interstellar medium (e.g., bubbles and superbubbles blown by young star clusters). Analysis of the four deep IPC exposures in the Wing of the SMC will also expand the number of known sources in that galaxy.

Only two X-ray satellites capable of obtaining data on the Cloud sources are now in orbit. TENMA, the Japanese timing and spectroscopy mission could obtain some data on the bright Cloud binaries, but little of startling import is likely to result. The ESA satellite, EXOSAT, has experienced some difficulties since its launch in June, but its high resolution imaging detectors are still operating and could be used, albeit with rather long exposure times, to obtain ~5" positions for as many of the remaining unidentified sources as time allows; observation of all 12 unidentified sources in the complete, flux-limited sample of the LMC is an achievable goal. Other than a handful of rocket flights and Shuttle missions with maximum experiment durations of minutes to hours, there will be no other opportunity for further X-ray studies of the Clouds until ~1988 when the German ROSAT mission will give us a soft X-ray (0.03 to 2 keV) imaging capability similar to that of Einstein with sensitivities higher by a factor of ~2.

6. PROSPECTUS FOR THE FUTURE (1988-   )

The next major advance planned in X-ray astronomy instrumentation is the Advanced X-ray Astronomy Facility (AXAF), currently scheduled for launch in the early 1990s. It will provide arcsecond resolution with a sensitivity ~10^2 times that of Einstein over a broader, 0.1 to ~8 keV band. High sensitivity, high resolution spectroscopy and X-ray polarimetry will also be possible. The advances such an observatory would bring to Magellanic Cloud research are legion. Surveys similar to those carried out on Einstein with perhaps a factor of 3 increase in exposure time (the scheduled mission life is five times that of Einstein) could:

1. Detect all main sequence and post-main sequence stars individually for types earlier than O6 in the LMC
2. Detect most Wolf-Rayet stars in both Clouds
3. Detect all of the predicted ~200 Be stars with collapsed companions in their quiescent state
4. Detect the brightest ~10% of cataclysmic variable systems in both Clouds

5. Determine the frequency of low luminosity X-ray sources in Magellanic clusters
6. Map all SNR through a substantial portion of their radiative phases (a factor of ~4 older than currently possible) yielding a sample of perhaps 200 remnants as the definitive test of SNR evolution models.
7. Detect many of the ~100 radio pulsars with ages $\lesssim 5 \times 10^4$ yr via their surrounding synchrotron nebulae
8. Map, on all scales, the structure of the hot phase of the LMC interstellar medium.

Clearly, this partial list of discoveries would go far beyond the important questions of X-ray source population studies we are addressing in the current era; it would represent a contribution of fundamental importance in problems of stellar and galactic evolution.

The author wishes to acknowledge the support of the National Aeronautics and Space Administration under contract NAS 8-30753 and the United States Air Force under grant AFOSR 82-0014. This is Columbia Astrophysics Laboratory Contribution No. 255.

REFERENCES

Borken, R. J. 1976, paper presented at meeting of HEAD-AAS, Cambridge, Mass.
Bunner, A., Sanders, W., and Nousek, J. 1979, Ap. J. (Letters), 228, L29.
Clark, D.H., Tuohy, I.R., Long, K.S., Szymkowiak, A.E., Dopita, M.A., Mathewson, D.S., and Culhane, J.L. 1982, Ap. J., 225, 440-443.
Clark, G., Doxsey, R., Li, F., Jernigan, J. G., and van Paradijs, J. 1978, Ap. J. (Letters), 221, L37-L41.
Clark, G., Li, F., and van Paradijis, J. 1979, Ap. J., 229, 54.
Cowley, A.P., Crampton, D., Hutchings, J.B., Thorstensen, J.R., Charles, P.A., Helfand, D.J., and Hamilton, T.T. 1983, Ap. J., submitted.
Crampton, D., Hutchings, J. B., and Cowley, A. P. 1978, Ap. J., 223, L79.
Dopita, M.A., Tuohy, I.R., and Mathewson, D. S. 1981, Ap. J. (Letters), 248, pp. L105.
Giacconi, R. et al., 1979, Ap. J., 230, 540.
Gorenstein, P., and Tucker, W. 1976, Ann. Rev. Astr. Ap., p. 373.
Griffiths, R. E., and Seward, F. D. 1977, M.N.R.A.S., 180, 75p-79p.
Hamilton, T. T., and Helfand, D. J., in preparation.
Inoue, H., Koyama, K., and Tanaka, Y. 1983, "Supernova Remnants and Their X-Ray Emission," Proc. IAU Symposium No. 101, P. Gorenstein and J. Danziger (Dordrecht: Reidel).
——————————————, 1984, Ap. J., submitted.
Johnston, M. D., Bradt, H. V., and Doxsey, R. E. 1979, Ap. J., 233, pp. 514.
Johnston, M. D., Bradt, H.V., Doxsey, R. E., Griffiths, R. E., Schwartz, D. A. and Schwartz, J. 1979, Ap. J. (Letters), 230, pp. L11.

Kelley, R. L., Jernigan, J. G., Levine, A., Petro, L. D., and Rappaport, S. 1982, Ap. J., 264, pp. 568-574.
Lang, F. L. et al. 1981, Ap. J. (Letters), 246, pp. L21.
Leong, C., Kellog, E., Gursky, H., Tananbaum, H., and Giacconi, R. 1971, Ap. J. (Letters), 170, L67-L71.
Li, F., Rappaport, S., and Epstein, A. 1978, Nature, 271, p. 37.
Long, K. S., Agrawal, P. C., and Garmire, G. G. 1976, Ap. J., 206, pp. 411.
Long, K.S., Helfand, D.J., and Grabelsky, D.A. 1981, Ap. J., 248, 925.
Lucke, R., Yentis, D., Friedman, H., Fritz, G., and Shulman, S., 1976, Ap. J. (Letters), 206, pp. L25.
Mark, H., Price, R. E., Rodriques, R.M., Seward, F.D., and Swift, C.D. 1969, Ap. J. (Letters), 155, L143-144.
Markert, T. H., and Clark, G. W. 1975, Ap. J. (Letters), 196, L55.
Mathewson, D. S., Ford, V.L., Dopita, M.A., Tuohy, I.R., Long, K.S., and Helfand, D.J., 1983, Ap. J., Suppl., 51, 345.
Mathewson, D.S., Dopita, M.A., Tuohy, I.R., and Ford, V.L. 1980, Ap. J. (Letters), 242, L73.
McCammon, D., Meyer, S., Sanders, W., and Williamson, F. 1976, Ap. J., 209, pp. 46.
McCammon, D., Bunner, A., Coleman, P., and Kraushaar, W. 1971, Ap. J. (Letters), 168, pp. L33.
McKee, J. D., Fritz, G., Cruddace, R. G., Shulman, S., and Friedman, H. 1980, Ap. J., 238, pp. 93.
Mills, B.Y., Little, A.G., Durdin, J.M., and Kesteven, M.J. 1982, M.N.R.A.S., 200, pp. 1007.
Price, R. E., Groves, D. J., Rodriques, R.M., Seward, F.D., Swift, C.D., and Toor, A. 1971, Ap. J. (Letters), 168, pp. L7-L9.
Rappaport, S., Levine, A., Doxsey, R., and Bradt, H. V. 1975, Ap. J. (Letters), 196, pp. L15.
Sanduleak, N., and Philip, A.G.D. 1968, Ap. J., 73, 566.
Schreier, E., Giacconi, R., Gursky, H., Kellogg, E., and Tananbaum, H. 1972, Ap. J. (Letters), 178, pp. L71.
Seward, F. D., and Mitchell, M. 1981, Ap. J., 243, pp. 736.
Skinner, G. K. et al. 1982, Nature, 297, pp. 568.
Tuohy, I.R., Dopita, M.A., Mathewson, D.S., Long, K.S., and Helfand, D.J. 1982, Ap. J., 261, 473.
White, N. E., and Carpenter, G. F. 1978, M.N.R.A.S., 183, pp. 11.

SPECTROSCOPY OF STELLAR X-RAY SOURCES IN THE MAGELLANIC CLOUDS

J.B. Hutchings
Dominion Astrophysical Observatory
Victoria, B.C., Canada, V8X 4M6

ABSTRACT. In the Magellanic Clouds, about 75 candidate stellar X-ray sources have been detected. Most of these positions have now been investigated and optical identifications made for ~ 50%. The majority of sources are foreground dwarf stars or background active galaxies. Detailed investigations exist for 3 SMC sources and 6 LMC sources. It is possible to make a preliminary comparison with the population of galactic X-ray sources. The Magellanic Cloud X-ray binaries have a number of unique or remarkable properties and the most important ones are presented and discussed. These include the most rapid pulsars (SMC X-1, 0538-66), the possible precessing disk in LMC X-4, and the black hole candidates LMC X-3, LMC X-1. The properties of these objects relate to the evolution of stars in the Magellanic Clouds and how it differs from the Galaxy.

INTRODUCTION AND OVERVIEW

The two Magellanic Clouds offer us our best opportunity of studying complete populations of stellar X-ray sources. They are close enough for X-ray positions to yield good optical identifications, and do not suffer the large and variable optical and UV extinction found in galactic sources. We may also hope to identify optical counterparts of the principal types of known sources: OB stars, Sco X-1 like systems, and globular clusters. The X-ray properties of the Cloud sources are presumably described elsewhere in this volume (Helfand review). However, it is necessary to look briefly at the overall X-ray data base to start our discussion.

There are surveys of both Clouds, most completely made with the Einstein observatory. Seward and Mitchell (1981) report on 26 SMC field sources in ~ 40 square degrees with limiting luminosity 3×10^{35} erg s^{-1}, complete to 10^{36} erg s^{-1}. Significantly, they did not detect SMC X-2, X-3 which are two very luminous transient sources. The LMC has been surveyed by Long, Helfand and Grabelsky (1981), recently updated, with optical work, by Cowley et al. (1983). This survey covers the same area of sky and is complete to luminosities of 5×10^{35} erg s^{-1} in the LMC, and includes about 75 sources. The rough statistics of these surveys and their optical identifications to date, are shown in Table 1, together with some numbers for the galaxy and M31 (Crampton et al. 1983) for comparison. This table contains in parentheses my present best estimate of the final count, which will not be known for some years. Actual known numbers are given without parentheses. Estimates are all based on consideration of probable incompleteness and may be wrong by up to a factor two. With these uncertainties, not much can be said about the types and populations of the three basic source types, except that they appear to be consistent with the same basic fractions in all four galaxies, i.e. so far, no gross differences are apparent. The high X-ray luminosity of the MC sources is well known, so that the effects of the luminosity limit of the surveys is not easy to assess. If we regard the surveys as picking out the highest luminosity sources in each galaxy, the numbers suggest that this population is complete at ~ 10^{36} erg s^{-1} and that fainter sources do not occur in great numbers until say < 10^{34} erg s^{-1}. This again is consistent with what we find in the Galaxy.

TABLE 1
Approximate X-ray Source Comparison for Four Galaxies

PROPERTY	SMC	LMC	GALAXY	M31
Galaxy mass	1/60	1/10	1	2
L_x limit (erg s^{-1})	5×10^{35}	5×10^{35}	5×10^{36}	5×10^{36}
Survey area(sq°)	40	40	-	3
# Sources detected	28	75	-	>200
Foreground	2(12)	12(15)	-	(3)
AGN	2(5)	5(7)	-	(3)
SNR	4(6)	25(35)	-	(100)
Binaries	3(5)[3]	6(16)[5]	(150)	(150)
Pop I binaries	3(3)	4(8)	16(50)	22(50)
Pop II binaries	0(2)	1(8)	30(75)	20(60)
Cluster binaries	0(0)	0(0)	12(25)	25(35)

Numbers in parentheses are author's present guesses at final counts. Numbers without parentheses are present counts of identified objects, or in the case of M31, candidate objects. Numbers in square brackets give counts of identified sources to L_x 5×10^{36} limit.

At this stage, the completion of source identifications is necessary, requiring more accurate positions for the ~ 25 unidentified sources, before we can make detailed comparison with the Galaxy. We therefore proceed to discussion of the nine Cloud binary sources that have been identified and studied in detail. These binaries are remarkable in a number of ways, which may well relate to basic differences between the Clouds and our Galaxy. One, and quite possibly two of the 6 LMC sources contains a black hole, compared with less than one (possible) in a similar volume of Galaxy. Does this imply that stellar evolution is different in the Clouds? The cloud sources have much higher X-ray luminosities than galactic ones, as originally noted by Clark et al. (1978,9). Their supposition that the accretion rate and hence X-ray luminosity is governed by heavy element abundances which affect the opacity of an accreting cloud of gas about a compact object, seems very reasonable. The two fastest X-ray pulsars (SMC X-1 and 0538-66) are found in the Clouds, again possibly suggesting a more rapid accretion rate. The best evidence for the existence of accretion disks in massive X-ray binaries (SMC X-1, LMC X-3, LMC X-4) is found in the Cloud sources, once more suggesting high accretion rates. Finally, there are two unique systems in the LMC, the transient 0538-66, and CAL 83, which are described in the sections below.

INDIVIDUAL SOURCES

SMC X-1. This binary has all the requirements for complete mass determination - X-ray pulses, eclipses and optical radial velocities (Hutchings et al. 1977; Primini, Rappaport and Joss 1977). Rappaport (1982) summarizes Monte Carlo analysis of the basic parameters, and these are shown in Figure 1. The excellent X-ray pulse timing orbit shows up several interesting optical phenomena, which are probably general. The orbit is very circular, yet the optical radial velocities are best fit with appreciable eccentricity. Some of this can be modelled with X-ray heating and gravity darkening of the primary, but not all, and only in hindsight. Thus, the determination of orbital elements for such stars from optical spectroscopy is good to a first approximation only. Secondly, the ubiquitous He II λ 4686 emission is found to originate near to, but not at the X-ray star (see Fig. 2). From such studies, the accretion processes can be modelled, and in Fig. 2 it is seen that the entire roche lobe of the X-ray star may be occupied by an accretion disk. Van Paradijs and Zuiderwijk (1977) and Hutchings (1982) find evidence for a disk in analysis of the optical light curve. Similar conclusions are drawn from the UV light curve by Van der Klis et al. (1982). Are there other differences from galactic binaries? Not to a first approximation. The stellar masses are similar and the X-ray ionization of the stellar wind appears similar (Hutchings 1982,3) (Fig. 3). More detailed study of the UV resonance lines will probably show up stellar wind differences from the galaxy (Hammerschlag-Hensberge, Kallman and Howarth 1983), but will need careful work and, possibly, ST spectrographs.

SMC X-2,3. Very little has been done on these two sources, seen only once as very bright transients (Clark et al. 1978). The two proposed optical counterparts were observed by Crampton, Hutchings and Cowley (1978). They found both stars to be SMC members of type late O with weak lines, and weak He II emission. These sources need better positions to confirm the optical identifications, and the stars need better and more extensive observation if they are indeed the right ones.

LMC X-1. The identification of this source has been difficult

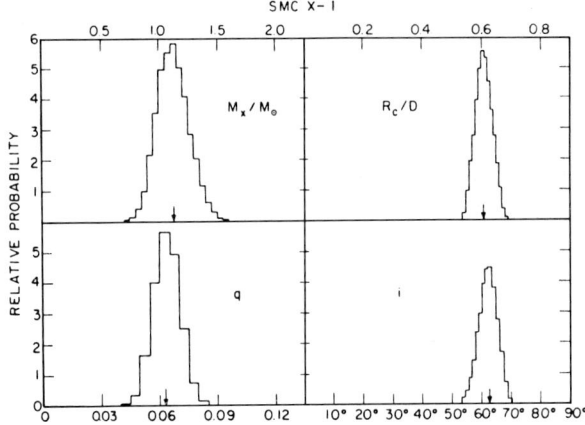

Fig. 1. Monte Carlo simulation of SMC X-1 binary parameters by Rappaport (1982).
Rc = primary radius
D = separation
q = mass ratio

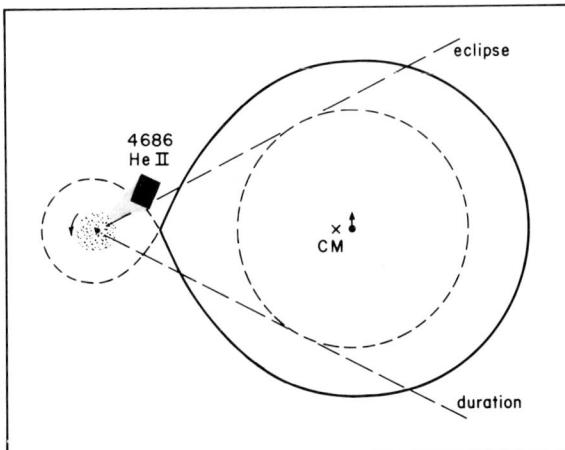

Fig. 2. Sketch in orbit plane of SMC X-1, typical of pop I sources. Note He II emission region offset from X-ray star.

Fig. 3. Modulation of UV wind resonance lines with orbit phase due to X-ray ionisation.

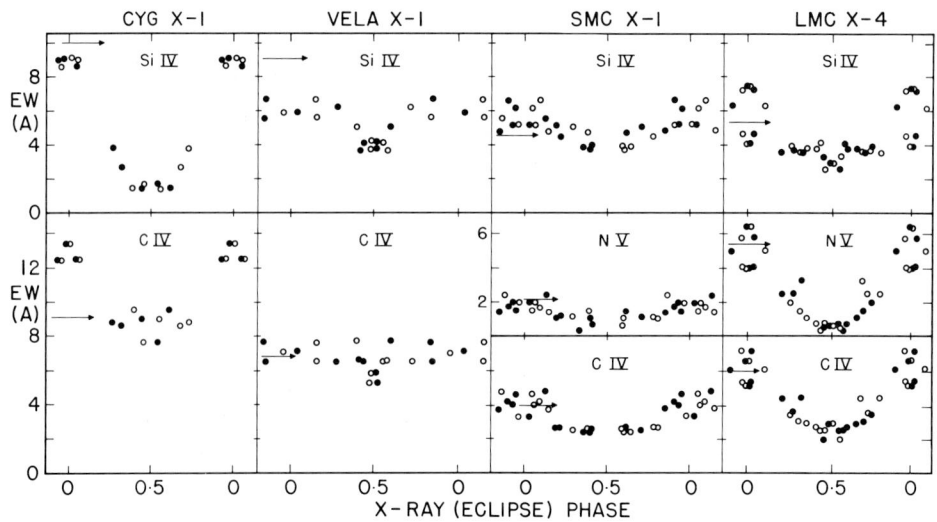

because two excellent candidate stars lie within ~ 3 arcsec of the X-ray position. Both have variable velocity and lie in a nebula with λ 4686 emission. Very recently (Hutchings, Crampton and Cowley 1983) the fainter star has been found to be an O-type binary of ~ 4 day period, with antiphased N III λ 4640 emission (λ 4686 is overlaid by nebular emission). The brighter candidate star is a supergiant B star, with velocity variations typical of supergiants and no discernable period. Assuming the fainter star to be the correct one, the companion appears to have a mass between 2.5 and 8 $M_\odot$ (Fig. 4), making it a black hole candidate of very similar characteristics to Cyg X-1.

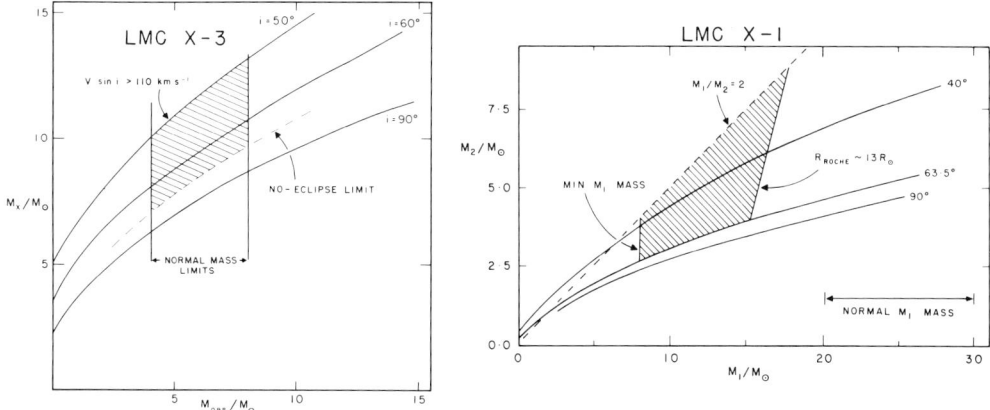

Fig. 4. Component masses in two black hole binary candidate systems in LMC, based on optical spectra. Shaded areas are allowed locations.

LMC X-2. This is identified with a faint blue star (~ 18m) with He II emission (Pakull and Swings 1979). Its velocity is somewhat low for the LMC and it is not completely certain that it is not a foreground object in the galactic halo. Little is known about it to date.

LMC X-3. This source is identified with a B type star, in a binary orbit of 1.7 day period with a very large (235 km s^{-1}) orbital velocity. The X-ray star in this system has a mass in the range 7-13 $M_\odot$ (Cowley, et al. 1983) and appears to be the strongest case yet found for a stellar black hole (Fig. 4). The overall weakness of the B star absorption lines and the large photometric variation of the system are evidence for the existence of a luminous accretion disk in this system. Unfortunately the star is too faint to be observed in the UV until ST is launched.

LMC X-4. This is another well-studied binary, as the X-ray data show pulses, eclipses, and a 30-day cycle similar to the Her X-1 precession cycle, and thought to be the same phenomenon (Li, Rappaport, Epstein 1978, White 1978, Chevalier and Ilovaisky 1977, Lang et al. 1981, Kelley et al. 1982, Van der Klis et al. 1982). The UV and optical light curves show time changes (Fig. 5) which are probably related to the 30-day cycle and can be modelled as precession of a luminous disk.

Optical orbit determinations by Hutchings, Crampton and Cowley (1978) and Petro and Hiltner (1982) show some disagreement in orbital amplitudes which may relate to variable "complications" connected with a disk precession, although they are not obviously 30-day phase related. This system is the only massive binary in which the three clocks are present (pulse, orbit and precession) and it offers a unique laboratory for studies of variable X-ray illumination, stellar wind modification, and disk behaviour, and should be the subject of detailed spectroscopy with ST.

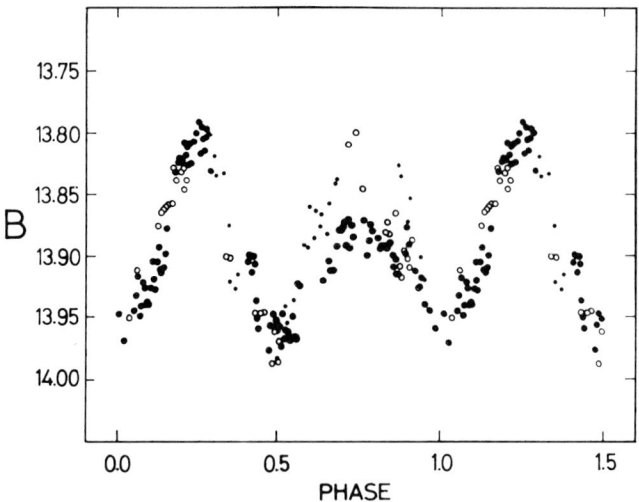

Fig. 5. Light curve of LMC X-4 showing 30-day phase variation, probably due to disk precession and changing X-ray heating (from Chevalier and Ilovaisky 1977).

0538-66. This is perhaps the strangest of the Cloud binaries. It is very variable, peaking at $\sim 10^{39}$ erg s^{-1}, and has the very rapid pulse period of 69 milliseconds (Skinner et al. 1982). The system contains a B star, whose spectrum so far has been rarely seen, as the binary has been in an extended state of optical emission for a period of some two years. During this time the spectrum has pronounced P Cygni characteristics, modulated very strongly on the known 16.5 day X-ray intensity period (Figure 6). This long period, presumably that of the binary orbit, makes for difficult observing, but limited pulse arrival timing data suggest orbital motions. Charles et al. (1982) and Howarth et al. (1983) discuss much of the spectroscopic evidence and possible models for the system. It seems generally accepted that the primary in quiescence is a somewhat inactive B star or rapid rotator (Be), and that the X-ray activity arises from a very eccentric orbit of the pulsar, which brings it close to the B star surface. During the extended high states, the B star photosphere becomes very extended and possibly kept that way by strong X-ray heating. Mass-loss from the system must be very high at such times, and both the high X-ray luminosity and the very rapid pulse period indicate a high accretion rate by the pulsar. An alternative possibility is that a massive accretion disk builds up and dominates the optical luminosity. Until orbital parameters are derived for this system, and the full range of its activity is studied, many questions will remain unanswered. So far, the system is without parallel in other known X-ray binaries.

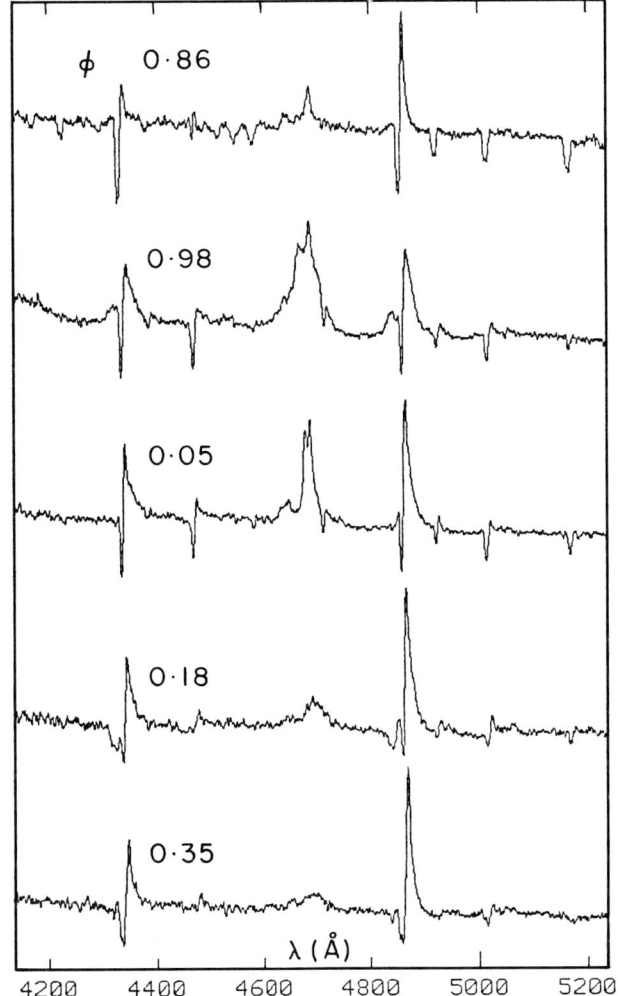

Fig. 6. Sequence of spectra of 0538-66 showing modulation of P Cygni type spectrum through X-ray maximum phase. All data are from the same cycle. (from I.B. Thompson, private communication)

CAL 83. This final identified source is faint and also unique. The star has a very broad emission complex in the $\lambda 4600-4700$ region, and this is the principal line feature. In this respect it is perhaps similar to the galactic burst source 1735-44 (Hutchings, Cowley, Crampton, 1982). Velocity measures so far have suggested periods in the range of a few days or near one day, but the amplitude is so low that unlikely masses or orbital indications are required. It seems more likely that the velocities are not purely orbital, perhaps arising between the stars in the system. The $\lambda 4686$ profile shows very rapid profile variations. The object has a luminosity similar to Sco X-1 and on this basis it is probably of low mass. Further study is needed to examine the unique properties of this object.

CONCLUSION

To conclude, it is clear that the full identification and study of Cloud X-ray sources is far from complete yet. Work to date has discovered a small number of very important binary sources, whose study has already suggested that a single parameter - the heavy element abundance - may account for a number of differences between Cloud and Galaxy sources. (However, we need to consider carefully whether the galactic halo sources should not also show these differences). A full understanding of these differences will almost certainly relate strongly to our understanding of stellar evolution and history in the Magellanic Clouds. Aside from the questions and needs already stated, I feel it is also important to study a number of normal binaries in the clouds, and at present we have little idea as to how fundamental normal stellar parameters may differ from those in the Galaxy.

I wish to acknowledge discussions and information from a number of colleagues in the preparation of this overview, particularly Anne Cowley and David Crampton.

REFERENCES

Charles, P.A. et al.: 1982, MNRAS **202**, 657.
Chevalier, C., Ilovaisky, S.A.: 1977, A & Ap. **59**, L9.
Clark, G., et al.: 1978, Ap.J. **221**, L37.
Clark, G., Li, F., Van Paradijs, J.: 1979, Ap.J. **227**, 54.
Cowley, A.P., et al.: 1983, Ap.J. in press.
Cowley, A.P., et al.: 1983, in preparation.
Crampton, D., Hutchings, J.B., and Cowley, A.P.: 1978, Ap.J. **223**, L79.
Crampton, D., et al.: 1983, preprint.
Hammerschlag-Hensberge, G., Kallman, T.R., Howarth, I.D.: 1983, preprint
Howarth, I.D. et al.: 1983, preprint.
Hutchings, J.B. et al.: 1977, Ap.J. **217**, 188.
Hutchings, J.B., 1982: Galactic X-ray Sources: Wiley pl.
_____. 1983, Adv. Space Res. **2**, 75.
Hutchings, J.B., Crampton, D., and Cowley, A.P.: 1983, Ap.J. preprint.
Hutchings, J.B., Cowley, A.P., and Crampton, D.: 1982, PASP **95**, 23.
Hutchings, J.B., Crampton, D., Cowley, A.P.: 1978, Ap.J. **248**, 925.
Kelley, R.L., et al.: 1982, Ap.J. **264**, 568.
Lang F.L. et al.: 1981, Ap.J. **246**, L21.
Li, F., Rappaport, S., and Epstein, A.: 1978, Nature **271**, 37.
Long, K., Helfand, D.J., and Grabelsky, D.S.: 1981, Ap.J. **248**, 925.
Pakull, M., and Swings, J.P.: 1979, IAUC **3318**.
Petro, L.D., and Hiltner, W.A.: 1982, A. J. preprint.
Primini, F., Rappaport, S., Joss, P.C.: 1977, Ap.J. **217**, 543.
Rappaport, S.: 1982, Galactic X-ray Sources: Wiley p. 171.
Seward, F.D., Mitchell, M.: 1981, Ap.J. **243**, 736.
Skinner, G.K., et al.: 1982, Nature **297**, 568.
van der Klis, M., et al.: 1982, A & Ap. **106**, 339.
van Paradijs, J., and Zuiderwijk, F.: 1977, A & Ap. **61**, L19.
White, N.E.: 1978, Nature, **271**, 38.

RADIO CONTINUUM EMISSION FROM MAGELLANIC-TYPE AND DWARF IRREGULAR GALAXIES

Ulrich Klein[1], Roland Gräve[1] and Rainer Beck[2]
[1] Max-Planck-Institut für Radioastronomie, Bonn, FRG
[2] Max-Planck-Institut für Kernphysik, Heidelberg, FRG

High-frequency radio continuum observations of a small sample of irregular and dwarf galaxies (Klein et al., 1983) have revealed that their integrated radio continuum spectra tend to be significantly flatter than those found for normal spirals (Gioia et al., 1982). This indicates a higher relative amount of thermal (free-free) emission.

The common feature of Magellanic Cloud-type galaxies seems to be a general lack of synchrotron disk radio emission at high frequencies if compared to normal spirals. The distribution of high-frequency radio emission is governed by strong thermal radiation from large HII complexes (like e.g. 30 Doradus in the LMC) which are situated at the termination of the bar-like stellar disk. There is evidence that dwarf galaxies with the lowest average mass surface densities exhibit the weakest nonthermal disk radio emission (like e.g. the Local Group member NGC6822, see Fig.1).

It is suggested that this deficiency in non-thermal disk emission of dwarf irregular galaxies is due to the absence of an energetic non-thermal cosmic rays via supernova-induced shock fronts takes place. The typical average mass surface densities of dwarf irregular galaxies are probably too low to stabilize a thick disk against internal pressure. Thus cosmic ray electrons quickly lose their energy and diffuse out into intergalactic space.

Models of stochastic self-propagating star formation predict that star formation is a rather short-lived phenomenon in dwarf galaxies (Gerola et al., 1980). This further inhibits the formation of a nonthermal thick disk. Finally, the low shear in irregular dwarf galaxies might not be sufficient to drive a dynamo that is required in order to generate and maintain a galactic magnetic field that is strong enough to produce detectable synchrotron emission at high frequencies.

REFERENCES

Gerola, H., Seiden, P.E., Schulman, L.S.: 1980, Astrophys. J. 242, pp. 517-527
Gioia, I.M., Gregorini, L., Klein, U.: 1982, Astron. Astrophys. 116, pp. 164-174
Klein, U., Gräve, R., Wielebinski, R.: 1983, Astron. Astrophys. 117, pp. 332-342

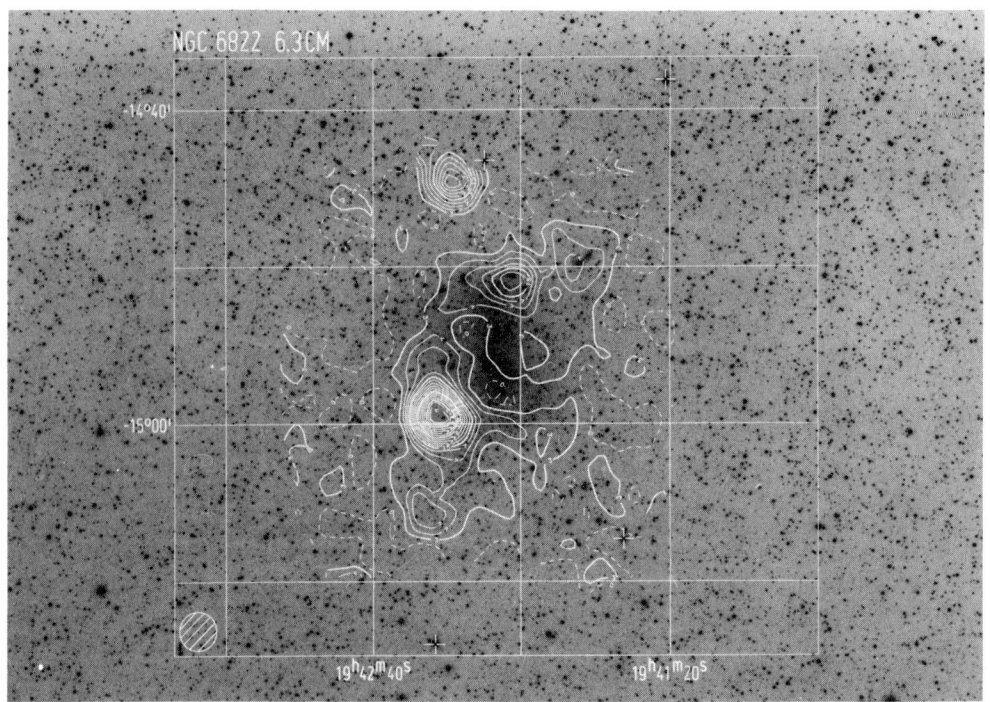

Figure 1

Radio map of NGC6822 at 4.75 GHz obtained with the Bonn 100-m telescope (HPBW 2.47 arcmin), superimposed onto a POSS plate. Contours are: 3,6,... mJy/beam area. The strongest source in the field has a rather steep spectrum ($S_\nu \sim \nu^{-1.02}$) and is most likely a background source. At the northern end of the bar-like stellar body a chain of HII complexes emits strong thermal emission. The radio emission from the disk of NGC 6822 is rather weak at this frequency.

ARE THE SUPERNOVA REMNANTS OF THE LMC IN THE ADIABATIC PHASE?

R. Fusco-Femiano and A. Preite-Martinez
Istituto di Astrofisica Spaziale, Frascati, Italy.

1. INTRODUCTION. A catalog of supernova remnants (SNRs) observed in the soft X-ray band 0.15-4.5 KeV by the imaging instruments on board the Einstein Observatory has been recently published by Mathewson et al. (1983). One of the conclusions in their paper is that SNRs in the LMC appear to evolve to large diameters much faster than predicted by the adiabatic expansion theory (Sedov 1959). This is based on the observed cumulative number-radius (N-R) relation that for the remnants in the LMC turns out to be approximatively linear. As suggested by Mathewson et al. (1983), a linear N-R relation can be obtained assuming a population of SNRs in the free-expansion phase (phase I, Woltjer 1972) expanding in a homogeneous medium. In the following we show that the X-ray surface brightness as a function of linear radius (Σ-R diagram) and the N-R relation can be alternatively explained assuming that the SNRs in the LMC are in the adiabatic phase.

2. THE Σ-R DIAGRAM. In Figure 1 we show the surface brightness-radius relations derived from our NE-hydrodynamical models. Each evolutionary track is characterized by the value of the fundamental parameter $\eta = n_o^2 E_{51}$. For the sake of simplicity, tracks are labelled with the value of the interstellar density n_o. T_s ranges from about 10 to 0.2 keV. The η=1 track has been followed down to 0.02 keV. The hydrodynamical models are allowed to radiate since the beginning of their evolution. The surface brightness has been computed using metal abundances $\{M\} = 0.3 \{M\}_\odot$. As discussed in detail by Fusco-Femiano and Preite-Martinez (1983), it is possible to define from a theoretical point of view the expected location in the Σ-R plane of adiabatic remnants expanding in IS media of different densities. No adiabatic remnant can be found to the left of the line M=10 in Figure 1, no remnant, either in the adiabatic or in the radiative phase, can be found to the right of the line labeled $(\Sigma-R)_{max}$. The overlap between the observational and the theoretical strip in Figure 1 can be taken as an indi-

cation that (i) the remnants are indeed in adiabatic or later evolutionary phases, and (ii) in the LMC there is a variety of possible environments in wich SN explosions can occur, with a large spread of IS densities (Long et al. 1981), possibly larger than that in our set of models.

3. THE N-R RELATION.

The cumulative number-radius relation has been redefined including all the sources listed in the survey except the three crab-like remnants (Figure 2). The N-R relation expected from a population of SNRs in adiabatic expansion in a homogeneous medium, $N \propto R^{2.5}$ (dashed line in Figure 2), is clearly inconsistent with the observed data. If we consider, though, that the ISM is not homogeneous, we are able to satisfactorily reproduce the shape of the observed N-R relation. Indeed, applying the Sedov's similarity solution, the number of remnants up to radius R is also an explicit function of the ambient density n : $N(R)=t(R,n)/\Delta t$ where Δt is the time interval between SN explosions, and $E_{51}=1$. We further introduce a possible dependence of Δt on n : $\Delta t \propto n^{-\alpha}$. As shown in Figure 2, even with $\alpha=0$ the N-R relation is much flatter than the relation expected for SNRs expanding in a homogeneous ISM. A slight dependence of Δt on the density (α in the range 0.0-0.5) can explain even the flattening at large radii (R>20pc).

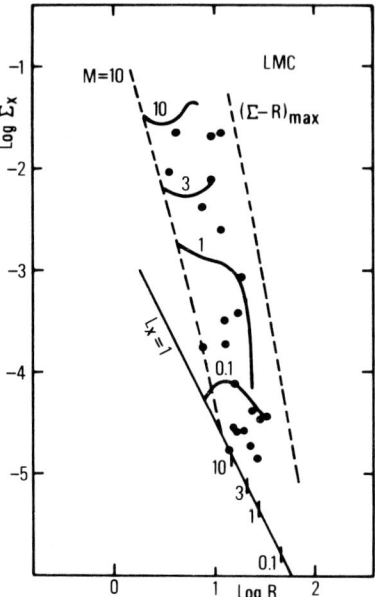

Figure 1. Surface brightness versus radius.

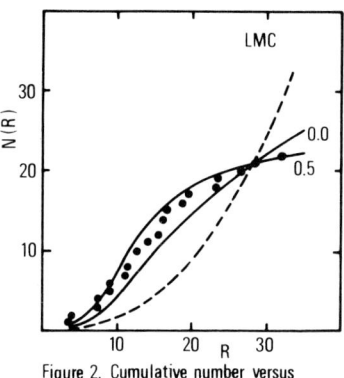

Figure 2. Cumulative number versus radius.

If this flattening is a real effect, it cannot be easily explained assuming that remnants are in free expansion. In other words, the shape of the N-R relation could turn out to be the stronger argument in favour of the adiabatic-expansion hypothesis.

References.
Fusco-Femiano, R., and Preite-Martinez, A. 1983, submitted to Ap.J.
Long, K.S., Helfand, D.J., and Grabelsky, D.A. 1981, Ap.J., 248, 925.
Mathewson, D.S., Ford, V.L., Dopita, M.A., Tuohy, I.R., Long, K.S., and Helfand, D.J. 1983, Ap.J.Suppl., 51, 345.
Sedov, L.I. 1959, Similarity and Dimensional Methods in Mechanics.
Woltjer, L. 1972, Ann.Rev.Astr.Ap., 10, 129.

COMPACT X-RAY SOURCES IN THE LMC: OPTICAL IDENTIFICATIONS*

Manfred W. Pakull
Institut für Astronomie und Astrophysik
Technische Universität Berlin, F.R.G.

Summary: In addition to the bright sources LMC X-1 - X-4 and A0538-66 fifteen EINSTEIN X-ray sources in the field of the LMC appear to be non-extended and have been accurately positioned with the HRI to make optical follow-up studies feasible. At least twelve of these are not associated with the LMC, including one Seyfert galaxy at redshift 0.076 and eleven late-type galactic foreground stars, some showing photometric variability typical for RS CVn and BY Dra stars.

The sources 1E 0524.5-7013 and 0543.8-6823 are identified with variable faint blue stellar objects, the latter has been observed to exhibit variable He II and Balmer emission lines being reminiscent of the low mass X-ray binary LMC X-2. The optical counterpart of LMC X-1 is a highly reddened (V=14.5, E(B-V)=0.61) O 7-9 type giant with variable He II and N III emission. The IPC source 1E 0501.6-7037 is probably associated with the LMC OB emission line star SK-70 36 which shows ellipsoidal light variations with a period of 6.94 days.

Further possible identifications of IPC sources with stellar objects and globular clusters in the LMC are discussed. The results indicate that only a small fraction of the non-extended X-ray sources are associated with the LMC and that a gap in the X-ray binary luminosity function exists between 10^{36} and 10^{38} erg/s.

A full account of this work will be published elsewhere.

---

* Based on observations collected at the European Southern Observatory, La Silla, Chile.

MOLECULES AND DUST IN THE MAGELLANIC CLOUDS

F.P. Israel
Astronomy Division, Space Science Department ESA, ESTEC
Noordwijk, The Netherlands.

1. INTRODUCTION.

A variety of studies over the last decade has shown molecular hydrogen to be a major constituent of the interstellar medium both in our Galaxy and in other spiral galaxies (Morris and Rickard, 1982). Our Galaxy contains roughly $M(H_2) = 4 \times 10^9 M_\odot$; between $R = 2$ kpc and $R = 10$ kpc the H_2 mass is one to three times that of HI; at the solar circle about 12 per cent of the total disk mass is in the form of H_2; most of this mass is in the form of several thousand giant molecular cloud complexes (GMCs) with sizes $d > 20$ pc and masses $M(H_2) > 10^5 M_\odot$ (Cohen et al, 1980; Sanders, 1981; Dame, 1983). These GMCs mainly consist of clumps with much smaller scales of order a few pc or less (e.g. Bally and Israel, 1983). Apart from their contribution to the total mass of the galactic interstellar medium, molecular clouds are also important as they are the major birthsite of massive early-type stars (see the review by Habing and Israel, 1979).

At least the disks of most late-type spiral galaxies observed thus far appear to share the molecular cloud characteristics of the Galaxy (Morris and Rickard, 1982; Scoville, 1984). In contrast, irregular dwarf galaxies of the Magellanic type are different: they either contain relatively little molecular material, or their molecular clouds are different (Elmegreen et al, 1980). Because of their proximity, the Large and Small Magellanic Clouds offer a unique possibility to study the similarities and differences of molecular clouds in spiral and irregular galaxies. In particular, such studies can provide insight in star formation mechanisms in these galaxies and in the process of formation and destruction of molecular material.

2. OBSERVING DUST AND MOLECULAR CLOUDS.

The most abundant molecule, H_2, can be observed directly only in particular cases (see the review by Shull and Beckwith, 1982). UV absorption lines yield information on molecular hydrogen in relatively unreddened clouds; it is observed only to an effective upper limit of order 10^{21} H_2 cm^{-2}. Near-IR emission lines yield information on dense molecular clouds, but only in relatively small zones of shocked molecular gas. Thus, most of the present knowledge of molecular clouds is derived from more easily observed tracers such as dust or less abundant molecules (isotopes of CO, OH, H_2CO, for instance).

Dust is a good indicator for the presence of molecules, because it plays a key role in both the production of molecules (grain surface formation of e.g. H_2) and the shielding of molecules (such as CO) against dissociation by the interstellar radiation field. Dust can in principle be detected and measured directly, by the extinction it causes or by its thermal emission (see Hildebrand, 1983). In the Galaxy, molecular emission is associated with reddening $E(B-V) > 0.05$ mag (Spitzer et al, 1973) and the column density of molecular hydrogen is usually taken to be $N(H_2)/E(B-V) = 3.75 \times 10^{21}$ cm^{-2} mag^{-1} at least for $A_V < 1.3$ mag (Jenkins and Savage, 1974; Dickman, 1978), although it is not yet clear that this relation is universally valid.

More quantitative information is supplied by tracer molecules, of which CO is the most extensively observed in its J=1-0 transitions at 115 GHz (^{12}CO) and 110 GHz (^{13}CO). The extrapolation of the properties of these tracer molecules to those of molecular hydrogen depends usually on somewhat uncertain assumptions regarding, for instance, radiative transfer, abundance ratios and excitation conditions. A good review of the problems associated with these procedures is given by Evans (1980). In case of CO, it is usually assumed that the ratio $N(H_2)/N(^{13}CO) = 5 \times 10^5$ (Dickman, 1978), but reality appears to be more complicated (Frerking et al, 1982).

In the Galaxy, main-line (type I) OH masers and strong H_2O masers commonly indicate sites of recent star formation in molecular clouds (c.f. Habing and Israel, 1979). They do not provide direct information on the properties of the molecular cloud in which they are embedded, but they do mark the presence of molecular clouds engaged in star formation (see the review by Reid and Moran, 1981).

3. DUST IN THE MAGELLANIC CLOUDS.

Estimates of the amount of dust in the Magellanic Clouds are subject to controversy, but the pattern that emerges indicates a relatively low dust content for both LMC and SMC. In the past, three methods have mainly been used to determine the dust content of the Magellanic Clouds: 1. determination of the reddening of individual stars by photometry and spectroscopy; 2. determination of surface extinction values by galaxy counts; 3. mapping of individual dust clouds by tracing their silhouet against the stellar background. All three methods are subject to observational difficulties. The more promising method of tracing the dust distribution throughout the Cloud volume by observing its thermal emission in the mid and far infrared has only recently become practical. Because this emission is optically thin, essentially all dust can be observed.

3.1. Reddening of Individual Stars.

Deducing dust content from line-of-sight extinction measurements requires a priori knowledge of the intrinsic colours of the observed stars and the distribution of reddening values to a certain limiting magnitude (c.f. Isserstedt, 1977). Moreover, this method only measures dust between the star and the observer. Because of the distance of the Magellanic Clouds, there is a systematic bias in favour of the brightest stars, i.e. stars at the near side of the Clouds suffering relatively little extinction. Moreover, the presence of circumstellar shells and double stars complicates the interpretation of observed reddening values (leading to overestimates; c.f. Isserstedt, 1975).

The observations of stars in the LMC show E(B-V) values ranging from 0.1 to 0.4 mag, and in the SMC ranging from 0.1 to 0.2 mag (e.g. Dachs, 1970; Lucke, 1974; Brunet, 1975; Azzopardi and Vigneau, 1977; Nandy et al, 1979). This includes a Galactic foreground reddening of E(B-V) = 0.03 - 0.07 mag for the LMC and E(B-V) = 0.02 - 0.04 mag for the SMC (McNamara and Feltz (1980). Thus, bright stars in both the LMC and SMC show, on average, very little reddening. In an extensive study of interstellar reddening in the LMC, Isserstedt (1975) has shown that the distribution of interstellar dust in the LMC is dominated by a huge dust cloud complex surrounding 30 Doradus. Nevertheless, even there, few stars have E(B-V) > 0.4 mag. A rough, but instructive map of the E(B-V) distribution across the LMC based on OB association extinctions (Lucke, 1974) is given by Page and Carruthers (1981). The dark clouds around 30 Doradus stand out, as well as a cloud complex associated with the HII regions N79/N83/N91 WSW of the Bar. The presence of dust in HII regions, as determined by H-beta photography, was briefly discussed by Johnson (1973).

UV spectroscopy and photometry including the Lyman alpha absorption has been used fruitfully to estimate line-of-sight gas-to-dust, or more exactly gas-to-colour excess ratios. The results indicate moderate dust depletion (by a factor of four) in the LMC and strong dust depletion (by a factor of sixteen or more) in the SMC when compared to Solar Neighbourhood values (see the review by Koornneef, these proceedings; Lequeux et al, these proceedings). Some information on the nature of the dust particles in the LMC and SMC is provided by the shape of the interstellar reddening curve, which is now relatively well-determined (see the review by Nandy, these proceedings; Lequeux et al, these proceedings). In the near-infrared and the visual it is virtually identical to the Galactic reddening law (Brück et al, 1970; Koornneef, 1982; Borgman and Danks, 1977). This is not true in the UV. Compared to the Solar Neighbourhood, the LMC reddening law shows a distinctly weaker 2200Å feature, and significantly higher far-UV extinction; the SMC reddening law shows the same behaviour, but to an even more extreme degree (Koornneef and Code, 1981; Nandy et al, 1981; Lequeux et al, these proceedings). The differences can be explained by assuming a grain size distribution as in the Galaxy, but a progressive absence of graphite particles in the LMC and SMC and, in the case of the SMC, by silicon being underabundant by a factor of ten, and mostly locked up in grains (Bromage and Nandy, 1983; Lequeux et al, these proceedings). The depletion of graphite is consistent with the low C/O ratio in Magellanic Cloud HII regions (Dufour et al, 1982).

3.2. Extinction from Galaxy Counts.

Only for the SMC attempts have been made to derive the extinction through the Cloud by way of galaxy counts. Determination of the area integrated extinction by galaxy counts (Hodge, 1974 and references therein) are influenced by galaxy clustering, by confusion of Cloud nebulosities with background galaxies, and by count incompleteness in crowded star fields. Small-scale structure in the dust distribution (such as small clouds with high extinction) is washed out or missed.

Shapley (1951) concluded that the SMC is essentially transparent with A_V at most 0.3 mag. Wesselink (1961a, b) counted 240 galaxies over in total 5 square degrees, but found that A_V = 1.5 mag over a large fraction of the SMC with, however, a considerable uncertainty because of count incompleteness. The most extensive work to date is by Hodge (1974a) who counted 2545 galaxies over an area of 85 square degrees. He found A_V > 0.8 mag over most of the SMC, reaching a peak at the SW end of the Bar with A_V = 1.3 mag. He estimated a total dust mass of $M_d = 5 \times 10^5$ $M_\odot$, yielding an atomic gas-to-dust ratio $M(HI)/M_d$ = 300, or a factor of three higher than in the Galaxy. The smoothed contours of dust extinction in Hodge's model show some structure, but its reality is questionable (Hodge, 1974a).

3.3. Discrete Dust Clouds.

Identification of dust clouds by their absorption of background starlight is limited by systematical effects (Hodge, 1972) such as the brightness and clumpiness of background stars (and galaxies). Thus, only dark clouds larger than about one arcmin (or about 20 pc) can be identified, preferentially on the near side and in front of a bright background. Cloud catalogues and maps were published by Hodge (1972, 1974b) and Van den Bergh (1974). Striking composite color photographs of the LMC and SMC clearly showing several dust clouds and dark lanes were published by Madsen and Tarenghi (1982), and Dufour (available from the Hansen Planetarium). McGillivray (1975) found several dark clouds lying outside the main body of the SMC by galaxy counts.

The catalogues by Hodge (1972, 1974b) provide some quantitative data. For the LMC, 68 clouds are listed (mean size 4.2 arcmin or 67 pc) and for the SMC 45 clouds (mean size 2.1 arcmin or 49 pc). In the LMC, a major concentration of dark clouds is found in the area near 30 Doradus, whereas in the SMC this is the case in the Southwest Bar. The dark cloud distribution is summarized in Table 1. Equal numbers of clouds are found in the LMC Bar and the 30 Doradus region, in roughly equal areas; in the SMC the a smaller fraction of the surface is covered despite a higher density of clouds. In Table 1, we have also

Table 1. Discrete Dark Clouds in LMC and SMC.
(From Hodge, 1972, 1974b).

	n	%	Mean Size (')	Number Density (kpc^{-2})	Fraction Surface Covered	'Darkness' Mean per Region	% of Total
LMC-Total	68	100	4.2	--	----	9.5×10^3	100
30 Dor	30	44	4.9	17	0.08	13.5	63
Bar	30	44	3.5	15	0.04	5.7	27
Other	8	12	4.1	3	0.01	8.4	10
SMC-Total	45	100	2.1	--	----	4.7	100
SW Bar	21	47	1.9	31	0.03	4.8	48
Mid Bar	13	29	2.2	30	0.04	4.1	25
NE Bar	9	20	2.3	21	0.03	4.0	17
Wing	2:	4:	3.3:	2	0.01	11.3:	11:

defined an empirical 'darkness' as the product [opacity x d^2 (pc)] derived from angular sizes and opacities given by Hodge (1972, 1974b). The 30 Doradus clouds are on average more than twice as 'dark' as clouds elsewhere in the LMC, and they appear to represent almost two thirds of the dark material seen in the form of discrete clouds. The SMC Wing appears to be poor in dark clouds, but the less intense stellar background makes dark clouds more difficult to detect. There is little variation in cloud 'darkness' over the SMC (the high value for the Wing being uncertain because of the above-mentioned selection effect). The average SMC cloud 'darkness' is only slightly lower than that of clouds in the LMC Bar, but only a third of the dark clouds in the 30 Doradus region, thereby confirming the interstellar reddening results reviewed in section 3.1.

The linear sizes of LMC and SMC dark clouds are similar to those of Galactic giant molecular clouds. As shown in section 4.1, they do indicate the presence of molecular material, but it appears that on average they do not represent very large molecular masses. From star counts, Hodge (1972, 1974b) finds for the <u>cores</u> of these clouds an average visual extinction of order 0.5 mag. Moreover, the largest clouds cause the least extinction. If we assume that cloud core extinctions apply to the <u>whole</u> cloud, and if we take the Galactic value for $N(H_2)/A_V$ from section 3.1, we find mean cloud masses $M(H_2) = 3.5 \times 10^4 M_\odot$ for the LMC clouds and $0.7 \times 10^4 M_\odot$ for the SMC clouds, i.e. an order of magnitude less than one finds in Galactic GMCs. The catalogued dark clouds would then imply total H_2 masses of order $10^6 M_\odot$, or less than one per cent of the atomic hydrogen mass of the Magellanic Clouds (c.f. section 5).

This discrepancy might, in principle, be overcome in several ways. 1. The LMC and SMC dark clouds may have a high degree of clumping, so that large amounts of dense material would cause only a relatively small <u>average</u> extinction. This is not unlikely, but it would not increase our mass estimate by much, as we already took extinction values corresponding to the darkest parts as representative for the whole cloud. 2. If the clouds are deeply embedded in the LMC and SMC, star counts underestimate the true extinction. However, this would mean that most clouds present are, in fact, seen. 3. The ratio of dust to molecular gas could be lower in the Magellanic Clouds. If dust is depleted with respect to H_2 to the same extent as with respect to E(B-V) (see section 3.1), the mean dark cloud mass in the LMC would be 1.4×10^5 $M_\odot$ and in the SMC 1.1×10^5 $M_\odot$. These masses are still lower than those of Galactic GMCs, but not dramatically so.

Nevertheless, it is fair to conclude that optical observations, and in particular the apparent low extinctions of the Hodge clouds fail to provide unambiguous proof for the presence of Galactic-type giant molecular clouds on the near side of either the LMC or the SMC.

3.4. Infrared Emission from Dust Clouds.

The best way of measuring dust throughout the Magellanic Clouds is by mapping its thermal infrared emission. So far, there are few published results. Hoffman et al (1973) only obtained an upper limit at 100 microns for 30 Doradus, while in the AFGL survey Price and Walker (1976) found four sources in the LMC at 20 microns (GL 4050, 4055, 4056 and 4057) of which two coincide with the HII regions 30 Doradus and N159. Characteristically, no source was found in the SMC. Werner et al (1978) mapped thermal dust emission between 30 and 200 microns in 30 Doradus, and detected dust in three nearby HII regions (N158C, N160A and N159). Their results show that the dust emission in 30 Doradus arises primarily near the two major ionization fronts. They estimated a dust mass of $M_d = 10^3$ $M_\odot$, and conclude that the average dust density in the observed HII regions is lower by a factor of 2.5 to 5 than in similar Galactic HII regions. Near-IR searches by Gatley et al (1981, 1982) have resulted in the discovery of two heavily reddened objects: in the LMC-N159 and the SMC-N76B, presumably very young and embedded in dust. They found further evidence of small dust clouds with A_V = 5 - 10 mag associated with these HII regions from nearby, heavily reddened background stars.

However, these results will rapidly become superseded as soon as IRAS Magellanic Cloud observations become available; several sources of far-infrared emission have already been identified in both the LMC and the SMC (see Brown, 1983; Habing, private communication).

4. MOLECULES IN THE MAGELLANIC CLOUDS.

4.1. Searches for Molecular Clouds.

The present paucicity of molecular observations of LMC and SMC, in particular of systematic surveys, primarily reflects the lack of major millimeter-wave facilities in the southern hemisphere over most of the past period. In fact, most Magellanic Cloud CO observations published till now have been made with millimeter-wave detectors attached to optical telescopes. The situation is aggravated by the large angular extent of the Magellanic Clouds and their weak molecular emission necessitating long integration times at a large number of positions. The first detection of molecular emission in the Magellanic Clouds was that of CO(1-0) by Huggins et al (1975), using the AAT. So far, the only published systematic search for molecular clouds in both LMC and SMC is the one by Israel et al (1983a, b) in the CO(2-1) line at 230 GHz, covering only a small fraction of the Clouds and thus far from complete. A major step forward is the large-scale CO(1-0) survey undertaken by the New York group (Cohen et al, these proceedings; Rubio et al, these proceedings). The survey uses an 8 arcmin beam and aims at full coverage of at least the LMC. In the SMC, full coverage has been abandoned in view of the weakness of CO emission, and instead selected positions are observed with very long integration times.

Table 2 summarizes the presently available results of molecular line observations in the Magellanic Clouds; numbers in parentheses refer to the number of positions searched. The only observations of H_2 itself are those by Koornneef and Israel (1983) who detected weak emission at two microns in LMC-N159 and SMC-N81, but the data do not lend themselves to quantitative interpretation. In the survey by Israel et al (1983), CO is detected towards the known LMC and SMC masers (see also Gardner, these proceedings), most of the dark clouds and bright HII regions observed. Positions in the LMC and SMC Bars not associated with such objects yielded poor detection rates. The LMC HII region/maser source N159 has a CO intensity higher by a factor of two

Table 2. Summary of LMC and SMC Molecular Observations.

Molecule	N of Detections LMC	SMC	Reference
H_2	1(3)	1(2)	Koornneef & Israel, 1983
OH (abs)	1(1)	---	Whiteoak & Gardner, 1976b
	1(1)	---	Caswell & Haynes, 1981
	0(1)	---	Haynes & Caswell, 1981
HCO^+	1(1)	---	Batchelor et al, 1981
CO(1-0)	1(3)	---	Huggins et al, 1975
	3(4)	---	Gardner, these proceedings
	(many)	---	Cohen et al, these proceedings
	---	1(40)	Rubio et al, these proceedings
CO(2-1)	11(22)	5(15)	Israel et al, 1982; 1983
H_2CO (abs)	1(2)	---	Whiteoak & Gardner, 1976a
	0(1)	---	Haynes & Caswell, 1981
OH Maser	1(1)	---	Caswell & Haynes, 1981
	1(1)	---	Haynes & Caswell, 1981
H_2O Maser	2(17)	2(15)	Scalise & Braz, 1982
	3(10)	---	Whiteoak et al (1983)

than any other position in the LMC; a similar cloud is found towards the SMC H_2O maser/HII region N19. N159 is clearly the most succesful target for molecular observations: every species searched for has been found in this source. Israel et al (1982) argued that the N159 molecular cloud complex is the closest to a Galactic GMC yet found in the Magellanic Clouds. Its association with an OH/H_2O maser, shocked H_2 emission and an embedded infrared source shows it to be an active site of star formation (Gatley et al, 1981, 1982; Israel et al, 1982).

In the LMC, CO is concentrated in the region south of 30 Doradus and in Shapley's Constellation I (Israel et al, 1983a, b; Cohen et al, these proceedings), in good agreement with the dust distribution (sections 3.1 and 3.3). In the SMC, CO is detected almost exclusively in the southwestern part of the Bar (Israel et al, 1983a, b; Rubio et al, these proceedings). The outstanding characteristic of CO emission from the Magellanic Clouds is its almost uniformly low intensity as compared to that of the Galaxy. In the LMC, and in the SMC-SW Bar, CO intensities are lower than expected from an extrapolation of Galactic GMC's to Magellanic distances by a factor of two to four. In the remainder of the SMC the difference is a factor of six or more.

4.2. Searches for OH and H_2O Masers.

After unsuccessful attempts by Johnston et al (1971) and Kaufmann et al (1977) the first detection of a Magellanic Cloud maser was that of an H_2O maser in LMC-N159 (see section 4.1) by Scalise and Braz (1981). Caswell and Haynes (1981) then found an OH maser in the same source. This was followed by the discovery of a second OH/H_2O maser in LMC N105 (Haynes and Caswell, 1981; Scalise and Braz 1982), stimulating more extensive and still continuing H_2O maser surveys (Scalise and Braz, 1982; Whiteoak et al, 1983). In the SMC, 15 HII regions and dark clouds have been searched; there are two detections, one in the SW Bar, the other to the north, outside the SMC optical image, but coincident with a radio continuum source (McGee et al, 1976). In the LMC, a total of 22 HII regions and dark clouds have been searched with four detections, three in the region south of 30 Doradus. Only one detection (that of N105A) is common to both surveys. The luminosities of the H_2O masers discovered are comparable to those of H_2O masers (3×10^{29} erg s^{-1}) in the Galaxy but there is a conspicuous lack of masers comparable to the strongest (10^{33} erg s^{-1}) found in the Galaxy (Scalise and Braz, 1982; Whiteoak et al, 1983). In contrast, the N105 OH maser has a luminosity comparable to that of the strongest Galactic OH masers (Haynes and Caswell, 1981)

5. LACK OF DUST AND MOLECULES IN THE MAGELLANIC CLOUDS.

5.1. Lack of Dust.

Almost all studies reviewed in the preceding indicate that the Magellanic Clouds are lacking in dust as compared to the Galaxy. This lack of dust is particularly clear in determinations of the gas to dust ratio. Compared to the Solar Neighbourhood, the LMC appears to be underabundant in dust by a factor of four, and the SMC by a factor of seventeen (Koornneef, 1982; these proceedings; Lequeux et al, these proceedings). The lack of dust correlates well with the low abundances of heavy elements in the Magellanic Clouds, such as that of carbon (see the review by Dufour, these proceedings). It is particularly pronounced in view of the high atomic gas content of the Magellanic Clouds (section 5.3). Although the cause of these low abundances has not yet been established, it may well turn out to reflect a <u>mean</u> star formation rate significantly lower than the <u>present</u> star formation rate in the Magellanic Clouds or Galactic star formation rates. This would be consistent with models of stochastic star formation (Matteucci and Chiosi, 1983; Feitzinger et al, 1981).

5.2 <u>Lack of CO</u>.

CO is less widespread in most of the LMC and SMC than in the Galaxy, and where detected, it has intensities significantly less than that of comparable Galactic molecular cloud complexes (section 4.1). This result for the archetypical Magellanic Clouds is consistent with that obtained for other Magellanic dwarf irregulars by Elmegreen et al (1980). Most likely, the observed low CO intensities do not primarily reflect low CO_2 <u>abundances</u>, because the results have been obtained by observing the ^{12}CO J=1-0 and J=2-1 transitions which are expected to be optically thick. Measured instead is the product of brightness temperature and the fraction of the beam filled with CO clumps. The CO observations of the LMC, the SMC and other Magellanic galaxies thus indicate low brightness temperatures, few clumps per beam or both.

Elmegreen et al (1980) gave an extensive discussion of possible explanations for weak CO emission. They note the complex nature of CO formation, destruction and excitation mechanisms, and advance two categories of explanations. One is based on lower heating rates in irregular galaxies and the other on lower CO abundances. For instance, low heating rates of molecular clouds, and hence low CO excitation temperatures may result if cosmic ray fluxes are relatively low. Also, the CO formation rate might be lowered. This could be the case either because of low production rates or poor confinement of cosmic ray particles. However, the present supernova rate <u>per unit mass</u> of both the LMC and the SMC is three times higher than that of the Galaxy (Tammann 1982; Mathewson et al, 1983). In order to achieve a tenfold decrease in cosmic-ray fluxes (necessary to decrease CO excitation temperatures by a factor of two or three) confinement would have to be poor (see also the restraints on magnetic field strengths and energy densities derived from polarization measurements by Schmidt, 1970).

High luminous star formation rates and efficiencies may cause rapid desintegration of parent molecular clouds, shortening their mean lifetimes. This effect probably explains the lack of CO near giant HII regions such as 30 Doradus (Israel et al, 1982) but seems insufficient to explain the overall lack of CO in the Magellanic Clouds.

Discussing their CO observations, Israel et al (1983b) attempted to explore the consequences of the known low dust-to-gas ratio in the Magellanic Clouds. They argue that the lack of dust allows deeper penetration of individual gas clumps in a molecular cloud complex by a stronger interstellar UV radiation field, thus leading to higher CO destruction rates, hence to fewer CO clumps within the observing beam. Therefore, a low CO content would be the direct consequence of a low ratio of dust to gas. The results of a simple model show that the observed underabundance of dust may well explain the observed low CO antenna temperatures. The greater lack of dust in the SMC and the equally low heavy-element abundances (in particular the low C/O ratio) may conspire to create extremely low CO column densities, perhaps even leading to the unusual presence of a significant fraction of optically thin CO in the SMC. The best way of verifying this are measurements of the ^{13}CO isotope, but the line is expected to be very weak.

5.3 The H_2 Abundance.

Direct determination of H_2 column densities towards a few of the brightest stars in the Magellanic Clouds may, in the near future, come within reach of the high-resolution spectrograph of the Space Telescope. Such observations are of great importance, as they provide virtually the only way of determining the molecular hydrogen content of the Clouds. In the meantime, the different radiation field and dust particle properties suggested by UV observations of the LMC and the SMC complicate model calculations of H_2 formation, whereas in view of the above the value of the CO molecule as a quantitative tracer for H_2 is diminished. For instance, if H_2 formation and destruction is not influenced by the different physical conditions in the Magellanic Clouds, contrary to CO, the CO/H_2 ratio in the LMC would be lower than in the Galaxy by a factor of five or more, and in the SMC by a factor of ten or more, depending on CO destruction rates and on CO abundances. If the depletion of H_2 were to go linearly with that of dust, the effect on the CO/H_2 ratio would be less; probably not more than a factor of two for the LMC, and a factor of four or so for the SMC. Whatever the case, the use of the Galactic CO/H_2 ratio as given in section 2 is suspect and probably leads to underestimates for the molecular hydrogen content of the Magellanic Clouds.

One pertinent observation remains. The LMC and SMC have surprisingly high amounts of neutral atomic hydrogen. The LMC has $M(HI) = 5.4 \times 10^8 M_\odot$ (McGee and Milton, 1966) and the SMC has $M(HI) = 4.8 \times 10^8 M_\odot$ (Hindman, 1967), which must be contrasted with total (dynamic) masses $M(tot) = 6 \times 10^9 M_\odot$ for the LMC (McGee and Milton, 1966; Feitzinger, 1980 and references therein) and $M(tot) = 1.5 \times 10^9 M_\odot$ for the SMC (Hindman, 1967). The HI surface density is almost an order of magnitude higher in the SMC than in the LMC. Thus the atomic hydrogen mass fraction $M(HI)/M(tot)$ is 0.09 for the LMC and 0.32 for the SMC, as compared to about 0.01 for the Galaxy as a whole, and about 0.06 for the Solar Neighbourhood. Such large amounts of atomic hydrogen appear to preclude the possibility that an appreciable fraction of hydrogen is in molecular form. In the LMC and the SMC we are unlikely to find H_2 to HI ratios of one to three as CO observations are taken to indicate for the disks of the Galaxy and Sc galaxies (Sanders, 1981; Young and Scoville, 1982; Dame, 1983). We finally note that the very high M(HI)/M(tot) ratio of the SMC indicates the presence of even less molecular hydrogen than in the LMC, in line with the lower dust, CO and heavy-element abundances.

6. REMARKS ON STAR FORMATION IN THE MAGELLANIC CLOUDS.

The present (luminous) star formation rates per unit mass are higher than that of the Galaxy by factors of 2.7 for the LMC and 1.5 for the SMC (see the review by Lequeux, these proceedings; also Israel, 1980). The present supernova rates are consistent with these numbers (see section 5.2). Thus, the lack of dust, medium-heavy elements (C, N, O) and molecules does not appear to prohibit vigorous formation at least of massive ($M > 10\ M_\odot$) stars; if anything, the contrary is the case. Some relevant considerations are that a. formation of massive stars may be favoured at low metallicity levels because of lower cooling rates in the contraction phase and b. the molecular content of the densest and most massive clumps will be least affected by enhanced radiative destruction due to dust depletion, also favouring the formation of massive stars over those of low-mass.

The major site of molecules and dust in the LMC is the 30 Doradus complex; the northern part of this complex contains HII regions and OB associations that represent up to 30 per cent of all newly formed stars in the LMC. The morphology of this Greater Doradus complex suggests a sequence of star formation in a north-south direction, with star formation presently taking place in the middle (in N159). Less outstanding concentrations in the LMC are Constellation I and perhaps the outlying N79/N91 complex. Similarly, in the SMC current star formation appears to be limited mainly to the southwest Bar.

7. FUTURE PROSPECTS.

It is unlikely that optical as opposed to UV observations will add much to our present knowledge of the dust content of the Clouds, but IRAS far-IR observations will allow us to study in detail the dust component of the Magellanic Clouds. Further molecular observations, especially the completion of the ^{12}CO survey by the New York Group, and additional observations of e.g. ^{13}CO are urgently needed. If feasible, H_2 UV absorption measurements with the Space Telescope, would be of great importance. More generally, theoretical modelling of molecular formation and destruction processes as a function of the dust and radiation field properties appears worthwhile. Finally, the rapidly increasing data base on dust and molecules in the Magellanic Clouds should stimulate further theoretical investigations linking star formation processes to the evolutionary history of the Clouds.

REFERENCES

Azzopardi, M., Vigneau, J., 1977 Astron. Astrophys. $\underline{56}$, 15
Bally, J., Israel, F.P., 1983, in preparation
Batchelor, R.A., McCulloch, M.G., Whiteoak, J.B., 1981 M.N.R.A.S. $\underline{194}$, 911
Borgman, J., Danks, A.C., 1977 Astron. Astrophys. $\underline{54}$, 41
Bromage, G.E., Nandy, K., 1983 M.N.R.A.S. $\underline{204}$, 29P
Brown, D.A., 1983 Aviation Week Space Technology $\underline{118}$, No. 9, p.21
Brunet, J.A., 1975 Astron. Astrophys. $\underline{43}$, 345
Brück, M.T., Lawrence, L.C., Nandy, K.N., Thackeray, A.D., Wood, R., 1970 Nature $\underline{225}$, 531
Caswell, J.L., Haynes, R.F., 1981 M.N.R.A.S. $\underline{194}$, 33P
Cohen, R.S., Cong, H., Dame, T.M., Thaddeus, P., 1980 Ap. J. Lett. $\underline{239}$, L53
Dachs, J., 1970 Astron. Astrophys. $\underline{9}$, 95
Dame, T.M., 1983, Ph.D. Thesis Columbia University (USA)
Dickman, R.L., 1978 Ap. J. Suppl. $\underline{37}$, 407
Dufour R.J., Shields, G.A., Talbot, R.J., 1982 Ap. J. $\underline{252}$, 461
Elmegreen, B.G., Elmegreen, D.M., Morris, M., 1980 Ap. J. $\underline{240}$, 455
Evans, N.J., 1980 in: 'Interstellar Molecules', Ed. B.M. Andrew, IAU Symposium 87, Reidel Publ. Co., p. 1
Feitzinger, J.V., 1980 Space Sc. Rev. $\underline{27}$, 35
Feitzinger, J.V., Glassgold, A.E., Gerola, H., Seiden, P.E., 1981 Astron. Astrophys. $\underline{98}$, 371
Frerking, M.A., Langer, W.D., Wilson, R.W., 1982 Ap. J. $\underline{262}$, 590
Gatley, I., Becklin, E.E., Hyland, A.R., Jones. T.J., 1981 M.N.R.A.S. $\underline{197}$, 17P
Gatley, I., Hyland, A.R., Jones, T.J., 1982 M.N.R.A.S $\underline{200}$, 521
Habing, H.J., Israel, F.P., 1979 Ann. Rev. Astr. Ap. $\underline{17}$, 345
Haynes, R.F., Caswell, J.L., 1981 M.N.R.A.S. $\underline{197}$, 23P
Hildebrand, R.H., 1983 Quart. J. R.A.S. $\underline{24}$, 267
Hindman, J.V., 1967 Austral. J. Phys. $\underline{20}$, 147
Hodge, P.W., 1972 P.A.S.P. $\underline{84}$, 365
Hodge, P.W., 1974a P.A.S.P. $\underline{86}$ 263
Hodge, P.W., 1974b Ap. J. $\underline{192}$, 21
Hoffman, W.F., Frederick, C.L., Emery, R., 1973 B.A.A.S. $\underline{5}$, 31
Huggins, P.J., Gillespie, A.R., Phillips, T.G., Gardner, F., Knowles, S., 1975 M.N.R.A.S. $\underline{173}$, 69P
Israel, F.P., 1980 Astron. Astrophys. $\underline{90}$, 246
Israel, F.P., De Graauw, Th., Lidholm, S., Van de Stadt, H., De Vries, C.P., 1982 Ap. J. $\underline{262}$, 100
Israel, F.P., De Graauw, Th., Van de Stadt, H., De Vries, C.P., 1983a in: IAU 106 'The Milky Way as A Galaxy', in press.
Israel, F.P., De Graauw, Th., Van de Stadt, H., De Vries, C.P., 1983b Ap. J. submitted
Isserstedt, J., 1975 Astron. Astrophys. $\underline{41}$, 175
Isserstedt, H., 1977 Astron. Astrophys. $\underline{59}$, 167
Jenkins, E.B., Savage, B.D., 1974 Ap. J. $\underline{187}$, 243

Johnson, H.M., 1973 in: 'Interstellar Dust and Related Topics', Ed. J.M. Greenberg and H.C. van de Hulst, Reidel Publ. Co., p. 471
Johnston, K.J., Knowles, S.H., Sullivan, W.T., 1971 Ap. J. Lett. 167 L93
Kaufmann, P., Zisk, S., Scalise, E., Schaal, R.E., Gammon, B.R.H., 1977 Astron. J. 82, 577
Koornneef, J., 1982 Astron. Astrophys. 107, 247
Koornneef, J., Code, A.D., 1981 Ap. J., 247, 860
Koornneef, J., Israel, F.P., 1983 Ap. J. Lett. submitted
Lucke, P.B., 1974 Ap. J. Suppl. 28, 73
Madsen, C., Tarenghi, M., 1982 ESO Messenger No. 30, p. 15
Mathewson, D.S., Ford, V.L., Dopita, M., Tuohy, I.R., Long, K.S., Helfand, D.J., 1983 Astrophys. J. Suppl. 51, 345
Matteucci, F., Chiosi, C., 1983 Astron. Astrophys. 123, 121
McGee, R.X., Milton, J.A., 1966 Austral. J. Phys. 19 343
McGee, R.X., Newton, L.M., Butler, P.W., 1976 Austr. J. Phys. 29, 329
McGillivray, M.T., 1975 M.N.R.A.S 170, 241
McNamara, D.H., Feltz, K.A., 1980 P.A.S.P. 92, 587
Morris, M., Rickard, L.J, 1982 Ann. Rev. Astr. Ap. 20, 517
Nandy, K., Morgan, D.H., Carmochan, D.J., 1979 M.N.R.A.S. 186, 421
Nandy, K., Morgan, D.H., Willis, A.J., Wilson, R., Gondhalekar, P.M., 1981 M.N.R.A.S. 196, 955
Page, T., Carruthers, G.R., 1981 Ap. J. 248, 908
Price, S.D., Walker, R.G., 1976 AFGL-TR-76-0208, USAF Geophys. Lab., Hanscomb AFB, Mass. (USA)
Reid, M.J., Moran, J.M., 1981 Ann. Rev. Astr. Ap. 19, 231
Sanders, D.H., 1981 Ph.D. Thesis SUNY, Stony Brook (USA)
Scalise, E., Braz, M.A., 1981 Nature 290, 36
Scalise, E., Braz, M.A., 1982 Astron. J. 87, 528
Schmidt, Th., 1970 Astron. Astrophys. 6, 294
Scoville, N.Z., 1984 in: 'Galactic and Extragalactic Infrared Spectroscopy', Ed. M.F. Kessler and J.P. Phillips, Reidel Publ. Co, p. 167.
Shapley, H., 1951 Proc. Nat. Acad. Sc. 37, 136 (USA)
Shull, J.M., Beckwith, S., 1982 Ann. Rev. Astr. Ap. 20, 163
Smith, L.F., Biermann, P., Mezger, P.G., 1975 Astr. Ap. 66, 65
Spitzer, L., Drake, J.F., Jenkins, E.B., Morton, D.C., Roguson, J.B., York, D.G., 1973 Ap. J. Lett. 181, L116
Tammann, G.A., 1982 in: 'Supernovae, A Survey of Current Research', Ed. M.J. Rees and R.J. Stoneham, Reidel Publ. Co. p. 371
Van den Bergh, S., 1974 Ap. J. 193, 63
Werner, M.W., Becklin, E.E., Gatley, I., Ellis, M.J., Hyland, A.R., Robinson, G., Thomas, J.A., 1978 M.N.R.A.S. 184, 365
Wesselink, A.J., 1961a M.N.R.A.S. 122, 503
Wesselink, A.J., 1961b M.N.R.A.S. 122, 509
Whiteoak, J.B., Gardner, F.F., 1976a M.N.R.A.S. 174, 51P
Whiteoak, J.B., Gardner, F.F., 1976b M.N.R.A.S. 176, 25P
Whiteoak, J.B., Wellington, K.J., Jauncey, D.L., Gardner, F.F., Forster, J.R., Caswell, J.L., Batchelor, R.A., 1983 M.N.R.A.S. 205, in press
Young, J.S., Scoville, N.Z., 1982 Ap. J. Lett. 260, L11

DISCUSSION

van den Bergh: I have two comments:
1. In small fields it is dangerous to estimate absorption from counts of relatively bright galaxies. Because the distribution of galaxies is so clumpy the absence of galaxies may only imply that there are no galaxies.
2. It is tempting to assume that the difference in dust abundance between the LMC and the SMC is responsable for the fact that the Large Cloud HI has a much clumpier distribution than that of the Small Cloud gas.

Israel: I quite agree. With respect to your second point I think we should also take into account the fact that there will, in any case, not be very much molecular hydrogen either, because the M(HI)/M(tot) ratio is already so high.

Shull: If possible, please describe in more detail the structure of a giant molecular cloud. You said it was a swarm of clumps (1 pc in size) moving balistically. What are your guesses about densities (within and between clumps), distances between clumps, and so forth.
My interest in this is related to my investigation of the inhomogeneous environments in which SNs explode: I have SNR spectra showing clumps about 1 pc in size that have been accelerated outward by the blast wave.

Israel: ^{12}CO pictures of molecular cloud complexes are rather misleading, because they really only show the temperature distribution. If you look at molecular cloud complexes in ^{13}CO or NH_3, you see a fair number of clumps; typically of parsec size, moving with velocities of a few km/s with respect to one another; typical masses are a few hundreds of solar masses (in H_2!). The projected separation is of order of a few clump diameters. Quite possibly, there are HI clumps of lower mass and density in between the molecular clumps.

GAS-TO-DUST RATIOS IN THE MAGELLANIC CLOUDS

Jan Koornneef
Space Telescope Science Institute
Homewood Campus,
Baltimore, MD 21218

Abstract: From radio observations it has been known for some time that a relatively large fraction of the total mass of the Magellanic Clouds is in the form of atomic Hydrogen. In contrast, optical photometry shows that both galaxies suffer rather little obscuration by dust, except in regions where the stellar population indicates recent star formation. However, the possibilities of discussing gas-to-dust ratios have been limited by the poor spatial resolution of the 21 cm HI-observations and uncertainties with respect to the relative location of the various components (gas, dust and stars).

Over the last few years, observations of about thirty Cloud supergiants with the International Ultraviolet Explorer have yielded estimates of the neutral Hydrogen column densities from measurements of the Lyman α absorption profile in the ultraviolet spectra which were also used to study the wavelength dependence of the extinction due to the intervening dust. The extinction laws have been shown to be strongly anomalous in both of the MCs. It now appears certain that the gas-to-dust ratios in the MCs are significantly larger than those in our Galaxy. If expressed by the parameter N_H/E_{B-V}, the LMC ratio is approximately four times larger than the Galactic value, whereas the SMC value is about seventeen times the Galactic ratio.

The presently available data fit into the following scheme: Galaxy $\longrightarrow$ LMC $\longrightarrow$ SMC: decreasing heavy-element abundances, decreasing strength of the 2200 Å dust-feature, increasing gas-to-dust ratio and increasing far-ultraviolet extinction.

RADIO OBSERVATIONS

Neutral hydrogen observations of the Magellanic Clouds have been available for quite some time. In 1966, McGee and Milton published the 21 cm Parkes Survey of the LMC which has a spatial resolution of 15'. A similar HI map for the SMC was published by Hindman (1967). The SMC data were re-examined by McGee and Newton (1981) and the same authors added new SMC-HI observations in 1982. Total hydrogen masses from these studies are roughly $5\ 10^8\ M_\odot$ for each of the galaxies, which translates into a typical column density of $N_{HI} = 4.5\ 10^{21}\ cm^{-2}$ for the SMC (Azzopardi and Vigneau 1977) and somewhat smaller values for the LMC because of its larger size. The local gas profile near zero velocity gives a galactic foreground column density in the range of $N_{HI} = 3.2 \pm 1.3\ 10^{20}\ cm^{-2}$ in the direction of the LMC (McGee, Newton and Morton, 1983). Towards the SMC, local hydrogen is mostly between 2.7 and $3.5\ 10^{20}$ (McGee, priv. comm.). In both MCs, the HI profiles contain a wealth of velocity information indicative of the existence of very large scale structures. In the context of the present review, this information has so far remained largely unexplored, partly because the low spatial resolution of these surveys is poorly matched to the much finer structure sampled when observing individual stars. But the relevance of hydrogen velocity profiles has already been demonstrated in the specific case of the immediate surroundings of R136 in 30 Doradus (Blades, 1980; Blades and Meaburn, 1980; de Boer, Koornneef and Savage, 1980) and in the study of the SMC supergiant Sk143 (Lequeux et al. 1982).

THE DUST CONTENT

The mean colour excess in both MCs is rather low. For the SMC, the total mean colour excess is $E_{B-V} = 0.07 \pm 0.04$ of which 0.04 ± 0.03 is due to the foreground (Azzopardi and Vigneau, 1976 and references therein). The main absorption lane in the SMC, as defined by de Vaucouleurs (1955), has an "exceptionally high" average $E_{B-V} = 0.11$ (in SMC reddening). Similar studies for the LMC, also based on the study of supergiants (Isserstedt, 1975; Brunet, 1975), give $E_{B-V} = 0.07 \pm 0.04$ for the foreground with an additional 0.03 to 0.11 within the LMC, dependent on the region.

Very few stars are reddened by more than $E_{B-V} = 0.20$ in LMC-dust (Isserstedt, 1975), which is also the upper limit for the effective reddening in the LMC-OB associations measured with OAO-2 (Koornneef, 1984). Reddening in excess of $E_{B-V} \simeq 0.30$ has been claimed by Madore (1976, 1982) for some LMC and SMC Cepheids, but the intrinsic colours of Cepheid variables are rather controversial (e.g. see McNamera and Feltz, 1980; Feast, 1984 and Stift, 1984). Isserstedt (1975) gives a list of LMC supergiants which might have $E_{B-V} > 0.30$. Often however, such candidates for high reddening turn out to have been either misclassified, are double or have emission lines (e.g. R 108 in the SMC: Hutchings, 1982; HD37974 in the LMC: Koornneef and Code, 1981).

Three candidates for high reddening in the SMC are BBB275 and 277 (Martin and Thackeray, 1973) and Sk143 ($E_{B-V} = 0.27$, Lequeux et al. 1982). Sk 108-69 in the LMC has $E_{B-V} \simeq 0.45$ according to Nandy and Morgan (1978). The dust in front of these last two stars has been shown by these authors to have ultraviolet extinction properties which are different from the average laws in the respective galaxies.

The statistics of the colour excesses suffer from selection effects as the more heavily reddened stars are naturally fainter and thus easily escape detection. It is thus easy to underestimate the mean reddening as most of the observed stars are only slightly reddened "even in those systems that are optically thick" (Wesselink, 1961). Other proponents of a high ($A_v \gtrsim 1$ mag) total dust content (in the SMC) are Hodge (1974) and MacGillivray (1975) who base their findings on background galaxy counts. However, crowding in dense fields might easily lead to an underestimate of the number of background galaxies (e.g. Westerlund, 1974). We will argue later in this review that the actual global gas-to-dust ratios in the MC's are indeed substantially in excess of those derived by the use of galaxy counts. But high dust concentrations are likely to exist locally. IRAS observations should be able to pinpoint their location in addition to providing good estimates of the total mass and distribution of the MC dust.

GLOBAL GAS-TO-DUST RATIOS

As demonstrated in the last section, the evaluation of the mean dust content of the MCs differs from study to study. In addition, there is some uncertainty with respect to the interpretation of 21 cm observations. Although clearly penetrating the whole of the external galaxies, the conversion of the measured antenna temperature to column densities can be readily done only under the assumption that the gas is optically thin. Nevertheless, the range in the published estimates in the gas-to-dust ratios is surprisingly large, although all authors who base their values on large scale HI maps and global estimates of the mean colour excesses at least agree that both MCs have more gas relative to dust than our Galaxy (see table 1).

Table 1. Gas to dust ratios relative to the Galactic value

SMC	van den Bergh (1968)	$\gtrsim 10$
	van Genderen (1969)	5-12
	Dachs (1970)	$\gtrsim 100$*
	Hodge (1974)	3
	MacGillivray (1975)	3
	Azzopardi and Vigneau (1977)	15 (9-24)
LMC	van Genderen (1970)	4
	Brunet (1975)	20*
	Martin et al. (1976)	9*

* these values have been derived by dividing the <u>total</u> HI column densities by the <u>mean</u> extinction and should therefore be systematically too large.

GAS TO DUST RATIOS USING HYDROGEN COLUMN DENSITIES FROM Lyα ABSORPTION IN INDIVIDUAL SIGHT-LINES

The average galactic gas-to-dust ratio most frequently quoted ($N_{HI}/E_{B-V} = 4.8 \; 10^{21}$ atoms cm^{-2} mag^{-1}; Bohlin, Savage and Drake, 1978) is derived from the measurement of the interstellar Lyman α atomic hydrogen absorption profile. The Copernicus satellite used by these authors was not sensitive enough for similar measurements of individual stars in external galaxies, but the International Ultraviolet Explorer has now been successfully used for this purpose in both MCs.

A study of Lyα absorption in the IUE spectra of eleven LMC early type supergiants has been published by Koornneef (1982). The total (LMC plus local) colour excesses of these stars range from $E_{B-V} = 0.06 - 0.26$. Under the assumption that the galactic foreground component is the same for each of these stars, a LMC gas to dust ratio of $N_{HI}/E_{B-V} = 2 \; 10^{22}$ atoms cm^{-2} mag^{-1} is derived.

A similar study for SMC supergiants is being made by Lequeux et al. (1983). Due to the lower mean reddening in the SMC, it is even more difficult to derive accurate gas-to-dust ratios for individual stars than in the LMC. But all of the fifteen stars they are studying have new ground based photometry as well as accurate spectral classifications and Lequeux et al. (1984) conclude that the "normal" SMC gas to dust ratio is approximately 17 times the galactic value.

EXCEPTIONS TO "NORMAL" GAS TO DUST

Three stars in the SMC sample studied by Lequeux et al. (1984) show gas to dust ratios closer to the average galactic value. They consider them exceptions to the rule and point out that the extinction curves for these stars tend to be more like the average Galactic law. One specific example is Sk143 (Lequeux et al. 1982). Its spectrum shows a significant depression around 2200Å in contrast with the "average" SMC curve which shows no 2200Å bump (Rocca-Volmerange et al. 1981, Hutchings 1982, Nandy et al. 1982). Sk143 is the star with the highest colour excess of all objects studied in the SMC. Its "counterpart" in the LMC is Sk108-69 (Nandy and Morgan, 1978) which, with $E_{B-V} = 0.45$, suffers more extinction than all other LMC supergiants which have so far been observed with IUE. Its extinction curve is rather more similar to the galactic law than to the "average" LMC curve (Koornneef and Code 1981; Nandy et al. 1981), but regrettably, no gas to dust ratio has so far been derived for this star.

DOES E_{B-V} MEASURE THE AMOUNT OF DUST?

The gas to dust ratios quoted so far are all normalized to unit extinction between B and V, and should thus be better described as gas to colour excess ratios. One way of explaining the high ratios reported for the MCs is by arguing that the "missing" dust is locked up in grains large enough to have a low visual reddening efficiency. This in fact is the probable explanation of the high N_{HI}/E_{B-V} ratio in the Ophiuchus dark cloud (Carrasco, Strom and Strom, 1973). In that region, Whittet and van Breda (1975) also report a high value for the ratio R of total to selective absorption, thus confirming the presence of larger grains. Determinations of R are notoriously difficult and unreliable, but near infrared (JHK) studies in the LMC by Koornneef (1982) and Morgan and Nandy (1982) have not revealed any obvious anomalies in the large particle size distribution function. A similar conclusion has been reached by Clayton and Martin (1984) on the basis of observations of optical linear polarization. For the SMC, Isserstedt (1976, 1980) has derived a surprisingly low value of $R \simeq 2$. This, if confirmed, would indicate a lower than normal number of large particles and this would tend to make the quoted SMC gas-to-dust ratios ever higher if they were expressed as a ratio by mass. Note, however, that Stift (1984) has now argued that the value of R in the SMC substantially exceeds the Galactic value (of $R \simeq 3.1$).

The very pronounced differences of the ultraviolet part of the extinction curves between the Galaxy, the LMC and the SMC could at least be partly due to changes in the particle size distribution function. Their effect on the normalization of the observed gas to dust ratios deserves further investigation.

SOME LIMITATIONS OF THE PRESENT ULTRAVIOLET DATA

As the IUE low dispersion Ly α observations do not provide any possibility to discriminate between local and MC gas due to saturation of the profiles, the assumption has to be made that the spatial (i.e. star to star) variations in the foreground gas and dust are very much smaller than their MC counterparts. This might not be quite true as the mean internal reddening in the external galaxies is of the same order of magnitude as the Galactic reddening. If the foreground is not constant it is likely, statistically, that the heavily reddened stars in the sample will have a relatively large foreground component, whereas the stars with the lowest E_{B-V} are likely to suffer little foreground extinction. This would bias the derived gas-to-dust ratios in the direction of the galactic value, i.e. the observed ratios would degrade into a lower limit (a similar argument has been made for MC extinction laws by Koornneef and Code, (1981)).

Some concern should also be expressed with respect to the accuracies of the colour excess values as derived from UBV-photometry and spectral types. Most of the MC-supergiants studied so far are intrinsically brighter than their Galactic counterparts and their (ultraviolet) intrinsic colours are likely to be luminosity dependent (e.g. Brunet, 1975; Isserstedt, 1982). Potential errors in this area can be minimized by restricting the range in luminosity and spectral type of the sample. This has been carefully considered by Koornneef (1982) and Lequeux et al. (1983) so that their principal results should not be affected. But the study of star-to-star variations of gas-to-dust ratios (as well as extinction laws) requires more accurate intrinsic colours of early type supergiants as a function of spectral type, luminosity and metallicity. More information is also needed with respect to the relative locations of gas, dust and star along the total (local plus MC) line of sight, both in order to suppress the local effects and to further the study of the multiple components known to exist within the MCs. High resolution optical and ultraviolet interstellar line studies are likely to make important contributions in this area in the near future. It will be 4-5 years, however, before the Australian Synthesis Telescope will start supplying high spatial resolution radio observations.

CONCLUSIONS

The gas to dust ratios in both Magellanic Clouds are higher than measured in our galaxy by factors of approximately 4 in the LMC and 17 in the SMC. These factors are closely similar to the underabundance of Carbon in the respective Clouds (Dufour, 1983). But the abundances of Si and O, which are needed for the formation of silicates, are down with respect to the solar neighbourhood by much smaller factors (1.8 in the LMC and 5.8 in the SMC,; see Lequeux et al. 1979). Empirically, there appears to be a coupling between the gas-to-dust ratios, the metallicity and the characteristics of the ultraviolet interstellar extinction.

Acknowledgements: It is a pleasure to acknowledge very valuable comments by Dr. J. Chris Blades. Ms. Jean Engelke typed and edited the manuscript and Ms. Sarah Stevens-Rayburn provided competent bibliographical assistence.

REFERENCES

Azzopardi, M., Vigneau, J. 1977, Astron. Astrophys. 56, p.151.
Boer, K.S. de, Koornneef, J., Savage, B.D., 1980, Astrophys. J. 236, p. 769.
Blades, J.C. 1980, Mon. Not. Roy. Astr. Soc. 190, p. 33.
Blades, J.C., Meaburn, J. 1980, Mon. Not. Roy. Astr. Soc. 190, p 59P.
Bergh, S. van den 1968, J. Roy. Astron. Soc. Canada, 62, p. 145.
Bohlin, R.C., Savage, B.D., Drake, J.F., 1978, Astrophys. J. 224, p. 132.
Brunet, J.P. 1975, Astron. Astrophys. 43, p. 345.
Carrasco, L., Strom, S.E., Strom, K.M., 1973, Astrophys. J., 182, p. 95.
Clayton, G.C., Martin, P.G. 1984, this volume, p. 403.
Dachs, J. 1970, Astron. Astrophys. 9, p. 95.
Dufour, R.J. 1984, this volume, p. 353.

Feast, M.W. 1984, this volume, p. 157.
FitzGerald, M.P., 1970, Astron. Astrophys. 4, p. 234.
Genderen, A.M. van, 1969, Bull. Astron. Inst. Neth. Suppl. 3, p. 299.
Genderen, A.M. van, 1970, Astron. Astrophys. 7, p. 49.
Hindman, J.V. 1967, Australian J. Phys. 20, p.147.
Hodge, P.W., 1974, Astrophys. J. 192, p. 21.
Hutchings, J.B. 1980, Astrophys. J. 237, p. 285.
Hutchings, J.B. 1982, Astrophys, J. 255, p. 70.
Isserstedt, J. 1975, Astron. Astrophys. 41, p. 175.
Isserstedt, J. 1976, Astron. Astrophys. 47, p. 463.
Isserstedt, J. 1980, Astron. Astrophys. 83, p. 322.
Isserstedt, J. 1982, Astron. Astrophys. 115, p. 97.
Koornneef, J., Code, A.D. 1981, Astrophys. J. 247, p. 860.
Koornneef, J. 1982, Astron. & Astrophys., 107, p. 247.
Koornneef, J. 1984, this volume, p. 105.
Lequeux, J., Peimbert, M., Rayo, J.F., Serrano, A., Torres-Peimbert, S., 1979, Astron. Astrophys. 80, p. 155.
Lequeux, J., Maurice, E., Prevot-Burnichon, M-L., Prevot, L., Rocca-Volmerange, B. 1982, Astron. Astrophys. 113, p. L15.
Lequeux, J., Maurice, E., Prevot, L., Prevot-Burnichon, M-L., Rocca-Volmerange, B. 1984, this volume, p. 405.
MacGillivray, H.T. 1975, Mon. Not. Roy. Astr. Soc. 170, p. 241.
Madore, B.F. 1976, Mon. Not. Roy. Astr. Soc. 177, p. 215.
Madore, B.F. 1982, Astrophys. J. 253, p. 575.
Martin, N., Prévot, L., Rebeirot, E., Rousseau, J. 1976, Astron. Astrophys. 51, p.31.
Martin, W.L., Thackeray, A.D. 1973, Mon. Not. Roy. Astr. Soc. 161, p. 5P.
McGee, R.X., Milton, J.A., 1966, Australian J. Phys. 19, p. 343.
McGee, R.X., Newton, L.M., 1981, Proc. Astron. Soc. Aust. 4, p.189.
McGee, R.X., Newton, L.M., 1982, Proc. Astron. Soc. Aust. 4, p.308.
McGee, R.X., Newton, L.M., Morton, D.C. 1983, Mon. Not. Roy. Astr. Soc., submitted.
McNamara, D.H., Feltz, K.A. 1980, Publ. Astron. Soc. Pac. 92, p. 587.
Morgan, D.H., Nandy, K. 1982, Mon. Not. Roy. Astr. Soc. 199, p. 979.
Nandy, K., Morgan, D.H. 1978, Nature 276, p. 478.
Nandy, K., Morgan, D.H., Willis, A.J., Wilson, R., Gondhalekar, P.M., 1981, Monthly Not. Royal Astron. Soc. 196, p. 955.
Nandy, K., McLachlan, A., Thompson, G.I., Morgan, D.H., Willis, A.J., Wilson, R., Gondhalekar, P.M., Houziaux, L. 1982, Mon. Not. Roy. Astr. Soc. 201, p. 1p.
Rocca-Volmerange, B., Prévot, L., Ferlet, R., Lequeux, J., Prévot-Burnichon, M.L. 1981, Astron. & Astrophys. 99, p. L5.
Stift, M.J. 1984, this volume, p. 229.
Vaucouleurs, G. de 1955, Astron. J. 60, p. 219.
Wesselink, A.J. 1961, Mon. Not. Roy. Astr. Soc. 122, p. 509.
Westerlund, B.E. 1974, Proc. 1st European Astr. Meeting. Vol. III, p. 39.
Whittet, D.C.B., van Breda, I.G. 1975, Astrophys. Space Sci. 38, L3.
Whittet, D.C.B., van Breda, I.G., 1978, Astron. Astrophys. 66, p. 57.

INTERSTELLAR EXTINCTION IN THE MAGELLANIC CLOUDS

K. Nandy
Royal Observatory, Edinburgh

ABSTRACT

The extinction properties of interstellar dust in the Large and Small Magellanic Clouds have been systematically investigated, using recent UV observations of early type Cloud members along with complementary visible data. The extinction curves differ systematically from the standard Galactic curve. The latter shows a broad absorption feature centred near 2200Å in virtually all sight lines but this is absent or only weakly present in the SMC; also the SMC extinction in the far UV is the largest known relative to E_{B-V}. Dust in the LMC appears to be intermediate in extinction properties between the SMC and normal Galactic material. However, exceptions from the average extinction curves have been found in both Clouds.

Model computations show that the range of grain sizes and their number distribution law may not be significantly different in the Clouds and the Galaxy; the difference in extinction laws can be accounted for by varying the graphite contribution relative to silicate.

1. INTRODUCTION

Extinction due to scattering and absorption occurs when starlight passes through a medium containing small particles. The wavelength dependence of extinction over a long wavelength range is a diagnostic of the grain-constituents and their sizes and size distributions. In the Magellanic Clouds, the nearest extragalactic systems, we have the opportunity of studying the interstellar extinction from comparison of individual spectra of reddened and unreddened stars of similar types.

It is well established that the abundances of heavy elements relative to H are significantly lower in the Magellanic Clouds than in the Sun and nearby Galactic HII regions (Pagel et al. 1978, Lequeux et al. 1979, Dufour et al. 1982). A probable site of grain formation is the atmosphere of cool stars. It may be that the underabundance of heavy elements in stellar atmosphere could affect the production and compostion of dust grains. A systematic investigation of extinction properties of dust in the Magellanic Clouds would, therefore, yield important information on the question of their origin and the chemical enrichment of the interstellar medium in these galaxies. An accurately determined extinction curve in the LMC and SMC would also enable us to derive the intrinsic flux distributions of the Magellanic Cloud members. Their distances being well determined, their absolute flux distributions would provide important stellar physical parameters such as angular diameters, radii and effective temperatures.

Earlier work of Brück et al. (1970) on the measurement of interstellar extinction in the visible range indicated that the ratio of UV slope to BV slope is not significantly different from the Galactic mean value. The first positive suggestion of differences in the UV extinction shortward of 3000Å between the Galaxy and the LMC was made by Borgman et al. (1975) from measurements of the surface brightness distribution using the ANS data in the 30 Doradus region. Because of many assumptions involved, the analysis of this surface photometry led to discordant results (Borgman 1978, Koornneef 1978, Nandy et al. 1979). A survey on a systematic basis using the conventional method of comparing the flux distributions of individual stars with different extinctions, was undertaken by several authors after the launch of the International Ultraviolet Explorer. In this article the current observational situation is summarised. The available data are examined in order to look for regional variations, if any, in the extinction curves and to determine mean LMC and SMC extinction curves. Specific grain models which might account for the observations, and some possible explanations thereof are discussed.

2. OBSERVATIONS

The survey of OB stars in the Magellanic Clouds currently available (Rousseau et al. 1978, Azzopardi et al. 1975, Ardeberg et al. 1977) is limited to V ∼ 14.00. At the distance of the Magellanic Clouds most of these must be supergiants. The average reddening E_{B-V} of these supergiants is not large, about ∼ 0.1 in the LMC and ∼ 0.07 in the SMC; the foreground reddening is ∼ 0.05. However, in the Large Magellanic Cloud there are several moderately reddened supergiants with E_{B-V} > 0.2; the majority being located in the 30 Doradus region. In the SMC only a small sample of supergiants have E_{B-V} > 0.2; they are located near the core and Wing. (A few OB stars with large B-V values turned out to be unreddened and the large value of B-V is due to the presence of a red companion.)

The UV observations (1150 - 3350Å) were obtained with the International Ultraviolet Explorer (Boggess et al. 1978) in both long and short wavelength channels in low dispersion mode (Δλ ∼ 6Å) through the large aperture (Koornneef et al. 1981, Nandy et al. 1980, 1981, Rocca-Volmerange et al. 1981, Lequeux et al. 1982). Complementary visible spectra of a representative sample of LMC members were obtained with the ESO 3.6m telescope and the SAAO 1.9m telescope; observations at J (1.2μm), H (1.65μm) and K (2.20μm) were made by Koornneef (1981) and Morgan and Nandy (1982). The ground based coverage of the SMC members is limited to published UBV photometry (Ardeberg and Maurice 1977; Azzopardi and Vigneau 1975).

3. REDUCTION

The extinction has been derived from comparison of the moderately reddened and little reddened stars in the Clouds. It is important that they be of similar spectral types and luminosity classes. The principal stellar features observed in the low dispersion UV spectra of early type supergiants are the CIVλ1550 and SiIVλ1400 doublets. They are detected in the supergiants as late as B5-B6. The strengths of these lines (equivalent width in Å) as a function of spectral types for the MC members have been studied by Hutchings (1982) and Houziaux et al. (this symposium). But the accuracy of spectral types determined from these lines is not better than ± 2 subclasses (Nandy et al. in preparation). Within this uncertainty the quoted spectral types and the spectral type determined from UBV photometry of the sample of the Cloud members studied here are consistent with the strength of CIV and SiIV absorption.

Also dominant in the spectra of B supergiants are the spectral lines due to FeIII causing a broad depression about 200Å wide in the continuum near 1920Å first detected in the S2/68 spectra of the Galactic supergiants. This feature is luminosity sensitive, but its strength can vary significantly amongst the supergiants and the difference does not necessarily imply a mismatch of temperature. Any mismatch of this feature in reddened and comparison stars would produce an apparent feature near 1920Å in the extinction curve derived by the comparison method. The effect of this mismatch was estimated and the extinction curve in the spectral region 2000-1800Å was corrected (for the details of the method see Koornneef & Code 1981, Nandy et al. 1981).

The visible spectra were reduced to give data points averaged over ∼ 50Å intervals from λ^{-1} = 1.66μm^{-1} to λ^{-1} = 2.54μm^{-1}. In order to compare the extinction curves derived from different pairs of reddened and little reddened Magellanic Cloud members, the curves were normalised to A_V = 0 and E_{B-V} = 1 (we have adopted λ^{-1} = 1.83μm^{-1} for V-band, 2.30μm^{-1} for B-band and 2.90μm^{-1} for U-band).

Photometric error and error due to spectral type mismatch are the major sources of the errors of the extinction curves derived from the comparison method. The error arising from spectral type mismatch is wavelength dependent. For example, for a B1 star an error of one subclass causes an error of ±0.02 near the V-band, ∼ ±0.0.1 near 3000Å rising to ∼ ±0.3 near 1500Å. Photometric accuracy is, on average, ∼ ±0.02 in the visible and ∼ ±0.05 in the wavelength range 2900-2400Å for fluxes averaged over 10Å, but is much lower, ∼ ±0.1 near 2200Å (2400-1900Å) and longward of 2900Å. The error of $\Delta m(\lambda)$ due to the uncertainty of the wavelength determination is < .03.

It is known that many early-type supergiants have long term variability. According to Appenzeller (1972) this may range up to 0.03 mag in V. Nothing is known about the possible variations in the stars studied here. Since the visible and ultraviolet observations are not simultaneous, the possibility exists that there may be uncertainty in V magnitude of about the same amount.

4. RESULTS

4.1 Interstellar extinction in the Large Magellanic Cloud

About half of our samples are located around the outer portion of the 30 Doradus nebula (within 1 kpc from the core), but the rest are well outside. It is, therefore, possible to look for differences in extinction properties between the 30 Doradus region and the rest of the LMC, and we have divided the data into two parts accordingly.

Figure 1. Comparison of infrared and (U-B) colour excesses between the 30 Doradus region and the rest of the Cloud. Larger symbols are given weight = 2.

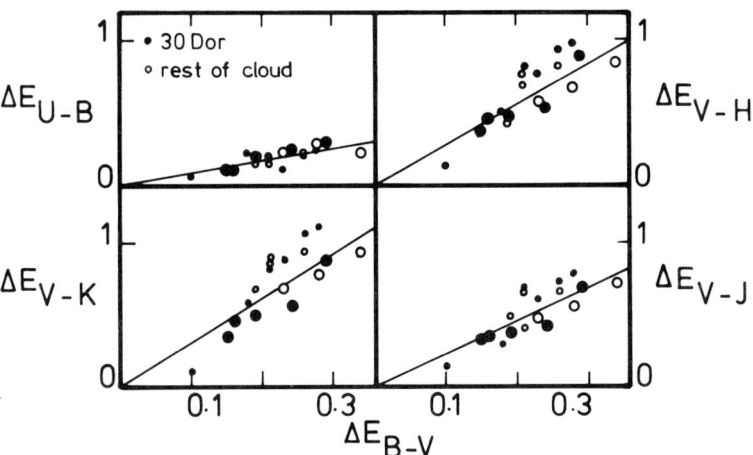

The values of extinction in the infrared and at the wavelength of the U band have been re-determined in the following way. The colour excesses ΔE_{U-B}, ΔE_{V-J}, ΔE_{V-H}, ΔE_{V-K} as a function of ΔE_{B-V} are shown in Fig. 1. Δ indicates the difference between reddened and comparison stars and different symbols are used for the 30 Doradus region and the rest of the Cloud. UBV data and spectral types are taken from Feast et al. (1960), Rousseau et al. (1978) and Walborn (1980). Wherever possible, more than one comparison star has been used; the mean values are shown in Fig. 1 by larger symbols and are given higher weights (weight = 2).

Since no systematic difference exists between the two sets of data, all the data points have been used to obtain the mean extinction at J, H, K and U bands normalised to $A_V = 0$ and $E_{B-V} = 1$. These values are obtained from the slope of best fit line passing through the origin. (They are not significantly different from the values obtained by Koornneef (1981) and Nandy and Morgan (1981, 1982)).

The visible spectra were obtained for 5 reddened stars (3 of them being located in the 30 Doradus region) and a similar number of comparison stars. Extinction in the visible wavelength range has been measured in the following way:- A set of extinction curves was derived for each of the reddened stars, using more than one comparison star, and a mean extinction curve was determined for each reddened star. This reduced the errors introduced by possible spectral type mismatch. The extinction in magnitude, $\Delta m(\lambda)$ was normalised to $\Delta m = 0$ at $1/\lambda = 1.82 \mu m^{-1}$ and $\Delta m = 1$ at $1/\lambda = 2.30 \mu m^{-1}$. Within the observational errors the individual extinction curves in the visible agreed well, thereby allowing a mean extinction curve to be constructed.

Individual extinction curves and colour excesses $E(\lambda-V)$ vs $E(B-V)$ in the UV wavelength range for the sample of LMC members discussed here have been published (see Nandy et al. 1980, 1981; Koornneef and Code 1981). Except for SK 69-108, a heavily reddened star located in the Bar, the extinction curves for all stars are similar within the errors. There appears to be no difference in the extinction law between the 30 Doradus region and the other parts of the LMC.

It should be noted that a number of reddened stars in the sample of Nandy et al. (1980, 1981) are more luminous than the comparison stars by up to 2 magnitudes. The fact that the luminosity sensitive 1920Å feature is also weak in the comparison stars raises the question of whether the luminosity effect is present in the extinction curves shortward of 1700Å, since the continuum fluxes of the luminous stars shortward of 1700Å could be intrinsically weaker than those of the less luminous stars (Humphries et al. 1975). However, this is not the case for the following reasons:

(a) Two reddened O stars which do not show 1920Å features have the same high UV extinction as others.

(b) The mean extinction curves derived from reddened stars with $-8.5 < M_V < -7.5$, and from those with $-7.5 < M_V < -6.5$ agree within the errors of observations.

(c) The difference in absolute magnitudes between the reddened and comparison stars in the sample of Koornneef et al. (1981) is slight, yet their mean extinction curve is not significantly different from that of Nandy et al. (1981).

All the available UV observations excluding SK 69-108 are combined to obtain a mean extinction curve for the LMC which is shown in Fig. 2; the comparison of the LMC extinction law with the mean Galactic law (Seaton 1979) is also shown. The Galactic extinction curve shows a change in slope in the blue region near $1/\lambda \sim 2.3 \mu m^{-1}$. This is also observed in the LMC curve. In the UV range the LMC curve differs from the Galactic curve in respect of the $\lambda 2200$ feature and far UV extinction (see Section 5).

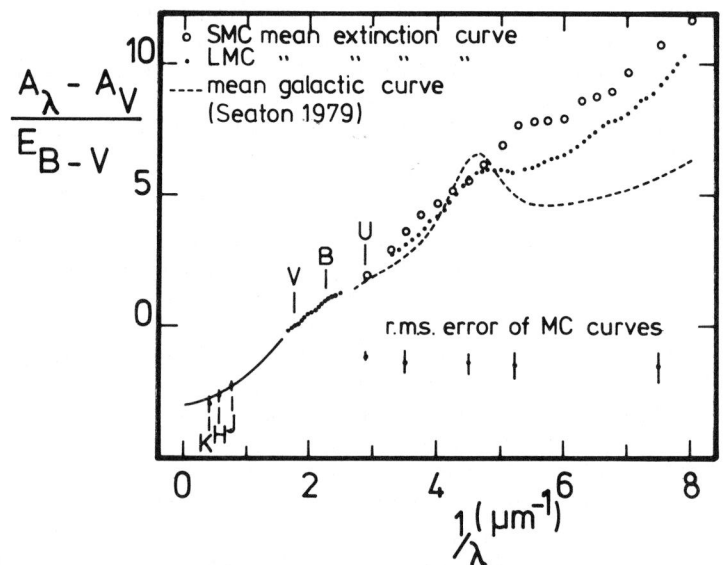

Figure 2. The comparison of the mean LMC and SMC extinction curves with the mean Galactic curve of Seaton (1979). The solid line is van de Hulst curve No. 15 (1949).

Although there is no evidence of systematic difference in extinction curves between the 30 Doradus region and the rest of the Cloud, exceptions have been found due to local anomalies. SK 69-108 which is heavily obscured and located in a region south of the LMC Bar exhibits an extinction curve similar to the Galactic mean curve. Koornneef and Mathis (1981) and Savage (private communication) have reported that the UV extinction properties of dust near the core of the 30 Doradus region are not as extreme as the mean LMC law in respect of the 2200Å feature and far UV extinction. They have suggested that the dust-to-gas ratio might affect the grain properties. Clearly there is a need for further observations.

4.2 The ratio of total-to-selective extinction for the LMC

R, the ratio of total-to-selective extinction measures the amount of dust present causing visible extinction and is an important quantity in many theories of star formation. The value of R is determined by extrapolating the extinction curve to $1/\lambda = 0$. This extrapolation

to infinite wavelength depends on the model. It is assumed that the
extinction in the visible has been mainly caused by dielectric particles (see Section 5). For $2\pi a/\lambda \ll 1$ the extinction efficiency of
the particles varies as λ^{-4} and the shape of the extinction curve
between $\lambda = 2.2\mu m$ and $\lambda \to \infty$ is not strongly dependent on species.
Using van de Hulst curve no. 15 (1949) the ratio R is given by R =
$1.10(E_{V-K}/E_{B-V})$. With the value of 3.0 ± 0.2 for E_{V-K}/E_{B-V}, the ratio
R for the LMC is found to be 3.3 with the uncertainty of $\sim \pm 0.2$.
Within this error, the value of R is the same as the mean value of R
for the Galaxy (Whittet and van Breda 1980). There is no guarantee
that this extrapolation is valid and further observations at longer
wavelengths are desirable. However, recent study of the wavelength
dependence of interstellar polarisation for a sample of 22 LMC members
over a range of E_{B-V} by Clayton et al. (1983) shows that the average
value of λ_{max}, the wavelength where maximum polarisation occurs is
$0.58\mu m \pm 0.05$. The value of R derived for the LMC is close to that
given by the empirical relation $R = 5.5 < \lambda_{max} >$ for Galactic stars
found by Serkowski et al. (1975).

4.3 Interstellar extinction in the Small Magellanic Cloud

The first ultraviolet extinction curves for the SMC were obtained
by Rocca-Volmerange et al. (1981) and Hutchings (1982) who reported
that the extinction curves do not show the $\lambda 2200$ feature and the far
UV extinction is even higher than in the LMC. The results were based
on a very small sample of reddened and comparison stars with a small
difference in colour excess. Further observations of considerably
reddened stars located near the core and the Wing of the SMC and
comparison stars were obtained by Nandy et al. 1982. The extinction
curves obtained from new pairs confirm earlier results. Irrespective
of the location of the reddened stars in the SMC, the individual
extinction curves are similar, although exceptions have been found due
to local anomalies [possible examples are SK143 (Lequeux et al. 1982)
and BBB338 (Nandy et al. 1982)]. In order to derive a mean extinction
curve for the SMC we have used the results of Rocca-Volmerange et al.
(1981) for SK13 and SK124, the pairs BBB280 comp. star SK32, HD4976
comp. star SK57 (Nandy et al. 1982) and additional pairs SK85
comp. star SK32, AZZ393 comp. star SK32. The comparison of the mean
LMC, SMC and Galactic extinction curves is shown in Fig. 2. The error
of the mean SMC curve may well be larger than those plotted in Fig. 3;
since there are still some uncertainties about cancellation of
stellar FeIII lines near 1920Å. These are to be carefully reinvestigated in future, but do not affect the main conclusions, viz. the
$\lambda 2200$ feature, so dominant in our Galaxy, is conspicuous by its
absence in the SMC, and the far UV extinction is significantly larger
than in the LMC.

5. DISCUSSION

The results presented here show that there is an overall similarity of individual extinction curves in both Clouds suggesting that

the "normal" LMC and SMC extinction curves are essentially defined (although anomalies have been found as discussed earlier). The comparison of the LMC extinction with the mean Galactic curve for the solar neighbourhood (Seaton 1979) shows two important differences:-

(a) the 2200Å feature is weaker in the LMC (by a factor $\sim$ 2);

(b) the extinction shortward of 2000Å is considerably higher in the LMC than in the Galaxy.

At the wavelength of the U-band and between 3300Å and 2700Å the LMC extinction values are higher than the Galactic values. This difference was noted by Koornneef and Code (1981) but is well within 2σ and may not be significant. In the visible and infrared wavelength range the LMC and Galactic extinction curves are similar.

The SMC curve is "extreme" in the sense of showing no $\lambda 2200$ feature (or very weak) and the largest known far ultraviolet extinction relative to E_{B-V}. There is almost a $1/\lambda$ dependence of $(A_\lambda - A_V)/E_{B-V}$ over the whole range from 1.8 to 8.5µm^{-1}.

As a first approximation to the composition of dust in the Magellanic Clouds we adopt the model proposed by Mathis and co-workers (1977, 1981). According to these authors the simplest dust model that accurately predicts the normal Galactic interstellar extinction observations is a mixture of uncoated graphite and silicate grains with a power law size distribution function $n(a) \propto a^{-3.5}$. Bierman and Harwit (1980) have supported the power law distribution on physical grounds. The model of Mathis et al. (1977) fits the general constraints imposed by element abundances; the particle radii range from $\sim$ 0.25µm to 0.01µm for silicate, and from 0.25µm to 0.005µm for graphite. Small uncoated graphite provides the "2200" feature in this model.

It is shown in Fig. 3 that this type of model fits the "normal" SMC extinction curve, the only difference being that the graphite contribution to Q_{ext} is at least a factor of seven weaker in the SMC than in the Galaxy (Bromage and Nandy 1983). A fit can be obtained to the LMC curve by increasing the graphite contribution. For example, as shown in Fig. 3 the "normal" LMC extinction curve can be approximated with 35% of the graphite:silicate ratio of the Galactic case (Bromage and Nandy in preparation).

It should be emphasised that the dust model derived by fitting to the extinction curve is not unique; it may be possible to obtain good fits by arbitrarily changing the optical constants, and compensating for this by changes in the size distribution function, or by addition of extra particale types. However, Fig. 3 indicates that the range of grain sizes and their number distribution law as proposed by Mathis et al. (1977) for the Galaxy are not necessarily different in the Magellanic Clouds, only the basic graphite:silicate ratio has to change. The apparently low value of the graphite:silicate ratio in the Clouds may have underlying important significance. Bromage and Nandy (1983) have considered several possibilities for this low ratio and have argued that graphite grain formation may have a low efficiency in the Clouds due to a low overall C/O abundance ratio. Carbon abundance in the Clouds has been derived by Dufour et al. (1982) from the UV and ground based observations of the HII regions. They conclude that C/O is very low, ~ 1/7 in SMC HII regions and ~ 1/3 in the LMC HII regions. Condensation of graphite grain from stellar atmospheres would be very unlikely in the SMC if C/O << 1. The fact that the extinction properties of the LMC dust are intermediate between the Galactic and SMC ones might be connected with the intermediate C/O ratio in this Cloud.

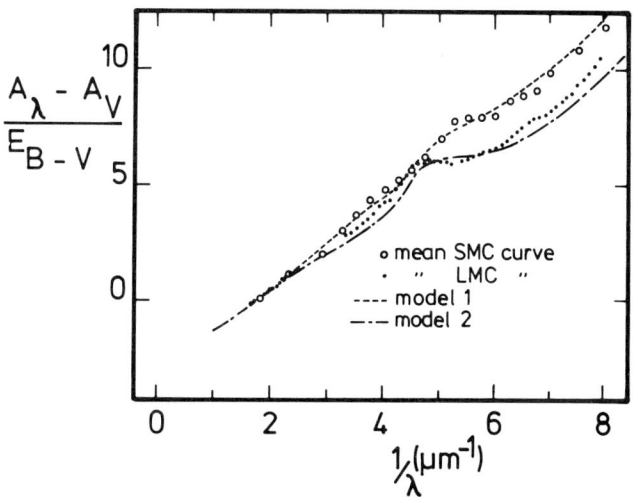

Figure 3. Model fitting to the mean LMC and SMC extinction curves. Model 1 - graphite contribution = nil; Model 2 - graphite contribution = 35% of Galactic value.

However it is known that the accuracy of carbon abundance measurement is limited by IUE detector sensitivity (see for example Maslen et al. (1982). Further work is necessary to establish a possible link between the apparent absence or underabundance of graphite grains and gas-phase depletion in the Clouds.

ACKNOWLEDGEMENT

I would like to thank The Deutsche Forschungsgemeinschaft (DFG) and IAU for providing support for travel to this symposium.

REFERENCES

Appenzeller, I., 1972. Publ.Astr.Soc., Japan 24, 483.
Ardeberg, A. & Maurice, E., 1977. Astron.Astrophys.Suppl. 30, 261.
Azzopardi, M. & Vigneau, J., 1975. Astron.Astrophys.Suppl. 22, 285.

Bierman, P. & Harwit, M., 1980. Astrophys.J. 241, L105.
Boggess, A. et al.., 1978. Nature 275, 372.
Borgman, J., van Duinen, R.J. & Koornneef, J., 1975. Astron.Astrophys. 40, 461.
Borgman, J., 1978. Astron.Astrophys. 69, 245.
Bromage, G.E. & Nandy, K., 1983. Mon.Not.R.astr.Soc. 204, 29P.
Brück, M.T., Lawrence, L.C., Nandy, K., Thackeray, A.D. & Wood, R., 1970. Nature 225, 531.
Clayton, C.G. & Martin, P.G., 1983. Astrophys.J. 265, 194.
Dufour, R.J., Shields, G.A. & Talbot, R.J., 1982. Astrophys.J. 252, 461.
Feast, M.W., Thackeray, A.D. & Wesselink, A.J., 1960. Mon.Not.R.astr.Soc. 121, 337.
Humphries, C.M., Nandy, K. & Kontizas, E., 1975. Astrophys.J. 195, 111.
Hutchings, J.B., 1982. Astrophys.J. 255, 70.
Koornneef, J., 1978. Astron.Astrophys. 67, 179.
Koornneef, J. & Code, A.D., 1981. Astrophys.J. 247, 860.
Koornneef, J., 1981. ESO Sci.Prepr., No. 162.
Koornneef, J. & Mathis, J.S., 1981. Astrophys.J. 245, 49.
Lequeux, J., Peimbert, M., Rayo, J.F., Serrano, A. Torres-Peimbert, S., 1979. Astron.Astrophys. 80, 35.
Maslen, D., Willis, A.J., Wilson, R., Nandy, K. & Blades, J.C., 1982. Proc. 3rd Eur.IUE Conf., ESA-SP 176, p. 431.
Mathis, J.S., Rumpl, W. & Nordsieck, K.H., 1977. Astrophys.J. 217, 425.
Mathis, J.S. & Wallenhorst, S.G. 1981. Astrophys.J. 244, 483.
Morgan, D.H. & Nandy, K., 1982. Mon.Not.R.astr.Soc. 199, 979.
Nandy, K., Morgan, D.H. & Carnochan, D., 1979. Mon.Not.R.astr.Soc., 186, 421.
Nandy, K., Morgan, D.H., Willis, A.J., Wilson, R., Gondhalekar, P.M. & Houziaux, L., 1980. Nature 283, 275.
Nandy, K., Morgan, D.H., Willis, A.J., Wilson, R. & Gondhalekar, P.M., 1981. Mon.Not.R.astr.Soc. 196, 955.
Nandy, K., McLachlan, A., Thompson. G.I., Morgan, D.H., Willis, A.J., Wilson, R., Gondhalekar, P.M. & Houziaux, L., 1982. Mon.Not.R.astr.Soc. 201, 1P.
Pagel, B.E.J., Edmunds, M.G., Fosbury, R.A.E., & Webster, B.L., 1978. Mon.Not.R.astr.Soc. 184, 569.
Rocca-Volmerange, B., Prevot, L., Ferlet, R., Lequeux, J. & Prevot-Burnichon, M.L., 1981. Astron.Astrophys. 99, L5.
Rousseau, J., Martin, N., Prevot, L., Rebeirot. E., Robin, A. & Brunet, J.P., 1978. Astron.Astrophys.Suppl.Ser. 31, 243.
Sanduleak, N., 1969a. Cerro Tololo Inter-American Obs.Contr.No. 89.
Sanduleak, N., 1968. Astron.J. 73, 246.
Sanduleak, N., 1969b. Astron.J. 74, 877.
Seaton, M.J., 1979. Mon.Not.R.astr.Soc. 187, 73P.
Serkowski, K., Mathewson, D.S. & Ford, V.L., 1975. Astrophys.J. 196, 261.
van de Hulst, H.C., 1949. Rech.astr.Obs.Utrecht 11, Part 1, 2.
Walborn, N.R., 1980. Astrophys.J. 215, 53.
Whittet, D.C.B.W. & van Breda, I.G., 1980. Mon.Not.R.astr.Soc. 192, 467.

DISCUSSION

Searle: The variations in the UV extinction law that you discuss may be one aspect of a more general phenomenon. I have recently determined the reddeneing law in M31 from spectrophotometry of its globular clusters. There is a radial change in this reddening law in the wavelength range 3300 to 10,000Å, across the disk of M31. At a galactocentric distance of 12 kpc, the M31 reddening law resembles that of the solar neighborhood. The ultraviolet extinction, for a given E(B-V), increases with decreasing galactocentric distance.

Hutchings: IUE spectra of M33 objects by Massey and myself show that an extinction law similar to that in the LMC applies there.

Davidson: A parenthetical remark: Several years ago there was a slightly silly dispute about the possible reddening of quasars. Some people insisted that such reddening was negligible because the 2200Å feature was not seen. But there is some evidence that reddening perceptibly affects the UV emission lines, at least, of quasars. So this combination of attributes may indicate something about carbon/oxygen ratios in quasars - a poorly known topic.

de Boer: It is worthwhile, I think, to draw attention to the work of Blair Savage and his collaborators (Meyer and Savage, ApJ 248, 545) on extinction in the UV. From UV photometry with the ANS, and later with the IUE satellite, they showed that the extinction in the UV differs a lot from star to star in the Milky Way. Actually, one cannot trust any value of E(B-V) that is derived entirely from UV observations. It is a pity that people working on UV spectra of extragalactic objects are not yet sufficiently aware of the problems due to differing extinction curves.

Dopita: What happens at wavelengths smaller than 912Å?

Nandy: That I do not know in detail. The laboratory data are conflicting.

Koornneef: What happens to the value on R if you take the graphite out of the Mathis Rumpl Nordsieck-mixture?

Nandy: A MRN-mixture with graphite-contribution = nil gives the value of $R \simeq 2.6$. At present we don't know its value for the SMC.

Savage: In reply to the question of M.Dopita, I recall that graphite has a second but stronger extinction bump at about 800Å. Amplifying what Klaas de Boer just said, indeed I like to caution everybody about using average extinction curves in correcting their UV data for the presence of dust. Studies of stars within the Milky Way and now in the LMC and SMC have shown that the difference in extinction from region to region can be enormous. Within the Milky Way $E(\lambda-V)/E(B-V)$ has been shown to vary from about 4 to 10 at 1300Å. The only safe approach is to obtain information about extinction along the line of sight to stars near to the object of interest. An example of this approach is contained in the paper by Savage and Fitzpatrick (this symposium).

THE COMPOSITION OF H II REGIONS IN THE MAGELLANIC CLOUDS

Reginald J. Dufour
Rice University

ABSTRACT

The state of our knowledge concerning the chemical composition of H II regions in the Magellanic Clouds is reviewed. New abundances derived from all modern published spectroscopy are presented. Some of the implications of the results regarding the nucleosynthesis of the elements and galactic chemical evolution are noted.

1. RECENT OBSERVATIONAL STUDIES AND BASIC RESULTS

Studies of the physical properties of H II regions in the Magellanic Clouds began in the late 1950s (Johnson 1959) and has continued at an active pace since. A discussion of the studies made prior to 1974 can be found in the review by Dufour (1976). The "modern age" of abundance studies of H II regions in the Clouds began in the mid 1970s (see Table 1) when four investigations were published by Peimbert and Torres-Peimbert (1974, 1976), Aller et al. (1974), and Dufour (1975). Compared to previous studies, these were notable in (a) the relatively high quality of the spectral data, (b) the direct measurement of electron temperatures and densities in most of the numerous nebulae observed, (c) the use of modern diagnostic techniques and atomic data in the abundance calculations, and (d) the discovery of significant abundance differences between the LMC, SMC, and the Galaxy. Collectively, these studies provided a picture of the physical conditions and chemical composition of the H II regions in the Magellanic Clouds that has changes relatively little since.

The four studies showed conclusively that the H II regions in the LMC and SMC had pronounced deficiencies in the gaseous-phase total elemental abundances of N, O, and Ne compared to the sun and nearby galactic H II regions, with those in the SMC showing the larger deficiencies in all elements studied. Compared to the sun, O and Ne were found to be deficient (relative to H) by factors of about 2.5 (-0.4 dex) in the LMC and 6 (-0.8 dex) in the SMC. Of all of the elements studied, N showed the largest deficiency: factors of about 6 (-0.8 dex) in the LMC and 25 (-1.4 dex) in the SMC. Relative to the Orion Nebula, these deficiencies were smaller but no less significant. In addition, Aller et al. noted that the deficiencies of S, Ar, and Cl (again relative to H) in the LMC are roughly similar to those of O and Ne — a result further substantiated by Dufour for both clouds using deeper photographic spectra. None of the investigations found significant evidence of radial (or other) composition gradients in the Clouds. Peimbert and Torres-Peimbert, using a more detailed analysis and smaller corrections for neutral He, found $N(He)/N(H) = 0.084 \pm 0.005$ for the LMC and $N(He)/N(H) = 0.078 \pm 0.005$ for the SMC.

These interesting findings concerning the composition of the H II regions stimulated additional spectroscopic (and other) studies of emission nebulae in the Clouds that included planetary nebulae and supernova remnants as well (cf. the reviews by Peimbert and Dopita in this volume). The studies of PN by Webster (1976), Osmer (1976), and Dufour and Killen (1977) included several small semi-stellar H II regions, as well as PN, and the results on these generally confirmed the findings of the previous investigations of the larger H II regions. Dufour and Harlow reported photoelectrically measured line strengths for 13 emission lines in 10 SMC H II regions, most notably the $\lambda\lambda 4471$, 5876, and 6678 He I lines that are crucial for determining accurate He abundances. Pagel et al. reported new observations of 6 LMC and 17 SMC H II regions made with the IDS and IPCS spectrometers at the AAT covering the $\lambda\lambda 3575$-8500 wavelength region, and used the new and previously published data to analyze the composition of a total of 19 LMC and 23 SMC H II regions. These new studies resulted in improved abundances for N, O, and Ne in the LMC and SMC, but not significantly different from those previously published. Dufour and Harlow's study is most notable in that it derived probably the most representative value for the He abundance of the SMC at the time: $N(He)/N(H) = 0.081 \pm 0.005$ (based on 10 nebulae), which was slightly larger than that found by

Table 1. Major Abundance Studies of H II Regions in the Magellanic Clouds

Reference	Instrument	WLrange	#LMC	#SMC	Elements
Peimbert & Torres-Peimbert 1974	scanner	3727-7330	4	0	He, N, O, Ne
Aller et al. 1974	scanner + image-tube	3727-7330	20	10	He, N, O, Ne, S, Ar
Dufour 1975	scanner + image-tube	3727-7136	11	3	He, N, O, Ne, S, Cl, Ar
Peimbert & Torres-Peimbert 1976	scanner	3727-7330	0	3	He, N, O, Ne
Dufour & Harlow 1977	scanner	3727-6731	0	10	He, N, O, Ne, S, Ar
Pagel et al. 1978	IDS + IPCS	3575-8500	19	23	He, N, O, Ne, S, Cl, Ar
Aller et al. 1979	scanner + image-tube	3727-7330	16	6	He, N, O, Ne, S, Cl, Ar
Dufour et al. 1982	IUE vidicon + scanner	1200-7330	4	3	He, C, N, O, Ne, S, Ne, S, Cl, Ar

Peimbert and Torres-Peimbert (1976): 0.078 ± 0.005 (based on 3 nebulae). Pagel et al.'s study did not impact the He abundance problem, but did contribute significantly to the accuracy of the determination of representative abundances of the other elements, particularly S, Cl, and Ar. More recently, Dennefeld and Stasinska (1983) published a reanalysis of the S abundances in several LMC, SMC, and MWG H II regions, based on new observations of the IR lines of [S III], which they derived S/H ratios for the LMC and SMC nebulae lower by a factor of about 2 compared to most previous studies.

In essentially all of the previously cited studies, the investigators adopted the ionization correction factor (ICF) approach first applied to galactic H II regions by Peimbert and Costero (1969). In this approach, the ionization corrections for various observable ions are largely estimated from the O^+/O^{+2} ratio based on comparison of ionization potentials of O^+, O^{+2}, and the ion. A potentially more accurate method of deriving total elemental abundances in H II regions is through the use of sophisticated nebular models based on matching the observed spectrum and physical conditions of a nebula. Such procedures have recently been applied to H II regins in the Magellanic Clouds by Aller et al. (1979) and by Dufour et al. (1981, 1982). Not only do these procedures result in more accurate determination of ionization corrections to use in deriving elemental abundances from the observable ions, but they also give an estimate of the extent of the temperature fluctuations in a nebula and the temperatures appropriate to zones for each ion.

In addition to the application of model analysis techniques, the study of Dufour et al. (1982) presented new observational data on the ultraviolet spectra of 4 LMC and 3 SMC H Ii regions using the IUE satellite. This permitted the first determination of the gaseous-phase C abundances in the H II regions of the Magellanic Clouds. In a prior preliminary paper, Dufour et al. (1981) found the C/H ratio in 3 SMC H II regions to be a factor of 8 (-0.9 dex) lower than in the sun. Later, based on additional IUE data, Dufour et al. (1982, cf. also Dufour and Shields 1982) found C/H lower in the LMC and SMC by factors of 6 (-0.75 dex) and 30 (-1.49 dex), respectively, compared to the sun. Recent radio studies of CO in the Clouds (cf. the review by Israel in this volume) and the weakness of the $\lambda 2200$ graphite feature in the UV extinction curves for the LMC and SMC (cf. the review by Nandy in this volume) further substantiate the existence of a C deficiency in the ISM of the Clouds.

2. SUMMARY OF RESULTS AND "RECOMMENDED ABUNDANCES" FOR THE LMC AND SMC

In Table 2 the averaged abundance results of the major recent studies of H II regions in the LMC and SMC are presented (note that those of the Peimberts have been rederived using smaller temperature fluctuations and more modern atomic data as noted in Lequeux et al. 1979). At the bottom of the table are newly calculated "recommended abundances" for the LMC and SMC derived by the author using the model-based analysis techniques of Dufour et al. (1982) applied to all of the spectral data from the various

Table 2. Abundance Results for H II Regions in the Magellanic Clouds

Reference	He	12 + log N(X)/N(H) (except He)						
		C	N	O	Ne	S	Cl	Ar
Large Magellanic Cloud								
PTP 1974(t^2 = 0.035)*	0.084		7.21	8.50	7.64			
(t^2 = 0.000)	0.084		7.03	8.34	7.44			
Aller et al. 1974	0.098		6.94	8.46	7.83	7.2		
Dufour 1975	0.102		6.80	8.43	7.64	7.15	5.01	7.10
Pagel et al. 1978			6.88	8.39	7.61	6.8	4.9	6.35
Aller et al. 1979			7.02	8.43	7.77	6.90		6.35
Dufour et al. 1982	0.083	7.90	6.94	8.38	7.68	7.01		6.10
Small Magellanic Cloud								
Aller et al. 1974	0.100		6.28	7.97	7.40	6.5		6.0
Dufour 1975	0.093		6.49	8.05	7.18	7.15		7.02
PTP 1976 (t^2 = 0.035)*	0.078		6.49	7.99	7.13			
(t^2 = 0.000)	0.078		6.41	7.89	7.03			
Dufour and Harlow 1977	0.081		6.48	8.02	7.29	6.4		
Pagel et al. 1978			6.41	7.98	7.16	6.4	4.5	5.94
Aller et al. 1979			6.45	8.10	7.58	6.29		5.86
Dufour et al. 1982	0.083	7.16	6.60	8.05	7.34	6.61		5.77
LMC Recommended	0.085	7.90	6.97	8.43	7.64	6.85	4.84	6.20
(rms error)	±0.004	±0.15	±0.10	±0.08	±0.10	±0.11	±0.20	±0.06
SMC Recommended	0.080	7.16	6.46	8.02	7.22	6.49	4.7:	5.78
	±0.003	±0.04	±0.12	±0.08	±0.12	±0.14		±0.12

*Note: The PTP 1974 and 1976 abundances shown are those rederived in Lequeux et al. (1979).

studies (and including the model results of Aller et al. 1979).

A number of comments regarding the new and old results and some of the associated problems involved in the calculations for the different elements are in order: <u>Helium</u> — Evaluation of the He/H abundance in the H II regions of the Clouds essentially reduces to a problem of determining and using only the best quality data on the He I lines and estimating the magnitude of the corrections for neutral He. The results are relatively insensitive to temperatures adopted. Values of He^+/H^+ in 14 LMC and 15 SMC H II regions were calculated using published spectrophotometry that contained measurements of at least two of the ratios: $I(4471)/I(H\gamma)$, $I(5876)/I(H\alpha)$, and $I(6678)/I(H\alpha)$ — and the results retained only if at least two He^+/H^+ values for the groups of LMC and SMC H II regions showed little indication of depending on ionization level (most had $O^{+2}/O^+ > 0.5$). This observation, coupled with recent models, suggest that the correction for He^0 in these large extended nebulae is negligible in those with ionization fractions $X(O^{+2}) > 0.6$. Consequently, the He^+/H^+ values 8 LMC and 10 SMC nebulae meeting these requirements were averaged to get the final He/H results for the LMC and SMC shown in Table 2. The errors (rms standard deviations) in the averages are smaller than in the previous studies of the Peimberts and of Dufour and Harlow. <u>Carbon</u> — The recommended C abundances for the LMC and SMC were taken straight from the study by Dufour et al. (1982) based on IUE observations of the $\lambda 1909$ [C III] lines in 4 LMC and 3 SMC H II regions and ionization corrections made by model calculations. Newer IUE data recently obtained from additional nebulae confirm these results. <u>Nitrogen</u> — Since N^+ is usually a very minor fraction of the total N abundance in the high ionization H II regions of the Clouds, and charge exchange with H is probably different for the ions of O and N, it is potentially dangerous to use the usual ICF approach based on the assumption that $N/O = N^+/O^+$. However, the models in Dufour et al. (1982) suggest that $X(N^+) \approx 0.96\ X(O^+)$ in the high excitation low density H II regions of the Clouds, and therefore, I used the standard ICF approach applied to data on 12 LMC and 16 SMC H II regions that had good [O III] temperatures ($\pm 10\%$ accuracy) to derive average N/H values for each nebula; which were then combined with the model results of Aller et al. (1979) to get the final galaxy averages based on a total of 20 LMC and 16 SMC H II regions. <u>Oxygen</u> — Since essentially all of the O in the H II regions of the Clouds is in the form of O^+ and O^{+2}, the accuracy of the oxygen abundances in the H II regions primarily depend on the quality of the two ionic abundances — and this to a large extent depends on the quality of the temperature determinations. While the models suggest that slightly different temperatures should be used for O^+ and O^{+2}, the effect is small, so I just averaged all data on O/H abundances for LMC and SMC H II regions that were based on accurate electron temperatures observed by the various investigations (including the data from Aller et al. 1979). <u>Neon</u> — The situation for Ne is similar to N in that charge exchange rate differences between the various ions of Ne and O make the standard ICF approach dangerous. The models in Dufour et al. (1982) suggest that $X(Ne^{+2}) \approx 1.1\ X(O^{+2})$, so I used this ICF to rederive Ne/H in the various H Ii regions and combined the data with the results of Aller et al. (1979). <u>Sulphur, Chlorine, and Argon</u> — The ICF's and the atomic data for the observable ions of these are subjects of much debate at present. Pagel et al. (1978) gives a good discussion of the problems involved with previous ICF approaches, which usually result in the S/O, Cl/O, and Ar/O ratios for individual H II regions in each Cloud varying with ionization level. A more critical study of the ICF problem and model inferences regarding S have been made by Dennefeld and Stasinska (1983), in which they utilized new observations of the strong IR [S III] lines in several LMC and SMC H II regions in their analysis. While their results were more consistent with ionization level, the final S/H values for the Clouds were about a factor of 2 lower than previous studies. I calculated S/O, Cl/O, and Ar/O with the available data using the model-inferred ICF's derived in Dufour et al. (1982, based on Model L), and these showed little variation with ionization level. However, the resulting S/H ratios were significantly larger than found by Dennefeld and Stasinska, and probably larger than in reality. Therefore, as a compromise, these S/H values were averaged with the results of Dennefeld and Stasinska for nebulae in common to get the recommended S/H results given in Table II. The results for Cl and Ar in Table 2 are based solely on the Dufour et al. model-inferred ICFs.

3. DISCUSSION: SOME IMPLICATIONS OF THE ABUNDANCE DEFICIENCIES

In Table 3 below, the "recommended abundances" derived for the LMC and SMC are compared with those of the sun (Ross and Aller 1976, Lambert 1978) and galactic H II regions in the solar neighborhood (Shaver et al. 1983, Dufour et al. 1982). Some of the implications of the well established abundance deficiencies in the LMC and SMC compared to the solar neighborhood will be noted in the remainder of this paper. A good discussion of these and other implications of the Clouds' results, along with those from stellar and nebular studies of other galaxies, can be found in the excellent review of Pagel and Edmunds (1981).

3.1. Applicability of the Simple Evolutionary Models to the Clouds

Pagel et al. (1978) made a detailed comparison between the abundance results for the LMC and SMC H II and the expectations of simple chemical evolution models. They noted that the apparent lack of significant abundance gradients in the Clouds suggested that they are well mixed systems which (along with many other metal-poor irregulars) follow the expectations of the simple "closed box" model with instantaneous recycling and no infall. Such a model predicts that the mass fraction of heavy elements, Z, in a galaxy is related to the fraction of gas mass to total mass, μ, by the relation: $Z = -p \ln \mu$, where p is the "yield" of the elements. While they found that the individual H II regions did not follow the relation in spatial detail, the average Z (inferred from O, which is presumed to result totally from primary nucleosynthesis processes) for each Cloud varied with μ in a manner analogous to the outer regions of M101 and M33 for the above equation with a yield of $p \approx 0.003$ by mass. This global behavior of the Clouds was further substantiated by Lequeux et al. (1979) using the Clouds' data with those of six other irregulars. They found a yield of 0.004 ± 0.001 which agreed best with those computed by Chiosi et al. (1978) with mass loss and a power law IMF with a slope of 2. Dufour et al. (1982) derived yields for the presumably primary elements C and O individually and they found the results consistent with massive star nucleosynthesis models for an IMF exponent $7/3 < \alpha < 10/3$ if mass loss is negligible, or with $4/3 < \alpha < 7/3$ if mass loss is significant. They also noted that C/H did not follow the expectations of the simple model and

Table 3. Comparative Abundances

Object	[He]	[C]	[N]	[O]	[Ne]	[S]	[Cl]	[Ar]	[C/O]	[N/O]
sun	11.03	8.65	7.96	8.87	8.05	7.23	5.5:	6.57	-0.22	-0.91
MWG H II	11.00	8.46	7.57	8.70	7.90	7.06	5.16	6.42	-0.24	-1.13
[LMC/sun]	-0.10	-0.75	-0.99	-0.44	-0.41	-0.38	-0.7:	-0.37	-0.31	-0.55
[LMC/H II]	-0.07	-0.56	-0.60	-0.27	-0.26	-0.21	-0.32	-0.22	-0.29	-0.33
[SMC/sun]	-0.13	-1.49	-1.50	-0.85	-0.83	-0.74	-0.80	-0.79	-0.64	-0.65
[SMC/H II]	-0.10	-1.30	-1.11	-0.68	-0.68	-0.57	-0.5:	-0.64	-0.62	-0.43

that deriving Z for metal-poor galaxies based only on the observed O/H value is dangerous since C/O and Fe/O may be much lower than solar values.

3.2. CNO Nucleosynthesis

The result that, until recently, N was observed to show the largest deficiency of all of the elements studied in the H II regions of the Clouds, coupled with similar behavior observed in other galaxies of not following the simple model, prompted speculation that most of the N resulted from secondary nucleosynthesis processes (e.g., Truran 1977), such as processing of C by the CNO cycle in H-burning shells of stars. This N would then be injected into the ISM from intermediate mass stars via planetary nebulae and novae. However, the similarity of the observed N/O ratio in the Clouds and in many other very metal-poor irregulars have led others to propose that a substantial primary component of N exists (cf. Edmunds and Pagel 1978 and references therein). If N is purely a secondary element from C, then the simple model predicts that N/C would be proportional to C/H. The C abundances of Dufour et al. (1982) for the Clouds show that, at least for these metal-poor systems, N/C decreases with increased C/H. They interpret this result as suggesting that most of the N in metal-poor systems is primary and that most of the C enrichment arises from less massive stars than those which produce most of the primary N. However, they point out that in evolved systems, N/C would increase as the secondary production of N from planetary nebulae and novae begin to dominate the chemical enrichment.

3.3. The Pre-Galactic He Abundance and Big Bang Nucleosynthesis

Peimbert and Torres-Peimbert (1974, 1976) showed that the small, but statistically significant He deficiencies in the Clouds coupled with those of the other primary elements suggested a chemical enrichment scenario for galaxies characterized by $\Delta Y/\Delta Z \approx 2.7$. They also derived a pre-galactic He abundance value (by mass) of $Y_p = 0.228 \pm 0.014$ and noted that this implied an open universe for $H_o > 16$ km s^{-1} Mpc^{-1} from standard big bang models. Using the Peimbert's He/H results for the Clouds with those of other metal-poor irregulars, Lequeux et al. (1979) derived very similar results. Several subsequent studies of abundances in spiral and irregular galaxies suggest $\Delta Y/\Delta Z \approx 3$ and $Y_p \approx 0.23$ (Serrano and Peimbert 1981).

4. CONCLUDING REMARKS

While we have come a long way in the study of the composition of the H II regions in the Magellanic Clouds and have apparently accurate information on the abundances of He, C, N, O, Ne, and probably Ar, much can still be done. Deeper optical spectra are needed to improve the determinations of S and Cl, as well as possibly obtaining Fe via the very weak [Fe II] and [Fe III] lines. Observatins in the UV, IR, and radio wavelength regions offer great potential in improving our understanding of physical conditions and determning the abundances of other elements like Mg and Si. The H II regions in the Magellanic Clouds are the most accessible example of the large low density H II regions which dominate the ISM in more distant galaxies, and therefore an understanding of those in the Clouds is necessary before the true potential of extragalactic H II region studies can be fully exploited.

REFERENCES

Aller, L. H., Czyzak, S. J., Keyes, C. D., and Boeshaar, G.: 1974, Proc. Nat. Acad. Sci. USA 71, 4496.
Aller, L. H., Keyes, C. D., and Czyzak, S. J.: 1979, Proc. Nat. Acad. Sci. USA 76, 1525.
Chiosi, C., Nasi, E., and Sreenivasan, S. R.: 1978, Astron. Astrophys. 63, 103.
Dennefeld, M., and Stasinska, G.: 1983, Astron. Astrophys. 118, 234.
Dufour, R. J.: 1975, Astrophys. J. 195, 315.
Dufour, R. J.: 1976, Earth Extraterrestrial Sci. 2, 245.
Dufour, R. J., and Harlow, W. V.: 1977, Astrophys. J. 216, 706.
Dufour, R. J., and Killen, R. M.: 1977, Astrophys. J. 211, 68.
Dufour, R. J., and Shields, G. A.: 1977 in Y. Kondo, J. Mead, and R. D. Chapman (eds.), "Advances in Ultraviolet Astronomy: Four Years of IUE Research," NASA Conf. Pub. 2238, Washington, D.C., p. 385.
Dufour, R. J., Shields, G. A., and Talbot, R. J.: 1982, Astrophys. J. 252, 461.
Dufour, R. J., Talbot, R. J., and Shields, G. A.: 1981, in R. D. Chapman (ed.), "The Universe at Ultraviolet Wavelengths: The First Two Years of the International Ultraviolet Explorer," NASA Conf. Pub. 2171, Washington, D.C., p. 671.
Edmunds, M. G., and Pagel, G. E. J.: 1978, Monthly Notices Roy. Astron. Soc. 185, 77p.
Johnson, H. M.: 1959, Pub. Astron. Soc. Pacific 71, 425.
Lambert, D. L.: 1978, Monthly Notices Roy. Astron. Soc. 182, 249.
Lequeux, J., Peimbert, M., Rayo, J. F., Serrano, A., and Torres-Peimbert, S.: 1979, Astron. Astrophys. 80, 155.
Osmer, P. S.: 1976, Astrophys. J. 203, 352.
Pagel, B. E. J., and Edmunds, M. G.: 1981, Ann. Rev. Astron. Astrophys. 19, 77.
Pagel, B. E. J., Edmunds, M. G., Fosbury, R. A. E., and Webster, B. L.: 1978, Monthly Notices Roy. Astron. Soc. 184, 569.
Peimbert, M., and Costero, R.: 1969, Bol. Obs. Tonantzintla Tacubaya 5, 3.
Peimbert, M., and Torres-Peimbert, S.: 1974, Astrophys. J. 193, 327.
Peimbert, M., and Torres-Peimbert, S.: 1976, Astrophys. J. 203, 581.
Ross, J. E., and Aller, L. H.: 1976, Science 191, 1223.
Serrano, A., and Peimbert, M.: 1981, Rev. Mex. Astron. Astrophys. 5, 109.
Shaver, P., A., McGee, R. X., Newton, L. M., Danks, A. C., and Pottasch, S. R.: 1983, Monthly Notices Roy. Astron. Soc. 204, 53.
Truran, J. W.: 1977, in J. Audouze (ed.), "CNO Isotopes in Astrophysics," D. Reidel, Dordrecht, Holland.
Webster, B. L.: 1976, Monthly Notices Roy. Astron. Soc. 174, 513.

DISCUSSION

Graham: Do you think the reported abundance gradients in the LMC are marginal at best?
Dufour: The Pagel et al. (1978) result: d (log O/H) / d R = 0.03 ± 0.02 is definitely marginal. With these data I find a less obvious gradient in N/H and O/H, and none in N/O.
Graham: Would you or Peimbert comment on Bob Williams recent suggestion that there may be substantial nitrogen enrichment in the Magellanic Clouds from novae?
Dufour: Yes, probably Manuel Peimbert should comment; he has restudied Williams results and believes that novae are less important than planetary nebulae.
Peimbert: I think that is a very provocative idea. I find three problem areas that should be clarified before the importance of nitrogen enrichment by novae is established: a) the N produced by novae, according to stellar evolution models, is the isotope N^{15} while in the solar neighborhood N^{14}/N^{15} is larger than 100; b) the shell masses of novae might have been overestimated if the filling factor is very small, that is in the presence of spatial density fluctuations; and c) I think that we should try to obtain masses and chemical compositions of the novae in the LMC.
Dopita: It is interesting that the SNR's (see IAU Symp. 101) in Magellanic irregulars also give constant N/O ratios from galaxy to galaxy. The study of SNR's has the potential to give very accurate C, Si, Ca and Fe abundances from their extremely rich UV spectra, and should furnish a very good check on chemical evolution theories.
Dufour: I agree totally. In fact, Don Cox, John Raymond, and I are working on modelling the SNR's in IC1613 and NGC6822 using my SMC and LMC abundances with the aim of matching their spectra for abundance determinations. I believe that SNR's have great potential in this area, particularly for elements like Fe and Ca, and commend you on your pioneering efforts in the past few years.
Lequeux: I would like to caution about any derivation of the yield from the Z vs. M(gas)/M(tot) relation in irregular galaxies. At the time I published my paper with the Peimberts and Talent wrote his thesis, there seemed to be a nice Z vs. M(gas)/M(tot) relation, but more recent work demonstrated the existence of exceptions. The most striking cases are the blue compact galaxies IZw18 and IZw36, studied by Viallefond, Thuan and myself. These objects are very metal deficient but have a small M(gas)/M(tot) (of the order of 0.1 - 0.2). This points to complications like accretion of external gas, changes in the IMF or galactic winds, which may well occur also on most irregular galaxies.
Dufour: I believe that your point is most correct and important. We, for example, find He/H in IC1613 exceptionally large for its Z inferred from O/H. Since these small systems are sensitive to external and internal effects (e.g., infall, encounters, etc.), I am sure that a good percentage of them will have "abnormal" characteristics. However, I think the group of nearby irregulars studied by Talent in his thesis is an exceptionally good set for such comparisons since they are sufficiently resolved to obtain total masses by rotation curves, gas masses by HI fluxes, and accurate astration (μ) values. This is not true for most of the Zwicky and the compact systems because of their greater distances.

THE MAGELLANIC CLOUDS AND PLANETARY NEBULAE

Manuel Peimbert
Instituto de Astronomía, Universidad Nacional Autónoma de México

Abstract. A review of the statistics, emission line intensities, central star fluxes, radial velocities and chemical compositions of PN in the MC is given. From these data a discussion is made of: a) the distance scale, b) the envelope mass, c) the comparison between the observed chemical abundances and those predicted from stellar evolution models and, d) the effect that intermediate mass stars have on the chemical evolution of the MC and our galaxy.

I. INTRODUCTION

The Magellanic Clouds can provide us with a better understanding of PN, for example: a) the distribution of PN across the face of the nebula and their radial velocities can give us an indication of the mass range of the progenitor stars, b) the common distance of MC PN can permit us to derive the [O III] or Hα luminosity function and the mass distribution of their envelopes, c) the very low He, C, N, O abundances from which the progenitors of PN were formed can allow us to study the enrichment produced by stellar evolution and to check the stellar evolution models. Alternatively PN in the MC can tell us something about the star formation rate, the chemical evolution of the MC as a whole and the kinematical history relative to the formation and evolution of the MC. The purpose of this review is to study some of these relations. Previous reviews related to PN in the MC and in the Local Group are those by Webster (1978), Jacoby (1983) and Ford (1983) where several aspects related to MC PN as well as many references that will not be discussed here are presented. Special effort will be made to discuss the effect that intermediate mass stars, IMS, $1 < M/M_\odot < 8$ have on the chemical evolution of the MC, some overlap with the review by Dufour (1984) on chemical evolution is anticipated.

II. DISTRIBUTION AND KINEMATICS

Webster (1978), from an objective prism study done with the 1.2 m

Schmidt and covering the central 36 square degrees of the LMC, found that PN exhibited no concentration either to the Bar or to the regions of active star formation and concluded that the distribution of bright PN, identified up to then, originated from more massive, and consequently younger, stars than the bulk of galactic PN.

Freeman et al. (1983) compared the galactocentric velocities from Feitzinger and Weiss (1979) of 35 PN in the LMC with those of the HI/HII rotation curve and with the older cluster rotation curve, derived from objects with ages larger than 10^9 years, and found that PN do appear to show a wide range of kinematical properties; the PN within about 2° of the rotation center follow the HI/HII region rotation curve, as noted by Smith and Weedman (1972), while the outer ones show higher scatter and two of them lie near the older cluster rotation curve. These results are not in contradiction with the conclusions of Webster (1978) and the idea that most IMS become PN. A larger number of high accuracy radial velocity determinations coupled with abundance determinations are needed to be more specific.

III. [O III] LUMINOSITY FUNCTION AND NUMBER OF PN

Jacoby (1980) obtained on-line/off-line filter photographs in [O III] or Hα of four fields in the central regions of each of the clouds using the Cerro Tololo 4m prime-focus camera; from this survey he was able to detect planetary nebulae 250 times fainter than the brightest ones, improving the detection limit by three magnitudes relative to previous surveys. He established an [O III] luminosity function and from it estimated that the total numbers of PN in the SMC and LMC are 285±78 and 996±253 respectively. Previous estimates of the total number of PN are reviewed by Jacoby (1983).

Based on his MC PN luminosity function Jacoby (1980) estimated that, for local group galaxies, the average luminosity specific number is $6.1\pm2.2\times10^{-7}$ PN/$L_\odot$ and the mass specific number is $2.1\pm1.5\times10^{-7}$ PN/$M_\odot$, where the uncertainties are 1 standard deviation values derived from the averaging procedure. From the luminosity specific number he concludes that there are 10 000±4 000 PN in our Galaxy.

IV. MASS FUNCTION OF THE SHELL, PN DISTANCE SCALE, PN AS STANDARD CANDLES

Jacoby (1980) was able to spatially resolve 9 faint PN in the MC and to obtain diameters for the first time. Under the assumption that the PN are optically thin it follows that

$$M\epsilon^{-1/2} = K[I(H\alpha)\ d^5\ \phi^3]^{1/2} , \qquad (1)$$

where M is the mass of the shell, ε is the filling factor, K is a constant, I(Hα) is the intrinsic Hα flux, d is the distance to the MC, and φ is the angular radius in arcseconds. Jacoby obtained a spread of a factor of forty in $M\epsilon^{-1/2}$ which, under the assumption of a constant ϵ, implies that there is a spread of shell masses of a factor of 40 as observed in galactic PN. The spread in M might be smaller due to two reasons: a) LMC 18, which is the minimum mass object observed by Jacoby, might be optically thick and its real mass might be consequently higher; b) the filling factor ϵ might decrease with the size of the PN (e.g., Torres-Peimbert and Peimbert 1977), reducing the mass of the most massive objects which are the largest. To improve our knowledge of this problem spectra of the objects are needed: to establish if they are optically thin and to determine electron densities from forbidden lines to be able to derive a value of ϵ for each object.

Seaton (1968), under the assumption that the masses of optically thin nebula are the same in the Galaxy and the MC, obtained a distance scale for galactic PN that sometimes is referred as the Seaton-Webster distance scale and that has been widely used by Cahn and Kaler (1971). Seaton made three additional assumptions that can affect the distance scale: a) MC PN have $T_e = 8000°K$, b) MC PN have $N(He)/N(H) = 0.16$, and c) that the maximum values of S, the surface brightness, and E, the total Hβ emission, will occur at the time in which PN become optically thin. From recent observations of MC PN it has been found that $\langle T_e \rangle = 12000°K$ and that $\langle N(He)/N(H) \rangle = 0.11$ each of these two effects amount to about 20% in the distance scale but they go in opposite directions and consequently cancel each other; with respect to c) if the number of ionizing photons increases with time E will become maximum when the nebula reaches the transition between optically thick and optically thin, alternatively S might reach its maximum before the PN becomes optically thin since S is proportional to $N_e^2 R$ and even if within the ionized zone R is increasing with time, due to the advance of the ionization front, N_e is diminishing with time due to the PN expansion. Since Seaton compared MC E values with galactic S values, from the previous argument it follows that his distance scale becomes a lower limit to the real distance scale (if the average masses of optically thin PN are the same in the MC and the Galaxy). That the maximum S values are attained by galactic PN before they become optically thin has been argued by Cudworth (1974) and Pottasch (1980). The distance scales by Cudworth (1974) and Weidemann (1977) are 1.5 and 1.3 times larger than the Seaton scale; moreover the estimated total number of PN in the Galaxy also support the larger distance scales (Jacoby 1980; Alloin et al. 1976). If the Cudworth distance scale is adopted, and if the comparison between the MC and galactic PN is valid, it follows that when E is maximum S is two times fainter than its maximum value.

PN have been used as standard candles to determine extragalactic distances. Jacoby and Lesser (1980) have been using the intrinsically luminous PN in M31 and the MC to establish distances to galaxies whose Local Group membership is uncertain. Moreover Ford and Jenner (1979, see also Ford 1983) have obtained a preliminary distance to M81 of 2 to 3 Mpc based on 8 PN.

V. CHEMICAL COMPOSITION

Chemical abundance determinations of PN in the MC have been obtained by: Osmer (1976), Webster (1976,1977,1978), Dufour and Killen (1977), Aller et al.(1981a), Maran et al.(1982), Aller and Czyzak (1983), Barlow et al. (1983), and Aller (1983). In Table 1 we present some of these determinations as well as those of other objects for comparison. The accuracy of the determinations are typically of: 0.04 dex for He/H, 0.1 dex for O/H, 0.1 dex for N/O and 0.2 dex for C/O at the one sigma level. With the possible exception of O the elements presented in Table 1 are those expected to be affected by stellar evolution of the progenitor star.

PN apparently are produced by stars in the 0.8 to 8 $M_\odot$ mass range (e.g., Peimbert 1978; Torres-Peimbert 1984). Peimbert (1978) based on chemical composition and kinematics divided galactic PN in four types which roughly correspond to the following mass intervals of the progenitor stars: Type I (He-N rich), 2.4-8 $M_\odot$; Type II (intermediate population), 1.2-2.4 $M_\odot$; Type III (high velocity), 1-1.2 $M_\odot$; and Type IV (halo), 0.8 to 1.0 $M_\odot$. Those PN whose progenitors have masses larger than 1.4 $M_\odot$ on the main sequence are expected to show abundances very similar to

TABLE 1
Chemical Abundances

Object	He/H[a]	O/H[a]	N/O[b]	C/O[b]	Source
⟨PN LMC⟩	11.07	8.3	−0.8	+0.7	1,2
⟨He-N⟩ LMC	11.23	8.0	+0.4	...	1,3,4
N25 LMC	...	8.3:	−1.8	...	5,6
N97 LMC	11.26	8.4	+0.2	−0.8	4
⟨H II⟩ LMC	10.92	8.34	−1.31	−0.48	7,8,9
⟨PN⟩ SMC	11.02	8.1	−0.8	+0.6	2,10
N67 SMC	11.27	7.7	−0.1	...	3
⟨H II⟩ SMC	10.89	7.89	−1.48	−0.89	8,9,11
PN 6822	11.27	8.1	+0.7	...	12
⟨H II⟩6822	10.92	8.3	−1.7	...	8
Type I PN	11.18	8.6	+0.0	−0.3+0.5	13
Type II PN	11.04	8.7	−0.6	−0.3+0.5	13
Orion	11.00	8.62	−0.97	−0.10	14,15
Sun	...	8.92	−0.93	−0.25	16

a. Given in log N(A)/N(B)+12; b. Given in log N(A)/N(B);
1. Aller 1983; 2. Maran et al. 1982; 3. Dufour and Killen 1977;
4. Barlow et al. 1983; 5. Webster 1977; 6. Webster 1978;
7. Peimbert and Torres-Peimbert 1974; 8. Lequeux et al. 1979;
9. Dufour et al. 1982; 10. Aller et al. 1981a; 11. Peimbert and Torres-Peimbert 1976; 12. Dufour and Talent 1980; 13. Torres-Peimbert 1984; 14. Peimbert and Torres-Peimbert 1977;
15. Torres-Peimbert et al. 1980; 16. Lambert 1978.

those of H II regions for elements not affected by their stellar evolution, like Ar and S; those PN with smaller mass progenitors should show underabundances in these elements reflecting the chemical evolution of the interstellar medium from which they formed. Results of investigations of MC PN indicate that the Ne, Ar and S abundances are similar to those of H II regions in their respective galaxy, indicating that their progenitor stars formed relatively recently. From the previous argument it follows that most of the observed objects in the MC seem to be of Types I and II with the possible exception of N25 in the LMC (Webster 1977, 1978), which is 45 arc seconds from the center of the red globular cluster NGC 1852, that might have a progenitor of smaller mass. N67, N97 and N102 are He-N rich objects corresponding to those of Type I.

Stecher et al. (1982) have obtained masses of ~ 1 $M_\odot$ and T* $\sim 1 \times 10^5$ °K for the central stars of P40 in the LMC and N2 and N5 in the SMC. The central star masses are higher than those of average PN in the Galaxy (Schoenberner and Weidemann 1983) which might be due to: a) errors in the distance determinations of galactic PN, b) severe selection effects towards brighter objects with more massive progenitors, and c) stellar evolution differences between MC and galactic PN. The masses of the progenitor stars in the main sequence are expected to be around 4 $M_\odot$ and would correspond to He-N rich objects, which they are not.

VI. CHEMICAL ENRICHMENT AND STELLAR EVOLUTION MODELS

Renzini and Voli (1981) (see also: Iben and Truran 1978; Becker and Iben 1979, 1980; Iben and Renzini 1982a,1982b,1983) have reviewed the evolution of the surface abundances of He, C, N and O for IMS, from the main sequence until the ejection of the PN envelope, or until C ignition in the core. Torres-Peimbert (1984) has compared these predictions with observations of galactic PN, the agreement in general is good but she notices two important differences: a) Type I PN show an anticorrelation between O and N probably indicating that some O has been converted into N, and b) C enrichment is more efficient than predicted and extends to masses as low as 0.8 $M_\odot$.

Since the initial abundances of He, C, N and O relative to H are considerably smaller for the progenitors of MC PN than for those of galactic PN (see Table 1), the expected relative enrichments for MC objects are higher, therefore the observations provide stronger constraints to the stellar evolution models.

In Figure 1 we show the relative enrichment of N and He for MC PN (Table 1 and references therein) compared with galactic PN (Torres-Peimbert 1984 and references therein). The MC values are higher than those of galactic PN for several reasons: a) their original He/H values are smaller, and higher relative enrichments are expected for a given mass (Renzini and Voli 1981), b) some Type II galactic PN in the anticenter direction started with smaller N/O and He/H values than those of the Orion Nebula, which invalidates the region with [N/O] $\lesssim$ 0.3 and [He/H] $\lesssim$ 0.4. The N enrichment for He-N rich PN is higher than expected under the assumption of secondary production of N from C (see also Table 1), indicating that most of the N in these objects is of primary

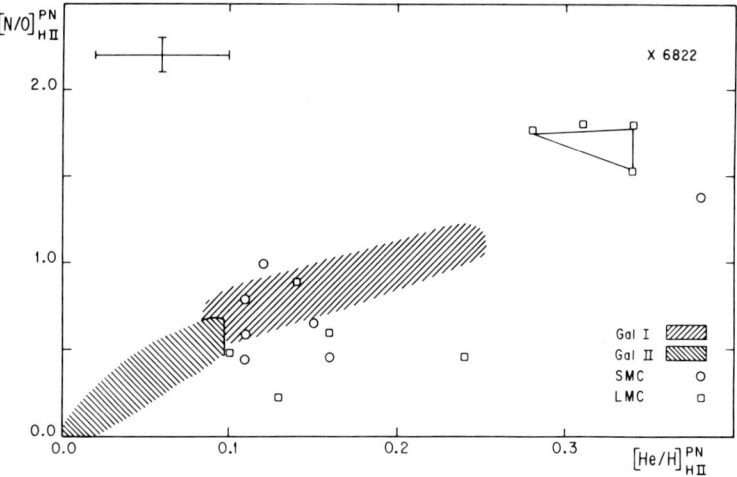

Fig. 1. Relative enrichment of N and He in PN where $[A/B]^{PN}_{HII} = \log(A/B)_{PN} - \log(A/B)_{HII}$. The shaded areas correspond to Type I and Type II galactic PN normalized to the Orion Nebula values. The PN in the local group galaxy NGC 6822 is included. The error bars in the upper left hand corner are typical of extragalactic determinations at the one sigma level. The three squares joined with straight lines correspond to N97 observed by Osmer (1976), Barlow et al. (1983), and Aller (1983).

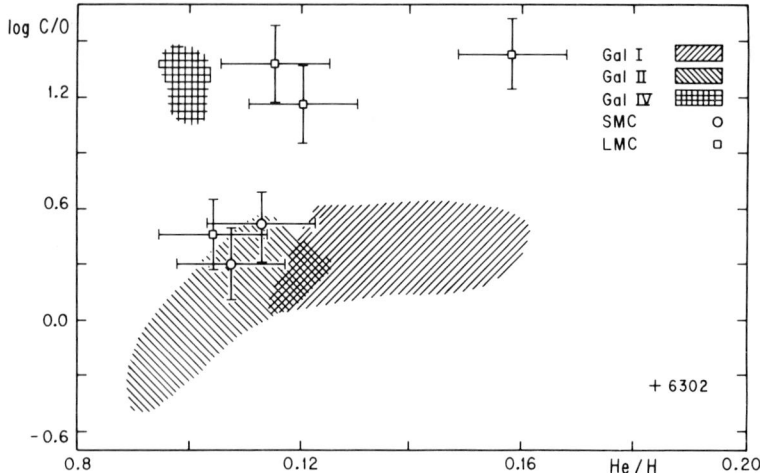

Fig. 2. $\log N(C)/N(O)$ versus $N(He)/N(H)$ values for MC and galactic PN.

origin, unless N is produced by secondary mechanisms from O instead of C, see below.

In Figure 2 the C/O versus He/H diagram is presented where the MC PN (Table 1 and references therein) are compared with galactic PN (Torres-Peimbert 1984 and references therein). Again the relative enrichment of C is considerably higher in the MC PN than in the galaxy, considering the underabundance of C present in the Clouds (see Table 1); this result indicates that indeed PN are producing C and again is in agreement with the predictions by Renzini and Voli (1981). The three values with C/O $\sim$ 1.0 were obtained from the $\lambda 4267$ C II recombination line and may correspond to upper limits due to probable overestimates of this very weak line, on the other hand the three values with log C/O $\sim$ 0.4 were derived from IUE data and may correspond to lower limits due to probable overestimates of T_e and to the possible effect of dust absorption on the $\lambda 1550$ C IV lines that was not considered. Under the assumption that C/O = C^{++}/O^{++} and from the 1909/1663 intensity ratios for NGC 6302 by Aller et al. (1981b) and Barral et al. (1982), it is obtained that log C/O = -0.35. These results are in agreement with models by Renzini and Voli (1981) with $\alpha \sim 2$ and $M_i \sim 8\ M_\odot$; Barlow et al. (1983) obtain for N97 similar abundances to those of NGC 6302 (see Table 1). These values support the results of Koester and Reimers (1981) and Reimers and Koester (1982) who found that white dwarfs do occur up to progenitor masses of $\sim 7\ M_\odot$.

In Figure 3 the relative enrichment of O versus N/O is presented for PN of Type I (Table 1 and references therein; Peimbert and Torres-Peimbert 1983). As it was noticed before (Aller 1983; Ford 1983; Peimbert and Torres-Peimbert 1983) there is a strong anticorrelation between [O/H] and N/O which implies that: a) there is a systematic effect in the abundance determinations, or b) the effect is real and there has been a substantial conversion of O into N. With the exception of the values by Barlow et al. (1983) for N97, all the other values have been obtained from optical data and the following equations:

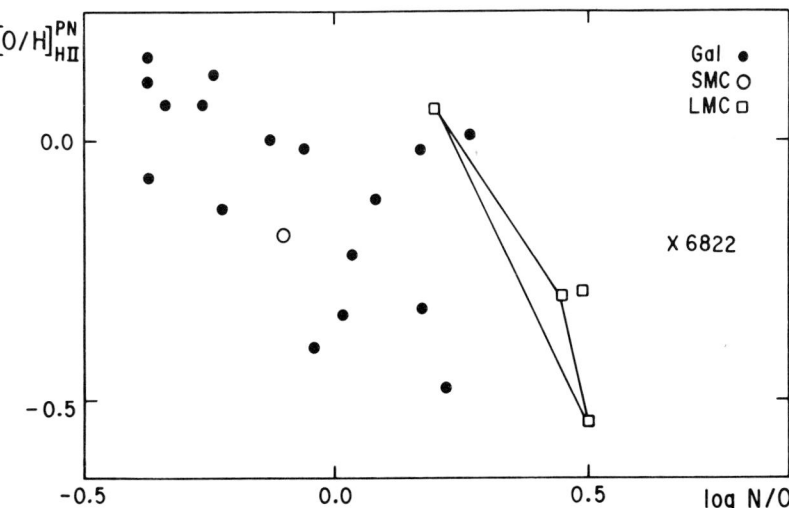

Fig. 3. Relative enrichment of O versus log N(N)/N(O), symbols as in Figure 1.

$$\frac{N(N)}{N(O)} = \frac{N(N^+)}{N(O^+)}, \qquad (2)$$

and

$$\frac{N(O)}{N(H)} = \frac{N(O^+) + N(O^{++})}{N(H^+)} \frac{N(He^+) + N(He^{++})}{N(He^+)}. \qquad (3)$$

Equation (2) is a valid approximation even for objects where $N(N^+)/N(N)$ ~ 0.01, like NGC 7662 for which Harrington et al. (1982) find that equation (2) is accurate within 10%. Equation (3), which is the crucial one, is in very good agreement with models for NGC 3918 by Torres-Peimbert et al. (1980) and models for NGC 7662 by Harrington et al. (1982). Alternatively Aller et al. (1981b) partially based on the O IV]1402 line derives for NGC 6302 an O/H value a factor of two higher than that given by equation (3). The O/H value by Barlow et al. (1983) for N97 is also partially based on the O IV]1402 line, but from the published results it is not possible to say if it is in disagreement with equation (3). Another possible source of systematic error is due to the $N(O^{++})/N(H^+)$ ratio which might be underestimated if there are regions of very high density contributing to $\lambda 4363$ and not to $\lambda 5007$ producing a spuriously high electron temperature.

If the O abundance determinations in Figure 3 are correct, the O depletion reaches factors of two to three in some objects which is not explained by present stellar evolution models where the O depletion is of at most a few percent (Renzini and Voli 1981). If there is no O depletion the excess N is mostly primary; alternatively if there is O depletion then the excess N is mostly secondary.

VII. GALACTIC CHEMICAL EVOLUTION

Tinsley (1978) on quite general grounds has shown that an overabundance X(PN)/X(HII) larger than a factor of seven would indicate that PN precursors are the main source of enrichment of the element considered. This estimate was made under the assumption of a shell mass of 0.3 $M_\odot$ for PN. Since the average mass of the progenitors in the main sequence is ~1.5 $M_\odot$ (Alloin et al. 1976) and the average mass of the PN central stars is ~ 0.6 $M_\odot$ (Shoenberner and Weidemann 1983) about 0.6 $M_\odot$ have been ejected prior to the PN formation, this material could already show some C and N overabundances thus lowering the estimate by Tinsley. From Figures 1 and 2 and Table 1 it follows that PN might be the main responsibles for C and N enrichment in the MC.

Based on earlier results Osmer (1976) reached the conclusion that PN are an inadequate source of N by a factor of ten; Dufour and Killen (1977) reached the conclusion that N enrichment of the Clouds interstellar medium is provided by sources other than PN; and Williams (1982) concluded that novae were the main contributors to the N enrichment in

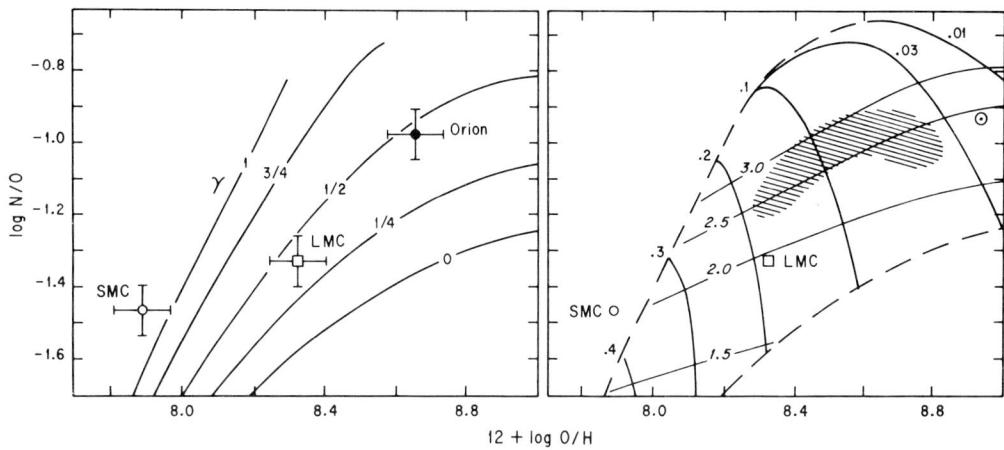

Fig. 4a. Galactic evolution models by Serrano and Peimbert (1983) where the following assumptions were made: a) secondary production of N, b) $\eta = 1.75$ a parameter proportional to the star formation rate, c) constant star formation rate, d) variable yield with $p_z = 0.002+0.6Z$. γ denotes the ratio of accretion to star formation rates.

Fig. 4b. Isochrones and lines of constant Mg/M_{tot} for the models in Figure 4a, the nearby horizontal lines are isochrones and are labelled by the value t/τ_N, where $\tau_N \sim 3 \times 10^9$ years. The nearby vertical lines are labelled by the value Mg/M_{tot}, which is constant along each time. The shaded area denotes galactic H II regions and the position of the sun is also reproduced in this diagram.

the Clouds, in this case most of the N would be of primary origin. Alternatively Aller et al. (1981a) concluded that the SMC PN contribute nearly enough N to enrich the ISM, but that massive stars may also be involved; moreover Serrano and Peimbert (1983) made models in which most of the N is of secondary origin and has been produced by IMS.

In Figure 4 models by Serrano and Peimbert (1983) are shown, these models are in agreement with the observed N/O versus O/H diagram of galactic and extragalactic H II regions. From this figure, it follows that in the SMC the star formation rate is almost equal to the accretion rate ($\gamma \sim 1$) and that in the LMC, γ, the ratio of star formation to accretion rates is about 1/2. The predicted M_{gas}/M_{tot} ratios are 0.17 and 0.38 for the LMC and the SMC respectively in good agreement with the observed values of 0.12 and 0.42 (Lequeux et al. 1979). From a $\tau_N \sim$ 3 Gyr it is obtained that the MC have an age of 5-6 Gyr in good agreement with the results of Cohen (1982) and Mould and Aaronson (1982).

Maran et al. (1982) by comparing the C abundances in PN to those of
HII regions (Table 1) concluded that most of the C enrichment in the MC
is due to PN. Serrano and Peimbert (1981) from chemical evolution models
of the solar neighborhood concluded that stars of $M \geq 10\ M_\odot$ contributed
with 0.25 and stars of $1 \leq M/M_\odot \leq 8$ (IMS) contributed with 0.28 to the
C/O ratio. The value of C/O = 0.13 for the SMC determined by Dufour
et al. (1982) implies that if the initial mass function is the same in
the SMC and the Galaxy, then stars with $M \geq 10\ M_\odot$ contribute at most with
0.13 to the C/O ratio in the solar neighborhood; consequently α, the
ratio of the convective mixing length to the pressure scale height, for
IMS has to be larger than 2. The C/O value in the solar neighborhood, the
C/O values in N97 and NGC 6302 and the very low C/O values in the MC all
indicate that α for the IMS has to be > 2.

Dufour and Shields (1982) have discussed several possible explanations for the very low C/O values in the MC. Support in favor of the
low C/O values comes from the shape of the interstellar extinction curve
where the 2200 A feature attributed to C also becomes fainter when
going from the Galaxy to the LMC and to the SMC (e.g., Seaton 1979;
Roca-Volmerange et al. 1981; Koornneef and Code 1981; Nandy et al. 1982).
It seems to me that a reasonable explanation for the low C/O values in
the MC is a delay in the C production, considering that the MC are
somewhat younger than the solar vicinity, which again indicates that
most of the C production is due to PN. The N/O behaviour indicates that
most of the C production takes place later than most of the N production,
which apparently is in contradiction with the idea that most of the N
is of secondary origin, this contradiction is not very strong due to
three reasons: a) at least part of the C comes from massive stars
(Arnett 1978), b) N/C is smaller than 1 even in the SMC, and c) part of
the N might be of secondary origin but produced by O instead of C.

It is a pleasure to acknowledge fruitful correspondence with L.H.
Aller and S. Maran as well as valuable discussions with J.F. Rayo,
A. Serrano and S. Torres-Peimbert.

REFERENCES

Aller, L.H.: 1983, Astrophys. J. 273, in press.
Aller, L.H., and Czyzak, S.J.: 1983, Proc. Natl. Acad. Sci. USA, 80,
    1764.
Aller, L.H., Keyes, C.D., Ross, J.E., and O'Mara, B.J.: 1981a, Monthly
    Notices Roy. Astron. Soc. 194, 613.
Aller, L.H., Keyes, C.D., Ross, J.E., and O'Mara, B.J.: 1981b, Monthly
    Notices Roy. Astron. Soc. 197, 95.
Alloin, D., Cruz-González, C., and Peimbert, M.: 1976, Astrophys. J.
    205, 74.
Arnett, W.D.: 1978, Astrophys. J. 219, 1008.
Barlow, M.J., Adams, S., Seaton, M.J., Willis, A.J., and Walker, A.R.:
    1983, in D.R. Flower (ed.), "Planetary Nebulae", IAU Symp. 103,

Dordrecht: Reidel, p. 538.
Barral, J.F., Cantó, J., Meaburn, J., and Walsh, J.R.: 1982, Monthly Notices Roy. Astron. Soc. 199, 817.
Becker, S.A. and Iben, I. Jr.: 1979, Astrophys. J. 232, 831.
Becker, S.A. and Iben, I. Jr.: 1980, Astrophys. J. 237, 111.
Cahn, J.H. and Kaler, J.B.: 1971, Astrophys. J. Suppl. 22, 319.
Cohen, J.G.: 1982, Astrophys. J. 258, 143.
Cudworth, K.M.: 1974, Astron. J. 79, 1384.
Dufour, R.J.: 1984, this book.
Dufour, R.J. and Killen, R.M.: 1977, Astrophys. J. 211, 68.
Dufour, R.J. and Shields, G.A.: 1982, in Y. Kondon, J.M. Mead and R.D. Chapman (eds.),"Advances in Ultraviolet Astronomy: Four years of IUE Research", NASA, p. 385.
Dufour, R.J., Shields, G.A. and Talbot, R.J. Jr.: 1982, Astrophys. J. 252, 461.
Dufour, R.J. and Talent, D.L.: 1980, Astrophys. J. 235, 22.
Feitzinger, J.V. and Weiss, G.: 1979, Astron. Astrophys. Suppl. 37, 575.
Freeman, K.C., Illingworth, G., and Oemler A. Jr.: 1983, preprint.
Ford, H.C.: 1983, in D.R. Flower (ed.),"Planetary Nebulae", IAU Symp. No. 103, Dordrecht: Reidel, p. 443.
Ford, H.C. and Jenner, D.C.: 1979, Bull. A. Astron. Soc., 10, No. 4, 665.
Harrington, J.P., Seaton, M.H., Adams, S., and Lutz, J.H.: 1982, Monthly Notices Roy. Astron. Soc. 199, 517.
Iben, I. Jr. and Renzini, A.: 1982a, Astrophys. J. (Letters) 259, L79.
Iben, I. Jr. and Renzini, A.: 1982b, Astrophys. J. (Letters) 263, L23.
Iben, I. Jr. and Renzini, A.:1983, Ann. Rev. Astron. Astrophys. in press.
Iben, I. Jr. and Truran, J.N.: 1978, Astrophys. J. 220, 980.
Jacoby, G.H.: 1980, Astrophys. J. Suppl. 42, 1.
Jacoby, G.H.: 1983, in D.R. Flower (ed.), "Planetary Nebulae", IAU Symp. No. 103, Dordrecht: Reidel, p. 427.
Jacoby, G.H. and Lesser, M.P.: 1981, Astron. J. 86, 185.
Koester, D. and Reimers, D.: 1981, Astron. Astrophys. 99, L8.
Koornneef, J. and Code, A.D.: 1981, Astrophys. J. 247, 860.
Lambert, D.L.: 1978, Monthly Notices Roy. Astron. Soc. 182, 249.
Lequeux, J., Peimbert, M., Rayo, J.F., Serrano, A., and Torres-Peimbert, S.: 1979, Astron. Astrophys. 80, 155.
Maran, S.P., Aller, L.H., Gull, T.R., and Stecher, T.P.:1982, Astrophys. J. 253, L43.
Mould, J. and Aaronson, M.: 1982, Astrophys. J. 263, 629.
Nandy, K., McLachlan, A., Thompson, G.I., Morgan, D.H., Willis, A.J., Wilson, R., Gondhalekar, P.M., and Houziaux, L.: 1982, Monthly Notices Roy. Astron. Soc. 201, 1.
Osmer, P.S.: 1976, Astrophys. J. 203, 352.
Peimbert, M.: 1978, in Y. Terzian (ed.), "Planetary Nebulae: Observations and Theory", IAU Symp. No. 76, Dordrecht: Reidel, p. 215.
Peimbert, M. and Torres-Peimbert, S.: 1974, Astrophys. J. 193, 327.
Peimbert, M. and Torres-Peimbert, S.: 1976, Astrophys. J. 203, 581.
Peimbert, M. and Torres-Peimbert, S.: 1977, Monthly Notices Roy. Astron. Soc. 179, 217.
Peimbert, M. and Torres-Peimbert, S.: 1983, in D.R. Flower (ed.), "Planetary Nebulae", IAU Symp. No. 103, Dordrecht: Reidel, p. 233.

Pottasch, S.R.: 1980, Astron. Astrophys. 89, 336.
Reimers, D. and Koester, D.: 1981, Astron. Astrophys. 99, L8.
Renzini, A. and Voli, M.: 1981, Astron. Astrophys. 94, 175.
Rocca-Volmerange, B., Prévot, L., Ferlet, R., Lequeux, J. and Prévot-Burnichon, M.L.: 1981, Astron. Astrophys. 99, L5.
Schoenberner, D. and Weidemann, V.: 1983, in D.R. Flower (ed.), "Planetary Nebulae", IAU Symp. 103, Dordrecht: Reidel, p. 359.
Seaton, M.J.: 1968, Astrophys. Letters, 2, 55.
Seaton, M.J.: 1979, Monthly Notices Roy. Astron. Soc. 187, 73p.
Serrano, A. and Peimbert, M.: 1981, Rev. Mexicana Astron. Astrof. 6,41.
Serrano, A. and Peimbert, M.: 1983, Rev. Mexicana Astron. Astrof. 8, in press.
Smith, M.G. and Weedman, D.W.: 1972, Astrophys. J. 177, 595.
Stecher, T.P., Maran, S.P., Gull, T.R., Aller, L.H., and Savedoff, M.P.: 1982, Astrophys. J. 262, L41.
Torres-Peimbert, S.: 1984, in C. Chiosi (ed.), "Stellar Nucleosynthesis", Dordrecht: Reidel, in press.
Torres-Peimbert, S. and Peimbert, M.: 1977, Rev. Mexicana Astron. Astrof. 2, 181.
Torres-Peimbert, S., Peimbert, M., and Daltabuit, E.: 1980, Astrophys. J. 238, 133.
Webster, B.L.: 1976, Monthly Notices Roy. Astron. Soc. 174, 513.
Webster, B.L.: 1977, Publ. Astron. Soc. Pac. 88, 669.
Webster, B.L.: 1978, in Y. Terzian (ed.), "Planetary Nebulae", IAU Symp. No. 76, Dordrecht: Reidel, p. 11.
Weidemann, V.: 1977, Astron. Astrophys. 61, L27.

DISCUSSION

Graham: Just a remark. The distribution of the planetaries as shown in Sanduleak's survey looks roughly similar to that of the novae. Jacoby's survey for faint planetary nebulae was made only in the Bar region of the LMC. His results do not necessarily imply that they concentrate towards the Bar.
Feast: A. Walker (S.A.A.O.) has doubled the number of planetaries with measured radial velocities in each cloud. A preliminary result of his is that the expanded sample in the SMC now shows no division into two velocity groups such as appeared in the earlier work (although the values of previous individual velocities are confirmed).
Peimbert: From the near encounter between the SMC and the LMC about 200 million years ago (see both Fujimoto and Mathewson in these proceedings) this result might indicate that the large majority (if not all) of the progenitors of the observed PN have masses smaller than 4 $M_\odot$.
Bessel: N abundances derived for a good sample of galactic G-K dwarfs by three groups (Mt Stromlo, Yale and Texas) show no variation of N/C with Fe/H or O/H. This indicates that N is mostly of primary origin at least over two decades in $[Fe/H]$.

CORONAE OF THE MAGELLANIC CLOUDS

Klaas S. de Boer
Astronomisches Institut Tübingen, D-7400 Tübingen, F.R.Germany

A hot gaseous corona was postulated by Spitzer (1956) for the Milky Way as a consequence of the detection of cool interstellar clouds (seen in CaII absorption) on paths to stars in the Milky Way halo. Halos around extra galactic systems were proposed by Bahcall and Spitzer (1969) as a possible explanation for the wealth of high redshift absorption line systems in the spectra of quasars. The recording with the International Ultraviolet Explorer (Boggess et al 1978) in 1978 of echelle spectra of R136 and HD38282 in the LMC showed strong absorption lines of the high ionization stage ions CIV and SiIV, due to material with velocities clearly pertaining to locations outside the Milky Way disk (Savage and de Boer 1979). Thus the reality of the galactic corona became established (review de Boer 1984). There is confusion with the words corona (gas - massive component) and halo (location - stars) and I (1984) proposed for the large mass from dynamics MASsive DArk Component, MASDAC.

From far-UV echelle spectra of stars in the Magellanic Clouds (MC), de Boer and Savage (1980; =dBS80) derived that the pattern of absorption at MC velocities by both neutral and highly ionized gas was similar to that of the Milky Way. This is based on the assumption that gas between us and a MC star has radial velocities correlated with distance from us. It suggested that the MC's may have coronae as well.

1. Observational Aspects

Ultraviolet absorption lines such as CII and MgII have large optical depths and are very sensitive probes for interstellar gas, much more than the CaII, the NaI or 21-cm HI lines. References to UV data papers can be found in Table 1. Observations of visual interstellar absorption lines to MC stars show absorption over widely differing velocity ranges, depending on the direction looked in and the instrumental sensitivity. Blades and Meaburn (1980) detected an extended blue wing in the CaII K profile seen toward R136. Songaila and York (1980) and Songaila (1981) observed MC stars to study the CaII K and NaI D absorption in the MCs. The Magellanic stream was probed by observations of extragalactic objects such as Fairall 9 (see York et al. 1982).

The low ion lines seen in LMC star spectra show a considerable range in absorption velocity, in particular up to 100 km/s less than the radial velocity of the stars (dBS80). Gas velocities more positive than the stellar ones are limited to 30 km/s. For the SMC stars the minimum absorption velocity cannot be determined because of blending by absorption of Milky Way gas, the positive limit is about 70 km/s. One cloud at 300 km/s LSR was found (dBS80; Fitzpatrick and Savage 1983).

The CIV lines pose an interpretational problem. The OB stars may produce some interstellar CIV ions in their immediate vicinity, and that absorption cannot be separated very well from absorption occurring on other portions of the line of sight. In Table 1 data are collected for the CIV absorption due to interstellar material in the MCs and, for comparison, data to Milky Way stars. The strength of CIV seen in the MCs is larger than that in Milky Way gas. Into directions with disturbed material (R136 in the LMC; maybe HD5980 in the SMC; e.g. the Cygnus Loop in the Milky Way) CIV lines are exceedingly strong. However, none of the here listed MC WR stars has a detectable ring nebula (Chu and Lasker 1980). The consistently large strength of the MC CIV lines suggests there is more CIV than from the HII regions or shells alone.

Detailed studies of UV absorption lines, including information from local nebular emission lines, were carried out for R136 by de Boer, Koornneef and Savage (1980), for HD36402 by de Boer and Nash (1982) and for HD5980 by Fitzpatrick and Savage (1983). The R136 region is clearly a-typical for the LMC and has no relevance here. The association to which HD36402 belongs would hardly be able to support a large amount of CIV, if it behaves like a normal (Dyson and de Vries 1972; Weaver et al 1977) interstellar bubble. In that case most of the detected CIV has to be outside the stellar environment. If the CIV absorption seen near the velocity of the nebular lines would be local to the star indeed, there remains otherwise unaccounted for CIV absorption near 220 km/s, a quite different velocity from that of the HD36402 velocity, thus requiring a different explanation. Toward HD5980 the CIV situation is rather confused due to severe velocity blending.

Table 1. Equivalent Widths for Interstellar CIV Lines in MC Stars and Comparable Milky Way Stars

MC star		Sp. Type	V(star)	V(CIV)	W(1550)	Ref	MW star	Sp. Type	W(1550)	Ref
Sk-67 18	--	O6-7+WN5	272	240:	200:	dBS	HD153919	O6f	90	BKM
Sk-67 5	HD268605	O9.7Ib	294	250:	>150:	dBS	HD213087	B0.5Ib	65	BDHR
Sk-67 104	HD 36402	WC5+OB	315	270	200	dBN	HD113904	WC6+O9BOI	80:	BKM
Sk-69 246	HD 38282	WN6	245:	210	230	dBS	HD192163	WN6	320	SWW,1
Sk-71 45	HD269676	O4-5III	229	220	270	GWMN	HD 46223	O4Vf	100	BDHR
Sk-69 243	HD 38268	R 136	245:	215	600	dBS	HD 37022	O6+B0.5V	260	FS,2
Sk 108	R31	O6.5+WN3	129	150	300	dBS	HD 93403	O6f+O7.5	150	BKM
Sk 80	---	O7Iaf+	---	150	280	dBS	HD 57060	O7f	30	BKM
Sk 78	HD 5980	OB?+WN3	---	140	410	dBS	HD190918	WN5+O9.5III	470	BKM,1
Sk 159	---	BOIa	---	170	60	P9	HD152667	BOIa	75	BKM

Velocities in km/s LSR. Equivalent widths in mÅ
Sk   = Sanduleak (LMC 1970, SMC 1968)
dBS = dBS80 and other papers (see de Boer and Savage 1983)
dBN = de Boer and Nash (1982)
GWMN= Gondhalekar et al (1980)
P9   = Prévot et al (1980); Fitzpatrick (1984)

BDHR= Black et al (1980)
BKM = Bruhweiler et al (1980)
FS = Franco and Savage (1982)
SWW = Smith et al (1980)
1: Star in direction of Cygnus SNR
2: V(HI)=-10 km/s, V(CIV)=-30 km/s

Interstellar NV, which essentially cannot be produced by stellar ionization, is seen in absorption in the direction of both HD36402 (de Boer and Nash) and HD5980 (Fitzpatrick and Savage). NV has been detected thusfar at Milky Way velocities only in the spectrum of HD5980. It is not impossible that X rays produce some NV, either directly influencing the temperature structure of the medium or by the Auger process. But most likely the NV exists outside the stellar environment and then indicates the presence of a region of coronal gas around each MC.

Doubt on the reality of the MC coronae proposed by dBS80 was cast by Prévot et al (1980) from the observational point of view. They had collected with the IUE one echelle spectrum of a star in the SMC and found that absorption due to CIV was weak, if present at all. They suggested that dBS80 had underestimated the contribution from the stellar vicinity to the CIV absorption present in the dBS80 spectra. Indeed, the CIV line to this star is weak (see Fitzpatrick, this symposium), but the HII region problem is not clear cut, and it was marked as a potential problem by dBS80.

Feitzinger and Schmidt-Kaler (1982a) compared LMC absorption line velocities available from published UV spectra with Feitzinger's (1980) dynamical model, which is based on all radial velocities available for the LMC. The model results in, among others, a rotation curve which is branched, and Feitzinger argues that there is gas outside the LMC disk, most likely as a warp (see also McGee and Milton 1964). Feitzinger and Schmidt-Kaler then state that all absorption seen in the UV spectra is due to gas either in the disk or in this warp. I think there is a problem with that suggestion. The absorption in the large optical depth lines of low ions occurrs between 200 and 300 km/s for the entire LMC (dBS80), hardly showing effects of rotation. In part the UV absorption velocities coincide (at Feitzingers $r<0$) with those from the rotation curve, but in other parts (at $r>0$) the absorption extends to velocities quite different from the rotation curve. The branching of the rotation curve is solely based on the 21-cm data of McGee and Milton (1964, 1966). The velocities producing the branching are from gas in the region south of 30Dor, the data points are projected onto the assumed major axis, and thus have little to do with lines of sight studied by dBS80 and in particular de Boer and Nash. For the SMC Feitzinger and Schmidt-Kaler (1982b) indicate that the absorption components detected coincide with those from 21-cm emission found by Hindman (1967) as well. Again, the UV absorption extends way outside the 21-cm velocities, thus indicating additional absorption from other locations. The CIV lines, hardly mentioned by Feitzinger and Schmidt-Kaler, at first sight follow the low ion lines. However, near the approaching side of the LMC the CIV absorbs at the velocity of the rotation, at the receeding side it absorbs at lesser velocities.

A relevant new development is that both observations (de Boer and Savage 1983) and model calculations for the rotation of the Milky Way halo (Feitzinger and Kreitschmann 1982) show that outside the plane of the Milky Way rotational velocities are much lower than in the disk (see also de Boer 1983). Although the rotation of the LMC cannot be compared directly with that of the Milky Way, one might expect a similar effect in the LMC. At the receeding (NW) side of the LMC one so would expect gas absorption at velocities less positive, at the approaching (SE) side gas absorption at velocities more positive than those of the main rotation curve. That is just what the UV absorption line data show, when compared to the rotation curve of Feitzinger and Schmidt-Kaler. In particular the CIV velocities are of the slow halo rotation type.

Quite extreme velocities were detected in HI 21-cm by McGee, Newton and Morton (1983). Following the call by Savage and de Boer (1981) radio measurements sensitive to very small HI column densities were carried out at Parkes. These show that at almost all positions neutral gas components are seen at a smallest velocity of about 175 km/s and a largest velocity of about 345 km/s LSR. Note that the smallest velocities usually were seen in (UV) absorption as well (see in particular the CII absorption to HD269357 displayed in Savage and de Boer 1981), whereas the largest 21-cm velocities were not seen in absorption. The lower velocity gas indeed is at the near side of the LMC. These data indicate that there is an envelope of hardly rotating HI gas (halo) around the LMC, in which is embedded a rotating "(HI) disk".

Summarizing the observational aspects: the absorption lines of both high and low ionization interstellar ions in the MC star spectra are strong, cover a large range in velocity, mostly at more negative velocities than the stars. New HI data reveal 180 km/s wide velocity ranges of emission from almost all positions of the LMC. The CIV lines are strong, on average stronger than those seen in Milky Way interstellar gas. The bubble theory predicts some CIV, but much less than detected. Absorption by CIV gas has velocities staying behind the main body rotation. The detection of NV is strong evidence for gas of an unusual nature around the MCs.

2. Theoretical Considerations

Following the recognition that much of the volume of interstellar space in the Milky Way is at temperatures of the order of a million K, Shapiro and Field (1976) speculated that stationary conditions may exist with convection of the hot phase of the matrix into the halo regions of the Milky Way. The full balance of heating, mass flux into the halo, and cooling was studied by Cox (1981). The Milky Way indeed may be just capable of maintaining a fountain with a mass flow of > 5 M☉ per year.

Whether or not a galaxy can maintain a fountain, and thus possibly possesses a hot gaseous corona, depends on the heating and cooling rate of the gas (Cox 1981). The heating is due to supernova (SN) explosions; other energy sources being only of minor importance. The total energy released in the ISM per unit volume is $h = 2.E_{sn}/V$ in erg cm^{-3} s^{-1}, with E_{sn} the energy per SN and V the volume in which this energy is dumped. The cooling is $l = L(T).n(T)^2$ in erg cm^{-3} s^{-1}, where $L(T)$ is the cooling function which is metallicity dependent, and $n(T)$ is the gas density at temperature T. A further requirement is that the ISM is porous, i.e. the hot matrix gas must have a filling factor q, such that the matrix stays hot between successive SN explosions (Cox and Smith 1974). Elaborating an earlier excercise (de Boer 1982) an attempt will be made here to compare the conditions in the LMC and SMC with those in the Milky Way to see if gaseous coronae can exist around the MCs.

The values of the relevant parameters for the Milky Way (MW) are set as follows. The average SN energy E= 10(+51) erg, and with a SN rate of about 1 per 30 years Esn= 10(+42) erg s^{-1}. The volume is taken as a cylinder V= $2\pi R^2$ H with R= 15kpc and scale height H= 100pc (Cox). The heating then is h(MW)= 2 10(+40) erg kpc^{-3} s^{-1}. For the cooling function the curve of Cox may be used. The gas density is derived from the pressure in the disk of 2 10(+4) K cm^{-3}, to be 2 cm^{-3} at 10(+4) K (the intercloud medium). The required porosity q=1 is fulfilled by the Milky Way due to its dependence on the SN influence radius and the disk thickness, q= Rsn/H (Cox).

For the Magellanic Clouds not all these parameters can be determined easily. The energy per SN may be taken the same as that for MW supernovae. The SN rate in the LMC is about 1 per 200 y (Long et al 1981), the rate for the SMC is about one per 1000 y from the fact that the number of SMC SNR is only about 10 (Seward and Mitchell 1981) compared to about 50 for the LMC (Long et al). So Esn(LMC)= 1.6 10(+41) and Esn(SMC)= 3 10(+40) erg s^{-1} (if the average energy per SN is the same as in the MW). The gas volumes and densities into which this energy goes are hard to determine. From the LMC SNR, Long et al estimate gas densities of 0.3 < n(H) < 3.1 cm^{-3}, and near HD36402 gas densities are 0.4 cm^{-3} in front and 1.6 cm^{-3} in the rear (de Boer and Nash). This suggests a typical n(H)= 1 cm^{-3}, at about 10(+4)K. The disk thickness is uncertain. Since the MCs are smaller but less compact than the MW, a value of 100pc will be adopted. Support for this value may be found from the detection of a pulsar in the LMC (McCulloch et al 1983). Its dispersion measure is 125 cm^{-3} pc which can be reduced to 80 cm^{-3} pc interior to the LMC. The pulsar is near the direction to HD36402 and using the n(e)=0.5 cm^{-3} found there the electron disk would be 150pc thick. For the Milky Way the electron layer is thicker than the HI layer, and thus the H= 100pc for the LMC seems reasonable. Since the entire LMC has SNR's, the entire volume has to be considered, V= 6kpc^3 based on R= 3kpc. For the SMC the radius is about 1.5kpc, and with equally H= 100pc, V= 1.4 kpc^3. Thus h(LMC)= 5 10(+40) and h(SMC)= 5 10(+40) erg kpc^{-3} s^{-1}. Considering now cooling, the metallicity of the LMC is about 0.5 that of the MW, for the SMC it is about 0.2 of the MW (see Lequeux et al 1979), reducing the L(T) by the same factor. For the LMC the pressure is about 10(+4) K cm^{-3}, thus a density of about half that of the MW. For the SMC H= 100pc was assumed and with generally double the column density N(H), n(SMC) may be double that of the LMC.

The heating can now be compared for the three systems. From above: h(LMC)= 2.5 h(MW) . (100/H) and h(SMC)= 2.5 h(MW) . (100/H), where the factor 100/H allows to adjust for any better value of the scale height H for each MC. These heating rates are, apart from the value of H, fairly accurate since they derive from reasonably well observed quantities. They suggest that the LMC and SMC have essentially equal heating capacity, but larger than that of the MW. The cooling is less well determined as: $l(LMC) = 0.5\ l(MW) \cdot (n(H,LMC)/n(H,MW))^2$ and $l(SMC) = 0.2\ l(MW) \cdot (n(H,SMC)/n(H,MW))^2$, where $n(MC)/n(MW)$ shows how the cooling depends on the gas density in each MC compared to the MW. Using the HD36402 area densities from above, $l(LMC) = 0.12\ l(MW)$ and from above $l(SMC) = 0.2\ l(MW)$. This seems to suggest that the LMC may cool less efficient than the SMC, and both less than the MW. Note, however, that the SMC may have too small a porosity, due to the low SN rate, to maintain matrix gas. But the densities are very uncertain and do not really allow any definite derivations.

Summarizing the theoretical considerations: With the larger rate at which the interstellar medium is heated in the LMC and the SMC, than in the MW, the vigour of the LMC and SMC fountains may be larger than that of the MW. The cooling rates are very poorly determined, basically because of very uncertain mean gas densities in the LMC and SMC; the lower metallicities all by themselves point to a more likely existence of coronae around the MC's than around the MW.

3. Summary and Outlook

Recent observations of interstellar absorption lines in the ultraviolet have given the long expected support for the old proposal that the Milky Way possesses a hot gaseous corona. Related observations in the Magellanic Clouds have shown that it is very likely that these smaller galaxies possess such coronae as well, but the proof is complicated due to blending of absorption from the regions in the immediate vicinity of the MC stars. An extended envelope containing neutral gas (a cool halo) follows from the 21-cm measurements. Theoretical models for the Milky Way interstellar medium including fountain-type flow have been developed only recently and are - of course, but also fortunately - in agreement with what is seen in the MW. The parameters for the MC's needed for similar models are insufficiently known, in particular the density of the gas in the MC's. No firm conclusions for the existence of coronae can be drawn from the energy balance calculations. The high heating rate (compared to the MW) from the known SN rate may point independently to relatively recent extensive star formation. Observing facilities with good photon sensitivity, with spectral resolution of better than 10 km/s, and capable of measuring wavelengths as small as 1000Å are needed. They will allow the detection of ion stages ranging from neutrals such as NaI, over dominant stages such as CII and SiII, to the high temperature ions SiIV, CIV, NV, and OVI, and thus help unravel the structure of the interstellar medium in and around the Magellanic Clouds.

References

Bahcall, J.N., Spitzer, L. 1969, Ap.J. **156**, L 63
Black, J.H., Dupree, A.K., Hartmann, L.W., Raymond, J.C. 1980, Ap.J. **239**, 502

Blades, J.C., Meaburn, J. 1980, M.N.R.A.S. **190**, 59p
Boggess, A., et 33 altera. 1978, Nature **275**, 377
Bruhweiler, F.C., Kondo, Y., McCluskey, G.E. 1980, Ap.J. **237**, 19
Chu, Y.-H., Lasker, B.M. 1980, P.A.S.P. **92**, 730
Cox, D.P. 1981, Ap.J. **245**, 534
Cox, D.P., Smith, B.W. 1974, Ap.J. **189**, L 105
de Boer, K.S. 1982, in "La Structure du Petit Nuage de Magellan", Comptes Rendus **4**, Strasbourg, p 50
de Boer, K.S. 1983, in "Highlights of Astronomy", 6, Ed. R.M. West, Reidel, p 657
de Boer, K.S. 1984, in "The Milky Way Galaxy", IAU Symp. **106**, Ed. H. van Woerden, W.B. Burton, R.J. Allen; Reidel, in press
de Boer, K.S., Koornneef, J., Savage, B.D. 1980, Ap.J. **236**, 769
de Boer, K.S., Nash, A.G. 1982, Ap.J. **255**, 447 (and **261**, 747)
de Boer, K.S., Savage, B.D. 1980, Ap.J. **238**, 86 (**dBS80**)
de Boer, K.S., Savage, B.D. 1983, Ap.J. **265**, 210
Dyson, J.E., de Vries, J. 1972, A.Ap. **20**, 223
Feitzinger, J.V. 1980, Space Science Rev. **27**, 35
Feitzinger, J.V., Kreitschmann, J. 1982, A.Ap. **111**, 255
Feitzinger, J.V., Schmidt-Kaler, T. 1982a, Ap.J. **257**, 587
Feitzinger, J.V., Schmidt-Kaler, T. 1982b, in "La Structure du Petit Nuage de Magellan", Comptes Rendus **4**, Strasbourg, p 45
Fitzpatrick, E.L. 1984, this Symposium, p. 407.
Fitzpatrick, E.L., Savage, B.D. 1983, Ap.J. **267**, 93
Franco, J., Savage, B.D. 1982, Ap.J. **255**, 541
Gondhalekar, P.M., Willis, A.J., Morgan, D.H., Nandy, K. 1980, M.N.R.A.S. **193**, 875
Hindman, J.V. 1967, Australian J. Phys. **20**, 147
Lequeux, J., Peimbert, M., Rayo, J.F., Serrano, A., Torres-Peimbert, S. 1979, A.Ap. **80**, 155
Long, K.S., Helfand, D.J., Grabelsky, D.A. 1981, Ap.J. **248**, 925
McCulloch, P.M., Hamilton, P.A., Ables, J.G., Hunt, A.J. 1983, Nature **303**, 307
McGee, R.X., Milton, J.A. 1964, in "The Galaxy and the Magellanic Clouds", IAU Symp. **20**, p 289; Ed. F.J. Kerr, A.W. Rodgers; Canberra
McGee, R.X., Milton, J.A. 1966, Australian J. Phys. **19**, 343
McGee, R.X., Newton, L.M., Morton, D.C. 1983, M.N.R.A.S. in press
Prévot, L., et 9 altera. 1980, A.Ap. **90**, L 13
Sanduleak, N. 1968, A.J. **73**, 246
Sanduleak, N. 1970, C.T.I.O. Contribution no 89
Savage, B.D., de Boer, K.S. 1979, Ap.J. **230**, L 77
Savage, B.D., de Boer, K.S. 1981, Ap.J. **243**, 460
Seward, F.D., Mitchell, M. 1981, Ap.J. **243**, 736
Shapiro, P.R., Field, G.B. 1976, Ap.J. **205**, 762
Smith, L.J., Willis, A.J., Wilson, R. 1980, M.N.R.A.S. **191**, 339
Songaila, A. 1981, Ap.J. **248**, 945
Songaila, A., York, D.G. 1980, Ap.J. **242**, 976
Spitzer, L. 1956, Ap.J. **124**, 20
Weaver, R., McCray, R., Castor, J., Shapiro, P., Moore, R. 1977, Ap.J. **218**, 377
York, D.G., Blades, J.C., Cowie, L.L., Morton, D.C., Songaila, A., Wu, C.-C. 1982, Ap.J. **255**, 467

DISCUSSION

Shull: What are the densities and temperatures in the corona?
de Boer: From the observations one can guess densities after assuming some velocity-distance relation for the gas in the corona. We did so for the Milky Way corona (Savage and de Boer 1979, 1981) but the results are highly uncertain. Temperatures were inferred from comparison with ionization models; the 10^5 K stated there followed when collisional ionization was assumed, but it is not clear at all that the gas is in that state (for a discussion of the Milky Way corona see de Boer 1983b). For the Magellanic Clouds nothing is known about the depth of the halo-corona and any density or temperature would be a wild guess.
Feitzinger: The rotation curve of the LMC derived from the new HI survey (Rohlfs, Kreitschmann, and Feitzinger, this Symposium) shows identical features to that of the rotation curve derived from the McGee and Milton data. Again, the velocity space covered by the HI and the absorption lines is the same.
de Boer: That is indeed so. Yet, from the 21-cm data one cannot infer which fraction of the gas is (which velocities come from) behind the stars. The UV data show that the interstellar absorption takes place at velocities mostly smaller than those of the stars. In particular the velocity of the CIV absorption deviates from the main-body rotation curve.
Mathewson: The stars you observed are amongst the hottest in the Magellanic Clouds so that the absorption lines you observe may be produced in their immediate vicinity. For example HD5980 in the SMC is in a nebulosity which is an X-ray source.
de Boer: That is correct and these very concerns were very carefully phrased by us (dBS80). HD5980 indeed is the prime case for such difficulties. Two important aspects of the data, recently more fully discussed by Fitzpatrick and Savage (1983) and by Fitzpatrick (1983 preprint; partly represented at this symposium) are: 1) the ratio of the column densities N(CIV)/N(SiIV) found in the Magellanic Clouds is very similar to the one of the Milky Way corona; and 2) the detection of CIV in Sk159, away from any nebulosity, with a column density small as expected from the lower metallicity of the SMC. Toward other stars there is of course the detection of NV, as discussed in the text.
Gondhalekar: Did you determine any abundances for the gas?
de Boer: For the Milky Way corona gas abundances are a bit lower than solar. For the Magellanic Cloud corona nothing could be derived. For completeness I mention, since you ask, that for the neutral gas in front of R136 some reasonable limiting numbers for Si and Fe could be derived, while the oxygen abundance was found to be a factor 1.7 below solar (de Boer and Nash, Table 2). The latter was derived from the 1356Å OI intercombination line seen in the sum of 5 IUE spectra available at Washburn Observatory at that time. The value agrees closely with the oxygen abundance from HII regions.

AGE DETERMINATION OF HII REGIONS OF THE LMC AND SMC

M.V.Copetti[†], H.A.Dottori, E.L.Bica[†], M.G.Pastoriza
Departamento de Astronomia, Instituto de Física,
UFRGS and CNPq, BRAZIL
[†]CNPq Fellows

ABSTRACT

HII region models were constructed which take into account: 1º) A burst for the formation of the ionizing association; 2º) Different Salpeter's initial mass function ($1 \leq \chi \leq 3$) and upper stellar mass limit ($30 \leq M_u/M_\odot \leq 120$); 3º) Models of stellar evolution with and without mass loss (Maeder, 1980, Hellings et al. 1981).

From these models the temporal evolution of the H_β emission line equivalent width (W_{H_β}), the ratio of the forbidden lines 4959, 5007 [OIII] to H_β, and of the He^+, H^+ zones volume ratio (R) was obtained. It was found that W_{H_β}, [OIII]/H_β and R decrease as a function of the time and consequently they are good age indicators. Some of the models appropriate for the LMC are shown in figures 1.

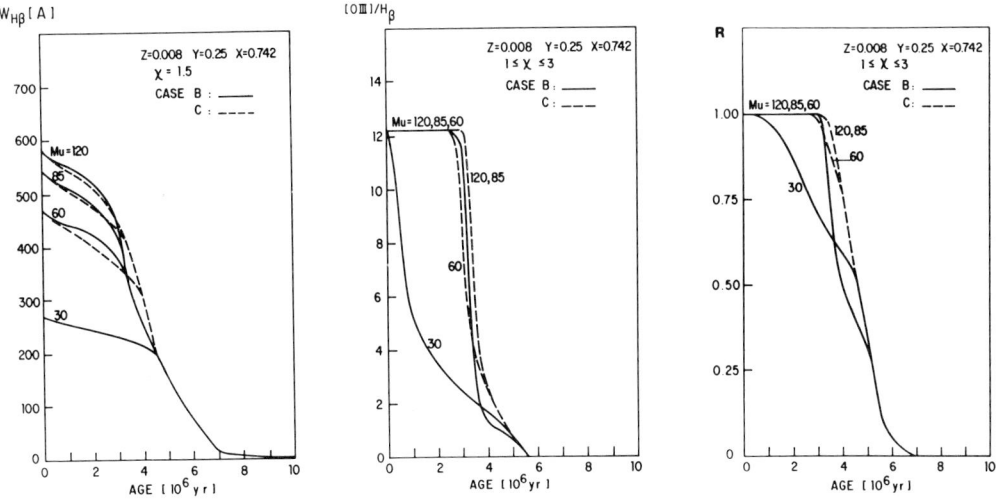

Figures 1: Evolution of W_{H_β}, [OIII]/H_β and R as a function of the time.

Measurements of $W_{H\beta}$ for 29 regions of the LMC (Dottori and Bica, 1981) were analized using the models. Models with $x = 3$ can be disregarded because they are incompatible with $W_{H\beta}$ higher than 90 Å, which are common in the LMC. For $x = 2.5$ the maximum $W_{H\beta}$ is around 200 Å and is not consistent with the values found in 30 Dor. and NGC 2032. Values of $1 \leq x \leq 2$ appear to be compatible with data from all regions and the choice of x within the interval does not significantly affect the scale of ages. For $W_{H\beta} \leq 200$ Å (26 of the regions) the M_u also do not influence the age determination. In figure 2 we plot the histogram of ages for $x = 1.5$ and $M_u = 60\ M_\odot$, which indicates a burst of star-formation with highest activity about 6.0 to 6.5×10^6 years ago, and with a duration of 1.5 to 2.0×10^6 years, measured at half maximum.

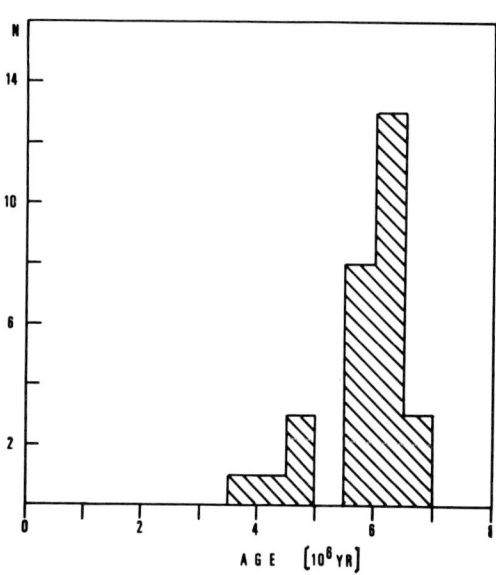

Figure 2: Histogram of formation of HII regions in the LMC.

Ratios [OIII]/H_β and R were obtained for 11 HII regions of the LMC from data of Dufour (1975). The mean age obtained through these two parameters are very similar (The limits are 2.2 and 4.0×10^6 years assuming $M_u = 30\ M_\odot$ and $60\ M_\odot$ respectively). These ages are smaller than those obtained from $W_{H\beta}$, which may be related to the fact that the two samples have a few objects in common.

For the SMC, data on [OIII]/H_β for 12 regions (Dufour and Harlow, 1977), lead to a mean limit of ages of 1.4 and 3.0×10^6 years for $M_u = 40\ M_\odot$ and $100\ M_\odot$ respectively. We emphasize that the dispersion around these ages is very small, suggesting that all the regions were formed simultaneously.

REFERENCES
Dottori, H.A.; Bica, E.L.D., 1981, Astron. Astrophys. 102, 245.
Dufour, R.J., 1975, Astrophys. J. 195, 315.
Dufour, R.J.; Harlow, W.V., 1977, Astrophys. J. 216, 706.
Hellings, P.; Vanbeveren, D., 1981, Astron. Astrophys. 95, 14.
Maeder, A., 1980, Astron. Astrophys. 92, 101.

THE EFFECT OF MASS-LOSS ON THE EVOLUTION OF HII REGIONS IN THE L.M.C.

I.R.G. Wilson[1],[2] and M. A. Dopita.[2]

1) Royal Greenwich Observatory, Herstmonceux Castle, Hailsham East Sussex, United Kingdom BN27 1RP

2) Mount Stromlo and Siding Spring Observatories, Private Bag Woden P.O. Canberra Australia 2606.

ABSTRACT

Empirical calibrations based on prior studies of Galactic OB Stars are used to determine the integrated stellar wind mechanical luminosity (L_W) and the integrated Lyman continuum photon luminosity ($S*$) for 10 OB clusters in the L.M.C. These values of L_W and $S*$, together with narrow band H-α surface photometry of an overlapping sample of 15 L.M.C. HII regions, are used to show that the large shell-like HII regions, in the L.M.C. are stellar wind bubbles 3 to 5 million years old. In order to reproduce the general properties of these HII shells, a thick cushion of shocked stellar wind gas must have been present on the inside of the shell for most of the life time of the nebula, and the shell itself must be ram pressure confined by the HI/molecular cloud out of which it formed.

THE TWO STELLAR-WIND MODELS

The stellar-winds of young OB associations sweep up a shell in the surrounding I.S.M., the properties of which are dependent on the mode of interaction of the wind with the shell. The two extreme possibilities are:

1) that there is a reverse-shock set up in the stellar-wind creating a pad of hot shocked gas between the free-flowing stellar-wind and the shell (Hot cushion model, HC; Weaver et. al. 1977).

2) that the wind impinges directly onto the shell due to a collapse of the hot pad of shocked gas early in the lifetime of the nebula (Direct-impingement model, DI; Steigman et. al. 1975).

The three major characteristics which can be used to distinguish the two models are:

1) the dependence of the nebula radius (R_{SH}) upon the stellar-wind parameters, stellar age and I.S.M. density ie:

$$R_{SH}(p.c.) = 427.9 \ L_{29}^{1/5} \ n_{II}^{-1/5} \ t_6^{3/5} \qquad \text{HC Model (1a)}$$

$$R_{SH}(p.c.) = 402.5 \ M_{23}^{1/4} \ n_{II}^{-1/4} \ t_6^{1/2} \qquad \text{DI Model (1b)}$$

where L_{29} (10^{29} Joules), M_{23} (10^{23} Newtons), n_{II} (m^{-3}) and t_6 (10^6 years) are the mechanical luminosity of the stellar wind, the wind momentum, the density of the surrounding media and the nebula age, respectively.

2) whether or not the shell nebula is ionization-bounded. It is possible for the forming shell to become sufficiently massive to absorb all of the ionizing photons from the central cluster. The conditions for this to occur prior to the shell stalling is given by:

$$n_{II} \ L_{29}^3 \ S_{48}^{-2} \geqslant 9.05 \times 10^4 \qquad \text{HC Model (2a)}$$

$$n_{II} \ M_{23} \ S_{48} \geqslant 1.07 \times 10^{10} \qquad \text{DI Model (2b)}$$

where S_{48} (10^{48} photons/sec) is the number of ionizing photons emitted by the OB cluster.

Taking typical values of L_{29} (13.0), S_{48} (40.0) and M_{23} (5.0) (Wilson 1983), we can see that the photons are cut off prior to stalling if, $n_{II} \geqslant 1.2 \times 10^6$ for the HC model, and $n_{II} \geqslant 1.4 \times 10^{11}$ for the DI model. This means that the HC will produce an ionization-limited shell-like HII region, in all except low density media, while the DI model will only produce this type of nebula if the surrounding media is exceptionally dense. In almost all cases the DI model will produce a HII shell stalled within a larger HII region.

3) different densities in the shells due to different pressures being applied to the inner surface of the shell ie.

$$n_{HC} = 2.80 \times 10^{10} \ L_{29} t_6 R_{SH}^{-3} \qquad \text{HC Model (3a)}$$

$$n_{DI} = 6.03 \times 10^7 \ M_{23} R_{SH}^{-2} \qquad \text{DI Model (3b)}$$

OBSERVATIONS

The M.K. spectral types of 25 OB stars in 10 clusters associated with the L.M.C. HII regions were determined (Wilson 1983). These classifications were used to derive Teff, L_*, R_* and M_* for each of the stars, which in turn were used to determine the stellar mass-loss rates ($\dot{M}$) via an empirical calibration relating $\dot{M}$ to the stellar parameters (Wilson and Dopita 1983). These rates were then integrated to obtain L_{29} and M_{23} for each of the clusters, assuming $V_\infty = 3 V_{esc}$ (Abbott 1978). The integrated S_{48} for each cluster was also obtained from an empirical expression relating S_{48} to the spectral type of the star (Morton 1969, Georgelin et. al. 1975).

15 narrow band H-α images of L.M.C. HII regions, taken with an S.E.C. Vidicon T.V. photometer, were used to determine S_{48} for each nebula. The nebula surface brightnesses were also used to determine the r.m.s. densities of the HII regions through the adoption of simple geometric models.

DISCUSSION AND CONCLUSIONS

Figure 1a shows that the observed integrated H-α flux emitted by the HII regions (and hence the Lyman photon emission rate S_{48}) is equal to or larger than the integrated H-α flux (S_{48}) expected to be produced by the ionizing flux of the exciting stars. This is confirmed by a larger sample of data plotted in figure 1b, which shows that there is a general relationship between an HII region's integrated H-α flux and its diameter. Figure 1c shows a similar plot for the integrated H-α flux expected to be produced by the ionizing cluster. Comparing figures 1b and 1c we see that almost all of the HII regions appear to absorb a greater number of ionizing photons (0.2-0.4 in $\log(S_{48})$) than that expected from the ionizing cluster. This difference can be explained by the decreased line blanketing in the L.M.C. due to reduced metalicity, which makes a star of a given spectral type appear bluer and more luminous, and hence increases S_{48}. In order to produce agreement between the expected and observed S_{48} (ie for the HII region to be marginally ionization bounded) the Lyman continuum spectral distribution of L.M.C. OB stars would have to mimic that of a star with a T_{eff} 2-4,000 K hotter than that of a corresponding star in the Galaxy. This is the largest increase in U.V. colour temperature we could reasonably expect, thus figures 1a-1c show conclusively that the L.M.C. HII regions observed are in general ionization bounded. This conclusion, coupled with the fact that all the larger nebulae show a distinct shell morphology, strongly favours the HC model.

Table 1 shows the observed r.m.s. densities for the HII shells (column 6) which can be compared with mean shell densities predicted the HC and DI models (columns 4 and 5 respectively). The observed densities also strongly support the HC model.

The HC model can successfully reproduce the observed shell morphologies, the fact that the nebulae are ionization bounded and the shell densities. Taking this model as being applicable to the nebulae, we can substitute t_6, R_{SH} and L_{29} into equation 1a to determine the I.S.M. densities. The densities obtained (column 7) are much higher than what is typically expected in the I.S.M., though they compare favourably with the densities expected in the outerlying parts of massive gas clouds (~ $10^5 M_\odot$) which are in the latter stages of isothermal collapse (column 8; Wilson 1983).

TABLE 1

NEBULA	R(pc)	t_6	n_{HC}	n_{DI}	n_{OB}	n_{II}	n_C
301	53.0	3.6	8.7	0.10	5.8	13.4	20.6
25	43.5	3.3	6.0	0.08	4.6	13.2	17.8
137	88.5	–		0.06	2.4	4.7	5.7
229	66.1	3.4	9.5	0.72	6.2	10.1	12.9
196	47.5	3.2	7.8	0.12	6.0	14.8	18.1
106	53.0	5.6	9.6	0.12	–	15.6	54.9
11	24.4	2.8	28.9	0.20	–	58.9	194.8
226	31.0	3.0	21.9	0.22	–	40.4	105.0
235	39.0	3.8	7.3	0.08	–	18.5	35.5
31	71.5	2.4	4.8	0.12	–	4.2	2.7

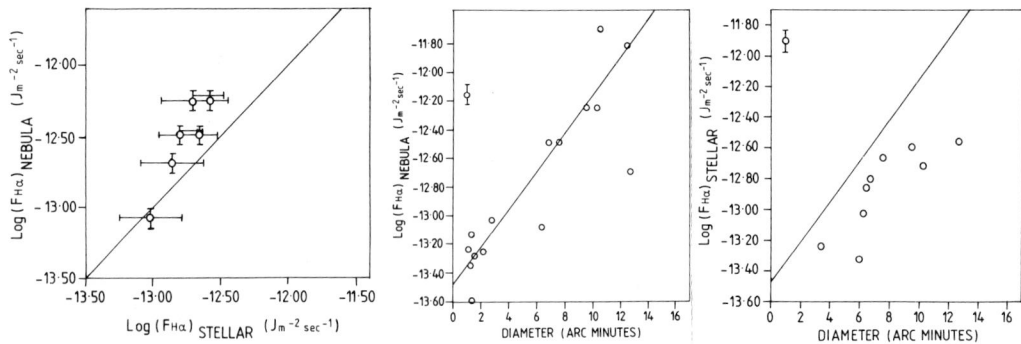

Figure 1a: The integrated dereddened H-α flux ($F_{H\alpha}$) for 6 nebulae compared to the $F_{H\alpha}$ produced by the exciting cluster.
Figures 1b and 1c: $F_{H\alpha}$ plotted against the nebula diameter.

Thus large shell-like HII regions in the L.M.C. are stellar-wind bubbles, 3-5 million years old, which have been formed by a thick pad of shocked stellar wind gas and are ram-pressure confined by the infalling gas of the massive cloud out of which they formed.

REFERENCES

Abbott, D. C. 1978. Astrophys. J. 225, 893.
Georgelin, Y. M., Lortet-Zuckermann, M. C., Monnet, G. 1975, Astron. Astrophys. 42, 273.
Morton, D. C. 1969, Astrophys. J. 158, 629.
Steigman, G., Strittmatter, P. A., Williams, R. E. 1975, Astrophys. J., 198, 575.
Weaver, R., McCray, R., Castor, J.I., Shapiro, P., Moore, R. 1977. Astrophys. J., 218, 377.
Wilson, I. R. G. 1983, Thesis, Australian National University.
Wilson, I. R. G., Dopita, M. A. 1983, in preparation.

ABSOLUTE Hα AND Hβ PHOTOMETRY OF LMC HII REGIONS

J. Caplan and L. Deharveng
Observatoire de Marseille

A nebular photometer equipped with a scanning Fabry-Perot interferometer has been used at the ESO 50-cm telescope to observe about 50 HII regions of the LMC (Caplan and Deharveng, 1983). This instrumentation gives the profiles of the observed lines, thus allowing discrimination against unwanted continuum and line emission. The integrated fluxes were measured with a 4.9' diaphragm, to facilitate comparison with the 6-cm radio continuum fluxes obtained by McGee et al. (1972) with a similar resolution. Nebular reddening, from the Hα/Hβ ratio, is greatest for the 30 Doradus Nebula. Relatively high reddening was also observed for the following zones: N159 (at the border of the largest HI-molecular complex of the LMC) and N160; N79, N81 and N83; N48; N64. Reddening is rather low towards other HII regions, especially for those in the bar and for large bubble-shaped regions.

In addition to our measured Hα and Hβ fluxes, we have assembled other data for these same HII regions. Radio continuum fluxes at 6 cm are from McGee et al. (1972); HI column densities are from McGee and Milton (1964, 1966); B-V colour excesses of associated stars have been supplied by N. Martin, L. Prévot, E. Rebeirot and J. Rousseau (private communication). Among the possible comparisons are the following:

1. The nebular extinction as measured by the S(6 cm)/Hβ ratio is (as is usually found to be the case) higher than that estimated from the Hα/Hβ reddening using a standard interstellar extinction law. This fact suggests the presence either of dust inside the HII regions or of clumpy external dust.

2. Assuming reasonable HI/dust ratios (Koornneef, 1982), the HI column densities are too low to account for the extinction, even if one assumes that all of the neutral hydrogen observed in the direction of an HII region lies in front of the region. This effect could be caused by extinction due to dust associated with molecular material or within the HII region. Most likely, however, the HI column density is underestimated because of line saturation and/or the smearing effect of the radio lobe. In any case, it is certainly dangerous to use HI column densities

to estimate the extinction of an HII region and of its associated cluster.

3. The colour excesses of the stars observed in the direction of or near the HII regions are too low to account for the nebular extinction. This may be due to a selection effect in the stellar observations. We think that the amount of <u>interstellar</u> extinction is generally correctly estimated from the colour excesses of the outlying stars, but that additional reddening in the Balmer lines results from internal dust. The amount of internal dust needed to explain this additional reddening is much greater than in the interstellar case, and corresponds to an A_V of several magnitudes. The most highly reddened stars inside each region would not be observed, and this would lead to an underestimate of the extinction. Because of this effect, stellar colour excesses should be used with caution when estimating the extinction of HII regions.

4. The colour excesses of the outlying stars in each HII region allow us, however, to correct the nebular extinction and reddening measurements (see § 1 above) for the effects of the <u>interstellar</u> dust. The residual nebular extinction tends to be large for the rather small amount of nebular reddening. This can be explained by clumpy external dust or, as discussed in § 3, by internal dust. Points plotted in a nebular extinction-vs-reddening diagram are consistent with an $H\alpha$ albedo smaller than that for $H\beta$, according to model calculations of Mathis (1983) and of P. Cox (private communication) for HII regions with internal dust.

A catalogue-atlas of our observations is in preparation.

Similar observations have been undertaken for about 25 HII regions of the SMC.

REFERENCES

Caplan,J., Deharveng,L. 1983, The Messenger No. 32,3
Koornneef,J. 1982, Astron. Astrophys. 107,247
Mathis,J.S. 1983, Astrophys. J. 267,119
McGee,R.X., Milton,J.A. 1964, IAU Symp. No. 20, The Galaxy and the Magellanic Clouds, eds. F.J. Kerr and A.W. Rodgers, 289
McGee,R.X., Milton,J.A. 1966, Aust. J. Phys. 19,343
McGee,R.X., Brooks,J.W., Batchelor,R.A. 1972, Aust. J. Phys. 25,581

KINEMATICS OF THE HII REGION N11 IN THE LMC

A. Hänel
Observatorium Hoher List der Universitätssternwarte Bonn
D-5568 Daun, Fed. Rep. Germany

INTRODUCTION

Four mechanisms have been proposed to explain the structures and large scale mass motions of HII regions: the Champagne model, the 'classical' picture of an expanding Strömgren sphere, interaction of stellar winds and supernova blast waves with the surrounding medium. Especially the last three models have been used to explain the LMC shell structures. Braunsfurth et al. (1983) showed that it is not possible to distinguish the driving mechanisms for individual regions from their global appearance (dimension). Detailed studies of individual shells however, have been limited to the largest and brightest (30 Dor) or the more symmetric ones (e.g. N70, rings around WR stars).

The N11 complex (DEM 34, MC 18) is the most massive LMC HII region, has the largest excitation parameter, highest Hα and FUV flux besides 30 Dor (Braunsfurth et al., 1983, Israel, 1980). Meaburn (1978) described the shell like structure around the OB association LH 9, while Henize (1956) catalogued several bright HII condensations, which dominate this complex. Only the brightest of these (B) has been studied in detail to identify the excitation sources (Heydari-Malayari et al., 1983).

OBSERVATIONS

A Fabry-Perot etalon (interference order=1482, free spectral range =4.4 Å= 200 km/s) has been used with a focal reducer at the 1m telescope of the European Southern Observatory to study the radial velocities of the Hα line. 440 velocity points have been measured on 6 plates using a modified Abbe comparator. From these, mean values have been derived over a grid of 1'=16 pc side length, centered on the bright WC5 star HD 32228 (R64, FD6, Br9...). The internal error is about 3 km/s for well exposed ring segments, while it can be considerably higher for weak ring segments. The 1☐' mean values were used to derive a velocity field which was checked with the individual velocity points to position contour lines more exactly. This was however possible only for the brighter regions, where the number and accuracy of individual data points is large enough.

To study the exciting sources of the nebulae, CMDs were derived from published photographic magnitudes and colours (Woolley, 1963, Lucke, 1972). After applying a correction of 0.2mag to Woolley's V and B-V data, the values of both authors agree very well for LH 9. The stars of the associations LH 10 and LH 13 which are embedded in the nebulae N11 B and C respectively, are too red to fit the main sequence. If this is not due to observational difficulties (nebular background, multiple stars), reddening of about $E(B-V)=0.5$mag and absorption of $A_V=3 \cdot E(B-V)=1.5$mag must be assumed to fit these stars to the main sequence. With the distance modulus of 18.6 mag and $E(B-V)=0.1$mag for LH 9 a $(B-V)_0-M_V$ diagram can be constructed. Some of the bluest and brightest stars are either compact HII regions or multiple stars or have smaller reddening values. The high absorption would allow to identify most of the exciting sources of the nebulae. Fitting isochrones to the data yields an age of about 4.5 for LH 9 and about 3.5 million years for the clusters embedded in the nebulae. This age sequence is supported by the ages derived for the HII regions (Dottori et al., 1981), though their ages are older.

DISCUSSION

The velocity fields of N11B and C vary from about 285km/s at the southwest edge to about 295(B) and 290(C) km/s at the NW edge. These two condensations are at the SW border of a large (0.5x1kpc) HI cloud with a radial velocity of 293km/s. A geometric model could explain the two HII condensations as cavities (blisters) at the edge of the HI cloud with flows of ionized gas into the direction of the observer.

The region around the association LH 9 appears even on long exposure plates devoid of Hα emission, while the surrounding nebulae show an ellipsoidal ring structure of 70x100pc inner diameter. Dominant object of this association is something like a compact cluster around the WC5 star HD 32228. The galactic counterparts of these stars have high mass loss rates with wind powers of about $1.5 \cdot 10^{38}$erg/s. Interaction of this wind with the neutral cloud (density 2.5 cm^{-3}) would form a bubble of the observed dimensions after about 700 000 years. Due to the faint hydrogen emission around the association, radial velocities are difficult to measure. On one interferogram however, split lines are marginally visible, indicating an expansion velocity of about 40 ± 20 km/s, which could also be explained by a wind-driven bubble.

Sequential star formation appears attractive to explain the age difference and the spatial arrangement of the N11 complex: first LH 9 has been formed and the stellar wind and ionization-shock front expanding into the HI cloud might have triggered star formation in N11B and C.

REFERENCES:
Braunsfurth,E., Feitzinger,J.V.: 1983, Astron. Astrophys., in press
Dottori,H.A., Bica,E.L.D.: 1981, Astron. Astrophys. 102, 245
Henize,K.G.: 1956, Astrophys. J. Suppl. Ser. 2, 315
Heydari-Malayari,M., Testor,G.: 1983, Astron. Astrophys. 118, 116
Israel,F.P.: 1980, Astron. Astrophys. 90, 246
Lucke,P.: 1972, Thesis, University of Washington
Meaburn,J.: 1978, Astrophys. Space Sci. 59, 193
Woolley,R.v.d.R.: 1963, Roy. Obs. Bull. Ser. E, 66, 263

THE MULTIPLE-PHASE STRUCTURE OF THE INTERSTELLAR MEDIUM IN THE LMC

Peter Shull Jr.
Max-Planck-Institut für Astronomie

Physical properties of the ISM near SNRs in the LMC can be deduced from their very high-resolution, two-dimensional spectra. The ISM around the N49 and N63A SNRs apparently has as many as three different phases of varying density, clumpiness, and spatial distribution.

1. OBSERVATIONS AND RESULTS

The N49 and N63A SNRs were observed with the 4 m echelle spectrograph at Cerro Tololo. The optical emission lines of H I, [O III], [N II] and [S II] were imaged. Angular resolution was 1".9, and the night-sky line was 10 km s^{-1} wide (HWHM). Further details concerning this and following sections are in Shull (1983) and references therein.

Each emission "line" consists of a sharp spike (at rest in the LMC reference frame) flanked by spectral features called narrow bands and broad bands. N49's features indicate spherically symmetric expansion, while those in N63A show only asymmetrical blueshifting. Characteristic velocity widths, surface brightnesses, and velocity shifts (relative to the spikes) of these spectral features are summarized in Table 1.

TABLE 1.   Properties of Spikes and Bands in N49 and N63A.

SNR	Feature	Δv (km s^{-1}, HWHM)	v shift (km s^{-1})	Surf. Brightness
N49	Spike	7-16	0	...
"	Narrow Band	30-60	100-140	Low
"	Broad Band	80-130	20-70	High
N63A	Spike	9-17	0	...
"	Narrow Band	30-50	20-50	High
"	Broad Band	70-110	100-140	Low

2. INTERPRETATION OF THE RESULTS

N49 and N63A both have round X-ray disks about 1' in diameter (Mathewson et al. 1983). Optically, N63A is much smaller than N49 (which is almost coextensive with its X-ray disk). Since both SNRs are in the "radiabatic" phase, simply meaning that radiative clouds and the adiabatic blast wave simultaneously exist, the first inference is that radiating clouds are less plentiful in N63A than in N49.

The broad and narrow bands are produced by shocked gas, while the spikes are produced by photoionized gas. Therefore it is reasonable to identify the velocity shifts of the bands with the bulk outward motion of gas clouds accelerated by the high-speed SNR blast wave.

Clouds are accelerated by momentum transfer and the pressure change across the shock front. Equations by Cox (1979) indicate that the resulting velocity increase is proportional to (matrix density / cloud density) as measured before the collision, to the blast velocity at impact, and to a geometrical factor. Using measured expansion velocities, it is therefore possible to (1) estimate the density ratio between the intercloud matrix and a given cloud species and (2) to show that the density ratio between various cloud species is inversely proportional to their expansion velocities.

Thus, the second inference for N49 and N63A is that the density ratios for the band and matrix gas are about 100:30:1. For N49, the mean preshock density as determined from X-ray measurements is 1 cm^{-3}, which means that the two cloud populations have mean densities of ~ 30 and ~ 100 cm^{-3}. Interestingly, preshock cloud densities that Dopita et al. (1977) infer from line intensities cluster near 30 and 100 cm^{-3}. Uncertainties in these ratios due to variations in blast wave impact speed, finite cloud acceleration times, velocity projection effects, and cloud deceleration through mass accumulation are small.

Finally, Table 1 shows that surface brightness and velocity width of the bands are correlated differently in N49 than in N63A. Let us assume that cloud brightness is proportional to mean density ρ, and that shocks propagate within clumpy clouds so as to conserve kinetic energy. The third inference is then that the internal rms density fluctuations vary as $\rho^{1.5}$ or more in N49, and less strongly than this in N63A.

The author acknowledges grants NASA NAS 09-15940 and NSF AST 80-07540 to Rice University and the grant of CTIO observing time.

REFERENCES

Cox, D.P.: 1979, Ap.J., 234, p. 863.
Dopita, M.A., Mathewson, D.S., and Ford, V.L.: 1977, Ap.J., 214, p.179.
Mathewson, D.S., Ford, V.L., Dopita, M.A., Tuohy, I.R., Long, K.S., and Helfand, D.J.: 1983, Ap.J.Suppl., 51, p. 345.
Shull, P., Jr.: 1983, Ap.J., 275, Dec. 15, in press.

A NEW 21-CM LINE SURVEY OF THE LMC

K. Rohlfs, J. Kreitschmann and J.V. Feitzinger
Astronomisches Institut der Ruhr-Universität,
FRG - 4630 Bochum 1

The measurements were made in Feb. 1982 with the Parkes 64 m telescope using a corrugated waveguide horn with total half-power beam width of 15', the first sidelobes being 19 dB down, resulting in an aperture efficiency $\eta_A = 0.53 \pm 0.007$, a main beam efficiency of $\eta_{mb} = 0.80 \pm 0.005$ and a ratio of source flux to antenna temperature of $\Gamma = 0.62 \pm 0.1$ K/Jy (Murray, priv. comm.). A cooled two channel FET frontend used in the frequency switching mode with $\Delta\nu = 2$ MHz resulted in a system noise temperature at zenith of $T_{syst} = 40$ K for one channel and $T_{syst} = 50$ K for the other. Each frontend channel received a single polarization mode, and this radiation was then further analysed in a 2 x 512 channel autocorrelation spectrometer set at a channel separation of 3.906 KHz corresponding to a velocity resolution of V = 0.824 km s^{-1}. Hanning smoothed this resulted in a $\sigma_T = 0.05$ K for the average of both polarization.
A field of $-2°.8 \leq X \leq 3°.4$, $-2°.6 \leq Y \leq 4°.0$ in the standard tangential coordinates of the LMC (Isserstedt 1975) corresponding to roughly $4^h54^m < \alpha < 6^h00^m$, $-72°12' < \delta < -65°48'$ have been sampled with a grid spacing of $0°.2$ both in α and in δ. Some results of this survey are shown in Fig. 1 to Fig. 4. Fig. 1 gives the distribution of N_{HI}, while Fig. 2 shows the velocity field in the gas. All radial velocities in this diagram have been referred to the galactic center using a rotational velocity for the LSR of 225 km s^{-1}. The position angle for the major axis of the system as shown by both the velocity field and by the distribution of the gas in the individual channel maps is 28° in contrast to 168° as given by Feitzinger (1980). A transversal velocity can change the kinematical value at most by 10° - 15°.
Many of the profiles are double peaked or strongly asymmetric, Fig. 3 shows the distribution of $V_{mean} - V_{mode}$, where V_{mean} is the average radial velocity of a line profile, while V_{mode} is the velocity for the largest peak. $V_{mean} - V_{mode}$ is clearly oriented to the geometry of the system.
Fig. 4 finally shows the velocity field (V_{mean}) along the major axis. In the central region a velocity perturbation probably caused by the bar is clearly visible.

References:
Feitzinger, J.V., 1980, Space Sci Rev. <u>27</u>, 35
Isserstedt, J., 1975, Astron. Astrophys. <u>41</u>, 21

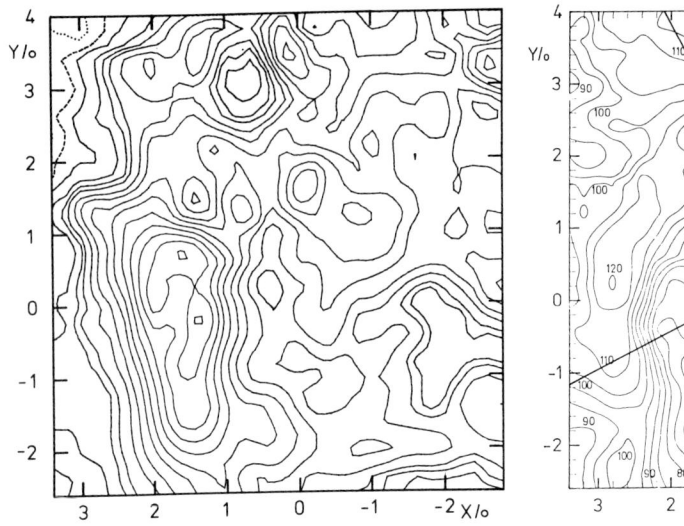

Fig. 1 The distribution of total neutral hydrogen over the LMC. Contour levels at .3, .5, 1, 2, 3, 4, 6, 8, 10, 12, 15, 20, 30 × 10^{21} atoms/cm^2.

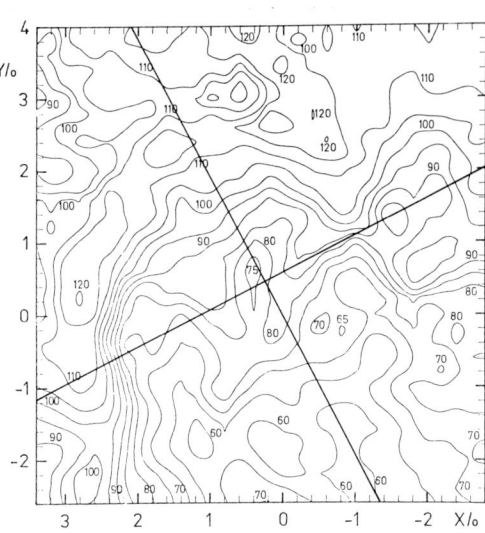

Fig. 2 Iso-velocity contours of LMC. The major axis at $\Phi = 28°$ and the minor axis at $\Phi = 118°$ (NESW) are indicated.

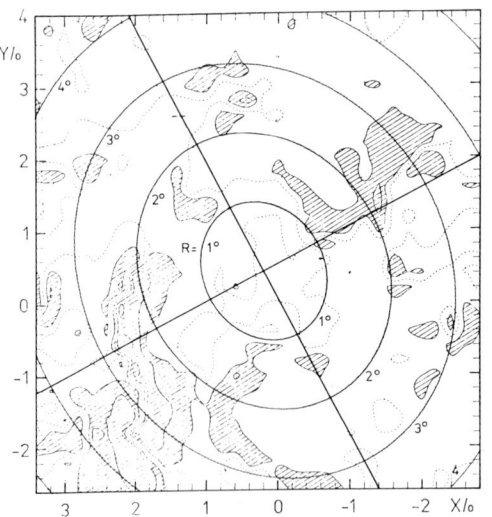

Fig. 3 Colour line of equal $V_{mean} - V_{mode}$ and circles of r = const. from the kinematical center in the plane of the system ($i = 33°$).

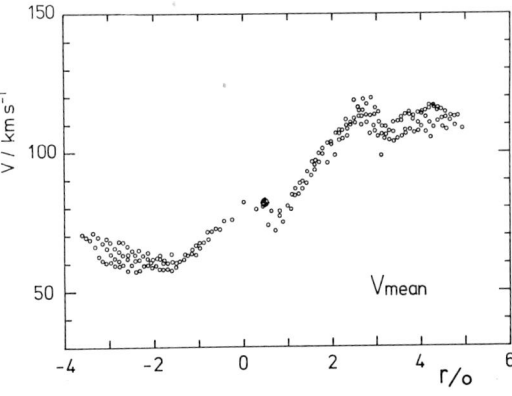

Fig. 4 The velocity field along the major axis, the kinematical center is marked.

MOLECULAR LINE OBSERVATIONS IN THE LARGE MAGELLANIC CLOUD

F.F. Gardner
Division of Radiophysics, CSIRO, Sydney, Australia

This is a preliminary report on on-going observing programs in the LMC using CSIRO facilities. Molecular-line observations made with the 64-m Parkes telescope include OH at 18 cm (Whiteoak & Gardner 1976b; Caswell & Haynes 1981), H_2CO at 6 cm (Whiteoak & Gardner 1976a) CH at 9 cm (Whiteoak et al. 1980) and H_2O at 1.3 cm with beamwidths between 13' and 1'.7 arc; observations of HCO^+ at 3.4 mm (Batchelor et al. 1981) and CO at 2.6 mm have been made with the 4-m Epping telescope.

The CO survey with the 4-m telescope is still preliminary, and this note will only discuss results for positions where other molecules have been detected. At the 4 H_2O maser positions, N105A, 30 Dor, N160A and N159 (Whiteoak et al. 1983; Scalise & Braz 1981), CO was detected with corrected antenna temperature values (T_a^*) between 0.6 and 1.5 K and with widths of 5.5-8.0 km s^{-1}. It was not found (<0.25 K) at a dark cloud position 4'.5 NW. of N160A, where Israel et al. (1982) reported 1.3 mm CO emission with T_a^* = 1.1 K. CO, H_2O and radio recombination velocities for the HII regions (the peaks of which were within ~1' of the maser positions) were in close agreement. For N159 observations made with positional offsets of ±1'.5 in right ascension and declination showed that the cloud size was 2'.1±1'.0 and the corrected peak intensity ~4 K.

In his review talk 'Molecules and Dust in the Magellanic Clouds', F.P. Israel demonstrated the generally low intensity of CO emission from the Clouds as compared to the Galaxy and suggested that this might result from the observed low dust/gas ratio, which is probably a consequence of the lower metal abundances of the Clouds. For the other molecules the information available is more limited. H_2O masers have been found in 4 and OH masers in 2 of the sources, N105A and N159; the other lines have been found in 1 source only, N159. The comparison of the masers in the LMC and the Galaxy is limited by the small numbers and selection effects. Whiteoak et al. (1983) note that the luminosities of their 3 H_2O masers (out of 10 HII regions observed) were similar to the average for H_2O masers in the Galaxy (~$10^{29.5}$ erg). The stronger OH 1665 MHz source N105A had an observed intensity of 0.76 Jy equivalent to 2×10^3 Jy for a source 1 kpc distant. Haynes & Caswell (1981) note that only 2 (from a sample of 40 galactic Type I masers) have an intrinsic luminosity greater than the source in N105.

The results for the non-masering lines are summarized in Table 1. A comparison with the Galaxy is limited by uncertainties in the beam dilution for N159. With the most favourable estimates (see references given earlier) the N159 values for OH, H_2CO and CH could equal the average values in the Galaxy. For HCO^+ the ratio $T_a^*(HCO^+)/T_a^*(CO)$ observed with similar beams, $0.18/1.45 = 0.12$, is close to the average ratio in the Galaxy. Since N159 must be one of the most massive molecular clouds in the LMC (as discussed in detail by Israel et al. 1982), it is likely that the LMC is generally deficient in the 4 molecules in Table 1. However, the fact that the abundances of these molecules relative to CO are similar for the LMC and the Galaxy would argue against CO abundance being determined primarily by the UV intensity. As discussed by Israel (this volume), CO destruction by UV should be considerably greater in the LMC because of the greater UV penetration of the clouds in the LMC as compared with the solar neighbourhood. However, it is probable that molecular destruction by the UV would be more effective for non-CO molecules than for CO.

Table 1. Molecular Line Observations towards N159

Molecular Line	Beamwidth ('arc)	N159 Value[a]	Typical Galactic Value[a]
OH: 1.667 GHz	12.4	L/C = −0.013	L/C ∼−0.1−0.2
H_2CO: 4.83 GHz	4.3	L/C = −0.014	L/C ∼−0.07
CH: 3.26 GHz	6.4	0.013 Jy; L/C = 0.006	L/C ∼0.025
3.33 GHz		" "	
HCO^+: 89 GHz	3.3	0.18 K T_a^*	∼2.0 K T_a^*

[a] L/C, ratio of line/continuum intensity.

A comparison of molecular concentrations and excitation for the two galaxies will clearly assist our understanding of molecular formation and destruction processes. There is an obvious need for improved and considerably more extensive molecular data for the Magellanic Clouds.

REFERENCES
Batchelor, R.A., McCulloch, M.G. & Whiteoak, J.B.: 1981, Mon. Not. R. Astron. Soc. 194, 911.
Caswell, J.L. & Haynes, R.F.: 1981, Mon. Not. R. Astron. Soc. 194, 33P.
Haynes, R.F. & Caswell, J.L.: 1981, Mon. Not. R. Astron. Soc. 197, 23P.
Israel, F.P. et al.: 1982, Astrophys. J. 262, 100.
Scalise, E. & Braz, M.A.: 1981, Nature 290, 36.
Whiteoak, J.B. & Gardner, F.F.: 1976a, Mon. Not. R. Astron. Soc. 174, 51P.
Whiteoak, J.B. & Gardner, F.F.: 1976b, Mon. Not. R. Astron. Soc. 176, 25P.
Whiteoak, J.B. et al.: 1983, Mon. Not. R. Astron. Soc. (in press).
Whiteoak, J.B., Gardner, F.F. & Höglund, B.: 1980, Mon. Not. R. Astron. Soc. 190, 17P.

CO OBSERVATIONS IN THE SMALL MAGELLANIC CLOUD

M. Rubio
Universidad de Chile, Departamento de Astronomía

R. Cohen and J. Montani
Columbia University

The dwarf Magellanic irregular galaxies apparently have a very low molecular content compared to the Milky Way. In the LMC, molecular clouds are fairly common, but the ratio of molecular to atomic gas is at least 5 times lower than in the Galaxy (Cohen et al. 1984). Elmegreen et al. (1980) searched for CO in 6 dwarf galaxies and failed to detect any emission even though their sensitivity was adequate to detect galactic giant molecular clouds placed at the distance of these galaxies. Israel (1984) observed the J=2→1 transition of CO at 15 points in the Small Magellanic Cloud and detected CO emission from five of them, but at a level two to six times lower than typical galactic values.

Since January 1983, Columbia University and the University of Chile have been making a more extensive survey of CO in the SMC, using the J=1→0 line at 2.6 mm. The observations are being made with the Columbia Southern Millimeter-Wave Telescope at Cerro Tololo, Chile, a 1.2 m Cassegrain with an 8 arcminute beam (137 pc at the SMC).

Observations are now under way. We have observed 51 positions to date: 32 dark clouds, 11 optical HII regions, 5 WR star positions, 1 continuum source, 1 infrared source, and 1 H_2O maser. These positions cover a total of 0.5 square degree in the bar and 0.2 square degree in the wing. We have detected CO in only one position, Hodge's (1974) dark cloud 4 ($\alpha = 0^h 44^m 32^s$, $\delta = -73°39'$, 1950.0). The spectrum is shown in figure 1. We have also found possible CO emission at five other positions, four of them in the southwest part of the bar and one in the central part of the bar. We are now re-observing these positions to obtain a lower noise level.

The low molecular content is confirmed by our observations. Typical galactic giant molecular clouds at the distance of the SMC would produce a CO antenna temperature of about 0.5 K. This is 15 times our noise level and 6 times the intensity of the detected CO line. If the ratio of H_2 mass to CO luminosity is the same as in our galaxy (Lebrun et al. 1983), then the H_2 mass at the position of our CO detection is 7×10^4 $M_\odot$, only 7% of the HI mass within the telescope beam (McGee and

Newton 1981). In contrast, our galaxy contains roughly equal quantities of atomic and molecular hydrogen. It is quite likely, however, that because the abundance of metals, and especially of carbon, is so low in the SMC, that the CO luminosity is also low, and that we have therefore underestimated the H_2 mass.

Observations of young objects in the SMC will continue during 1984. During the next few years we plan to make a fully sampled map of the entire cloud.

REFERENCES

Cohen, R.S., Montani, J., Rubio, M. 1984, this volume, p. 401.
Elmegreen, B.G., Elmegreen, D.M., Morris, M. 1980, Astrophys. J. <u>240</u>, 455.
Hodge, P.W. 1974, Publ. Astron. Soc. Pac. <u>86</u>, 263.
Israel, F.P. 1984, this volume, p. 319.
Lebrun, F. <u>et al</u>. 1983, Ap. J., in press.
McGee, R.X., Newton, L.M. 1981, Proc. Astron. Soc. Australia, <u>4</u>, 189.

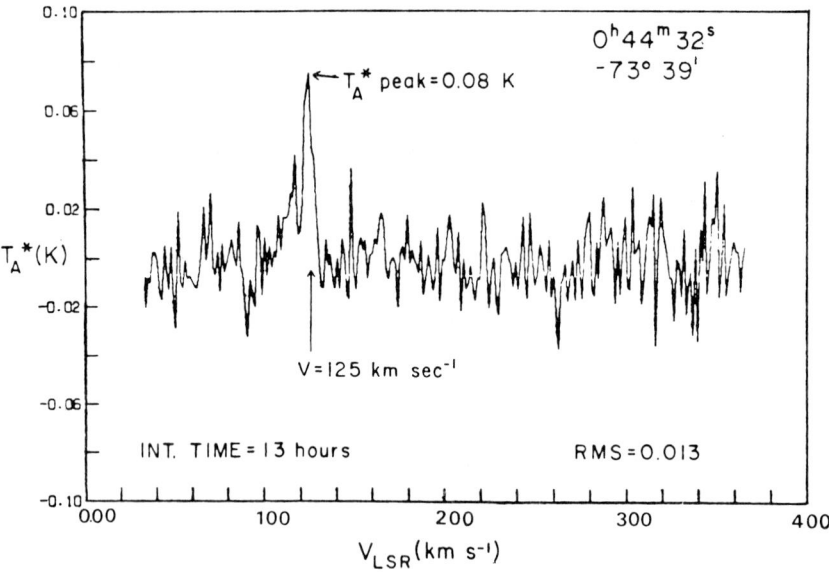

Figure 1 Spectrum of CO at Hodge dark cloud 4. As with all our spectra, only a linear baseline has been subtracted. At most positions, we integrated for 3 hours to achieve a noise level of 0.3K (RMS). Here, because the line is so weak, the time was increased to 13 hours.

CO IN THE LARGE MAGELLANIC CLOUD

R. Cohen and J. Montani
Columbia University

M. Rubio
Universidad de Chile

In our galaxy molecular clouds account for about half the total interstellar gas and are probably the sites of all star formation. The high gas content and widespread star formation in the Large Magellanic Cloud would therefore suggest a high molecular content. Very little however is actually known about molecules in the LMC. The most extensive previous survey (see Israel in this volume) found CO in half of the 22 points observed but covered less than 10^{-4} of the LMC area.

When the Columbia Southern Millimeter-Wave Telescope began operation in January 1983, we therefore immediately started a fully sampled survey of 2.6 mm CO emission from the LMC. The telescope, a close copy of the Columbia telescope in New York, is a 1.2 meter Cassegrain with a very accurate surface capable of working well into the sub-millimeter. It has a liquid nitrogen cooled receiver with a single sideband noise temperature of 385 K and a 256 channel filter bank spectrometer with a resolution of 500 kHz (1.3 km sec^{-1} at 2.6 mm). The beam of the telescope, 8' or 120 pc at the LMC, is large enough to make a complete LMC map practical -- the survey will require several thousand hours -- but is still able to distinguish objects about the size of galactic giant molecular clouds.

The final survey will cover virtually the entire LMC (at least 6° x 6°) with points spaced every beamwidth and each point observed for about 30 minutes to obtain a noise of 0.07 K RMS per channel. Although the survey is now only about 30% complete (Figure 1), it already shows that molecules are more abundant than in other megallanic irregulars (Elmegreen et al. 1980) and can be detected in many regions of star formation. In particular, there is a ridge of emission starting south of 30 Doradus at N159 and extending over one kiloparsec further south. Assuming the standard galactic H_2 mass to CO luminosity relation (Lebrun et al. 1983) applies in the LMC, this ridge contains a molecular mass of 4×10^6 $M_\odot$, about the same as a galactic giant molecular cloud but spread over a much larger area. The HI mass in the same region is 50×10^6 $M_\odot$. This ratio seems to be typical of the LMC, with calculated column densities for H_2 generally about 5 to 10 times lower than for HI.

It is possible, however, that the galactic mass-to-luminosity does not apply because of the low metalicity, and hence CO abundance, in the LMC.

REFERENCES
Elmegreen, B.G., Elmegreen, D.M., and Morris, M. 1980, Ap.J., 240, 455.
Lebrun, F. et al. 1983, Ap.J., in press.

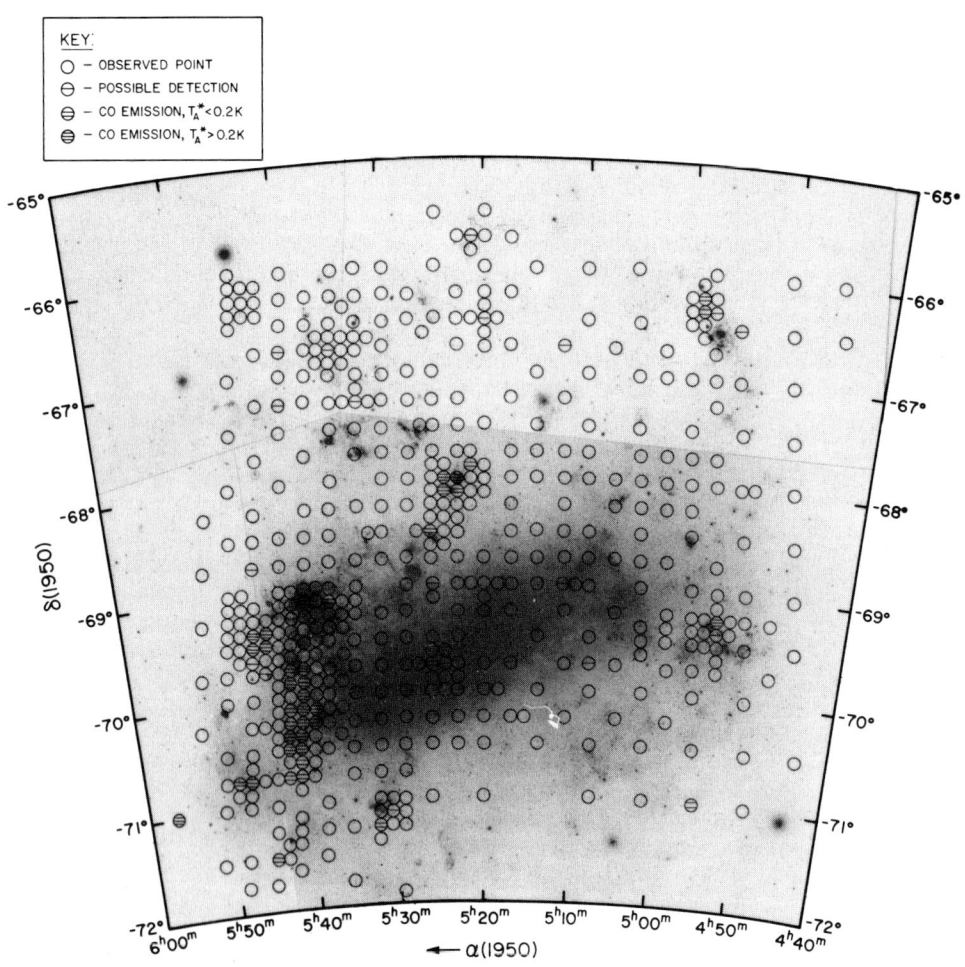

Figure 1. LMC CO observations superimposed on the ESO/SRC Atlas IIIaJ plate.

INTERSTELLAR DUST IN THE LMC

Geoffrey C. Clayton and Peter G. Martin
Department of Astronomy, University of Toronto

ABSTRACT

New IUE observations confirm the differences between the Galactic and LMC ultraviolet extinction curves and show some evidence for variations within the LMC. Visual and infrared photometry and polarimetry show that the anomalous properties of the LMC dust do not extend to longer wavelengths. Despite the much different dust to gas ratios in the Galaxy and the Magellanic clouds, the dust formation efficiency is similar when the abundances are included.

1. EXTINCTION AND POLARIZATION

Significantly reddened stars are rare in the LMC but extinction curves were constructed for 12 stars from 0.13 to 2.2 μm. The wavelength dependence of interstellar linear polarization was studied from 0.35 to 0.84 μm for 18 stars.

The differences in the average extinction curves of the two galaxies, seen in the ultraviolet, are confirmed in the present study. In the visual and infrared the curves are very similar. The polarization characteristics are also very similar. Typically, the Serkowski curve can be fit to the LMC data with the same parameters used to fit Galactic polarization (Clayton, Martin, and Thompson 1983). The polarization efficiency is also comparable.

Apparently the relevant variables, namely grain size, shape and composition, magnetic field strength, and gas density and temperature all conspire to make the grain alignment and polarization efficiency the same in these two quite different galaxies. The similarities between the extinction and polarization properties of interstellar dust in the LMC and the Galaxy are more remarkable than the differences considering the large number of potentially free parameters involved in modelling the dust.

2. VARIATION OF EXTINCTION WITH LOCATION IN THE LMC

The extinction curves measured for individual lines of sight within the Galaxy are often quite different from the average curve. Similar variations were searched for in the LMC data but any variations present are no larger than the estimated errors. To reduce these errors, the stars were grouped into 2 bins; one containing stars with a projected distance from the centre of 30 Dor of less than one degree and the other containing those lying at a distance greater than one degree. The average extinction curves for the 2 bins are different. The 30 Dor bin has a smaller 2200 Å bump and higher far ultraviolet extinction. However, these differences are only marginally significant.

If these regional differences are real, then it is possible that the grains near 30 Dor, a highly active region, have been modified. One possibility is sputtering and grain-grain collisions caused by shock waves. Massa and Savage (1983) have shown that a possible result of shocks is a steepening of the far ultraviolet extinction and a smaller 2200 Å bump. This is the trend seen in the binned data.

3. ABUNDANCES AND THE DUST-TO-GAS RATIO

The dust-to-gas ratio, $E(B-V)/N_H$, in the LMC is several times lower than the Galactic value (Koornneef 1982) which could be interpreted as evidence of less efficient dust production in the LMC. However, the amount of dust produced is limited by the abundance of condensable species. Taking into account the lower CNO abundances found for the LMC (Dufour et al. 1982), the dust to gas ratio, $E(B-V)/N_{CNO}$, is within a factor of two of the galactic value. For the SMC, which has even lower CNO abundances and $E(B-V)/N_H$ of one-tenth the Galactic value, the ratio $E(B-V)/N_{CNO}$ is in the same range as the LMC and the Galaxy. Therefore, it can be concluded that the dust formation efficiency is rather similar in these 3 galaxies.

It is interesting that the average extinction curve in the SMC is even more extreme than the LMC curve: the far ultraviolet extinction is higher and the bump is almost non-existent (Bromage and Nandy 1983). Therefore, at least for this small sample of 3 galaxies, there is an apparent correlation between the CNO abundances and the ultraviolet extinction characteristics.

REFERENCES

Bromage, G.E., and Nandy, K. 1983, Monthly Notices Roy. Astron. Soc. 204, pp. 29p.
Clayton, G.C., Martin, P.G., and Thompson, I. 1983, Astrophys. J. 265, 194
Dufour, R.J., Shields, G.A., and Talbot, R.J. 1982, Astrophys. J. 252, pp. 461.
Koornneef, J. 1982, Astron. Astrophys. 107, 247
Massa, D., and Savage, B.D. 1983, preprint.

SMC : UV EXTINCTION CURVES, GAS TO COLOR-EXCESS RATIOS*

J. Lequeux[1], E. Maurice[2], L. Prévot[1], M-L. Prévot-Burnichon[1], B. Rocca-Volmerange[3]
[1] Observatoire de Marseille, France
[2] European Southern Observatory, Chile
[3] Institut d'Astrophysique, Paris, France

We have made an extensive study of the far-UV extinction in SMC stars using IUE spectra mostly obtained by us. The studied stars are Sk 7, 13, 18, 32, 76, 82, 85, 103, 119, 124, 142, 143, 159, 191 and AV 398. A few other stars have also been observed but are unsuitable for extinction studies. We have obtained at ESO new photometry and spectral classifications for all the studied stars. The far-UV extinction curves derived from IUE spectra of matched star pairs are all similar, except for Sk 143 (Lequeux et al., 1982), Sk 124 and 191. The mean curve, which supersedes that of Rocca-Volmerange et al.(1981) is plotted Fig.1 with the label SMC "normal", as well as the curves for the three "anomalous" stars (full lines) For comparison, we also plot the LMC "normal" curve and that for the anomalous star Sk-69°108 (dotted-dashed curves), the "normal" galactic curve and the curves for two stars in Scorpius-Ophiuchus.

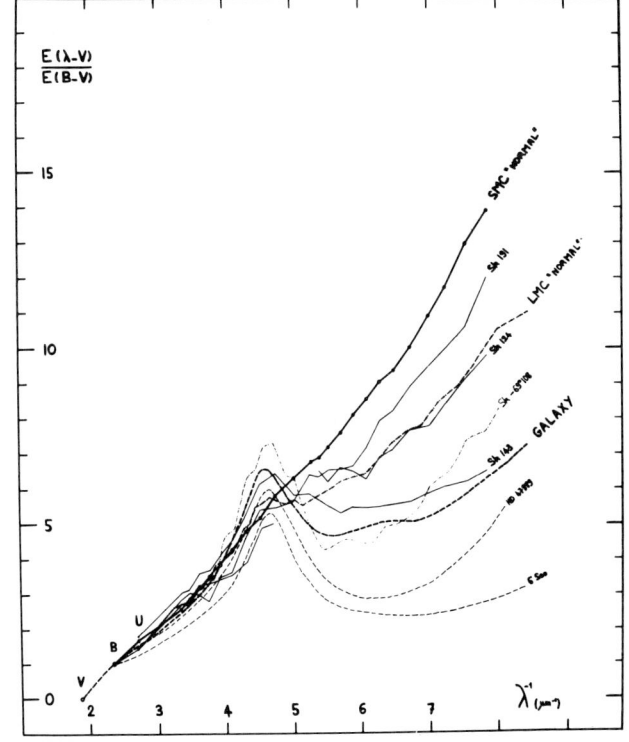

Fig. 1. UV extinction curves for SMC and other stars, normalized by E(B-V).

*Based on observations with the International Ultraviolet Explorer collected at the ESA Villafranca station, and at ESO, Chile.

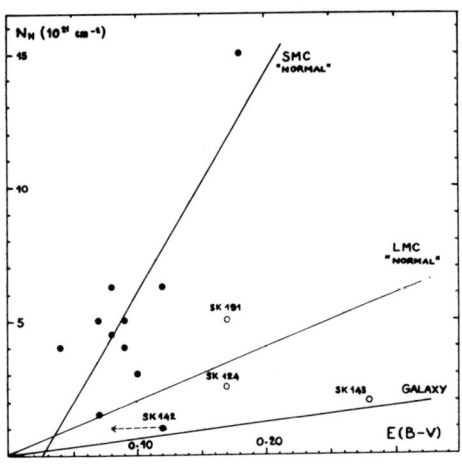

Fig. 2. Hydrogen column density vs. color excess for SMC stars (black dots). Three deviating stars are indicated by circles. The average errors are about 0.03 in color excess and $1.5\ 10^{21}$ in column density. The LMC and galactic mean relations are indicated for comparison.

Note the similar *relative* behaviour of the "anomalous" stars. In Fig.2 we plot the hydrogen column density N_H as derived from the interstellar Lyman α profile in the IUE spectra, vs. the color excess. Note that the three stars which deviate from the mean relation are *the same* that show an anomalous extinction curve. The intercept of the mean SMC relation on the E(B-V) axis gives approximately the galactic foreground color excess, about 0.03.

CONCLUSIONS : 1) The SMC, LMC and galactic "normal" UV extinction curves differ. The SMC curve is nearly linear in λ^{-1} and can be accounted for by silicate grains alone with a size distribution slightly steeper than the galactic one. The absence of graphite can be related to the large underabundance of carbon in the SMC.

2) The "normal" SMC $N_H/E(B-V)$ ratio is 10^{23} cm^{-2}mag^{-1}, 17 times the galactic value (the LMC one is 4 times galactic). If extinction is due to silicates, the ratio Si(in grains)/H is about 1/10 solar ; given the underabundance of the SMC in heavy elements, this is consistent with most of the Si being in grains.

3) Sk 143 and to a lesser extent Sk 124 and 191 show *both* UV extinction curves *and* $N_H/E(B-V)$ ratios closer to the galactic ones. A similar relative behaviour of UV extinction can be seen for Sk-69°108 in the LMC and for several Sco-Oph stars in the Galaxy. Graphite adds to the silicates. The $N_H/E(B-V)$ ratio for Sk 143 cannot be accounted for even if all Si and C in SMC abundances are in the grains. Either the extinction is circumstellar (but Sk 143 has little wind), or most of the hydrogen is molecular (but this does not explain the extinction curve), or the interstellar gas in front of Sk 143 has been enriched by recent star formation and not yet mixed with the general medium. A galactic extinction is excluded by our interstellar radial velocities.

REFERENCES

Lequeux, J., Maurice, E., Prévot-Burnichon, M-L., Prévot, L., Rocca-Volmerange, B.: 1982, *Astron. Astrophys.* 113, L 15.
Rocca-Volmerange, B., Prévot, L., Ferlet, R., Lequeux, J., Prévot-Burnichon, M.L.: 1981, *Astron. Astrophys.* 99, L 5.

ULTRAVIOLET INTERSTELLAR ABSORPTION TOWARDS SK159 IN THE SMALL MAGELLANIC CLOUD

Edward L. Fitzpatrick
Washburn Observatory
University of Wisconsin-Madison

Multiple high-dispersion <u>IUE</u> spectra of SK159 in the Small Magellanic Cloud (SMC) have been obtained and averaged in order to study interstellar gas associated with the SMC and, in particular, to search for evidence of SMC halo gas. A single high-dispersion <u>IUE</u> spectrum of SK159 was analyzed by Prévot et al. (1980). They reported the absence of C IV and only the marginal detection of Si IV at SMC velocities. The higher signal-to-noise ratio in our data allow a more definitive investigation of the absorption characteristics of these ions. SK159 is relatively cool, spectral type B0.5 Iaw (Walborn 1983) and is located in a region free of strong nebulosity. Thus the interstellar line spectrum will be less contaminated by H II region absorption than in previous studies of SMC stars. This favorable circumstance is somewhat offset by the great complexity of the underlying stellar spectrum and the resultant blending of stellar and interstellar lines. The main conclusions from our study of the SK159 spectrum are summarized as follows:

1.) The Milky Way disk and halo absorption features towards SK159, except for Al III, are very similar to those seen towards HD5980, also in the SMC. The Al III lines may be enhanced by photo-ionization from stars in the Milky Way disk.

2.) SMC H I 21-cm emission components in the SK159 direction are detected at 140 km s^{-1}, 175 km s^{-1}, and 201 km s^{-1}, corresponding to H I column densities of 1.4×10^{21} cm^{-2}, 3.3×10^{21} cm^{-2}, and 1.9×10^{20} cm^{-2}, respectively (McGee 1979; McGee and Newton 1982). From comparison with H I Lyα absorption and the other ultraviolet absorption lines, we determine the 140 km s^{-1} and 201 km s^{-1} H I clouds to be foreground to SK159 and the 175 km s^{-1} cloud to be beyond SK159.

3.) SMC ultraviolet absorption occurs at four principal velocities: 100 km s^{-1}, 150 km s^{-1}, 165 km s^{-1}, and 215 km s^{-1}.

4.) The 150 km s^{-1} absorption, seen in the low-ionization stage species, is associated with the 140 km s^{-1} 21-cm emission peak and originates in the SMC interstellar medium. Column density estimates suggest that S, Zn, and Mg are present in roughly their relative solar abundance ratios, but are deficient by a factor of ≤10 with respect to H. The gas-phase Si/S ratio may be a factor of four less than the solar value. Optical data suggest that the Na/H ratio is deficient by a factor of ~10, with respect to solar abundances, and that the Ca/Na gas-phase ratio is a factor of ~35 less than the solar ratio.

5.) The gas-to-dust ratio in the 150 km s^{-1} gas is between three and eight times greater than the galactic value. The uncertainty is mainly due to uncertainty in the SMC contribution to the color excess of SK159.

6.) The 165 km s^{-1} absorption corresponds to the weak H II region surrounding SK159. Only C II*, Al III, and Si IV absorption lines are resolved at 165 km s^{-1}. The strengths of the Si IV lines are consistent with a stellar photo-ionization origin. Nebular absorption is present in the strong lines of Si III, Si II, and C II, but not resolved.

7.) The low-ionization ultraviolet absorption seen near 100 km s^{-1} and 215 km s^{-1} may be associated with global SMC H I masses detected by McGee and Newton (1981). The 100 km s^{-1} H I mass is centered on the SMC. The 200 km s^{-1} H I mass is associated with the LMC-SMC H I "bridge". The presence of ultraviolet absorption shows that neither of these H I masses is composed of primordial material.

8.) C IV absorption, with a column density of ~3.6x10^{13} cm^{-2}, is detected near 100 km s^{-1}. It is not clear if there is an exact velocity coincidence with the low-ionization absorption near 100 km s^{-1}.

9.) Si IV absorption may be present near 100 km s^{-1} with a column density of ~3x10^{12} cm^{-2}. The C IV/Si IV ratio, ≳10, is similar to values found in the galactic halo and is not characteristic of stellar photo-ionized gas.

10.) Our data do not rule out the possibility of an SMC halo. To determine whether the 100 km s^{-1} C IV absorption has its origin in SMC halo gas, observations of stars in a variety of locations in the SMC are required. "Coolish" stars like SK159, located in quiescent regions, are most desirable for this because contamination from nebular gas will be minimal.

REFERENCES

McGee, R.X.:1979, private communication.
McGee, R.X., and Newton, L.M.: 1981, Proc.A.S.A., 4, pp. 189.
────────────────────────────── 1982, Proc.A.S.A., 4, pp. 308.
Prévot, L., et al.: 1980, Astr.Ap., 90, L13.
Walborn, N.R.: 1983, Ap.J., 265, pp. 716.

FUTURE DIRECTIONS OF RESEARCH IN THE MAGELLANIC CLOUDS
Panel Discussion

Chairman: B. E. Westerlund
Astronomical Observatory, Uppsala, Sweden

Members of the panel: I. J. Danziger, K. C. Freeman, J. A. Frogel,
J. A. Graham, M. Grewing, D. J. Helfand.

The aim of the panel discussion was to identify the most important remaining gaps in our knowledge of the Magellanic Clouds, and to find optimal ways of filling them. Particular attention has been paid to instrumentation available now and in the near future.

The following summary of the statements made by the members of the panel has been written by the Chairman on the basis of a tape recording of the discussion. The contributions by other participants have had to be left out as many parts could not be transcribed with sufficient accuracy from the tape. The same applies, unfortunately, also to parts of the statements by the members of the panel. Nevertheless, the summary should give a reasonably good idea of the main topics considered by the panel.

I. J. Danziger
European Southern Observatory, Garching bei München, F.R.G.

A truly important difference at the epoch of this conference as compared with the previous one on the Magellanic Clouds is that a good deal of work has been done now by theoreticians on the chemical evolution of galaxies, in particular on galaxies like the LMC. However, this conference has been dominated by the observers. It would have been important to have heard more from the theoreticians.

There are a number of things to be done to help those who do theoretical calculations on chemical evolution. Firstly, it is important to carry out direct measurements of the metallicity of individual stars. With the detectors now available much more can be done in that area. We do not necessarily have to work on very faint stars in the Clouds. It is, however, highly desirable to get good spectra of stars in the Bars of the Clouds. In addition to observing individual stars one should also aim at obtaining spectra of integrated parts of the bars. Some astronomers are working along this line, but we have not heard much about it here.

It will be particularly important to establish a photometric or a spectroscopic system that will make it possible to look at individual elements and say something about various groups of elements that are formed by different synthetic processes.

There has been some work reported here on old globular clusters in the SMC and their H-R diagrams. Since these are the most metal-poor of the systems it appears important to pursue that work as far as possible.

Another item regarding the chemical evolution of galaxies concerns the dependence of the slope of the Initial Mass Function on the metal abundance. Melnick and Terlevitch have made the case that this is so. It could be tested in the Magellanic Clouds. It is important, if it is correct, not only from the point-of-view of the chemical evolution, but also because it is telling us something about the physics of star formation.

Another point of importance: We lack any fundamental information about mass loss, particularly for late-type stars. Can anything be done about that in the Magellanic Clouds?

We were all struck by Mathewson's discussion of the SMC, dividing it into two. This suggestion is of interest even if it should not be true. Others at this meeting have talked about encounters between the SMC and the LMC. If that has happened, obviously - or is it obvious? - one galaxy could have borrowed some stars from the other, and some clusters and other material, too, at least until the next encounter when they might borrow it back. If this is true, it complicates the interpretation of the cluster contents of the Clouds, as they have then been mixed with each other. It is important to get this straight.

Finally, I wish to bring up a small item for the use of the Space Telescope. It should be used for a look at the SNRs in order to see what is produced in the supernova and what comes from the interstellar medium. The Space Telescope will be particularly advantageous for abundance work in the SNRs on elements and lines in wavelength regions not accessible from the Earth. The International Ultraviolet Explorer has not been reaching sufficiently faint. The Space Telescope also offers a better spatial resolution.

K. C. Freeman
Mount Stromlo and Siding Spring Observatories, Canberra,
Australia

We have now in the general area of kinematics and dynamics some marvelous new data on the kinematics of the whole Magellanic system. We have seen some very complex velocities in the bridge and in the LMC and the apparent disruption of the SMC itself. All of this is well over our understanding. The question is what needs to be done theoretically, so that we can make use of what we have actually seen.

We need a hydrodynamic treatment of the dynamics of first of all the whole LMC - SMC system, and secondly of the LMC and the SMC themselves. This is within the realm of dynamical possibilities. It is not just a dynamics exercise, because it could lead to some understanding of the flow fields. These flow fields affect the star formation distribution. Two particular areas of interest are the Constellation III and the 30 Doradus region in the LMC. You may think that they are peculiar for the LMC. It is not so. Similar areas of star formation and similar processes are seen in other magellanic systems. It appears to have something to do with LMC-type systems. The kind of studies that should be carried out are: Start with Fujimoto's orbits and find out what happens to the whole system. Then, assume also a hyperbolic orbit and look at what happens to the Clouds during a single pass about our Galaxy. It is important to treat the Magellanic Clouds as a whole, because this makes the system more bound.

An important question is whether there exists a halo population in the LMC. I say it does not exist. John Graham says that I and others agreeing with me are misled. We should look at the RR Lyr stars. In the globular clusters there are some RR Lyr stars. In the Galaxy we find the so called halo objects in the halo, and this is a spheroidal system. When we look at edge-on Magellanic-Cloud-type systems we do not see any spheroidal components, so it would be surprising to find a halo in the Clouds. However, in our Galaxy we do see stars of quite low mass scattered all over the disk. It would not be too surprising to see similar stars in the Clouds.

What are actually the RR Lyr stars? They may, according to some astronomers, not be so old. It would be very useful to have this question settled.

With regard to the globular clusters an important question concerns their formation in the LMC. We know that the young blue populous clusters are globular clusters in every way except in age, which for them may be 10^7 years or even lower. This has been mentioned several times during the meeting. The question is, why are they forming now? They

are abundant in the LMC; there may be a few in M 31, and, as far as we
know, there are none in the Galaxy. One particular aspect of this is
the question of binaries in globular clusters. There is a reasonable
amount of evidence that galactic globular clusters with giants of
about 0.8 solar masses have very few binaries. We do not yet know if
this is a general property of Extreme Population II or a general property of globular clusters. We have a great opportunity in the LMC
to study the stars in the young globular clusters. Their masses are
of the order of 8 - 10 solar masses. That puts them in the category
of Extreme Population I. These stars will then have to make up their
mind: Either they are of Extreme Population I and the clusters are
full of binaries, like this population in our Galaxy, or they are of
Extreme Population II, and there are no binaries. There is much to be
done about this problem, and the way to do it is obvious.

J. A. Frogel
Cerro Tololo Inter- American Observatory

I wish to make a few comments on the relevance of infrared (IR)
astronomy for Magellanic Cloud research and about what it has altered
in the picture. There are basically two areas where IR astronomy is
capable of making important contributions. The first area is in observing all stars where you measure the photospheric radiation of the star
itself. A group of us has been working on field stars and on cluster
stars, and we have done our best in exploiting this particular area of
IR astronomy.

The second area may be called more traditional IR astronomy. We
look at the dust at wavelengths longer than 2 µm, and at 5 -10 µm at
the longest. We try to find the thermal emission from the dust, that
is re-radiated starlight. Very little of this has been done in the
Magellanic Clouds, partly because it is rather much at the limit of the
instrumentation at present, particularly at 10 µm and with the existing
southern hemisphere telescopes. However, a lot is to be gained by
pursuing this research and by improving the instrumentation. As has
been pointed out at this meeting, when the results from IRAS become
available there will be a lot of new ideas and a number of objects
to look at. This will require a good deal of follow-up work from the
ground. In particular, there will be dust emission from HII regions
and molecular clouds and stars. Important is first of all that you can
find out something about mass loss rates from stars by looking at the
dust around them. If you can get spectra in the 10 µm region from the
stars, you can see whether the dust consists of silicates of
carboniferous material. This is clearly important for understanding
something about the composition of the Clouds. It is also important
to understand why the Clouds contain so much less dust than does the
Milky Way.

The ither aspect of dust emission concerns the birth of stars. A great deal of the infrared capacity during the past few years has been aimed at this question in the Galaxy. Important results have been derived by combining the IR observations with observations in the visual and radio bands, looking at molecules at mm wavelengths. This type of work, carried out in the Clouds, will yield very important, very exciting results on the star formation in the Clouds. It is a kind of a mystery there, because star formation is expected to occur in very dusty regions. In our Galaxy we find that star formation occurs in very extensive, very dusty regions. We do not see such extensive areas of dust in the Clouds, and yet, we know that star formation has been going on there in the relatively recent past. This is particularly evident in the LMC with its large number of supergiants and young clusters. Thus, we have to try to understand what is going on in the Clouds, and it looks likely that IR astronomy will provide the answer to this important question.

The imminent advent of high-resolution infrared spectroscopy is likely to yield important information about molecular abundance in late-type stars. Many molecules, in particular CO, have absorption bands only in the infrared. Thus, with the advent of multi-element spectroscopic detectors, and particularly if we can get a truly aperture-sized telescope in the southern hemisphere, great advances could be made in infrared spectroscopy. This would be important for understanding molecular abundances as well as for understanding the chemistry of the Clouds.

J. A. Graham
Cerro Tololo Inter-American Observatory, La Serena, Chile

The contrast between the research reported at this Symposium and at the IAU/URSI Symposium No. 20, held in Sydney 20 years ago, has been striking. In 1963 a number of surveys were reported, like the painstaking long-term spectroscopic work on supergiants in the Clouds by the Radcliffe astronomers, the radio surveys at Parkes, and the objective prism surveys with the Uppsala Schmidt telescope. Research is now carried out that was inconceivable at that time.

The Magellanic Clouds may be considered as laboratories for star formation and evolution and of extreme importance for our understanding of how galaxies form and evolve. We should appreciate that in the Clouds the globular clusters can still be resolved into individual stars. They are for us practically the last stage in studying the globular clusters in the Universe, before we go out into the very hazy realm of unresolved stellar systems and globular clusters.

The colour-magnitude diagrams of globular clusters presented 20 years ago could be interpreted in any number of ways. When we look at the most recently produced diagrams, they look unambiguous. It is essential that this work on clusters be continuously refined. Most important is that we take care of getting the zero-points right and that we avoid systematic errors in our data. In this way we diminish the possibilities of trouble for the theoretical interpretations.

It is important to obtain as precise data as possible for the globular clusters in the Clouds, in particular for those containing RR Lyr stars. One such cluster is NGC 121. It contains RR Lyr stars, but it is now said to contain also a kind of carbon star. Is it old? Old means more than 10^{10} years. I would not accept an age for the RR Lyr stars of 5×10^9 years, i.e. the age of the carbon stars. This does not agree with the situation in the general field. If anyone wants to know what RR Lyr stars really are or wants to study really old metal-poor giants he should observe the western part of the SMC. That is where you find field Population II.

Supergiant stars may be difficult to use for abundance determinations, but their studies are still important. The Magellanic Clouds are excellent for studying massive stars. They are all at the same, rather well-known distance and we can determine there, better than in the Galaxy, how massive stars will get. They can also be studied at a reasonable spectroscopic dispersion. Thus, we may find out what is going on in them, what holds them together, and what occasionally tears them apart.

Some topics were either not mentioned at this Symposium or covered rather unsufficiently. Little was said about abundances. Today, when we discuss abundances, we do not have to stay with the bright supergiants. We can push on in magnitude to the normal, better behaved stars and do work of the type that has been done on stars on the giant branch of the globular clusters. It is of great importance to get good abundances for cluster stars and for field stars so that we can tie things down for studies of more distant galaxies.

Another neglected item here is that of binaries. Everyone wants to get masses for X-ray binaries, but nobody seems to care about determining masses for ordinary binaries. There are many of the latter to be found in the Clouds. We badly need to know more about stellar masses. We also need to know if in the Magellanic Clouds the same stellar mass-luminosity relation applies as in the Galaxy.

Finally, when you observe today you frequently sit in a warm control-room with your cluster on the TV-screen. It is easy then to forget all about the atmospheric conditions outside and go on taking your CCD frames. Later on you apply a mean extinction coefficient to that night without thinking about the fact that the coefficient is just a mean and hardly applied to that observing night of yours. We must remember that only when we apply the new wonderful techniques with great care and make good use of them in every other way, will they produce physically meaningful results.

M. Grewing
Astronomical Institute, Tübingen, F.R. Germany

We have heard today about significant progress in the studies of the properties of both gas and dust in the Magellanic Clouds. The majority of these new data has been obtained with the International Ultraviolet Explorer (IUE) satellite, from low dispersion (6Å) spectra of a large number of stars in the Clouds, and from high resolution (0.2Å) observations for the brightest stars in the Clouds. These UV spectra are of particular importance since most of the resonance lines of the astrophysically relevant atoms and ions occur in this wavelength range and allow us to determine the physical state and chemical composition of the interstellar gas in the Clouds. Also, the slope of the extinction curve towards the ultraviolet and the strength of the 2200Å bump tell us something about the size distribution and composition of the interstellar dust.

In obtaining high resolution UV spectra of LMC and SMC stars, the IUE satellite was working at its sensitivity limit, i.e., only the brightest stars were accessible. These stars are known to greatly affect their circumstellar environment, both through their radiation and through their stellar winds. This introduces a potential bias into the interstellar medium studies. Also, due to their distribution accross the Clouds, they only allow probing the interstellar matter along very few lines of sight.

With the advent of the Space Telescope, this situation will change drastically. Basically all stars, which by their spectral type are suitable for interstellar medium studies, will then become accessible. The High Resolution Spectrograph (HRS) on the ST will also help to overcome, or at least lessen, another problem which has so far been hampering these studies; if the spectral resolution is not high enough to clearly separate individual line components arising from physically different regions along a particular line of sight (as is often the case with IUE observations), the quantitative determination of column densities and of element abundances becomes very uncertain if not impossible. Therefore, with the ST answers may be possible to questions like: (1) in what phases does the interstellar gas exist in the Clouds; do we find a variety of phases as in our Galaxy?, and (2) what are the chemical abundances and the abundance variations across the face of the Clouds which will reflect the amount of chemical processing and element mixing in these systems?

D.J. Helfand
Columbia Astrophysics Lab., New York

The planning for the next ten years is the following in X-ray astronomy: Now we have available EXOSAT. It was launched a couple of months ago. It is now said to have a lot of problems with the detectors,

Functioning is anyhow a HRI-type detector, like the Einstein Observatory HRI. It is, however, considerably less sensitive, and it has only 10 arcsec resolution. The Japanese have satellites for systematic observations and they will probably keep one up until the year 2000. But their satellites, at least before 1990, have no capability to observe anything but bright sources. There is a Russian experiment planned but not yet on mission.

There is a break until 1987, when ROSAT will be launched. It is a German national project with a telescope similar to Einstein. It will go to about 2.5 keV. Its IPC detectors should be considerably better than Einstein's and the overall sensitivity of the telescope should be higher. It will do an all-sky survey. This has some useful aspects as it will see the Magellanic Clouds. It will, however, not be much available, at least not during the first two years.

At around 1990 there are an Italian and a British mission proposed, both with good spectroscopic capabilities. They should be sensitive enough to observe SNRs in the Clouds. However, neither proposal is actually funded yet. Towards the end of the period AXAF (Advanced X-ray Astronomy Facility) may come. It is not inconceivable, however, that it will never be launched.

What does this give for the Magellanic Cloud research?

With EXOSAT, higher resolution positions will be obtained of many of the unidentified sources that we know exist. No new sources will be found in the Clouds with it, however, and it is possible that not all already known will be reached.

Between 1984 and 1988 some high-energy observations of the Magellanic Clouds may be carried out with 2 or 3 instruments on the Shuttle. Typically, one instrument may get 20 hours observing time. Perhaps 2 or 3 hours of this will be for the Clouds. So there will not be much of this work done during the next five years. Some of the experiments may, however, be able to distinguish between SNRs and point sources, not because of high spatial resolution but because of spectra of higher resolution.

In 1988-90 we have ROSAT and its all-sky survey to a level of 0.01 counts/sec. It will cover the Magellanic Clouds and it should see some SNRs. It should also measure high-precission positions for all the point sources that observing time can be obtained for. According to present plans it will not be Shuttle serviceable, so it may only last a couple of years. It should carry out a survey of the Clouds that is about about 10 times better than the Einstein survey. We should see sources that we do not know now.

A careful analysis of existing data may lead to the identification of at least some SNRs in the Clouds. The large-scale structure of the interstellar medium in the Clouds should be looked at, in particular the regions of hot gas, e.g. the surroundings of 30 Doradus.

The ROSAT should open the possibility to look at starforming regions and clusters in the Clouds. If the clusters are similar to clusters in our galaxy, ROSAT should see some sources of that kind and we should be able to find out how many there are in clusters. Apart from the neutron star binaries we should see all cataclismic variables.

If in 10 years time we have AXAF we will be able to see in the Clouds O stars and B stars and all SNRs. There should be at least 300 of the latter, i.e. about 10 times as many as known now. We should also see them better than in our Galaxy.

What has been detected so far are Population I binaries and SNRs and the hot interstellar medium. This is all extremely young population, less than 10^7 years old. The X-ray sources may be divided into two classes: the just mentioned sources and the old sources such as novae, cataclismic variables, population II binaries. We have not been able to do much about the latter so far. Once we do, we will have sources connected with two age groups in the Clouds, those associated with the extremely young population and those associated with a population older than 10^9 years.

INDEX

All entries implicitly refer to the Magellanic Clouds only. For galactic or (other) extragalactic topics see galactic- and extragalactic-.
nnn, subject referred to on several consecutive pages within an article.

Abundances	
gas	301, 333, 348, **353**, **366**, 382, 405, 415
radial gradient	361
(for stars and clusters see metallicity...)	
Adiabatic expansion	300, 315
AGB	80, 196
AGB stars (see stars...)	
Age	72, 99, 101, 103, 383, 391
of universe	12
Associations (OB-)	89, 93, 105, 385, 391
Asymptotic giant branch (see AGB...)	
Binaries (see stars...)	
Black hole	241, 301, 305
Bridge region	82, 103, 131, 411
Bubbles	89, 302
Carbon burning	367
Carbon production	196
Carbon stars (see stars...)	
Carina (η)	245
Cataclismic variables (see stars...)	
Centroids (see distribution...)	
Cepheids (see stars...)	
Chemical composition (see abundances...)	
Chemical evolution, history	19, 75, **107**, 409
Close encounter LMC-SMC	113, 121, 410
Clusters	1, 7, 13, 25-67, 71, 93, 172, 187
age	9, 29, 31, 43, 45, 47, 55, 64, 81, 189
age-ellipticity relation	5
age-metallicity relation	7, 17, 95
extinction in front of	5
formation	5, 7, 12, 66, 71, 411
hydrogen line strengths	15
integrated colors	**1**, 53
integrated spectra	**13**, 31, 53, 55, 57, 410
intermediate age	31, 199
kinematics	10, 107
mass-luminosity ratio	29
metallicity	10, 31, 33, 41, 43, 45, 47, 64
old	47, 75
open	10

Clusters (continued)
- photometric classification 13
- RR Lyrae 7, 37, 39
- spatial distribution (see distribution...)
- young 33, 35, 411

CO molecule (see interstellar...)
Collision Galaxy-MCs 103, 133
Core-helium burning 31, 217
Core rotation 206
Corona: There is confusion about the use and meaning of the words halo and corona (stars, gas, location, mass; see remark on page 375); in this index stars are under halo, gas is under corona.
Corona, galactic (see galactic...)
Corona
- gaseous **375**, 407
- massive dark component 111

Cosmic rays 313

Dark clouds 322
De Jager limit 151
Distance
- modulus, scale 192
- from Cepheids 165, 229
- from Novae 211
- from RR Lyrae 210
- from supergiants 145

Distribution of
- gas **115, 125,** 137, 319, 375, 395
- globular clusters 112, 128
- Novae 213
- planetary nebulae 113, 129, 231
- RR Lyrae 213
- supergiants 137

30 Doradus 105, 137, **245**, 255-261, **263**, 286, 297, 321, 344, 382, 411, 417

Dredge up mechanism 196
Dust (see interstellar...)
Dust shell 198

Electronography 41, 99
Ellipticity 1, 9, 27, 29
Emission nebulae (see HII regions...)
Enrichment of ISM 370
Evolutionary history (see history...)
Evolutionary tracks 68
Extinction (see interstellar...)
Extinction (atmospheric) 414
Extragalactic:
- blue compact galaxies 361
- Draco dwarf 39

INDEX

Extragalactic (continued)
    IC 1613   156, 361
    IZ18, 36   361
    irregulars   91, 95, 313
    M31   2, 20, 351
    M33   20
    M83   128
    M101   156
    NGC 1613   66
        3109   111
        4027   108
        4618/25   108
        5128   2
        6822   313

Fe II emission   235
Field stars   68, **79**, 97, 99, 101, 103, 184, 246
Flattening
    of clusters (see ellipticity...)
    of Magellanic systems   108

Galactic:
    chemical evolution   370
    corona (see gaseous... or mass...)
    fountain   378
    gaseous corona   131, 141, **375**
    (globular) clusters   33, 53, 59
    halo stars   309, 411
    high velocity clouds   115
    massive dark component   111, **115**, 134, 141, 375
        (see remark with corona)
    OB supergiants   233
    RR Lyrae   223
Gas (see interstellar...)
    dynamics   **271**
Gas-to-dust ratio   320, **333**, 403, 405, 407
Global properties (see large scale...)
Graphite   321, 348, 405
Gravitational potential   109
GRISM survey   184, 205

Hα flux   69, 89, 121, 364, 389, 391
HII regions   35, 75, 129, 286, **353**, 376, 382
                                           383-393
Halo
    (see remark with corona)
    population   64, 76, 82, 101, 113, 411
Heating, dynamical   108
Heating-cooling balance   378

History
    evolutionary   **1, 67, 107,** 134, 358
    of star formation   18, **79,** 185
Horizontal branch   18, 41, 43, 45, 47, 80, 99
Hubble-Sandage variables (see stars: S Dor...)
Hydrodynamic treatment LMC-SMC   411

Infrared space observations   324, 412
Initial mass function   **67, 79,** 329
Instability strip   227, 229
Inter Cloud region (see bridge region...)
Interacting galaxies   **107**
Intergalactic gas clouds   133
Interstellar
    absorption lines   53, 55, 116, 129, 319, **375,** 407
    abundances (see abundances...)
    dust   286, **319, 333, 341,** 389, 403, 412
    extinction   5, 59, 105, 255, 263, 296, 321,
                                           333, 341, 403, 405
        total-to-selective (R)   333, 341
    gas   72, **125,** 139, 141, 375, 395, 415
    molecules   **319,** 397
        CO   246, 325, 397, 399, 401
        hydrogen   **319,** 405
    phases of ISM   393, 415
Ionizing photons (see Lyman continuum...)
Isochrones   45, 84

K stars (see stars...)
Kinematics   76, **107, 115, 125,** 137

Large scale properties   **67, 125, 335,** 379, 395
LMC X (see X-ray sources...)
Local Group galaxies   134
Long-period variables (see stars...)
Luminosity function   76, 97, 105
Lyman continuum photons   69, 365, 387

M stars (see stars...)
Magellanic Stream   111, **115, 125,** 139, 141, 375
    distance of   116, 141
    stars in   103, 116
Magellanic type galaxies   2, **107,** 411
Maser   246, 320, **325,** 397
Mass loss   31, 35, 59, 151, **199,** 229, 239, 385
Mass transfer   202
Merger of galaxies   134
Metallicity
    of the MCs   91, 95, 409, 414
    of clusters (see clusters...)
    of Cepheids   157, 223, 225, 229

INDEX

Metallicity (continued)	
of RR Lyrae	209, 217
Mini Magellanic Cloud (MMC)	128
Mira stars (see stars...)	
Mixing	151, 198
MMC (see mini...)	
Molecules (see interstellar...)	
n-body simulations	27
Neutron star	295, 417

NGC (new data only)

121	57	2100	55, 59
330	57, 59	2121	31, 37
339	37	2136	33
416	37	2155	37
1466	47	2157	33
1783	31	2203	47
1786	37, 57	2209	31
1806	29	2210	37, 47
1818	29, 33, 55	2214	33
1866	33, 55	2231	31
1868	31	2257	39, 47
1962-70	35	Hodge 11	37, 41, 43, 47
1978	31	Kron 3	45
2004	55	LW4	47
2019	37	3603	**250**, 257, 261, 265
2070	(see 30 Dor)	6530	59

Nitrogen	
enrichment	281, 367
enrichment by Novae	212, 361
production	359
Novae	**207**, 359, 417
Nuclear burning (explosive)	274
Nucleosynthesis	359
OB (see associations... or stars...)	
OH/IR stars (see stars...)	
Orbit LMC-SMC	111, **115**, 131, 410
P Cygni profiles	150, 233, 235-239, 266
Period changes	227
Period-luminosity relation (see P-L-C...)	
Planetary nebulae	231, **354, 363**
(see also distribution...)	
P-L-C, P-L relation	**157**, 191, 202, 217, 221, 229
Polarization	347, 403
Pulsar	301, 305
dispersion measure	379

Pulsation mass   **174**

Quasar   284, 375

R136   66, 151, **245**, 255-262, **263**,
                                            375, 382
Radio continuum   **283**, 313, 389
Red variables (see stars...)
Reddening (see interstellar extinction...)
Reddening from Cepheids   167, 229, 334
Relaxation time   25, 27, 55
Ring nebulae   89, 376
Rotation curve LMC   110, 112, 377, 395
RR Lyrae (see stars... and clusters...)

Shell (gas)   35, 89, 385, 391
Shell burning   65
Shock waves   274, 393, 403
SMC depth   221
SMC remnant (RSMC)   129
SMC X (see X-ray sources...)
SNR (see supernova...)
Solar neighbourhood   68, 83, 105, 321, 372
Spatial distribution (see distribution...)
Spectral classification   **145**
s-process   177

Stars
    AGB   **171**, 217
    binaries   202, 305, 412
    carbon   14, 71, **172**, 184, **195**, 217
    cataclismic variables   302, 417
    Cepheid   71, **157**, 205, 219-229
    K   206, 225
    LPV   **171**, 206
    M   80, **172**, 185, 198, 217
    Mira   202, 217, 219
    O-B   35, 59, 70, 233, 237, 239, 264, 342
    OH/IR   181
    red giants   **183**, **195**
    red variables   **171**, 190, 217, 219
    RR Lyrae   64, **207**, 411, 414
    S   186, 198
    S Dor   145, 237
    supergiants   **145**, **243**, 414
        red, M   68, 79, 145, 246
        luminous   **145**, 201
        blue to red ratio   155
    WR (WN, WC)   153, 237, **245**, 266

Star counts   25, 29, 55, **79**, 97, 103

INDEX

Starformation	18, **67**, **79**, 91, 413
rate (SFR)	**67**, 197
propagation	71, 89
stochastic	72, 93, 95, 313, 327
Supergiants (see stars...)	
Supermassive stars	263
(see also R136...)	
Supernova rate	70, 327, 354, 379
Supernova remnant (SNR)	246, **271**, **286**, **293**, 315, 416
identifying procedure	271
Tarantula nebula (see 30 Doradus...)	
Tidal interaction	**115**, **125**
Transient X-ray source	295
UV observations (far-)	7, 53-59, 70, 91, 105, 233, 336, 341, 355, 375, 405, 407, 415
Warp	128
Wind, stellar	235, 237, 385, 415
WR stars (see stars...)	
X-ray sources	241, 246, 272, 286, **293**, **305**, 315, 317, 416